HVDC
POWER TRANSMISSION SYSTEMS

HVDC POWER TRANSMISSION SYSTEMS

K R Padiyar
Ex-Professor
IISC, Bangalore

An Imprint of

NEW AGE INTERNATIONAL (P) LIMITED, PUBLISHERS
LONDON • NEW DELHI • NAIROBI
Bangalore • Chennai • Cochin • Guwahati • Hyderabad • Kolkata • Lucknow • Mumbai
Visit us at **www.newagepublishers.com**

Published by New Age International (P) Ltd., Publishers
First Edition: 1990
Third Multi Colour Edition: 2015
Reprint: 2017

GLOBAL OFFICES

- **New Delhi** **NEW AGE INTERNATIONAL (P) LIMITED, PUBLISHERS**
  7/30 A, Daryaganj, New Delhi-110002, (INDIA)
  Tel.: (011) 23253771, 23253472, **Telefax:** 23267437, 43551305
  E-mail: contactus@newagepublishers.com • Visit us at www.newagepublishers.com
- **London** **NEW AGE INTERNATIONAL (UK) LTD.**
  27 Old Gloucester Street, London, WC1N 3AX, UK
  E-mail: info@newacademicscience.co.uk • Visit us at www.newacademicscience.co.uk
- **Nairobi** **NEW AGE GOLDEN (EAST AFRICA) LTD.**
  Ground Floor, Westlands Arcade, Chiromo Road (Next to Naivas Supermarket)
  Westlands, Nairobi, KENYA, **Tel.:** 00-254-713848772, 00-254-725700286
  E-mail: kenya@newagepublishers.com

BRANCHES

- **Bangalore** 37/10, 8th Cross (Near Hanuman Temple), Azad Nagar, Chamarajpet, Bangalore- 560 018
  Tel.: (080) 26756823, **Telefax:** 26756820, **E-mail: bangalore@newagepublishers.com**
- **Chennai** 26, Damodaran Street, T. Nagar, Chennai-600 017, **Tel.:** (044) 24353401, **Telefax:** 24351463
  E-mail: chennai@newagepublishers.com
- **Cochin** CC-39/1016, Carrier Station Road, Ernakulam South, Cochin-682 016
  Tel.: (0484) 2377303, **Telefax:** 4051304, **E-mail: cochin@newagepublishers.com**
- **Guwahati** Hemsen Complex, Mohd. Shah Road, Paltan Bazar, Near Starline Hotel, Guwahati-781 008
  Tel.: (0361) 2513881, **Telefax:** 2543669
  E-mail: guwahati@newagepublishers.com
- **Hyderabad** 105, 1st Floor, Madhiray Kaveri Tower, 3-2-19, Azam Jahi Road, Near Kumar Theater, Nimboliadda Kachiguda, Hyderabad-500 027, **Tel.:** (040) 24652456, **Telefax:** 24652457
  E-mail: hyderabad@newagepublishers.com
- **Kolkata** RDB Chambers (Formerly Lotus Cinema) 106A, 1st Floor, S N Banerjee Road, Kolkata-700 014
  Tel.: (033) 22273773, **Telefax:** 22275247
  E-mail: kolkata@newagepublishers.com
- **Lucknow** 16-A, Jopling Road, Lucknow-226 001, **Tel.:** (0522) 2209578, 4045297, **Telefax:** 2204098
  E-mail: lucknow@newagepublishers.com
- **Mumbai** 142C, Victor House, Ground Floor, N.M. Joshi Marg, Lower Parel, Mumbai-400 013
  Tel.: (022) 24927869, **Telefax:** 24915415, **E-mail: mumbai@newagepublishers.com**
- **New Delhi** 22, Golden House, Daryaganj, New Delhi-110 002, **Tel.:** (011) 23262368, 23262370
  Telefax: 43551305, **E-mail: sales@newagepublishers.com**

ISBN: 978-81-224-3785-0
C-17-02-10185

Printed in India at Goyal Offset Printers, Delhi.
Typeset at In-house, Delhi.

NEW AGE INTERNATIONAL (P) LIMITED, PUBLISHERS
7/30 A, Daryaganj, New Delhi-110002
Visit us at **www.newagepublishers.com**
(CIN: U74899DL1966PTC004618)

PREFACE TO THE THIRD EDITION

There have been several significant developments in the field of HVDC technology and applications since the first edition of the book was published in 1991. The major development involves the application of Voltage Source Converters (VSC) using IGBT devices for power levels up to 400 MW. VSC-HVDC has several advantages over the Line Commutated (current source) Converters (LCC) using thyristors. In particular, the problems of operating with low SCR (weak AC systems) are overcome due to absence of commutation failures and the fact that VSC can generate reactive power (within limits) to help regulate the converter bus voltage. The VSC converter can also supply passive loads with regulation of voltage and frequency. The use of IGBT devices permit pulse-width modulation (PWM) that helps to simplify harmonic filtering. The application of VSC goes hand in hand with the use of XLPE polymer extruded submarine/ land cables. With the development of HVDC breakers, it is anticipated that the application of VSC based MTDC systems will be a reality. The VSC-HVDC systems have found new applications in evacuation of power from off-shore wind farms, power supply to off-shore platforms (producing oil/gas), supply of controllable power to improve security, integration of dispersed or distributed generation.

The development of VSC-HVDC is unlikely to replace LCC based HVDC power transmission in the near future. This is due to several developments in the so called 'classical' technology—the introduction of Light Triggered Thyristors (LTT), increase in the voltage/current ratings of devices, the application of double/triple tuned passive filters and active (hybrid) filters, Capacitor Commutated Converters (CCC), optical current transducers, DSP controllers etc. The reliability has significantly improved in the recent times in view of these developments and LCC based HVDC is still economical at extra-high voltage, bulk power transmission. Recent trends are to develop ± 800 kV DC for long distance transmission of power in the range of 5000–6000 MW.

This edition takes care of the few errors that crept-up in the last edition. In particular, the emerging technology of VSC-HVDC links is described in detail. Instead of adding new chapters to present the new developments, the new material is added at appropriate places. For example, the analysis of VSC is presented in Chapters 2 and 3 along with the thyristor based Line Commutated Converters (LCC). Practically, in all chapters, there is discussion of VSC-HVDC. The book also presents other developments such as the application of hybrid active filters, Capacitor Commutated Converters, double and triple tuned filters etc. The Chapter 10 presents power flow analysis in AC-DC systems based on a novel approach. The modeling, simulation and study of interactions among AC-DC systems are covered in the last 2 chapters without missing any relevant topics. The appendices give details of thyristor and IGBT valves, transient simulation of converters and DC lines, synchronous generator modeling, SSR analysis and design of DC and AC voltage controls in BTB VSC-HVDC links. Several examples and case studies are included to illustrate the concepts. The new edition also gives an opportunity to correct errors in the first edition and add explanations wherever required for better understanding.

The new edition contains 12 chapters. The torsional and harmonic interactions are covered along with modeling and simulation in Chapter 11. Small signal (or dynamic) stability, power modulation, voltage and transient stability analysis are covered in Chapter 12. The bibliography given at the end of each chapter is fairly exhaustive and the reader can refer to them for more detailed study. The 7 Appendices at the end of the book contain many details on technology, modeling and analysis of interest to a diligent reader. Care is taken to ensure that the book is useful not only to students taking the first course on HVDC transmission, but also to graduate students pursuing research and practicing engineers interested in applying the HVDC technology. Thus, the book contains more material than what can be covered in a one-semester course.

K.R. Padiyar

PREFACE TO THE FIRST EDITION

Modern DC power transmission is relatively a new technology which made a modest beginning in 1954. The advent of thyristor valves and related technological improvements over the last 18 years has been responsible for the acceleration of the growth of HVDC systems. The HVDC technology is still undergoing many changes due to continuing innovations directed at improving reliability and reducing costs of converter stations. The latest development is the introduction of multi-terminal system operation which increases the scope of application of HVDC systems. However, the growth in the spread of knowledge on HVDC technology remains limited although several conferences/symposia are held regularly and many technical papers are being published. This is due to the fact that the published literature is generally directed at experts in the field.

When the number and size of DC systems are small, it was common to consider HVDC power transmission as too specialized and fit only to be taken up by manufacturers and consultants. With the growth of HVDC systems, there is now a greater awareness among engineers from utilities, regarding the potential of DC transmission from the point of view of interactions with AC systems. Some of these interactions are beneficial, while others may pose problems unless investigated thoroughly during the design stage and solutions incorporated to overcome the adverse effects. While it is true that the HVDC systems are quite reliable and converter control allows flexibility in the system operation, full realization of the potential benefits of HVDC transmission requires detailed studies during the planning and design stage. This necessitates proper modelling and analysis of DC systems. The dearth of books that tackle this subject in sufficient detail is also responsible for the slow growth in the knowledge which is essential for planning and evaluation of the performance of HVDC systems. As HVDC technology has reached a level of maturity and is contributing to the improved performance of AC systems (the development of static VAR compensators is a direct offshoot of HVDC technology)-there is a need to strengthen academic programmes in the universities by including HVDC power transmission in the curriculum.

The author has been offering a graduate course in HVDC power transmission (initially at Indian Institute of Technology, Kanpur and currently at Indian Institute of Science, Bangalore) since 1980 and has been involved with research and development in this area for over 15 years. He has also conducted intensive courses, workshops and lectured to practising engineers/researchers from industry and academic institutions. This book is the outgrowth of the notes and material used in lecturing to graduate students and others. It also includes material taken from the research work carried out by the author and his students.

It has been the author's experience that it is necessary to expose the beginner to HVDC technology before analysis and design can be taken up. Thus the book is divided into two parts—the first part dealing with the various components and techniques of the state-of-the-art in HVDC technology and the second part presenting several aspects of the interactions of AC/DC systems. The study of system interactions requires adequate modeling and analysis of HVDC systems. These topics are presented in sufficient detail and an effort has been made to bring together material that has been scattered in various journals and not available in books published so far. It is hoped that this book will meet the need for an up-to-date text book as well as a reference book for practising engineers and researchers from industry and academic institutions.

Part I of the book has 9 chapters which cover the components and the techniques of HVDC technology. The emphasis is on the unique components of HVDC systems, such as thyristor valves, converters, control, protection and harmonic filters. In describing these (particularly the controllers), the principles of operation are stressed and details of specific equipment (which are treated as proprietary information by manufacturers) are avoided. Reactive power control and multi-terminal DC system operations are also included as two separate chapters. Static VAR systems used for reactive power control in converter stations are also discussed.

Second part of the book deals with the modeling, analysis and simulation of AC/DC systems. Seven chapters are included in this part which cover component models, power flow, transient stability, dynamic stability and power modulation, harmonic and torsional interactions, simulation of converters and HVDC systems. The digital simulation is stressed as this has already become a complement to physical models (HVDC simulators). Also, in the author's view, digital simulation has the potential to become a powerful and inexpensive tool for the detailed study of converter controls.

It is to be understood that this book does not (and cannot) cover each and every aspect of HVDC technology and its applications. Obviously the choice of the topics and style of presentation are influenced by the author's predilections and experiences. However, effort is made to give a comprehensive and an up-to-date coverage of the topics presented. In the face of a changing technology, this effort can only be successful in a limited time frame.

I would like to acknowledge with gratitude the contributions of many individuals in the preparation of this book. I was introduced to the subject of HVDC transmission by Prof. John Reeve at University of Waterloo, Canada. Drs. Suresh C. Kapoor, Narain G. Hingorani, John J. Vithayathil, P.C.S. Krishnayya and Prof. R.M. Mathur have helped by providing much useful information. Prof. M.A. Pai at University of Illinois, Urbana (previously with Indian Institute of Technology, Kanpur) provided constant encouragement and support. I am also indebted to Dr. M. Ramamoorthy, Director General, Central Power Research Institute and Mr. M.S. Vasudeva, Additional Director, Department of Electronics, Government of India, for their interest and support of R&D programmes in HVDC technology in India. The support received by the Curriculum Development Cell of the Quality Improvement Programme of I.I.T., Kanpur is also gratefully acknowledged.

Several of my former and present graduate students have helped by contributions of their research work and assistance in the preparation of the manuscript. They are Drs. K. Gomathi, C. Radhakrishna, Sachchidanand, H.S.Y. Sastry, A.G. Kothari, Rambabu Adapa, P.K. Kalra, S. Bhattacharya, R.K. Verma, J. Senthil, Vijayan Immanuel and V. Kalyana Raman, Mr. Abraham at I.I.T. Kanpur and Mr. Vijayakumar at I.I.Sc. have typed the manuscript with painstaking care.

Finally, I would like to acknowledge the understanding and patience of my wife Usha during the long drawn out effort involved in completing the book.

K.R. Padiyar

ACKNOWLEDGEMENTS

I would like to acknowledge with gratitude, the contribution of several individuals. First and foremost, I wish to thank Dr. Narain Hingorani for inspiring me to continue the work on HVDC and FACTS and providing much useful knowledge.

Drs. John Vithayathil, P.C.S. Krishnayya, R.S. Thallam, Ram Adapa and Arunachalam have helped by useful discussions and information. I am indebted to Prof. M.A. Pai at University of Illinois, Urbana for his encouragement and inviting me to give lectures on the subject after the publication of the first edition.

The work of my graduate students at Indian Institute of Science has helped in the preparation of this edition. In particular, I wish to acknowledge Mr. Kalyana Raman, Drs. M.K. Geetha, Animesh Bhowmick, Suresh Rao, Nagesh Prabhu and Mr. Abhishek Tyagi. Dr. Anil Kulkarni provided constructive comments on the first edition and help with the literature survey. Mr. Babu Narayanan from CPRI provided the details of the second CIGRE Benchmark Model. BHEL provided the picture of the valve structure for National HVDC Project-Phase II, that is shown on the cover page.

Ms. Sudha Aithal has typed the manuscript with painstaking care. I also acknowledge the contribution of Ms. N. Jayanthi in preparation of the electronic version of the text in the first edition. Mr. Malleswara Rao, my graduate student, helped in the preparation of some figures. I wish to thank Mr. Saumya Gupta, Managing Director at New Age International Publishers for his keen interest in bringing out this multi colour edition. New Age team has done a tremendous job in the production of this book.

The financial assistance for the preparation of the manuscript provided by the Curriculum Development Cell, Indian Institute of Science, Bangalore, is gratefully acknowledged. I am deeply indebted to Indian Institute of Science for permitting me to pursue academic activities as an Honorary Professor from May 2003.

Last, but not the least, I appreciate the patience, understanding and support of my wife Usha, during the long hours of working on this book.

K.R. Padiyar

ABBREVIATIONS

AC	Alternating Current
AGC	Automatic Generation Control
AN	Audible Noise
AVR	Automatic Voltage Regulator
CACV	Constant AC Voltage (Control)
CC	Constant Current (Control)
CDCV	Constant DC Voltage (Control)
CEA	Constant Extinction Angle
CIGRE	Conference Internationale des Grands Reseaux Electriques a haute tension (International Conference on Large High-Voltage Electrical Systems), Paris
CRB	Current Reference Balancer
dB	decibel
DC	Direct Current
EMTP	Electro-Magnetic Transients Program
EPC	Equidistant Pulse Control
ESCR	Effective Short Circuit Ratio
FACT	Flexible AC Transmission System
GPG	Gate Pulse Generator
GPU	Gate Pulse Unit
GTO	Gate Turn-Off (Thyristor)
HP	High Pressure (Turbine)
HVDC	High Voltage Direct Current
Hz	Hertz
IGBT	Insulated Gate Bipolar Transistor
IGE	Induction Generator Effect

IGCT	Integrated Gate Commutated Thyristor
INV	Induced Noise Voltage
IP	Intermediate Pressure (Turbine)
IPC	Individual Phase Control
LCC	Line Commutated Converter
ln	Natural logarithm
LPF	Low Pass Filter
LP	Low Pressure (Turbine)
LTT	Light Triggered Thyristor
MAP	Maximum Available Power
MCT	Metal-oxide semiconductor Controlled Thyristor
mH	milliHenry
MIDC	Multi-Infeed DC
MOV	Metal Oxide Varistor
MRTB	Metallic Return Transfer Breaker
MTDC	Multi-Terminal DC
PFC	Pulse Frequency Control
PIV	Peak Inverse Voltage
PLC	Power Line Carrier
PLL	Phase-Locked Loop
PPC	Pulse Phase Control
PSS	Power System Stabilizer
PWM	Pulse Width Modulation
RF	Radio Frequency
RI	Radio Interference
rms	Root Mean Square
RoW	Right of Way
RTDS	Real Time Digital Simulator
SC	Synchronous Condenser
SCR	Short Circuit Ratio
S.C.R.	Silicon Controlled Rectifier
SIL	Surge Impedance Loading
SMC	Supplementary Modulation Controller
SSDC	Sub Synchronous Damping Controller
SSO	Sub Synchronous Oscillation
SSR	Sub Synchronous Resonance

STATCOM	Static synchronous Compensator
STATCON	Static Condenser
SVC	Static Var Compensator
TCR	Thyristor Controlled Reactor
THFF	Telephone Harmonic Form Factor
TIF	Telephone Influence Factor
TSC	Thyristor Switched Capacitor
T-G	Turbine Generator
TI	Torsional Interaction
UHV	Ultra High Voltage
VAR	Volt Amperes Reactive
VCO	Voltage Controlled Oscillator
VDCL	Voltage Dependent Current Limit
VDCOL	Voltage Dependent Current Order Limiter
VSC	Voltage Source Converter
VST	Voltage Setting Terminal
μF	microFarad
μs	microSecond

CONTENTS

3. ANALYSIS OF HVDC CONVERTERS 50–82

4. CONVERTER AND HVDC SYSTEM CONTROL 83–113

5. CONVERTER FAULTS AND PROTECTION 114–132

DC POWER TRANSMISSION TECHNOLOGY

1

1.1 INTRODUCTION

The industrial growth of a nation requires increased consumption of energy, particularly electrical energy. This has led to increase in the generation and transmission facilities to meet the increasing demand. In USA till the early seventies the demand doubled every ten years. In developing countries, like India, the demand doubles every seven years which requires considerable investment in electric power sector.

The imperatives of supplying energy at reasonable costs coupled with the depleting reserves of non-renewable energy sources has led to the establishment of remote generating stations–predominantly fossil-fuel fired thermal stations at pit head. Environmental considerations also dictate sometimes, siting of power stations at remote locations. Large hydro stations are invariably at distances of hundreds of kilometers from load centres. The need to economize on costly investments in generation reserves, sharing of benefits in utilizing variability in generation mixes and load patterns have given rise to interconnection of neighbouring systems and development of large power grids.

Remote generation and system interconnections lead to a search for efficient power transmission at increasing power levels. The increase in voltage levels is not always feasible. The problems of AC transmission particularly in long distance transmission, has led to the development of DC transmission. However, as generation and utilization of power remain at alternating current, the DC transmission requires conversion at two ends, from AC to DC at the sending end and back to AC at the receiving end. This conversion is done at converter stations – rectifier station at the sending end and inverter station at the receiving end. The converters are static—using high power thyristors connected in series to give the required voltage ratings. The physical process of conversion is such that the same station can switch from rectifier to inverter by simple control action, thus facilitating power reversal.

The HVDC transmission made a modest beginning in 1954 when a 100 kV, 20 MW DC link was established between Swedish mainland and the island of Gotland. Until 1970, the converter stations utilized mercury arc valves for rectification. The successful use of thyristors for power control in industrial drives encouraged its adoption in HVDC converters by development of high power semiconductor devices. The device voltage rating is now in the range of 10 kV, and current rating up to 5 kA (for 125 mm device). The highest transmission voltage reached is ± 600 kV.

The total installed HVDC capacity now exceeds 75000 MW. Several developments and innovations in HVDC technology, over the years, have made it attractive to consider HVDC transmission for a wide range of applications. This can cover a wide spectrum from bulk power transmission of thousands of megawatts over long distances to connecting small distributed electrical generators to a grid.

Deregulation or restructuring in the power sector has encouraged competition and private investments. The power transmission and distribution network is also viewed as a means for trading power among the competing generators and consumers of electrical energy. This introduces increased uncertainty in system operation and poses challenges in maintaining system security. The controllability of power flow in a transmission line becomes an important issue in planning of new lines.

The relative merits of AC and DC transmission are reviewed in the next section.

1.2 COMPARISON OF AC AND DC TRANSMISSION

The relative merits of the two modes of transmission (AC and DC) which need to be considered by a system planner are based on the following factors:

1. economics of power transmission
2. technical performance
3. reliability

A major feature of power systems is the continuous expansion necessitated by increasing power demand. This implies that the establishment of a particular line must be considered as a part of an overall long-term system planning.

1.2.1 Economics of Power Transmission

The cost of a transmission line includes the investment and operational costs. The investment includes costs of Right of Way (RoW), transmission towers, conductors, insulators and terminal equipment. The operational costs include mainly the cost of losses.

The characteristics of insulators vary with the type of voltage applied. For simplicity, if it is assumed that the insulator characteristics are similar for AC and DC and depend on the peak level of voltage applied with respect to ground, then it can be shown that for lines designed with the same insulation level, a DC line can carry as much power with two conductors (with positive and negative polarities with respect to ground) as an AC line with 3 conductors of the same size. (Note that $P_{AC} = \frac{3}{\sqrt{2}} V_p I \cos\phi$ and $P_{DC} = 2V_p I, V_p$ is the peak voltage of a conductor with respect to ground, $\cos\phi$ is the power factor and I is the effective current rating of a conductor. Thus, $P_{AC} \simeq P_{DC}$.) This implies that for a given power level, DC line requires less RoW, simpler and cheaper towers and reduced conductor and insulator costs. The power losses are also reduced with DC as there are only two conductors (about 67% of that for AC with same current carrying capacity of conductors). The absence of skin effect with DC is also beneficial in reducing power losses marginally. The dielectric losses in case of power cables is also very less for DC transmission.

The corona effects on DC conductors tend to be less significant than for AC and this also leads to the choice of economic size of conductors with DC transmission. The other factors that influence the line costs are the costs of compensation and terminal equipment. DC lines do not require compensation but the terminal equipment costs are increased due to the presence of converters and filters.

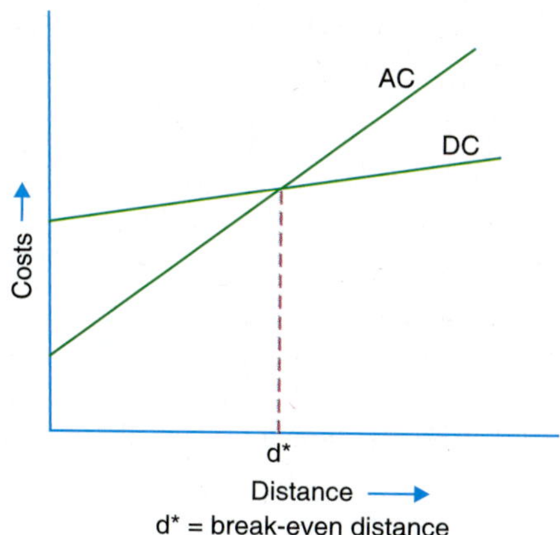

Fig. 1.1: Variation of costs with line length

Figure 1.1 shows the variation of costs of transmission with distance for AC and DC transmission. For distances less than 'break even' distance, AC tends to be economical than DC and costlier for longer distances. The break even distances can vary from 500 to 800 km in overhead lines depending on the per unit line costs.

1.2.2 Technical Performance

The DC transmission has some positive features which are lacking in AC transmission. These are mainly due to the fast controllability of power in DC lines through converter control. The advantages are:

1. full control over power transmitted
2. the ability to enhance transient and small signal stability in associated AC networks
3. fast control to limit fault currents in DC lines. This makes it feasible to avoid DC breakers in two terminal DC links.

In addition, the DC transmission overcomes some of the problems of AC transmission. These are described below:

Stability limits

The power transfer in AC lines is dependent on the angle difference between the voltage phasors at the two ends. For a given power level, this angle increases with distance. The maximum power transfer is limited by the considerations of steady state and transient stability. The power carrying capability of an AC line as a function of distance is shown in Fig. 1.2. The same figure also shows the power carrying capability of DC lines which is unaffected by the distance of transmission and is limited only by the current carrying capacity of the conductors (termed as 'thermal limit').

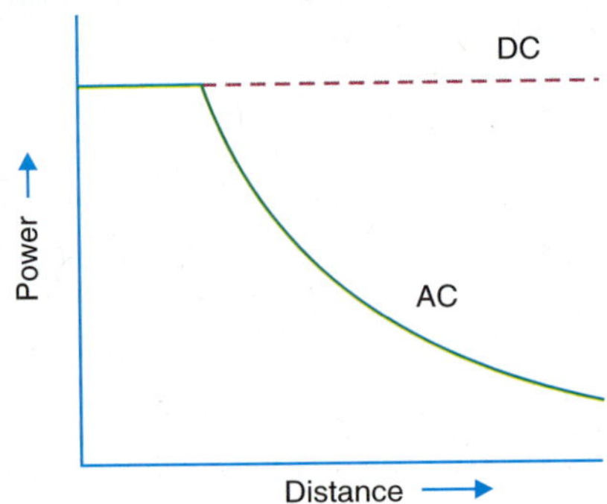

Fig. 1.2: Power transfer capability vs. distance

Voltage control

The voltage control in AC lines is complicated by the line charging and inductive voltage drops. The voltage profile in a AC line is relatively flat only for a fixed level of power transfer corresponding to surge impedance loading (SIL). The voltage profile varies with the line loading. For constant voltage at the line terminals, the midpoint voltage is reduced for line loading higher than SIL and increased for loadings less than SIL. This is shown in Fig. 1.3.

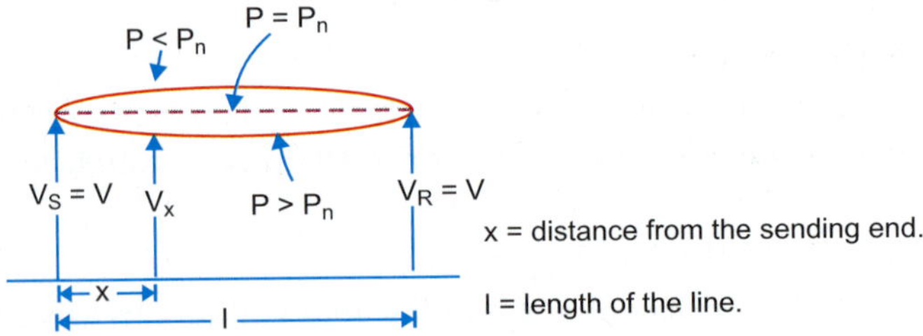

Fig. 1.3: Variation of voltage along the line

The maintenance of constant voltages at the two ends requires reactive power control from inductive to capacitive as the line loading is increased. The reactive power requirements increase with the increase in line lengths. Although DC converter stations with line commutated converters require reactive power related to the line loadings, the line itself does not require reactive power.

The steady-state charging currents in AC lines pose serious problems in cables. This puts the break-even distance for cable transmission around 40 km.

Line compensation

For reasons mentioned earlier, AC lines require shunt and series compensation in long distance transmission, mainly to overcome the problems of line charging and stability limitations. Series capacitors and shunt inductors are used for this purpose. The increase in power transfer and voltage control is also possible through the use of shunt connected Static Var Compensator (SVC).

In AC cable transmission, it is necessary to provide shunt compensation at regular intervals. This is a serious problem in underwater cables.

Problems of AC interconnection

When two power systems are connected through AC ties (synchronous interconnection), the automatic generation control of both systems have to be coordinated using tie-line power and frequency signals. Even with coordinated control of interconnected systems, the operation of AC ties can be problematic due to:

(*i*) presence of large power oscillations which can lead to frequent tripping

(*ii*) increase in fault levels

(*iii*) transmission of disturbances from one system to the other.

The controllability of power flow in DC lines eliminates all the above problems. In addition, for asynchronous DC ties, there is no need of coordinated control.

It is obvious that two systems which have different nominal frequencies cannot be interconnected directly with AC ties and require the use of DC links.

Ground impedance

In AC transmission, the existence of ground (zero sequence) currents cannot be permitted in steady-state due to high magnitudes of ground impedance which will not only affect efficient power transfer, but also result in telephone interference. The ground impedance is negligible for DC currents and a DC link can operate using one conductor with ground return (Monopolar operation). The ground return is objectionable only when buried metallic structures (such as pipes) are present and are subject to corrosion with DC current flow.

It is to be noted that even while operating in the monopolar mode, the AC network feeding the DC converter station operates with balanced voltages and currents. Hence, single pole operation of DC transmission systems is possible for extended periods, while in AC transmission, single phase operation (or any unbalanced operation) is not feasible for more than a second.

Disadvantages of DC transmission

The scope of application of DC transmission is limited by the following factors:

(*a*) the difficulty of breaking DC currents which results in high cost of DC breakers

(*b*) inability to use transformers to change voltage levels

(*c*) high cost of conversion equipment

(*d*) generation of harmonics which require AC and DC filters, adding to the cost of converter stations

(*e*) complexity of control.

Over the years there have been significant advances in DC technology, which have tried to overcome the disadvantages listed above except for (*b*). These are:

(*a*) development of DC breakers

(*b*) modular construction of thyristor valves

(*c*) increase in the ratings of thyristor cells that make up a valve

(*d*) twelve pulse operation of converters

(*e*) use of metal oxide, gapless arrestors

(*f*) application of digital electronics and fiber optics in control of converters.

Some of the above advances have resulted in improving the reliability and reduction of conversion costs in DC systems. It can be said without exaggeration that complexity of control does not pose a problem and can actually be used to provide reliable and fast control of power transmission not only under normal conditions but also under abnormal conditions such as line and converter faults. This has removed the need for DC current interruption in two terminal links. Even for multiterminal operation, the requirements of current ratings of DC breakers are modest due to effective converter control.

1.2.3 Reliability

The reliability of DC transmission systems is quite good and comparable to that of AC systems. An exhaustive record of existing HVDC links in the world is available from which the reliability statistics can be computed. It must be remembered that the performance of thyristor valves is much more reliable than mercury arc valves and further developments in devices, control and protection have improved the reliability level. For example, the development of direct light triggered thyristors (LTT) can improve reliability because of elimination of high voltage pulse transformers and auxiliary supplies for turning on the devices.

There are two measures of overall system reliability.

Energy availability

This is defined as

$$\text{Energy availability} = 100\left(1 - \frac{\text{equivalent outage time}}{\text{total time}}\right)\%$$

where equivalent outage time is the product of the actual outage time and the fraction of system capacity lost due to outage.

Transient reliability

This is a factor specifying the performance of HVDC systems during recordable faults on the associated AC systems

$$\text{Transient reliability} = \frac{100 \times \text{No. of times HVDC system performed as designed}}{\text{No. of recordable AC faults}}$$

Recordable AC system faults are those faults which cause one or more AC bus phase voltages to drop below 90% of the voltage prior to the fault. It is assumed that the short-circuit level after the fault is not below the minimum specified for satisfactory converter operation.

The energy availability or transient reliability of existing DC systems with thyristor valves is 95% or more.

The average failure rate of thyristors in a valve is less than 0.6% per operating year. It is common practice to provide redundant thyristors in the series string composing a HVDC valve, so that failed thyristors can be replaced during scheduled maintenance period (once or twice a year). The maintenance of thyristor valves is also much simpler than the earlier mercury arc valves.

Table 1.1: HVDC Outage Statistics

Equipment	MTTF (years)	MTTR (hours)
Thyristor Group	13.7	6.1
Converter Transformer	16.1	1700.0
Smoothing Reactor	76.8	1700.0
DC Filter	19.7	7.9
AC Filter	12.6	9.3
Master Control	25.0	6.9
Pole Control	9.0	8.6
Pole of Transmission Line	1.25/100 km	1.5
DC Line Switch	147.2	7.8

MTTF = Mean time to failure

MTTR = Mean time to repair.

Some of the HVDC outage statistics is given in Table 1.1. In comparing the reliability of various alternatives, it must be kept in mind that bipolar DC line can be as reliable as a double circuit AC line with same power capability. This is because of the fact that failure of one pole does not affect the operation of the other pole (with ground return). If the DC line conductor has adequate overload rating and if the converters on the failed pole can be paralleled with the converters on the healthy pole, the prefault power level can be maintained even with permanent outage of one pole.

1.3 APPLICATION OF DC TRANSMISSION

The detailed comparison of AC and DC transmission in terms of economics and technical performance, leads to the following areas of application for DC transmission:

1. long distance bulk power transmission
2. underground or underwater cables
3. asynchronous interconnection of AC systems operating at different frequencies or where independent control of systems is desired.
4. control and stabilization of power flows in AC ties in an integrated power system.

The first two applications are dictated primarily by the economic advantages of DC transmission, where the concept of break-even distance is important. To be realistic, one must also assign a monetary value for the technical advantages of DC (or penalty costs for the drawbacks of AC). The problem of evaluation of the economic benefits, is further complicated by the various alternatives that may be considered in solving problems of AC transmission—phase shifters, static var systems, series capacitors, single pole switching etc.

The technical superiority of DC transmission dictates its use for asynchronous interconnections, even when the transmission distances are negligible. Actually there are many 'back to back' (BTB) DC links in existence where the rectification and inversion are carried out in the same converter station with no DC lines. The advantage of such DC links lies in the reduction of the overall conversion costs and improving the reliability of DC system.

The alternative to DC ties may require strengthening of existing AC network near the boundary of the two systems. This cost can be prohibitive if the capacity of the tie required is moderate compared to the size of the systems interconnected.

In large interconnected systems, power flows in AC ties (particularly under disturbance conditions) can be uncontrolled and lead to overloads and stability problems, thus endangering system security. Strategically placed DC lines can overcome this problem due to the controllability of power. The planning of DC transmission in such applications requires detailed study to evaluate the benefits.

Presently the number of DC lines in a power grid is very small compared to the number of AC lines. This indicates that DC transmission is justified only for specific applications. Although advances in technology and introduction of multi-terminal DC (MTDC) systems are expected to increase the scope of application of DC transmission, it is not anticipated that AC grid will be replaced by DC power grid in future. There are two major reasons for this. Firstly, the control and protection of MTDC systems is very complex and the inability of voltage transformation in DC networks imposes economic penalties. Secondly, the advances in DC technology have resulted in the improvement of the performance of AC transmission, through introduction of FACTS (Flexible AC Transmission System) controllers.

The rate of growth of DC transmission was slow in the beginning. In over 16 years, only 6000 MW of DC systems were installed using mercury arc valves. The introduction of thyristor valves overcame some of the problems of system operation mainly due to the arc backs in mercury arc valves. Since then, the rate of growth of DC transmission capacity has reached an average of 2500 MW/year.

There are more than 92 HVDC projects worldwide with a total capacity exceeding 75000 MW. Initial application of HVDC transmission was for submarine transmission or frequency conversion where HVDC is the unique solution. Subsequently, HVDC has been applied for long distance bulk power transmission and for asynchronous BTB links.

Table 1.2 gives the list of BTB and line projects in India.

Table 1.2: List of BTB and Line Projects in India.

(a) Back to Back (BTB) Links

S. No.	System/Project	Year Commissioned	Supplier	Power Rating (MW)	Voltage (kV)
1.	Vindhyachal	1989	ABB	500	2 x 69.7
2.	Chandrapur–Ramagundam	1997/98	GEC Alsthom	1000	2 x 205
3.	Vizag–I	1999	GEC Alsthom	500	205
4.	Vizag–II	2005	ABB	500	±88
5.	Sasaram	2002	GEC Alsthom	500	205

(b) Line Projects

S. No.	System/ Project	Year Comm-issioned	Supplier	Power Rating (MW)	Voltage (kV)	Line Length (km)
1.	National HVDC Project–Stage–I	1989	BHEL	100	100	196
2.	NHVDC–Stage–II	2000	BHEL	100	200	196
3.	Rihand–Delhi	1991-92	ABB/BHEL	750/1500	± 500	814
4.	Chandrapur–Padghe	1998	ABB/BHEL	1500	± 500	736
5.	Talcher–Kolar	2003	Siemens	2000	± 500	1400
6.	Balia–Bhiwadi	2009	Siemens/BHEL	2500	± 500	780

A compilation of the worldwide HVDC projects titled "HVDC Projects List" is available at the website maintained by HVDC and FACTS subcommittee of IEEE/PES Transmission and Distribution committee [22].

1.4 DESCRIPTION OF DC TRANSMISSION SYSTEM

1.4.1 Types of DC Links

The DC links are classified into three types which are defined below:

1. Monopolar link [see Fig. 1.4 (*a*)] has one conductor usually of negative polarity and use ground or sea return. Sometimes metallic return is also used.
2. Bipolar link [see Fig. 1.4 (*b*)] has two conductors, one positive and the other negative. Each may be a bundled conductor in EHV lines. Each terminal has two sets of converters of identical ratings, connected in series on the DC side. The junction between the two sets of converters is grounded at one or both ends. Normally, both poles operate at equal currents and hence there is zero ground current flowing under these conditions.
3. Homopolar Link [see Fig. 1.4(*c*)] has two or more conductors all having the same polarity (usually negative) and always operated with ground or metallic return.

Because of the desirability of operating a DC link without ground return, bipolar links are most commonly used. Homopolar link has the advantage of reduced insulation costs, but the disadvantages of earth return outweigh the advantages. Incidentally, the corona effects in a DC line are substantially less with negative polarity of the conductor as compared to the positive polarity.

The monopolar operation is used in the first stage of the development of a bipolar line, as the investments on converters can be deferred until the growth of load which requires bipolar operation at double the capacity of a monopolar link.

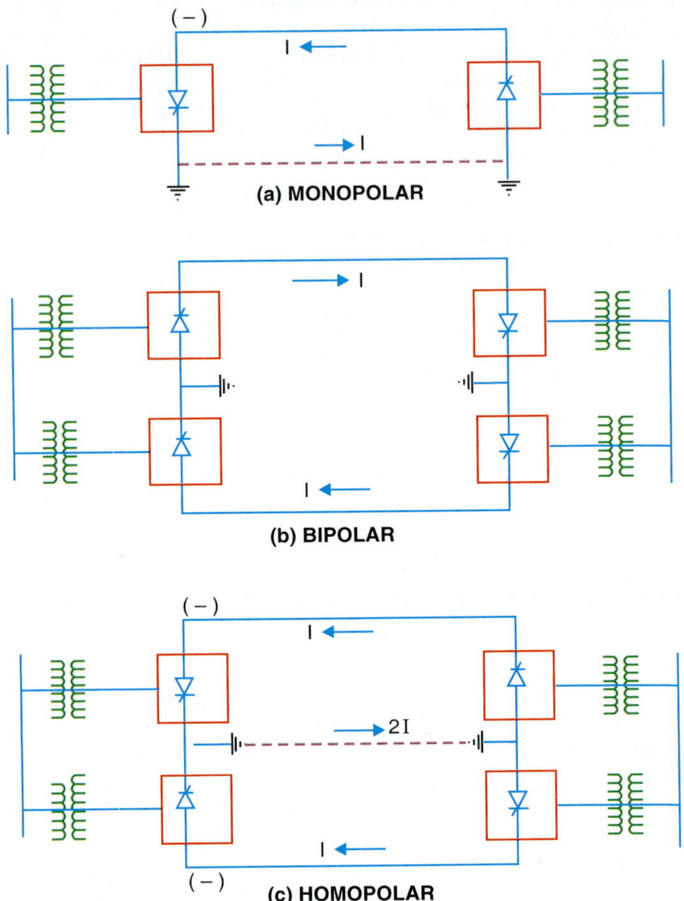

Fig. 1.4: DC link configurations

1.4.2 Converter Station

The major components of a HVDC transmission system are converter stations where conversions from AC to DC (Rectifier station) and from DC to AC (Inverter station) are performed. A point to point transmission requires two converter stations. The role of rectifier and inverter stations can be reversed (resulting in power reversals) by suitable converter control.

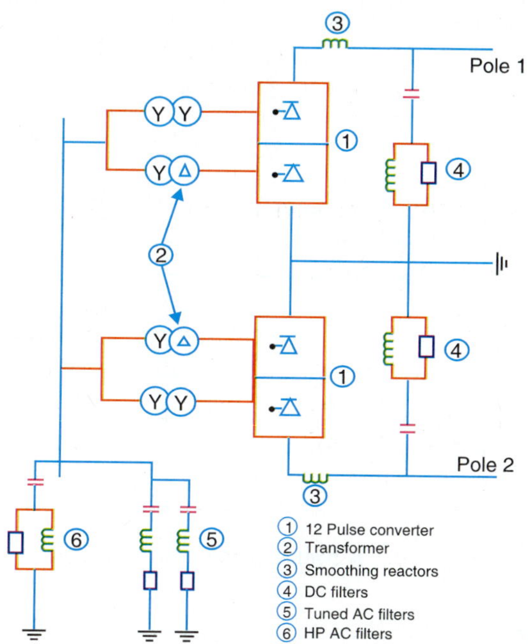

Fig. 1.5: Schematic diagram of a typical HVDC converter station

A typical converter station with one 12 pulse converter unit per pole, is shown in Fig. 1.5. The various components of a converter station are discussed below.

Converter unit

This usually consists of two three phase converter bridges connected in series to form a 12 pulse converter unit as shown in Fig.1.6. The total number of valves in such a unit are twelve. The valves can be packaged as a single valve, double valve or quadrivalve arrangements. Each valve is used to switch in a segment of an AC voltage waveform. The converter is fed by converter transformers connected in star/star and star/delta arrangements.

The valves can be cooled by air, oil, water or freon. Liquid cooling using deionized water is more efficient and results in the reduction of station losses. The ratings of a valve group are limited more by the permissible short circuit currents than steady state load requirements. The design of valves is based on the modular concept where each module contains a limited number of series connected thyristor levels.

Valve firing signals are generated in the converter control at ground potential and are transmitted to each of the thyristor in the valve through a fiber optic light guide system. The light signal received at the thyristor level is converted to an electrical signal using gate drive amplifiers with pulse transformers.

The valves are protected using snubber circuits, protective firing and gapless surge arrestors. Some of the details of the operation, control and protection of thyristor valves are given in Appendix A.

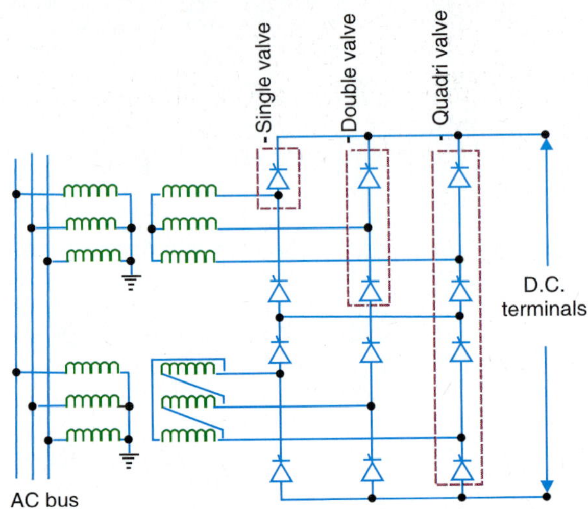

Fig. 1.6: A twelve pulse converter unit

Converter transformer

The converter transformer can have different configurations – (*i*) three phase, two winding, (*ii*) single phase, three winding, (*iii*) single phase two winding. The valve side windings are connected in star and delta with neutral point ungrounded. On the AC side, the transformers are connected in parallel with neutral grounded. The leakage reactance of the transformer is chosen to limit the short circuit currents through any valve.

The converter transformers are designed to withstand DC voltage stresses and increased eddy current losses due to harmonic currents. One problem that can arise is caused by the DC magnetization of the core due to unsymmetric firing of valves.

In back to back links, which are designed for low DC voltage levels, an extended delta configuration can result in identical transformers being used in twelve pulse converter units. This results in the reduction of the spare capacity required. However, the application of extended delta transformers is limited.

Filters

There are three types of filters used:

1. AC filters: These are passive circuits used to provide low impedance, shunt paths for AC harmonic currents. Both tuned and damped filter arrangements are used.
2. DC filters: These are similar to AC filters and are used for the filtering of DC harmonics.
3. High frequency (RF/PLC) filters: These are connected between the converter transformer and the station AC bus to suppress any high frequency currents. Sometimes such filters are provided on high-voltage DC bus connected between the DC filter and DC line and also on the neutral side.

Reactive power source

Converter stations require reactive power supply that is dependent on the active power loading (about 50 to 60% of the active power). This is due to the fact that current drawn by a Line Commutated (current source) Converter (LCC) can only lag the supply voltage. Fortunately, part of this reactive power requirement is provided by AC filters. In addition, shunt (switched) capacitors, synchronous condensers and static var systems (SVC or STATCOM) are used depending on the speed of control desired.

Smoothing reactor

A sufficiently large series reactor is used on DC side to smooth DC current and also for protection. The reactor is designed as a linear reactor and is connected on the line side, neutral side or at intermediate location.

DC switchgear

This is usually a modified AC equipment used to interrupt small DC currents (employed as disconnecting switches). DC breakers or metallic return transfer breakers (MRTB) are used, if required for interruption of rated load currents.

In addition to the equipment described above, AC switchgear and associated equipment for protection and measurement are also part of the converter station. This includes DC current and voltage transducers.

1.5 PLANNING FOR HVDC TRANSMISSION

The system planner must consider DC alternative in transmission expansion. The factors to be considered are: (*i*) cost, (*ii*) technical performance, and (*iii*) reliability.

Generally, the last two factors are considered as constraints to be met and the minimum cost option is selected among various alternatives that meet the specifications on technical performance and reliability.

For submarine, cable transmission and interconnecting two systems of different nominal frequencies, the choice of DC is obvious. In other cases, the choice is to be based on detailed techno-economic comparison.

The considerations in the planning for DC depends on the application. Two applications can be considered as representative. These are:

1. Long distance bulk power transmission
2. Interconnection between two adjacent systems.

In the first application, the DC and AC alternatives for the same level of system security and reliability are likely to have the same power carrying capability. Thus the cost comparisons would form the basis for the selection of the DC (or AC) alternative, if the requirements regarding technical performance are not critical.

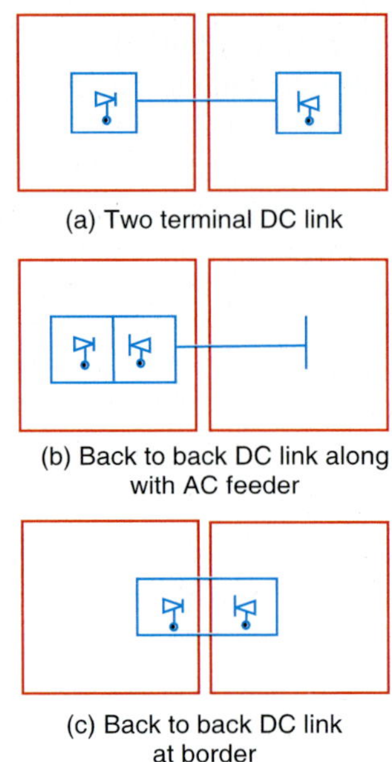

Fig. 1.7: Different configurations for asynchronous interconnection

In the second application, AC interconnection poses several problems in certain cases. For the same level of system security (and reliability), the capacity of AC interconnection will be much more than that for DC (even ignoring the beneficial aspects of DC power modulation). Thus the choice for DC interconnection will be based on the following considerations.

1. Small fluctuations in the voltage and frequency do not affect the power flow which can be set at any desired value.
2. The system security can be enhanced by fast control of DC power.

Having settled on the DC link for interconnection, there are three possible configurations for interconnection. These are:

1. A two terminal transmission where each terminal is located at a suitable place somewhere within the network and connected by a DC overhead line or cable.
2. A back to back HVDC station (also called HVDC coupling station) located somewhere within one of the network and an AC line from the other network to the common station.
3. A back to back station located close to the border between the two systems. This is a special case of the above.

These are illustrated in Fig. 1.7.

In the choice between the first and the second configuration, it is to be noted that converter costs are less for the common coupling station and the AC line costs are greater than the DC line costs. If the distances involved are less than 200 km, the second configuration is to be preferred. If the short circuit ratio (SCR) is acceptable then the third alternative will be the most economic.

The specifications and design of DC system require an understanding of the various interactions between the DC and AC systems. The interruption (or reduction) of power in a DC link can occur due to:

(*a*) DC line faults

(*b*) AC system faults

The speed of recovery from transient DC line faults is of concern in maintaining the integrity of the overall system. The power flow and stability studies are used in this context. The recovery of DC link from AC system faults is more complex. The depression of AC voltage at the inverter bus can lead to commutation failure and loss of DC power. The DC power is ramped up on the clearing of the fault. Too fast an increase in DC power output can lead to the reduction of AC voltage and failure of commutation (due to corresponding increase in the var demand). An optimum rate of increase in DC power can be determined from stability study. This is influenced by control strategy and system characteristics.

The following aspects also require a detailed study of the system interactions:

1. Var requirements of converter stations and voltage stability
2. Dynamic overvoltages
3. Harmonic generation and design of filters
4. Damping of low frequency and subsynchronous torsional oscillations
5. Carrier frequency interference caused by spiky currents in valves (at the beginning of conduction) due to the discharge of stray capacitances and snubber circuits.

The converter control plays a major role in these interactions and the control strategy should be such as to improve the overall system performance. Digital simulation and HVDC simulators are used for planning and design studies.

Choice of voltage level

For long distance bulk power transmission, the voltage level is chosen to minimize the total costs for a given power level (P). The total costs include investment (C_1) and cost of losses (C_2). The investment costs per unit length are modeled as:

$$C_1 = A_0 + A_1 nV + A_2 nq \tag{1.1}$$

where V is the voltage level with respect to ground

n is the number of conductors

q is the total cross-section of each conductor

A_0, A_1 and A_2 are constants.

The cost of losses per unit length is given by

$$C_2 = \frac{\left[n\left(\frac{P}{nV}\right)^2 \rho TLp \right]}{q} \tag{1.2}$$

where

ρ = conductor resistivity

T = total operation time in a year

L = loss load factor

p = cost per unit energy

C_2 can be simplified as

$$C_2 = \frac{\left[A_3\left(\frac{P}{V}\right)^2 \rho\right]}{nq} \tag{1.3}$$

where $A_3 = TLp$

By minimizing the sum of C_2 and the third term in C_1, we have,

$$nq = \sqrt{(A_3\rho/A_2)}.(P/V) \tag{1.4}$$

$$J = P/(nqV) = \sqrt{A_2/(A_3\rho)} \tag{1.5}$$

where J is the current density. The total costs, using (1.4) can be written as

$$C = C_1 + C_2 = A_0 + A_1nV + 2\sqrt{A_2A_3\rho}.(P/V) \tag{1.6}$$

The voltage level V is chosen to minimize C. The equation (1.6) ignores the variation of terminal costs with the voltage. Figure 1.8 shows the selection of optimum system voltage to minimize the sum of converter and line costs.

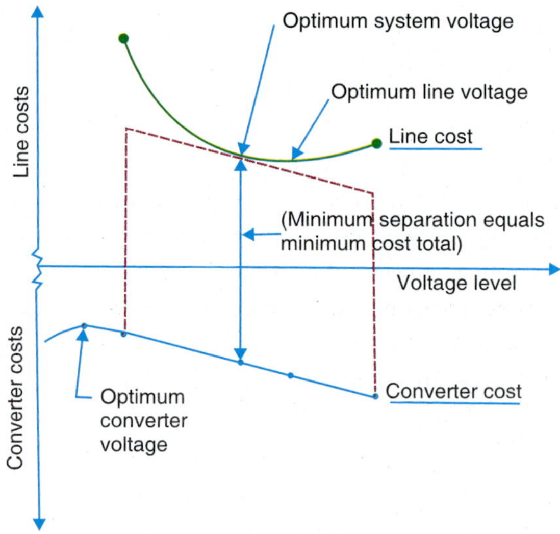

Fig. 1.8: Selection of optimum system voltage for a fixed power transfer

In case of back to back DC ties, the line costs are absent. Hence the voltage level is chosen to minimize converter costs. This level is generally much lower than that in the presence of an overhead line.

1.6 MODERN TRENDS IN HVDC TECHNOLOGY

The continuing technological developments in the areas of power semiconductor devices, digital electronics, adaptive control, DC protection equipment have increased the pace of application of DC transmission. The major contribution of these developments is to reduce the cost of converter stations while improving the reliability and performance.

Power semiconductors and valves

The cost of the converters can come down if the number of devices to be connected in series and parallel can be brought down. The size of the devices has gone up to 125 mm (in diameter) and there is no need for parallel connection. The increase in the current rating of the devices has made it possible to provide higher overload capability at reasonable costs and reduce the lower limits on transformer leakage impedance thereby improving the power factor. The voltage ratings are also on the increase. The development of direct light triggered thyristors (LTT) should also improve the reliability of converter operation. The cost of the valves is also reduced by the application of zinc oxide gapless arrestors and protective firing methods. Valve design with a suspended valve structure ensures safety in seismically active locations.

The power rating of thyristors is increased by better cooling methods. Deionized water cooling has now became a standard and results in reduced losses in cooling. The power rating of a 12 pulse converter unit has increased upto 3000 MW at 500 kV.

LTT valves rated at 125 kV, 2400 A are used at Sakuma frequency converter station in Japan. LTT valves rated at 133 kV, 2000 A have been in operation in the Celilo converter station of the 500 kV Pacific Intertie system in USA since 1997. Each valve is made up of 44 series connected thyristors arranged in two modules. Each thyristor is rated at 8 kV and requires only 40 mW of gate power to turn-on (compared to several watts for Electrically Triggred Thyristors (ETT)). To simplify thyristor electronics, the forward overvoltage protection is integrated in wafer.

As forced commutated converters operating at high voltages are uneconomic, the development of devices that can be turned off by application of a gate signal would be desirable. Gate turn off (GTO) thyristors are already available at 6 kV and 4000 A. However, the main disadvantage of GTO's is the large gate current needed to turn them off and low switching speeds which do not allow pulse-width modulation. In contrast, IGBT (Insulated Gate Bipolar Transistor) device requires much less power (to turn off) and has high switching speeds. IGBT converters are available at ±150 kV and 350 MVA rating. Switching speeds upto 2 kHz are feasible.

Converter control

The development of microcomputer based converter control equipment has now made it possible to design systems with completely redundant converter control with automatic transfer between systems in the case of a malfunction. Not only the forced outage rate of control equipment is reduced, it is also possible to perform scheduled preventive maintenance on the stand-by system when the converter is in operation. The use of a mini-simulator will make it feasible to check vital control and protection functions.

The microcomputer based control also has the flexibility to try adaptive control algorithms or even the use of expert systems for fault diagnosis and protection.

The control and protection requires the measurements of direct current. Traditionally the direct current is measured using transducers of zero-flux type based on magnetics, which are bulky and prone to problems during transients. Recently developed hybrid-optical measuring system uses an ohmic shunt, the voltage drop across which is measured and digitized by an electronic circuit located at high voltage potential. The measured signal is transmitted through optic link to ground potential. Future developments will employ simplified construction through the application of the fiber optic current measurement system based on Faraday's effect. According to this, linearly polarized light rotates by an angle directly proportional to the magnetic field produced by the current in the conductor passing through the optical sensor. The accuracy of the optical current transducers is better than 0.5 % and the bandwidth exceeds 5 kHz.

DC breakers

With the development and testing of prototype DC breakers, it will be possible to go in for tapping an existing DC link or the development of new MTDC systems. Parallel, rather than series operation of converters is likely as it allows certain flexibility in the planned growth of a system. The DC breaker ratings are not likely to exceed the full load ratings as the control intervention is expected to limit the fault current.

The control and protection of MTDC systems is not a straightforward extension of that used in the two-terminal DC systems. The possibility of decentralized control necessitated by communication failure, the coordination of control and protection are some of the issues, currently being studied.

Conversion of existing AC lines

The constraints on RoW are forcing some utilities to look into the option of converting existing AC circuits to DC in order to increase the power transfer limit. There could be some operational problems due to electromagnetic induction from AC circuits operating in the same RoW.

An experimental project of converting a single circuit of a double circuit 220 kV line was commissioned in 1989–90 in India between Lower Sileru and Barsoor (under the National HVDC Project). It is feasible to convert a double circuit AC line to a bipolar HVDC line with substantial power upgrading (about 3.5 times). [17]

Operation with weak AC systems

The strength of AC systems connected to the terminals of a DC link is measured in terms of short circuit ratio (SCR) which is defined as

$$\text{SCR} = \frac{\text{Short circuit level at the converter bus}}{\text{Rated DC power}}$$

If SCR is less than 3, the AC system is said to be weak. The conventional constant extinction angle control may not be satisfactory with weak AC system. The recovery of inverters following the clearing of fault in the connected AC system can also be problematic.

Constant reactive current control or AC voltage control have been suggested to overcome some of the problems with weak AC systems. The use of fast reactive power control at the converter bus by applying static var systems is another alternative. Limiting dynamic overvoltages through converter control during load rejection is becoming a standard practice.

The power modulation techniques used to improve dynamic stability of power systems will have to be modified in the presence of weak AC systems. Coordinated reactive and active power modulation has been suggested to overcome the problems of voltage variations that can limit the effectiveness of power modulation.

Active DC filter

In the nineties, a hybrid filter made up of an active filter (based on voltage source converter) in series with the passive filter has been developed to improve the filtering of harmonic currents flowing in the HVDC lines. The active filter can eliminate both characteristic and low frequency non-characteristic harmonics. Both Siemens and ABB have supplied active filters. In India, Chandrapur-Padghe HVDC project uses an active filter in each pole.

Capacitor commutated converter (CCC)

This involves the connection of commutation capacitors in series with the valve side windings of the converter transformer. This is not strictly a new technology and has been studied under the (erroneous) label of 'force commutated' converters. The advantages of CCC are—improved voltage stability when operating with weak AC systems, reduced risks of commutation failures, less load rejection overvoltages and reactive power requirements. The CCC can be combined with AC filters that have low reactive power ratings. These can be automatically tuned passive filters or active filters. The CCC technology was first applied in Garabi BTB project for power exchange between Argentina and Brazil. This is rated at 1000 MW and SCR can be as low as 1.5. Another project incorporating CCC is a 200 MW station at Rapid City, S. Dakota in U.S.A. commissioned in 2003. This is a BTB HVDC link that interconnects eastern U.S.A. system with western system. (There is no synchronous interconnection between the two grids.)

UHV DC transmission

Ultra High Voltage(UHV) DC transmission at +/– 800 kV is being considered for transmission of large blocks of power above 3000 MW for distances beyond 1500 km, in China, India, South Africa/Congo and Brazil where large hydro power resources available in remote areas are developed. A bipolar HVDC line operating at +/– 800 kV can transmit power in the range up to 5000 or 6000 MW with 2 twelve-pulse converters per pole. It is claimed that information currently available on the flashover and withstand characteristics of air gaps is adequate for the design of +/– 800 kV systems. Thyristor ratings may have to be increased by adopting 6 inch diameter wafers.

1.7 SOME OPERATING PROBLEMS

While the growth of HVDC technology over the years has been impressive and resulted in several new projects all over the world, some problems have surfaced that affected reliable operation and in few cases even led to catastrophic failures. These problems are being addressed and not likely to pose major difficulties in the adoption of HVDC technology by system planners.

1. Converter Transformers

CIERE AG 04 statistics reveal many failures in converter transformers have been occurring (although concentrated in a few HVDC projects). The major problem areas are: turn to turn failures in valve side windings, leads and cleats failures, DC bushing failures. Although the failure rates of converter transformers are not too far from the failure rate of AC transformers, the consequences of the HVDC converter transformer failures on the system operation are much more severe.

The effect of repetitive commutation transients has been identified as potentially harmful, particularly during high firing angle operations. This is aggravated by the effect of the Cu_2S depositions on the electric field distribution. The generation of Cu_2S is caused by high electrical stresses, which is under investigation.

It has been shown that the commutation failures in the upper rectifier bridge may result in voltage stresses in the lower bridge in excess of the stresses caused during the DC insulation tests.

In addition to the commutation transients, other high frequency transients generated by adjacent system (AC or DC) may cause internal winding resonances that can pose problems. These transients may be affected by substation design/layout, grounding system etc. Test specification may have to be changed to ensure that transformer functions satisfactorily in service.

2. Flashover Performance of HVDC Converter Station Insulators

Insulators such as wall bushings, capacitive voltage dividers and voltage transformers suffer external flashovers caused by rain, fog, snow and ice. At some converter stations, there were several flashovers at the beginning of their operation. The maximum flashover rate occurred between 1987 and 1989 involving about 20 flashovers per year. Most flashovers occurred in valve bushings followed by large-diameter insulators. The wall bushing flashovers occurred mainly in rain and at negative polarity.

The application of booster sheds, silicone grease coating, RTV silicone rubber coating etc. have resulted in reducing the flashovers per converter stations by a factor of ten to about 0.1 per station per year. Recently dry type bushings (with SF_6 insulation) have been developed which are also safer from fire risks.

3. Valve Hall Fires

In 1989, a converter in Itaipu HVDC station (in Brazil) was totally destroyed due to a fire, probably caused by a leakage in cooling system associated with corrosion. Subsequently, there were fires in valve halls in Rihand (India), Sylmar (USA) and Gezhouba (China).

In the Sylmar converter station of Pacific DC Intertie, the valves were completely destroyed in 1993. The adjacent valve hall had extensive smoke damage requiring much arduous clearing. Incidentally Sylmar station was also affected by the 1994 earthquake. The building having mercury arc valves was extensively damaged although the equipment damage was not as severe as in the 1970 earthquake. There were three fires in Nanqiao converter station (China) that were caused by thyristor electronics, water cooled resistor of the valve reactor and water leakage. The fire in Sylmar station is also ascribed to thyristor electronics. The fire in Gezhouba converter station in China was caused by a loose connection between the reactor winding and copper connecter.

4. Problems of Ground Return

Although ground return during monopolar operation cuts down power losses, it can also cause problems with buried metallic structures and three phase AC transformers with grounded neutral. The DC current flowing through the ground can enter the neutral and flow through the transformer windings and AC Lines. The DC current flowing through the three phase windings tend to saturate the magnetic cores unequally. The current magnitudes depend on the potential of the ground mats where the transformers are located. As the magnitudes increase, there is a distinct audible hum (at the fundamental frequency). The increase of the hum is noticeable when the DC current per phase exceeds the magnitude of the magnetizing current by a ratio of 1.2 to 1.5 (note that, typically, the magnetizing current is about 0.5% of the rated current). At higher currents, the transformers may overheat due to increase in iron losses.

The solutions to this problem are not simple. Blocking devices (capacitor in shunt with a protection arrestor) are generally connected between the neutral and ground mat. Sometimes, the neutrals can be grounded through resistors (of a few ohms). Alternately, capacitors may be connected in series with the three phases (of the line to which the transformer is connected).

The problems of the ground return can be avoided if we have:

(*a*) Monopolar operation with or without a dedicated metallic return path. The scheme without a dedicated metallic return path (and electrodes) is grounded only at one of the terminals and results in the lowest cost. In this case, monopolar operation is feasible during converter pole outage but not during the HVDC line conductor outage.

(*b*) Restriction on the ground current flowing during the monopolar operation.

Although problems have been experienced in the application of HVDC technology under various operating conditions, these have been tackled successfully. Hence they do not pose restrictions on the consideration of HVDC transmission solutions in planning the expansion of the power transmission network.

A major innovation in applying Voltage Source Converters (VSC) using IGBT valves, instead of Line Commuted Converters (LCC) using thyristor valves is an exciting development of recent times and expands the scope of application of HVDC technology. This is discussed in the next section.

1.8 HVDC TRANSMISSION BASED ON VOLTAGE SOURCE CONVERTERS

The first VSC based HVDC link was commissioned at Hellsjon in Sweden in 1997 with a modest rating of 3 MW. Since then, 8 projects have been commissioned, the last one in 2006 at Eastlink as part of Baltic ring in European Union that allows power exchange between Baltic countries (Estonia, Latvia and Lithuania) and the Nordel grid. It is rated at 350 MW, ±150 kV and involves sea crossing of 74 km in addition to land cable of 31 km. All projects so far have been executed by ABB, who call this technology as HVDC Light. They are also executing three more projects. One of the projects involves interconnection of largest offshore wind park in Germany, rated 400 MW. Another project in Namibia rated at 300 MW is the first time, the VSC–HVDC will be used for long overhead transmission lines. Siemens labels it as HVDC^{PLUS} where PLUS stands for 'Power Link Universal Systems'. The Transbay project in California connecting San Francisco to Pittsburg is rated at ± 200 kV and 400 MW, is being executed by Siemens and is likely to commission by 2010.

VSC based HVDC link can be viewed as two STATCOMs, at two terminals that are interconnected by a DC link. A BTB VSC based HVDC link can be operated as two parallel STATCOMs when their DC capacitors are not connected (see Fig. 1.9). By closing the switches S_1 and S_2 and opening switch S the configurations changes to a BTB HVDC link. While SVC was an offshoot of the conventional (LCC) HVDC technology using thyristor valves, VSC HVDC is an extension of the emerging FACTS controllers based on VSC.

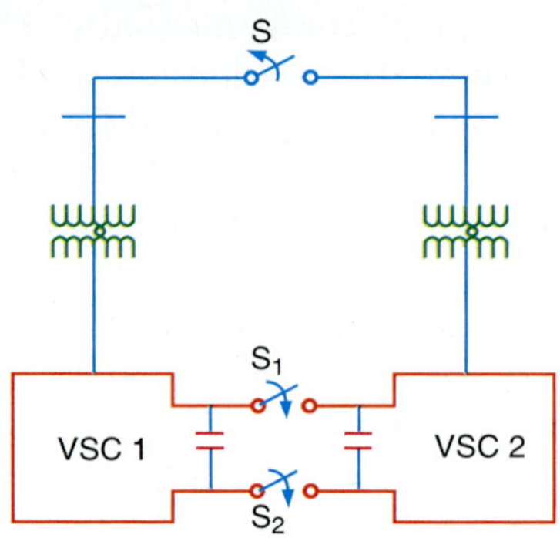

Fig. 1.9: Conversion of two STATCOMs to BTB VSC based HVDC link

VSC based HVDC links have several advantage over LCC based (or classical) HVDC links. Primarily there are:

1. A Voltage Source Converter (VSC) permits independent control over both active and reactive power outputs. In a LCC, the reactive power is a function of the active power drawn or supplied. In a classical (LCC based) HVDC link, reactive power is to be supplied at both rectifier and inverter stations. In a VSC based HVDC link, reactive power output can be positive or negative. A converter station can supply or absorb reactive power from the system and can help in regulating the bus voltage.

   In a two terminal, VSC based HVDC link, one of the terminal is used to control the power flow in the link, while the other regulates the voltage in the DC link.
2. Unlike in a LCC, a VSC does not require a voltage source for satisfactory operation. This implies, that a VSC based converter station can be used to supply passive loads. There are no problems of commutation failures in a VSC.
3. The harmonic filtering required is much simpler in a VSC as Pulse Width Modulation (PWM) is used to shift the frequency spectrum of the converter output. Using IGBT devices, the carrier frequency of 2 kHz is used for modulation.
4. Fast communication between the terminals is not required for control purposes.
5. The power reversal in the link does not require voltage reversal. This enables better control during multiterminal DC (MTDC) system operation.

VSC based HVDC links are equipped with extruded polymer insulated HVDC cables instead of oil filled, impregnated paper insulated cables.

VSC based HVDC links are also applied for power transfer from offshore wind power plants and supply to offshore platforms (in addition to asynchronous and cable transmission). A single line diagram of a VSC station is shown in Fig. 1.10. The major components at a converter station (excluding surge arrestors and circuit breakers) are shown in this diagram. The series reactor (typical value = 0.15 p.u) is used to isolate the injected voltage by the VSC from the converter bus (on the secondary side of the converter transformer) and limit the current harmonics. Pulse Width Modulation (PWM) ensures the current contains predominantly higher frequency components which are effectively eliminated by the action of series reactor and high pass AC filter. The converter transformer is the normal AC transformer with only sinusoidal voltages and currents in steady state. A separate series connected RI and PLC filters can be used to

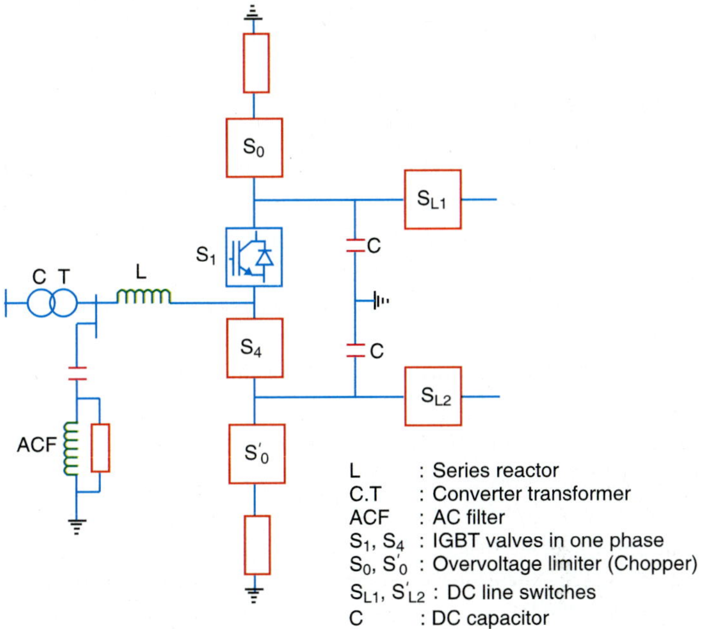

Fig. 1.10: A single line diagram of a VSC based HVDC convertor station

eliminate very high frequencies. Overvoltage limiter (chopper) is required for fast discharge of the DC capacitor if the DC voltage exceeds the maximum DC bus voltage for deblocked (conducting) converter. This function is realized by a fast switch and a resistor. DC line switches are required for fast isolation of the converter in the event of DC line faults. IGBT valves similar to the valves used in the Graetz bridge can be used as line switches and in chopper.

REFERENCES AND BIBLIOGRAPHY

1. Adamson C. and Hingorani N.G., *'High Voltage Direct Current Power Transmission'*, (London) Garraway, 1960.
2. Kimbark E.W., *'Direct Current Transmission'*, Vol. I (New York) John Wiley, 1971.
3. Uhlman E., *'Power Transmission by Direct Current'*, (Berlin) Springer-Verlag, 1975.
4. Arrillaga J., *'High Voltage Direct Current Transmission'*, Second Edition, (London) Peter Peregrinus, the Institution of Electrical Engineers, 1998.
5. Lava V.S., *'An Annotated Bibliography of High Voltage Direct Current Transmission'*, Pootland, Oregon: Bonneville Power Administration, 1966–68, 1969–1983.
6. Cory B.J., *'High Voltage Direct Current Power Converters and Systems'*, London: McDonald, 1965.
7. Meredith R.J., et al., *'Methodology for Integration of HVDC Links in Large AC Systems'*–Phase I: Reference Manual. Final Report EL-3004, RP 1964-1, Electric Power Research Institute, March 1983.

8. Hardy J.E., et al., '*AC or DC? One Utilities Approach*', IEE Conf. Publication No. 205, December 1981.

9. Carlsson L., '*Recent Developments in HVDC Converter Station Design*', IEEE Trans. Vol. PAS-103, No. 8, pp. 2166–71, August 1984.

10. Arrillaga J. and Smith B., '*AC-DC Power System Analysis*', London, The Institution of Electrical Engineers 1998.

11. Sood V.K., '*HVDC and FACTS Controllers—Application of Static Converters in Power Systems*', Boston, Kluwer Academic Publishers 2004.

12. Hingorani N.G., '*High Voltage DC Transmission–A Power Electronics Work Horse*', IEEE Spectrum, pp. 63–72 April 1996.

13. Asplund G., Eriksson K. and Stevenson K., '*DC Transmission Based on Voltage Source Converters*', CIGRE SC14 Colloquium, South Africa, 1997.

14. Hingorani H.G. and Gyugyi L., '*Understanding FACTS—Concepts and Technology of Flexible AC Transmission Systems*', New York, IEEE Press, 2000.

15. Long W. and Nilsson S. '*HVDC Transmission: Yesterday and Today*', IEEE Power & Energy Magazine, pp. 22–31, March/April 2007.

16. Padiyar K.R., '*FACTS Controllers in Power Transmission and Distribution*', New Delhi, New Age Int. Publishers, 2007.

17. Clerici A., Paris L. and Danfors P., '*HVDC Conversion of HVAC lines to provide substantial power upgrading*', IEEE Trans., Power Delivery, Vol. 6, pp. 324–333, 1991.

18. Lescale V.F., et al., '*Power Transmission with HVDC at 800 kV*', paper B4-106, CIGRE 2006.

19. Asplund G., et al., '*DC Transmission based on Voltage Source Converters*', paper 14–302, CIGRE, 1998.

20. Schettler F., Huang H. and Christl N., '*HVDC Transmission Systems using Voltage Sourced Converters-Design and Applications*', Summer Meeting, IEEE PES, 2000.

21. Agelidis V.G., Demetriades G.D. and Flourentzou N., '*Recent Advances in High-Voltage Direct-Current Power Transmission Systems*', National Power Electronics Conference, Xian, China, September 2006.

22. www.ece.uidaho.edu/HVDCfacts

23. Szectman M., Maruvada P.S. and Nayak R.N., '*800-kV HVDC on the Horizon*', IEEE Power and Energy, pp. 61–69, March/April 2007.

LINE COMMUTATED AND VOLTAGE SOURCE CONVERTERS

2

2.1 INTRODUCTION

In this chapter we will take up the study of converter circuits used in both Line Commutated (current source) Converters (LCC) and Voltage Source Converter (VSC) stations.

The basic configuration of a three phase converter (both LCC and VSC) is a bridge converter (also called Graetz Bridge) which can be fed from transformer windings connected in star or delta. The converter transformer feeding a Graetz bridge serves the objectives of providing

(i) galvanic separation between AC and DC sides

(ii) voltage transformation between AC and DC networks

(iii) adjustment of the applied AC voltage by On Load Tap Changer (OLTC).

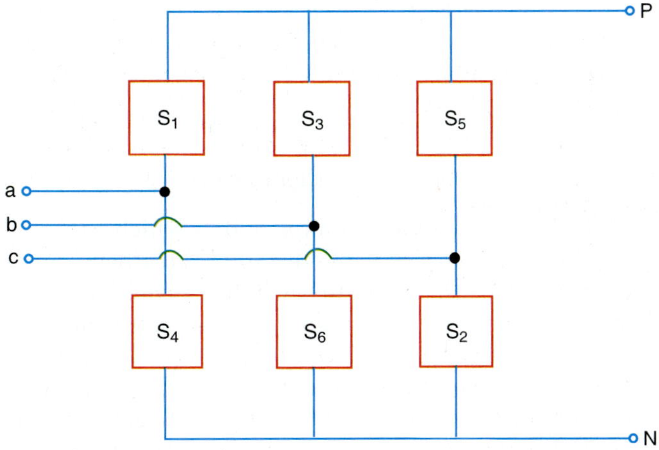

Fig. 2.1: A Graetz bridge

Note that an autotransformer can meet the last two objectives but cannot provide galvanic separation. The Graetz bridge circuit is shown in Fig. 2.1.

There are 6 switches numbered in the order in which they are turned on. In a LCC, the switches are made of thyristor valves, whereas in a VSC, they are made up of IGBT valves (a series connection of IGBT and anti-parallel connected diode). It is assumed that the phase sequence is 'abc'.

2.2 LINE COMMUTATED CONVERTER

2.2.1 Analysis of Graetz Bridge Neglecting Overlap

In a LCC, the Graetz Bridge is connected to three balanced sinusoidal voltage sources on the AC side and constant DC current source on the DC side as shown in Fig. 2.2.

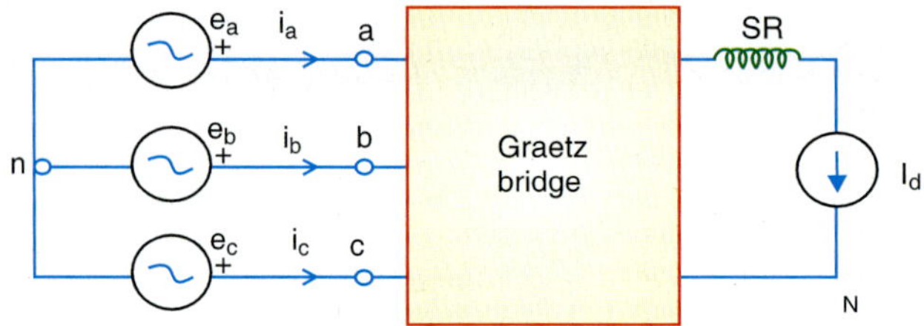

Fig. 2.2: Line commutated converter

Each of the switches S_1 to S_6 in the Graetz bridge are made of series connected thyristor devices (to provide the required voltage rating). The description of thyristor valve is given in Appendix A. The valve is denoted by the thyristor symbol shown in Fig. 2.3.

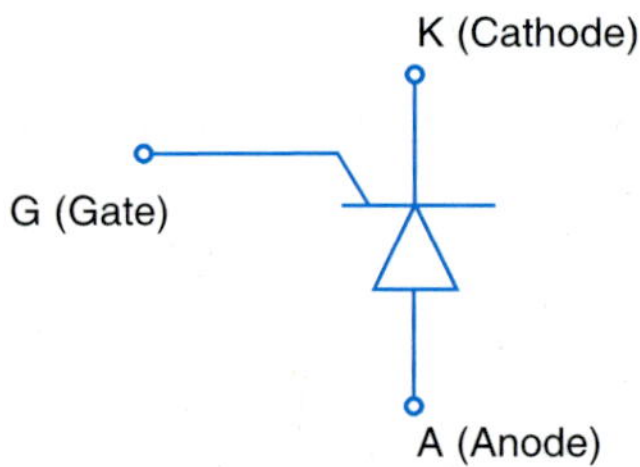

Fig. 2.3: Thyristor symbol

A thyristor valve (ideally) can be viewed as a switch which can be turned on when the terminal A (anode) is positive with respect to the terminal K (cathode) and a gate (firing) signal is provided at the terminal G. Once the thyristor switch is turned on, it can be turned off only when the current through it goes to zero and there is a minimum commutation margin (when the voltage across the thyristor is reverse biased – the voltage at terminal A is negative with respect to K). In modern high voltage thyristor device (rated at 8 kV and above), the margin required may be hundreds of microseconds. In general, when no gate pulse is present, the switch should withstand both positive and negative voltages. Neglecting losses in the thyristor valves, the voltage across the device is zero when it is turned on and the current through the devices is zero when it is not conducting. In other words, the valves can be viewed as ideal switches subject to the constraints mentioned above.

Note that the current can flow only in one direction through the thyristor switch (from anode to cathode) when it is on. Thus, the switches S_1, S_3 and S_5 carry current towards the terminal 'P' while the switches S_2, S_4 and S_6 carry current from the terminal 'N'. The direct current I_d has to flow through at least one valve in the upper group (S_1, S_3 and S_5) and one valve in the lower group (S_2, S_4 and S_6). Neglecting overlap between any two valves in a group, we can state that only one valve conducts in the upper group or lower group. Thus, in a cycle of the applied AC voltage, these are six equal intervals as shown in Fig. 2.4.

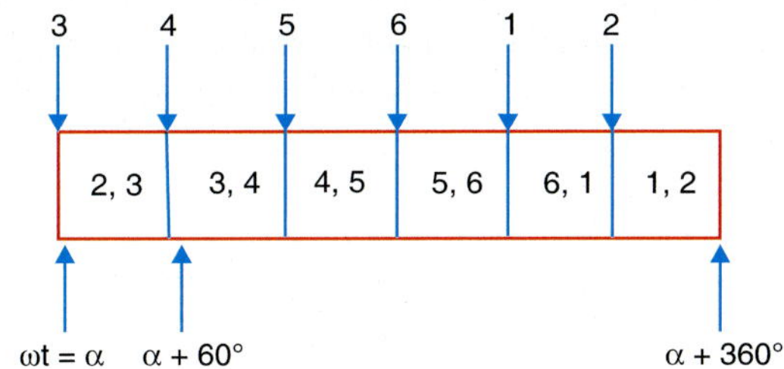

Fig. 2.4: Timing diagram

We assume that the switch S_3 is turned on at $\omega t = \alpha$. Prior to this, switches S_1 and S_2 are conducting and we also assume that the switch S_1 turns off as soon as the switch S_3 turns on (neglecting the overlap). Overlap between the switches S_1 and S_3 cannot be neglected if there are inductors in the circuit (due to leakage reactance of the converter transformer). We will analyze the converter circuit with overlap in the next chapter. Table 2.1 shows the DC voltages across the converter (V_{PN}) for each of the six intervals.

Table 2.1: DC Bus Voltage During Each Interval

Interval	I	II	III	IV	V	VI
Valves Conducting	2,3	3,4	4,5	5,6	6,1	1,2
Voltage (V_{PN})	e_{bc}	e_{ba}	e_{ca}	e_{cb}	e_{ab}	e_{ac}

Note that the voltage across the valve (switch S_3) is e_{ba} before it is fired (turned on). If we denote e_{ba}

$$e_{ba} = \sqrt{2}E_{LL} \sin \omega t \tag{2.1}$$

the voltage across the switch S_3 is negative when $\omega t < 0$ and becomes positive when $\omega t > 0$. For a thyristor valve, we can delay the turn on by a delay angle (α) by applying a gate pulse at $\omega t = \alpha$. It is assumed that the delay angle is constant in steady state.

The expression for e_{ba} given in Equation (2.1) implies that

$$\left.\begin{aligned} e_a(t) &= \sqrt{\frac{2}{3}}E_{LL} \sin(\omega t + 150^\circ) \\ e_b(t) &= \sqrt{\frac{2}{3}}E_{LL} \sin(\omega t + 30^\circ) \\ e_c(t) &= \sqrt{\frac{2}{3}}E_{LL} \sin(\omega t - 90^\circ) \end{aligned}\right\} \tag{2.2}$$

Note that we assume only positive sequence supply voltages. The expressions for the DC voltage (v_{PN}) and voltage across valve 1 for each interval, are shown in Table 2.2. The waveforms of these voltages for $\alpha = 0°$, 45°, 90°, 135° and 180° are shown in Fig. 2.5. The DC voltage waveform shows that there are six pulsations (cycles of ripple) per cycle of the AC voltage. The pulse number (which is defined as the ratio of the base frequency of the DC voltage ripple to the fundamental frequency of the AC voltage) for Graetz bridge is six. We also note that the average DC voltage reduces (and ripple content increases) as α is increased from 0° to 90°. On the other hand, as α is increased from 90° to 180°, the average DC voltage (in the negative direction) increases and the ripple content reduces.

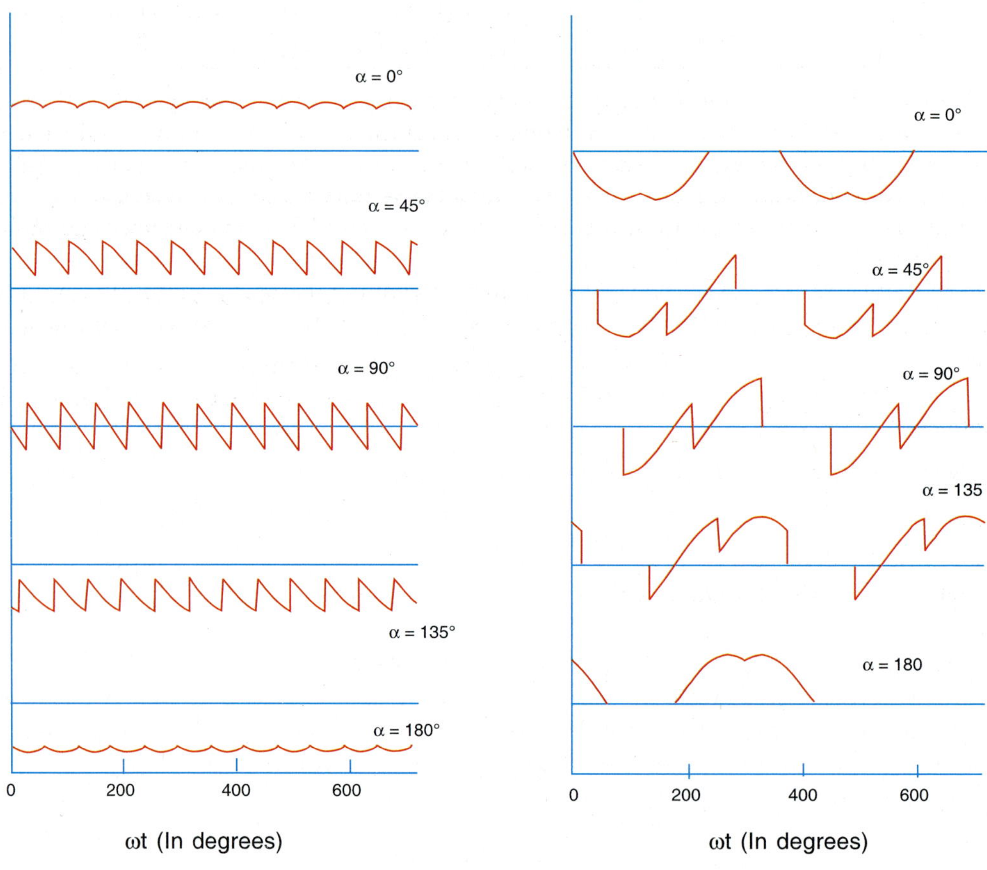

Fig. 2.5: DC and valve voltage waveforms

Table 2.2: DC and Valve Voltages

Interval No.	Conducting Valves	DC Voltage	Voltage Across Valve 1
I	2, 3	$\sqrt{2}E_{LL}\sin(\omega t+60°)$	$e_{ab}=-\sqrt{2}E_{LL}\sin(\omega t)$
II	3, 4	$\sqrt{2}E_{LL}\sin(\omega t)$	$e_{ab}=-\sqrt{2}E_{LL}\sin(\omega t)$
III	4, 5	$\sqrt{2}E_{LL}\sin(\omega t-60°)$	$e_{ac}=\sqrt{2}E_{LL}\sin(\omega t+120°)$
IV	5, 6	$\sqrt{2}E_{LL}\sin(\omega t-120°)$	$e_{ac}=\sqrt{2}E_{LL}\sin(\omega t+120°)$
V	6, 1	$\sqrt{2}E_{LL}\sin(\omega t-180°)$	0
VI	1, 2	$\sqrt{2}E_{LL}\sin(\omega t-240°)$	0

The DC voltage waveform repeats for each interval and hence average DC voltage can be found by taking the average of v_{PN} over one interval during which, a pair of valves conduct. This is based on the assumption that the applied voltages are balanced, sinusoidal (defined by Eq. 2.2) and E_{LL} remains constant (over a cycle of the supply frequency). We also assume that the firing of the valves is equidistant (at the interval of 60°). This also implies that each valve conducts for 120° duration.

Let us consider the interval corresponding to the conduction of valves 2 and 3. At the beginning of the interval, valve 3 is fired and at the end of the interval, valve 2 turns off.

Assuming the firing of valve 3 is delayed by an angle α, (that is $\alpha°$ after the zero crossing of the commutation voltage for valve 3 –voltage e_{ba}), the instantaneous DC voltage v_d during the interval is given by

$$v_d = v_{PN} = e_b - e_c = e_{bc}, \quad \alpha \le \omega t \le \alpha + 60°$$

If, $\quad e_{ba} = \sqrt{2}E_{LL}\sin\omega t$ from Eq. 2.1

then, $\quad e_{bc} = \sqrt{2}E_{LL}\sin(\omega t + 60°)$

$$\text{Average DC voltage} = V_d = \frac{3}{\pi}\int_{\alpha}^{\alpha+60°} \sqrt{2}E_{LL}\sin(\omega t + 60°)d\omega t$$

$$= \frac{3}{\pi}\sqrt{2}E_{LL}\left[\cos(\alpha+60°) - \cos(\alpha+120°)\right]$$

$$V_d = \frac{3\sqrt{2}}{\pi}E_{LL}\cos\alpha = 1.35E_{LL}\cos\alpha$$

$$= v_{do}\cos\alpha \tag{2.3}$$

Equation (2.3) indicates that for different values of α, V_d is variable. The range of α is 180° (from 0° to 180°) and correspondingly V_d can vary from V_{do} to $-V_{do}$. Thus the same converter can act as a rectifier or inverter depending upon whether the DC voltage is positive or negative.

It is to be noted that this is based on the assumption of continuous conduction of current by any valve over the 120° interval. Note that the DC current flow is unidirectional.

DC voltage harmonics

The DC voltage waveform contains a ripple whose fundamental frequency is six times the supply frequency. This can be analysed in Fourier series and contains harmonics of the order

$$h = 6n$$

where n is an integer.

The harmonic (rms value) in the DC voltage is given by

$$V_h = \frac{\sqrt{a_h^2 + b_h^2}}{\sqrt{2}}$$

where, $$a_h = \frac{1}{\pi}\int_{\alpha}^{\alpha+2\pi} V_d \cos(h\theta)d\theta,\ b_h = \frac{1}{\pi}\int_{\alpha}^{\alpha+2\pi} V_d \sin(h\theta)d\theta$$

It can be shown that the rms value of the hth order harmonic in DC voltage is given by

$$V_h = V_{do}\frac{\sqrt{2}}{h^2-1}\left[1+(h^2-1)\sin^2\alpha\right]^{1/2} \tag{2.4}$$

Valve voltage waveform

Figure 2.5 also shows the valve voltage waveform which shows three voltage jumps that arise due to the commutation from one valve to the next in the same commutation group. The valve voltage waveform can be obtained from the expressions for a valve voltage given in Table 2.2 (Note that the waveform is similar for all valves). The voltage jumps are all of the same magnitude given by

$$V_j = \sqrt{2}E_{LL}\sin\alpha = (\pi/3)V_{do}\sin\alpha \tag{2.5}$$

These voltage jumps contribute to the voltage stresses (dv/dt) across the valve. Valve damping circuits are provided to relieve these stresses.

Although α can vary from 0° to 180°, the full range cannot be utilized. In order to ensure the firing of all the series connected thyristors, it is necessary to provide a minimum limit of α greater than zero, (say 5°). Also, in order to allow for the turn-off time of a valve, it is necessary to provide an upper limit less than 180°. The delay angle α is not allowed to go beyond (180°–γ) where γ is called the extinction angle (sometimes also called the margin angle). The minimum value of the extinction angle is typically 10°, although in normal operation as an inverter, it is not allowed to go below 15° or 18°.

In the absence of forced commutation, α cannot be made negative.

AC current waveform

It is assumed that the direct current has no ripple (or harmonics). This is normally valid because of the smoothing reactor provided in series with the bridge circuit. The AC currents flowing through the valve (secondary) and primary windings of the converter transformer contain harmonics. The waveform of the current in a valve winding (say in phase b) is shown in Fig. 2.6.

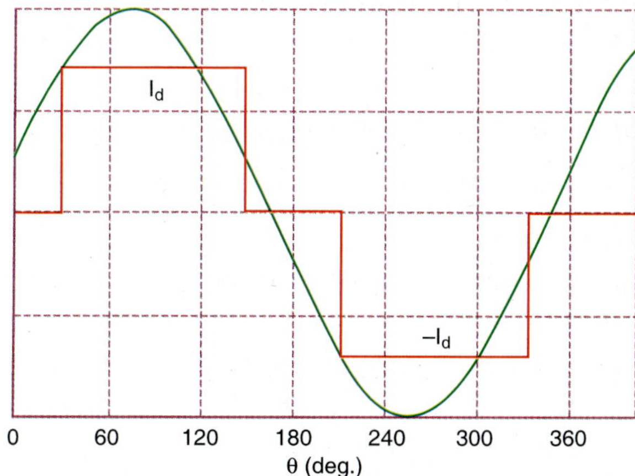

Fig. 2.6: Waveforms of i_b and e_b

The current in a valve winding is determined by the currents flowing in the valves connected to that winding. For example, the current in the phase 'b', i_b (see Fig. 2.2) is i_3 when valve 3 is conducting and $-i_6$ when valve 6 is conducting. Note that each valve conducts for 120° and valve 6 is turned on 180° after the firing of valve 3. Figure 2.6 also shows the applied voltage $e_b = \sqrt{2}E_{LL}\sin(\omega t + 30°)$. From equations (2.1) and (2.2), we have

$$i_b = i_3 = I_d,\ \alpha \le \omega t \le \alpha + 120°$$
$$= -i_6 = -I_d,\ \alpha + 180° \le \omega t \le \alpha + 300°$$

Defining $\theta = \omega t - \alpha - 60°$, the rms value of the fundamental component is given by

$$I_1 = \frac{1}{\sqrt{2}}\frac{2}{\pi}\int_{-\frac{\pi}{3}}^{\frac{\pi}{3}} I_d \cos\theta\, d\theta = \frac{\sqrt{6}}{\pi}I_d \tag{2.6}$$

whereas rms value of the current is

$$I = \sqrt{\frac{2}{3}}I_d \tag{2.7}$$

The harmonics contained in the waveform are of the order given by

$$h = 6n \pm 1$$

where n is an integer. Thus, the order of AC harmonics are 5, 7, 11, 13 and higher. These are filtered out by using tuned filters for each one of the first four harmonics and a high pass filter for the rest. The rms value of the hth harmonic is given by

$$I_h = \frac{I_1}{h} \tag{2.8}$$

The power factor

The AC power supplied to the converter is given by

$$P_{ac} = \sqrt{3}E_{LL}I_1\cos\phi$$

where ϕ is the (fundamental frequency) power factor.

The DC power must match the AC power ignoring the losses in the converter. Thus, we get

$$P_{ac} = P_{dc} = V_d I_d = \sqrt{3} E_{LL} I_1 \cos\phi \tag{2.9}$$

Substituting for V_d and I_1 from equations (2.3) and (2.6) in the above equation we obtain

$$\cos\phi = \cos\alpha \tag{2.10}$$

The reactive power requirements are increased as α is increased from 0 (or reduced from 180°). When $\alpha = 90°$, the power factor is zero and only reactive power is consumed.

2.2.2 Choice of Converter Configuration for any Pulse Number

We mentioned earlier that Graetz bridge is a six pulse converter for which the lowest DC voltage harmonic is sixth. Correspondingly, lowest AC current harmonics are 5th and 7th. To reduce the harmonic content in the AC current and DC voltage, it is desirable to use higher pulse numbers. In general, it can be stated that the characteristic harmonics (under ideal conditions) are of the order

$$h_{dc} = np, \; h_{ac} = np \pm 1 \tag{2.11}$$

where 'n' is an integer, 'p' is the pulse number.

There are several configurations for a converter of a specified pulse number. For $p = 6$, we have in addition to the Graetz bridge, six-phase diametral connection, cascade of three single phase full wave converters, cascade of two three phase converter, parallel connection with interphase transformer etc.

It is convenient to consider a 'p' pulse converter made up of series and parallel connections of a basic valve (commutation) group of 'q' valves or switches as shown in Fig. 2.7. Here, the switch is a thyristor valve whose firing can be delayed (from the instant when the valve voltage becomes positive). The voltage sources are actually obtained from the converter transformer windings. Neglecting overlap, only one valve conducts in a commutation group of 'q' valves.

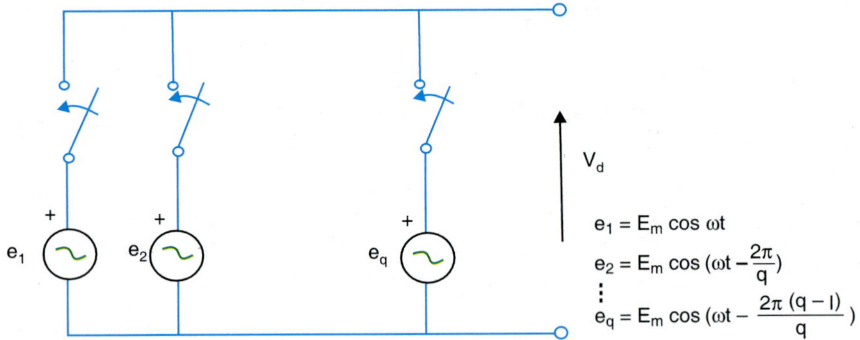

Fig. 2.7: A valve group with 'q' valves

If the converter is made up of a matrix of 's' valve groups in series and 'r' valve groups in parallel, then,

$$p = qrs \tag{2.12}$$

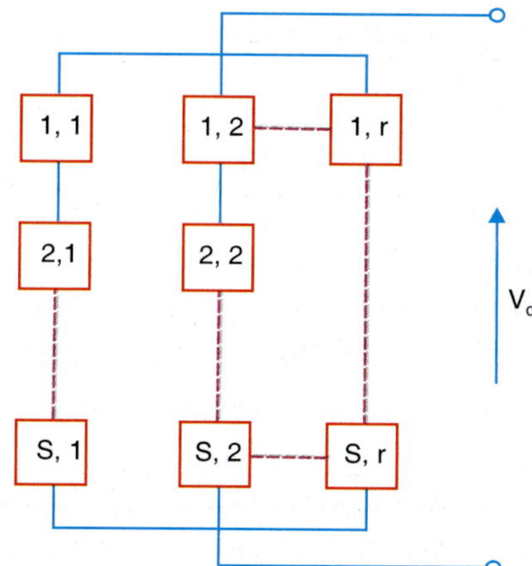

Fig. 2.8: Converter made up of series and parallel connection of communication groups

See Fig. 2.8 for the converter configuration. In general, there are '*p*' transformer windings. It will be shown later that sometimes windings can be combined (in particular for $q = 3$, $r = 1$, $s = 2$).

Valve rating

The valve voltage rating is specified in terms of peak inverse voltage (PIV) it has to withstand. The ratio of PIV to the average DC voltage is an index of the valve utilization. The average maximum DC voltage across the converter is given by

$$V_{do} = s\frac{q}{2\pi}\int_{-\frac{\pi}{q}}^{\frac{\pi}{q}} E_m \cos\omega t \, d\omega t$$

$$= \frac{sq}{\pi} E_m \sin\frac{\pi}{q} \qquad (2.13)$$

The peak inverse voltage (PIV) across a valve can be obtained as follows:

If '*q*' is even, then the maximum inverse voltage occurs when the valve with a phase displacement of π radian (180°) is conducting and this is given by

$$PIV = 2E_m$$

If '*q*' is odd, maximum inverse voltage occurs when the valve with a phase shift of $\pi \pm \pi/q$ is conducting. In this case,

$$PIV = 2E_m \cos\frac{\pi}{2q} \qquad (2.14)$$

The valve utilization factor is given by

$$\frac{PIV}{V_{do}} = \frac{2\pi}{sq\sin\frac{\pi}{q}} \text{ for } q \text{ even}$$

$$= \frac{\pi}{sq\sin\frac{\pi}{2q}} \text{ for } q \text{ odd}$$

Table 2.3 shows the valve utilization factor for different six pulse converter configurations. The best valve utilization is obtained for configurations 1 and 3.

Table 2.3: Valve Utilization Factor

Sl. No.	q	r	s	$\frac{PIV}{V_{do}}$
1	2	1	3	1.047
2	2	3	1	3.142
3	3	1	2	1.047
4	3	2	1	2.094
5	6	1	1	2.094

Transformer rating

The current rating of a valve (as well as transformer winding supplying it) is given by

$$I_v = \frac{I_d}{r\sqrt{q}} \tag{2.15}$$

where I_d is the DC current which is assumed to be a constant. The transformer rating on the valve side (in volt amperes) is given by

$$S_{tv} = p\frac{E_m}{\sqrt{2}}I_v$$

$$= p\frac{V_{do}\pi}{\sqrt{2}sq\sin\frac{\pi}{q}}\frac{I_d}{r\sqrt{q}}$$

$$= \frac{\pi}{\sqrt{2}}\frac{V_{do}I_d}{\sqrt{q}\sin\frac{\pi}{q}} \tag{2.16}$$

The transformer utilization factor $\frac{S_{tv}}{V_{do}I_d}$ is only a function of 'q'. The optimum value of q which results in maximum utilization is equal to 3. It is a fortunate coincidence that the AC power supply is 3 phase and the commutation group of 3 valves is easily arranged.

For $q = 3$,

$$\frac{S_{tv}}{V_{do}I_d} = 1.481 \tag{2.17}$$

The transformer utilization can be improved further if two valve groups can share a single transformer winding. In this case, the current (rms) rating of the winding can be increased by a factor of $\sqrt{2}$ while decreasing the number of windings by a factor of 2. For this case,

$$\frac{S_{tv}}{V_{do}I_d} = 1.047 \tag{2.18}$$

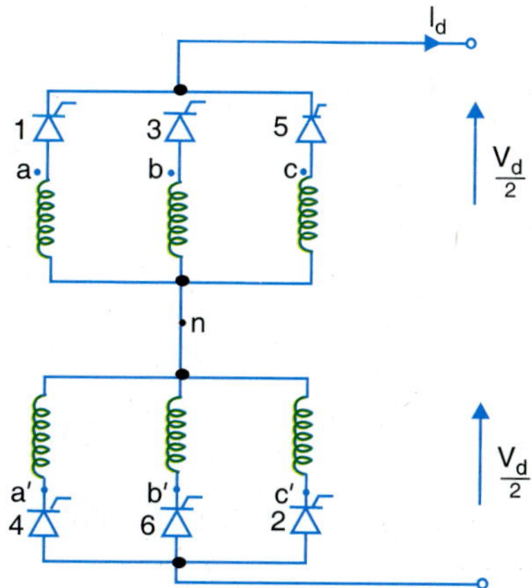

Fig. 2.9: Six pulse converter

For a 6 pulse converter, this can be easily arranged. The Graetz circuit shown in Fig. 2.1 is obtained when the two winding (shown in Fig. 2.9) are combined into one. This is possible since the terminals a and a' are at the same potential (with respect to the supply neutral point n). Thus, it is shown that both from valve and transformer utilization considerations, Graetz circuit is the best circuit for a six pulse converter.

Twelve pulse converter

In HVDC transmission, the series conduction of converter groups has been preferred because of the ease of control and protection as well as the requirements of high voltage rating. Thus a 12 pulse converter is obtained by the series connection of two bridges. The 30° phase displacement between the two sets of source voltages is achieved by the transformer connections, Y/Y for feeding one bridge and Y/Δ for feeding the second bridge.

The use of 12 pulse converter is preferable over the six pulse converter because of the reduced filtering requirements. However, increase in pulse number beyond 12 is not practical because the non-characteristic harmonics (which arise due to asymmetry in firing, imbalance in supply etc.) are not eliminated.

Example 1

For a 12 pulse converter with $q = 4$, $s = 3$, $r = 1$, calculate the maximum DC power and transformer ratings (valve windings) if PIV rating of the valve is V and the rms current rating is I. Rework the problem if $q = 3$, $s = 4$, $r = 1$.

Solution

With even value of q, $PIV = 2E_m$

$$V_{do} = \frac{q}{\pi} sE_m \sin\frac{\pi}{q} = \frac{12}{\pi}\frac{V}{2}\sin\frac{\pi}{4} = \frac{3\sqrt{2}}{\pi}V = 1.35V$$

$$I = \frac{I_d}{r\sqrt{q}} = \frac{I_d}{2} \Rightarrow I_d = 2I$$

$$P_{d\max} = V_{do} I_d = 2.70VI$$

$$S_{tv} = p\frac{E_m}{\sqrt{2}}I = \frac{12V}{2\sqrt{2}}I = 4.243VI$$

with $q = 3$, $s = 4$, $r = 1$

$$PIV = 2E_m \cos\frac{\pi}{2q} = 2E_m \cos 30^o = \sqrt{3}E_m$$

$$V_{do} = \frac{qs}{\pi}E_m \sin\frac{\pi}{q} = \frac{12}{\pi}\frac{V}{\sqrt{3}}\sin 60^o = \frac{6V}{\pi} = 1.91V$$

$$I_d = \sqrt{qr}I = \sqrt{3}I,\ P_{d\max} = V_{do}I_d = 1.91 \times \sqrt{3}VI = 3.31\,VI$$

$$S_{tv} = 12\frac{E_m}{\sqrt{2}}I = \frac{12VI}{\sqrt{6}} = 4.9VI$$

with only six windings on the valve side

$$S_{tv} = \frac{4.9}{\sqrt{2}}VI = 3.465VI$$

2.2.3 Analysis of a 12 Pulse Converters

To analyze the operation of a 12 pulse converter, consider the converter circuit diagram shown in Fig. 2.10. This shows two transformers with primary windings connected in star and connected to the common supply voltage. The currents supplied by the supply are the sums of the currents flowing in the two primary windings. Thus,

$$\left.\begin{aligned} i_A &= i_{AS} + i_{AD} \\ i_B &= i_{BS} + i_{BD} \\ i_C &= i_{CS} + i_{CD} \end{aligned}\right\} \qquad (2.19)$$

In determining the voltage and current waveforms, there is no loss of generality, in assuming that the turns ratio of the transformer feeding bridge 1 as 1:1 and the turns ratio of the transformer feeding the second bridge as $1:\sqrt{3}$. Assuming ideal transformers, we will get the identical magnitude of the AC voltages applied to the two bridges. Assuming e_A, e_B and e_C are defined as follows:

$$e_A = e_{aS} = \sqrt{\frac{2}{3}} E_{LL} \sin(\omega t + 150^\circ)$$

$$e_B = e_{bS} = \sqrt{\frac{2}{3}} E_{LL} \sin(\omega t + 30^\circ)$$

$$e_C = e_{cS} = \sqrt{\frac{2}{3}} E_{LL} \sin(\omega t - 90^\circ)$$

We get the line to line voltages applied to the second bridge as

$$e_{baD} = -\sqrt{3} e_A = \sqrt{2} E_{LL} \sin(\omega t - 30^\circ) \qquad (2.20)$$

Comparing the above equation with the Eq. (2.1) we note that the voltages applied to bridge 2 lag the voltages applied to the bridge 1 by 30°. Assuming the identical delay angle for both the bridges, valve 3 in the second bridge will be fired 30° after firing the valve 3 in bridge 1.

Thus, in a 12 pulse converter, there are 12 intervals in a cycle, each of duration 30°. In each interval, 4 valves (2 each from the two bridges) conduct. Table 2.4 shows the conducting valves from each bridge for all the intervals (note that the subscripts S and D denote the bridges with star connected and delta connected windings respectively).

Table 2.4: Bridge Voltages in Each Interval

<table>
<tr><td>Interval</td><td>I (a)</td><td>I (b)</td><td>II (a)</td><td>II (b)</td><td>III (a)</td><td>III (b)</td><td>IV (a)</td><td>IV (b)</td><td>V (a)</td><td>V (b)</td><td>VI (a)</td><td>VI (b)</td></tr>
<tr><td>Conducting Valves Bridge 1</td><td colspan="2">2, 3</td><td colspan="2">3, 4</td><td colspan="2">4, 5</td><td colspan="2">5, 6</td><td colspan="2">6, 1</td><td colspan="2">1, 2</td></tr>
<tr><td>v_{d1}</td><td colspan="2">e_{bcS}</td><td colspan="2">e_{baS}</td><td colspan="2">e_{caS}</td><td colspan="2">e_{cbS}</td><td colspan="2">e_{abS}</td><td colspan="2">e_{acS}</td></tr>
<tr><td>Conducting Valves Bridge 2</td><td>1, 2</td><td colspan="2">2, 3</td><td colspan="2">3, 4</td><td colspan="2">4, 5</td><td colspan="2">5, 6</td><td colspan="2">6, 1</td><td>1, 2</td></tr>
<tr><td>v_{d2}</td><td>e_{acD}</td><td colspan="2">e_{bcD}</td><td colspan="2">e_{baD}</td><td colspan="2">e_{caD}</td><td colspan="2">e_{cbD}</td><td colspan="2">e_{abD}</td><td>e_{acD}</td></tr>
</table>

The DC voltage V_{dc} for a 12 pulse converter is the sum of the DC voltages for the individual bridges. Thus,

$$v_d = v_{d1} + v_{d2} \qquad (2.21)$$

For the interval I(a), the R.H.S. can be expressed as (see Table 2.1)

$$\begin{aligned} v_d &= e_{bcS} + e_{acD} \\ &= \sqrt{2}E_{LL}\sin(\omega t + 60^\circ) - \sqrt{2}E_{LL}\sin(\omega t - 270^\circ) \\ &= \sqrt{2}E_{LL}\left[\sin(\omega t + 60^\circ) + \cos\omega t\right] \\ &= E_m \sin(\omega t + 75^\circ) \end{aligned} \qquad (2.22)$$

where E_m is defined as

$$E_m = \sqrt{2}E_{LL}(2\cos 15°) = \sqrt{2}E_{LL}(1.93185) \tag{2.23}$$

The average DC voltage (V_{dc}) is given by

$$V_{dc} = \frac{6}{\pi}\left[\int_{\alpha}^{\alpha+30°} E_m \sin(\omega t + 75°)d\omega t\right]$$
$$= 2V_{do}\cos\alpha \tag{2.24}$$

where $V_{do} = \frac{3}{\pi}\sqrt{2}E_{LL}$ (see Eq. 2.3)

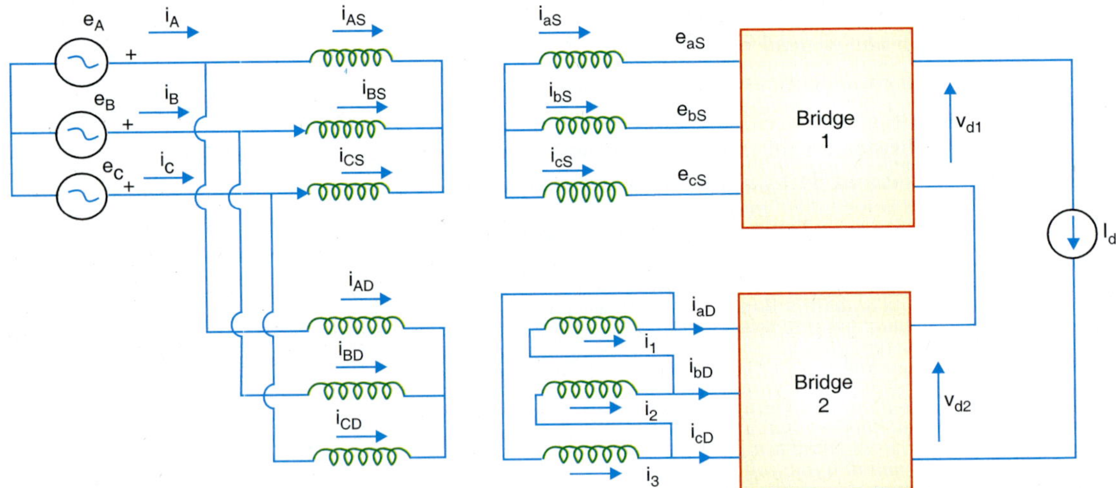

Fig. 2.10: A 12 pulse converter

The current i_{BD} flowing in the primary winding of the star/delta connected transformer (feeding bridge 2 in Fig. 2.10) is given by

$$i_{BD} = \sqrt{3}i_2 \tag{2.25}$$

where i_2 is the current flowing in the secondary winding that is coupled with phase B. Assuming that $i_1 + i_2 + i_3 = 0$, we can show that

$$i_{BD} = \frac{1}{\sqrt{3}}(i_{bD} - i_{cD}) \tag{2.26}$$

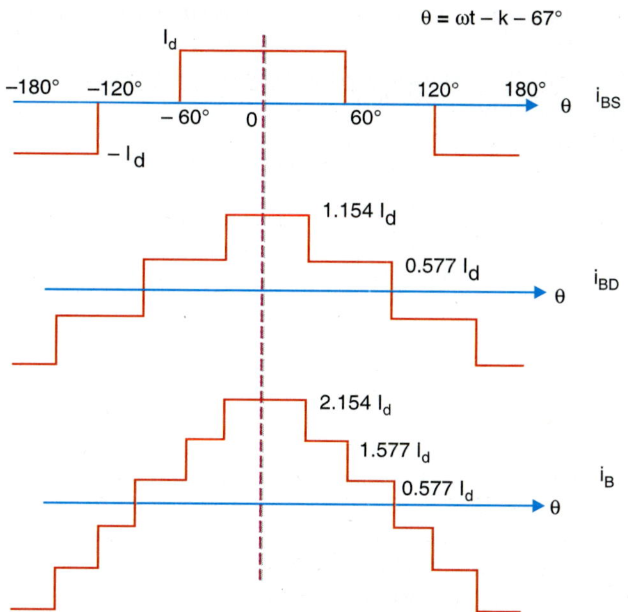

Fig. 2.11: Waveforms of current in 12 pulse converter

The waveform of i_{BS}, i_{BD} and i_B are shown in Fig. 2.11. In deriving the waveform of i_{BD}, it is to be noted that i_{bD} lags i_{bS} by 30° and i_{cD} lags i_{bD} by 120°.

It can be shown that the rms value of the fundamental component of the supply current in a 12 pulse converter is given by

$$I_1 = 2.\frac{\sqrt{6}}{\pi} I_d \tag{2.27}$$

The harmonic current I_h is given by

$$I_h = \frac{I_1}{h}$$

where $h = 12n \pm 1$, n is an integer

$= 6m \pm 1$, m is an even integer.

In comparing with the harmonic currents of a six pulse converter, it is obvious that the fifth and seventh harmonics flowing in the two transformers cancel each other. This can be generalized to the case whenever m is an odd integer.

2.2.4 Effect of Finite Smoothing Reactor

In the previous sections, it was assumed that I_d is constant which implies that the reactance of the smoothing reactor is very high (theoretically infinite). If the smoothing reactor is finite, then we cannot neglect the ripple in the DC current. The average DC current (I_d) should be above a minimum value, (corresponding to the peak value of the ripple) such that the current in the DC link remains continuous. Since valve currents can be only unidirectional, when the current goes to zero, there is discontinuity. The extinction of the link current can lead to overvoltages across the valves due to the lightly damped L-C circuit consisting of smoothing reactor and the line capacitance.

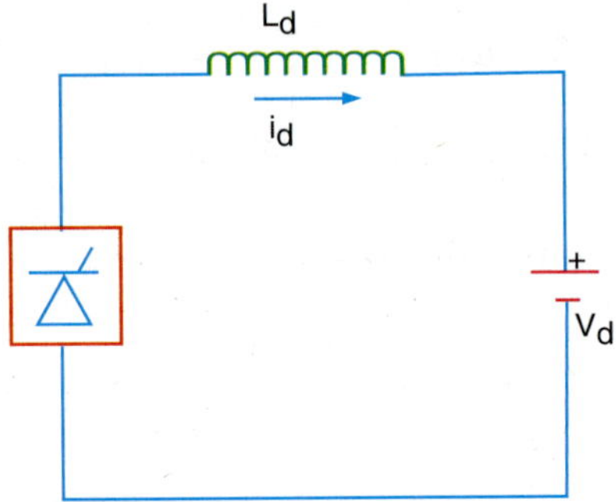

Fig. 2.12: Equivalent circuit for calculating ripple

To compute the ripple in the DC link current, consider the equivalent circuit in Fig. 2.12. Here L_d is the inductance of the reactor and V_d is the constant voltage source that represents the effect of the converter at the other terminal. The time varying voltage $v_d(t)$ across the converter is given by

$$v_d(t) = E_m \cos(\omega t), \quad \frac{-\pi}{p} + \alpha \le \omega t \le \frac{\pi}{p} + \alpha \tag{2.28}$$

Since, the losses in the circuit are neglected, we also have the constraint given by

$$V_d = \frac{p}{2\pi} \int_{\frac{-\pi}{p}+\alpha}^{\frac{\pi}{p}+\alpha} v_d(t) d\omega t \tag{2.29}$$

Substituting Eq. (2.28) in (2.29), we get

$$V_d = \frac{p}{\pi} E_m \sin\frac{\pi}{p} . \cos\alpha \tag{2.30}$$

$v_d(t)$ and V_d are shown in Fig. 2.13. It is obvious that current is minimum when $v_d(t) = V_d$ and is increasing. For $\alpha = 0$, the instant of minimum current occurs after the firing instant. However, it can be observed that as α increases, the current minimum occurs at the firing instant. The minimum value ($\bar{\alpha}$) for which the minimum current occurs at the firing instant, is given by the equation

$$V_d = E_m \cos\left(\bar{\alpha} - \frac{\pi}{p}\right) \tag{2.31}$$

Substituting Eq. (2.30) in Eq. (2.31), we can show that

$$\tan\bar{\alpha} = \frac{\pi}{p} - \cot\frac{\pi}{p} \tag{2.32}$$

For $p = 6$, $\bar{\alpha} \simeq 10°$ and for $p = 12$, $\bar{\alpha} \simeq 5°$. Since, α is normally greater than $\bar{\alpha}$ in steady state, we will compute the ripple current for this case.

The expression for the direct current, i_d in Fig. 2.12 is obtained as

$$i_d = \frac{1}{\omega L_d} \int_{\alpha-\frac{\pi}{p}}^{\omega t} (v_d - V_d) d\omega t + i_{d\,\min} \tag{2.33}$$

Substituting for v_d from Eq. 2.28, we get

$$i_d = \frac{1}{\omega L_d}\left[E_m \sin \omega t - E_m \sin(\alpha - \frac{\pi}{p}) - V_d(\omega t + \frac{\pi}{p} - \alpha)\right] + i_{d\,\min} \tag{2.34}$$

The average current I_d in the DC link is given by

$$I_d = \frac{p}{2\pi} \int_{\frac{-\pi}{p}+\alpha}^{\frac{\pi}{p}+\alpha} i_d d\omega t \tag{2.35}$$

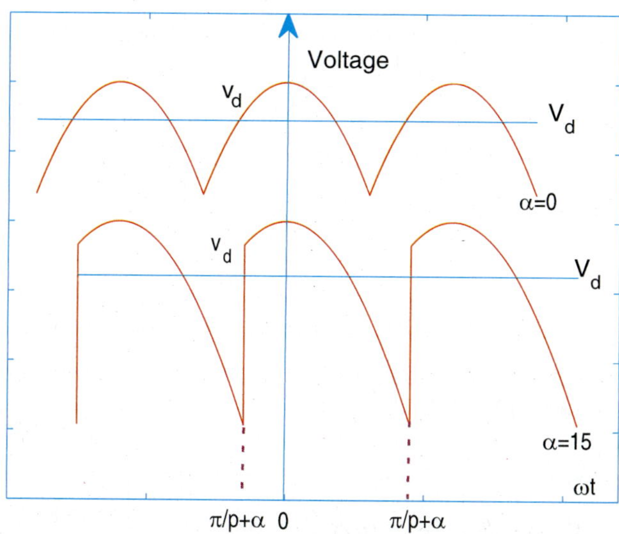

Fig. 2.13: Waveforms of v_d (ωt) and v_d

Substituting for i_d from Eq. (2.34) in Eq. (2.35), we can derive

$$I_d = \left(\frac{p}{\pi} E_m \sin\frac{\pi}{p}\right)\frac{1}{\omega L_d}\left(1 - \frac{\pi}{p}\cot\frac{\pi}{p}\right)\sin\alpha + i_{d\,\min}$$

$$= \frac{p}{6} \cdot \frac{V_{do}}{\omega L_d}\left(1 - \frac{\pi}{p} . \cot\frac{\pi}{p}\right)\sin\alpha + i_{d\,\min} \tag{2.36}$$

where $$V_{do} = \frac{3\sqrt{2}}{\pi} E_{LL} = \frac{6}{p}\left(\frac{p}{\pi} E_m \sin\frac{\pi}{p}\right)$$

For the current in the DC link to be continuous, $i_{d\,\min} \geq 0$. For the limiting case of the $i_{d\,\min} = 0$, we get

$$I_{d\min} = \frac{p}{6} \cdot \frac{V_{do}}{\omega L_d}\left(1 - \frac{\pi}{p}\cot\frac{\pi}{p}\right)\sin\alpha$$

The factor $\left(1 - \frac{\pi}{p}\cot\frac{\pi}{p}\right)$ has the value of 0.0931 for $p = 6$ and 0.0230 for $p = 12$.

2.3 VOLTAGE SOURCE CONVERTER

2.3.1 Basic Two Level (Graetz Bridge) Converter

The configuration of the Voltage Source Converter (VSC) with two levels is as shown in Fig. 2.1 with six switches. However, each switch consists of a self commutated IGBT device and an anti-parallel connected diode as shown in Fig. 2.14.

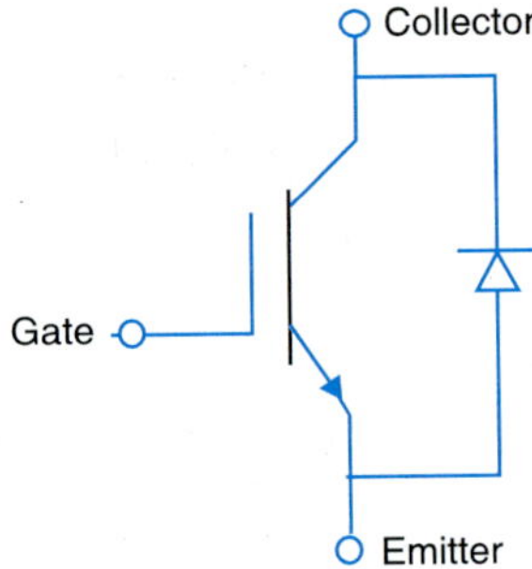

Fig. 2.14: Switch consisting of IGBT and diode

The DC side is connected to the capacitor and the AC side is connected to the source voltage through inductors as shown in Fig. 2.15.

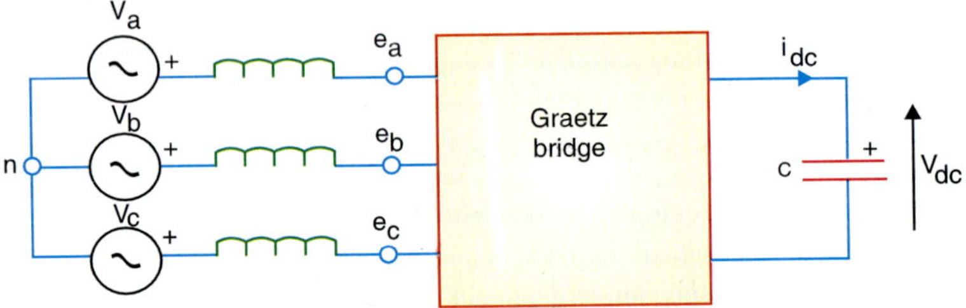

Fig. 2.15: A voltage source converter

Here, the voltage e_a, e_b and e_c are injected by the VSC and depend on V_{dc}, the voltage across the capacitor. The current 'i_{dc} flowing into the capacitor is a function of the source currents i_a, i_b and i_c.

In each leg of the converter, only one switch can be turned on at any given time to avoid short circuit across the capacitor. Each switch conducts for 180° and switch 1 must be turned off before switch 4 is turned on in the leg connected to phase '*a*'. Similar comments apply for switches in the legs connected to phases 'b' and 'c'. The waveform of the voltage e_a depends on the reference point. Figure 2.16 shows the leg connected to phase 'a'. With respect to the

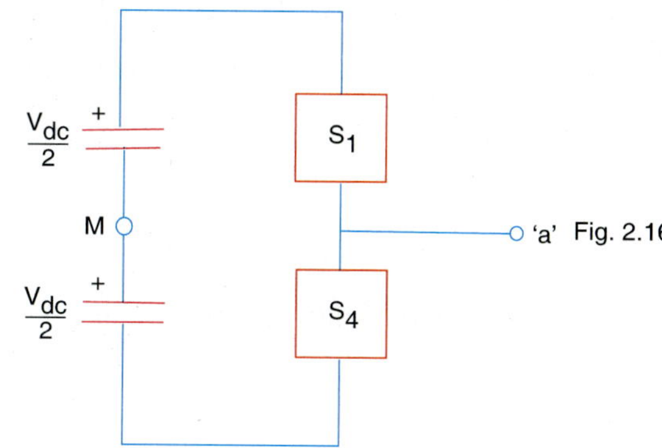

Fig. 2.16: A VSC leg connected to phase 'a'

midpoint 'M' of the capacitor, the waveform of the voltage e_{aM} is shown in Fig. 2.17(a). This voltage has only 2 levels $+\frac{V_{dc}}{2}$ and $-\frac{V_{dc}}{2}$. The voltages e_{bM} and e_{cM} lag e_{aM} by 120° and 240° respectively.

It can be shown that e_{Mn} (voltage of midpoint M with respect to the supply neutral) consists of only a zero sequence voltage with triplen harmonics [see Fig. 2.17(b)]. Subtracting e_{Mn} from e_{aM}, we get e_{an}(the injected voltage with respect to supply neutral). The waveform of e_{an} is shown in Fig. 2.17(c). The phase to phase voltage e_{ab} does not contain zero sequence component and its waveform is shown in Fig. 2.17(d). Both e_{an} and e_{ab} contain only harmonics of the order $h = 6k \pm 1$, k is an integer.

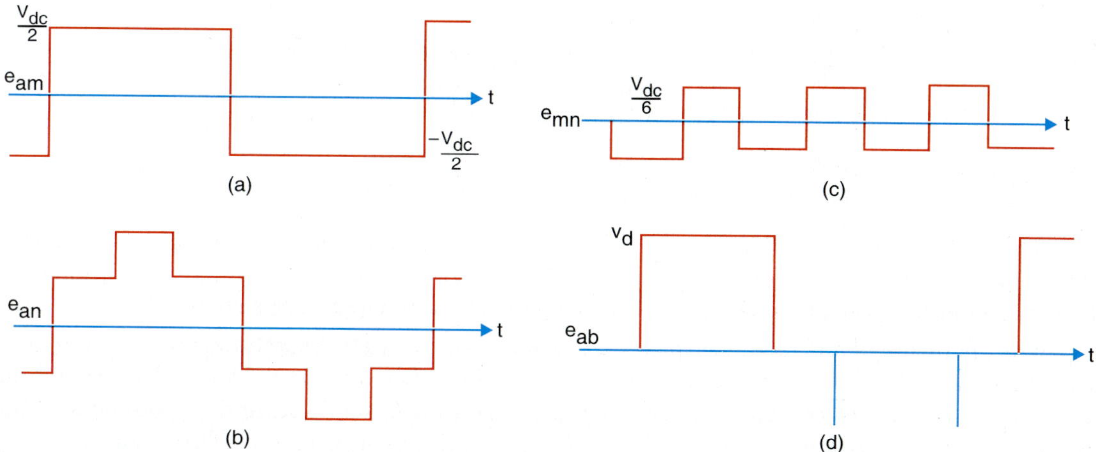

Fig. 2.17: Waveforms of e_{aM}, e_{Mn}, e_{an} and e_{ab}

The peak value of the fundamental components of the injected voltage e_a is given by

$$E_1 = \frac{4}{\pi}\left(\frac{V_{dc}}{2}\right) = \sqrt{2}(0.90)\left(\frac{V_{dc}}{2}\right) \tag{2.37}$$

0.90 is the rms value of the fundamental component of a square wave with an amplitude of 1.0. The harmonic voltage injected is given by

$$E_h = \frac{E_1}{h} \tag{2.38}$$

2.3.2 A Three Level Voltage Source Converter

A three level VSC (also termed as Neutral Point Clamped (NPC) converter) is a special case of multilevel diode-clamped converter. This has two capacitors on the DC bus and four valves (switches) in each phase leg as shown in Fig. 2.18.

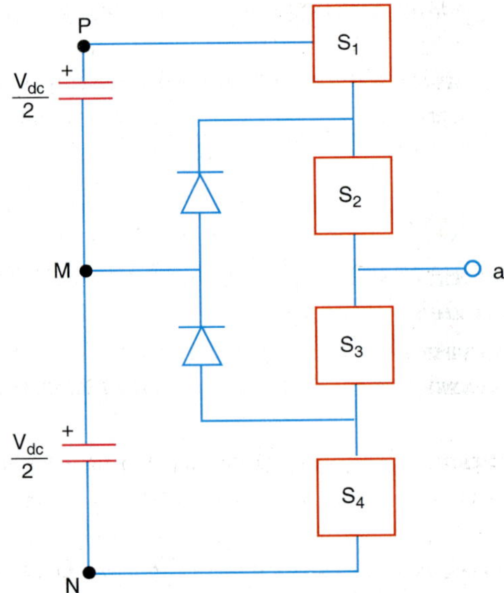

Fig. 2.18: A three level VSC leg connected to phase 'a'

There are 2 clamping diodes in each phase. Although the number of switches is doubled, the voltage rating is also doubled. For the same level of the capacitor voltage, the number of devices required per switch are halved (as compared to a two level converter).

The operation of a three level converter is determined by the switching rule that only two valves that are directly connected in series can be switched on at a time. In Fig. 2.18, switching on S_1 and S_2 connects the phase to the positive terminal *P*, while S_2 and S_3 connects to the midpoint *M* via clamping diodes, S_3 and S_4 connects to the terminal *N*. Thus, the resulting voltage injected (e_{aM}) has three levels (including zero level) as shown in Fig. 2.19.

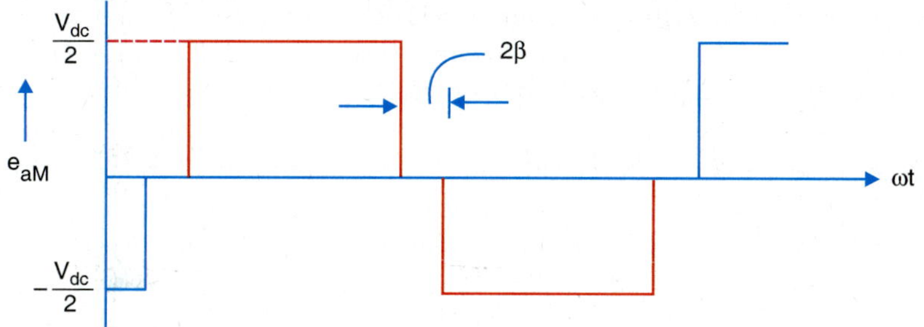

Fig. 2.19: Waveform of e_{aM} in a three level converter

It can be shown that the peak value of the fundamental components of the injected voltage e_a is given by

$$E_1 = \frac{4}{\pi}\left(\frac{V_{dc}}{2}\right)\cos\beta = \frac{2}{\pi}V_{dc}\cos\beta \tag{2.39}$$

The harmonic voltage (peak value) is given by

$$E_h = \frac{2}{h\pi}V_{dc}\cos(h\beta), h = 6k \pm 1 \tag{2.40}$$

where k is an integer.

For $\beta = 15°$, three level converter behaves like a quasi 12 pulse converter with low values of the 5th and 7th harmonics. The rms value of the fundamental component is $0.435V_{dc}$. (For a two level converter the rms value of the fundamental components is $0.45V_{dc}$).

In comparison with a two level converter, a three level converter provides better performance. However, it results in more complex layout resulting in larger footprint and higher costs.

2.3.3 Pulse Width Modulation

In order to control the power and reactive power, it is essential to control the magnitude and phase angle of the injected voltage by the converter. At the same time, it is desirable to regulate the DC bus voltage. Another requirement is to minimize the injection of low order harmonics by the converter. The Pulse Width Modulation (PWM) is utilized to achieve the following objectives:

1. To control the output AC voltage for a constant DC voltage.
2. To minimize the harmonics subject to the constraints on the switching losses, generation of noise etc.

   The frequency ratio

$$p = \frac{F}{f} \tag{2.41}$$

where F is the modulation frequency and f is the fundamental output frequency is an important parameter. p is also called as the pulse number. Higher the value of p, broader the spectrum of harmonics that can be eliminated. The switching frequency (F) is dependent on the devices used. The maximum switching frequency for GTO thyristors is only around 100 Hz. This is why PWM is not used in GTO based VSC employed in high power FACTS controllers. Multi-pulse or multilevel converters which require only one switching (on and off) in a cycle are used with GTO devices. On the other hand, IGBT devices can be operated up to 10 kHz. However, switching losses increase with frequency and are a limiting factor in the choice of F.

For $p > 20$, the modulating (carrier) wave and the reference (output) voltage need not be synchronized. It is not necessary that p be an integer. For comparing various modulation schemes, the modulation index (m) is normalized by taking the ratio of actual fundamental component of the modulated output voltage to that of the unmodulated output voltage, that is,

$$\bar{m} = \frac{E_{\text{mod}1}}{0.45V_{dc}} \tag{2.42}$$

where $E_{\text{mod}1}$ is the rms value of the fundamental component of the modulated output voltage.

Selective Harmonic Elimination (SHE)

This is also termed as the optimized PWM technique. By reversing the phase voltages a few times during each half cycle, it is possible to eliminate low order harmonics selectively. However, the high order harmonics may increase in magnitudes, but the current harmonics are not significantly affected due to the low pass filter characteristics of the AC system.

Figure 2.20 shows the waveforms of the phase voltage (e_{aM}) with four and six reversals within one half cycle. The voltage reversals are affected at chosen instants such that the notches (caused by the voltage reversals) are placed symmetrically about the centre line of each half cycle. If there are 'p' switching (voltage reversals) in a quarter cycle, the rms value of the nth harmonic voltage is given by

$$E_n = \pm\frac{1}{\sqrt{2}}\cdot\frac{4}{\pi}\cdot\frac{V_{dc}}{2}\left[-\int_0^{\alpha_1}\sin n\theta d\theta + \int_{\alpha_1}^{\alpha_2}\sin n\theta d\theta \ldots \pm \int_{\alpha_p}^{\pi/2}\sin n\theta d\theta\right]$$

$$= \pm\frac{0.45}{n}V_{dc}\left[2(\cos n\alpha_1 - \cos n\alpha_2 + \cos n\alpha_3 - \ldots) - 1\right] \tag{2.43}$$

where α_1, $\alpha_{2\ldots}$ α_p are switching angles within each quarter cycle. There are p degrees of freedom and are used to cancel (p – 1) harmonic components in the voltage and control the fundamental voltage. For example, if there are 3 switching at α_1, α_2 and α_3, we can eliminate the fifth and seventh harmonics in addition to controlling the fundamental voltage. We get 3 transcendental equations given by

$$\bar{m} = 2[\cos\alpha_1 - \cos\alpha_2 + \cos\alpha_3] - 1 \tag{2.44}$$

$$0 = 2[\cos 5\alpha_1 - \cos 5\alpha_2 + \cos 5\alpha_3] - 1 \tag{2.45}$$

$$0 = 2[\cos 7\alpha_1 - \cos 7\alpha_2 + \cos 7\alpha_3] - 1 \tag{2.46}$$

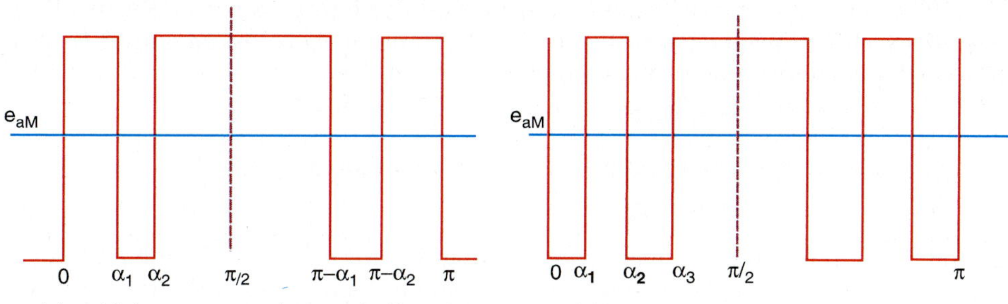

Fig. 2.20: Waveforms of e_{aM} with voltage reversals

Note that although the phase voltage contain triplen harmonics, the line to line voltage are free from them.

Example

If there are two voltage reversals in a quarter cycle, the values of α and α_2 are 16.2° and 22.0° such that the 5th and 7th voltage harmonics are zero. The modulation index ($\bar{m}$) is given by

$$\bar{m} = 1 - 2[\cos\alpha_1 - \cos\alpha_2] = 0.934$$

Sinusoidal PWM

The sinusoidal PWM technique is widely used for industrial converters. The method is also known as suboscillation method and uses a triangular carrier wave of frequency F to generate a sinusoidal reference voltage of frequency f. Actually, the modulated output voltage of the converter approximates a staircase shape in which each step is the average value over each cycle of the carrier wave.

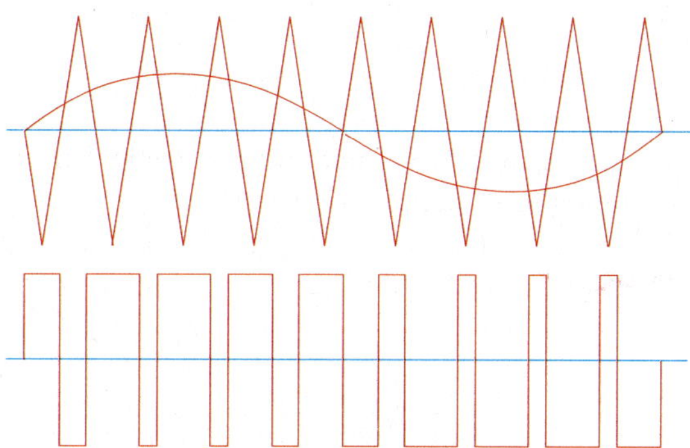

Fig. 2.21: Sinusoidal PWM

The sine PWM technique can be compared to the SHE PWM discussed earlier. Here, the instants of voltage reversals are determined by the intersection between a reference sinusoidal voltage of peak value m and a triangular voltage of amplitude 1. Figure 2.21 shows the operation sine PWM for $p = \frac{F}{f} = 9$ with m = 0.5. The modulation is said to be in the linear range, if $0 < m \leq 1$. The same carrier wave can be used for all three phases with reference voltages in each phase-phase shifted from each other by 120° (for generating balanced output voltages). The fundamental frequency voltage $e_1(t)$ is given by

$$e_1(t) = m\sin(2\pi ft) \tag{2.47}$$

The harmonics are present in output and have frequencies given by

$$f_h = MF \pm Nf \tag{2.48}$$

where M and N are integers. The harmonics are significant only around the frequency = MF. In general, the harmonics are determined by the frequency spectrum of the reference wave and that due to the modulation. With a sinusoidal reference wave, all harmonics are generated by the PWM process.

The reference wave can be divided into p intervals with each interval representing one cycle of the carrier wave. In each interval, the reference wave can be sampled by representing it as a constant value which is the mean of the reference wave within the interval. Thus, the reference wave is represented by a staircase waveform shown in Fig. 2.22.

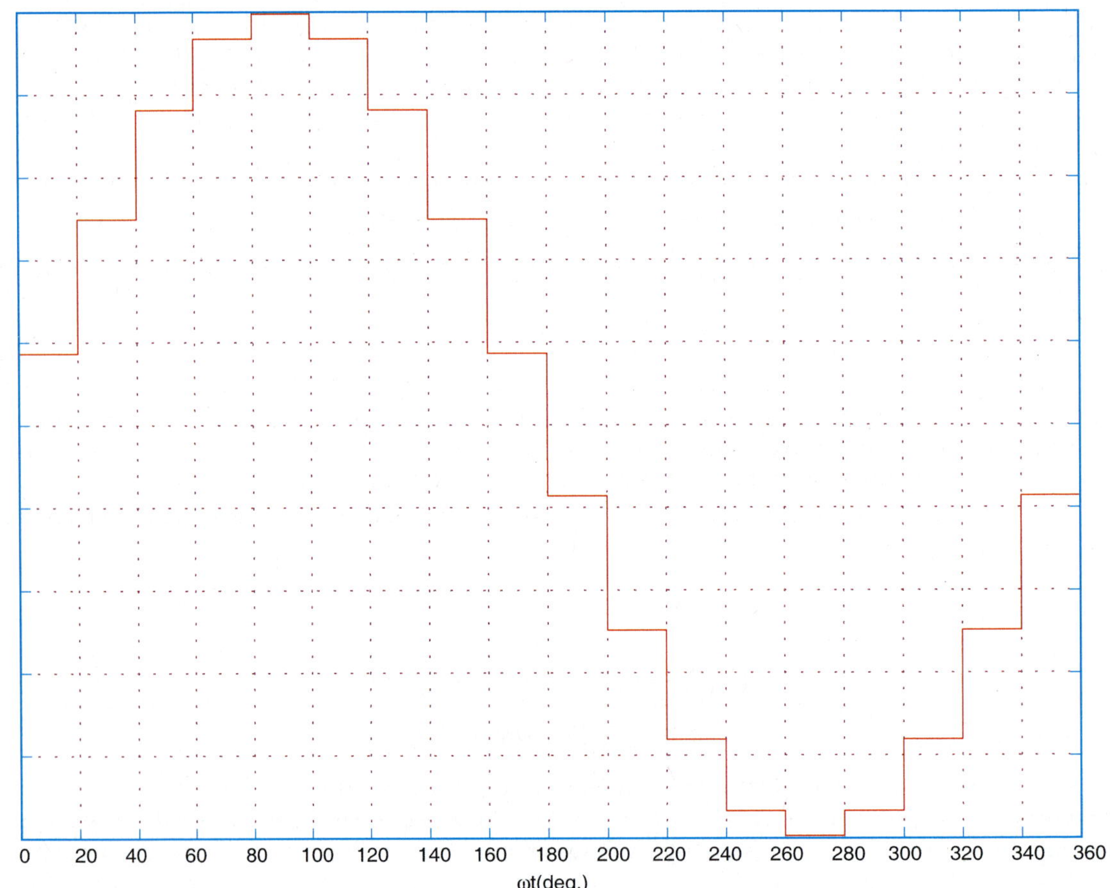

Fig. 2.22: Staircase waveform (After sampling)

Two sampled waveforms with identical mean values have approximately the same fundamental component and low order harmonics. Thus, the waveforms shown in Fig. 2.23 (*a*) and (*b*) which represent one stair and a notched waveform generated by the triangular wave shown in (*c*) have the same mean value (L). This shows that sine PWM results in the output voltage equivalent to a voltage having a staircase waveform. As the pulse number p is increased, the staircase waveform approximates closely the reference sine wave.

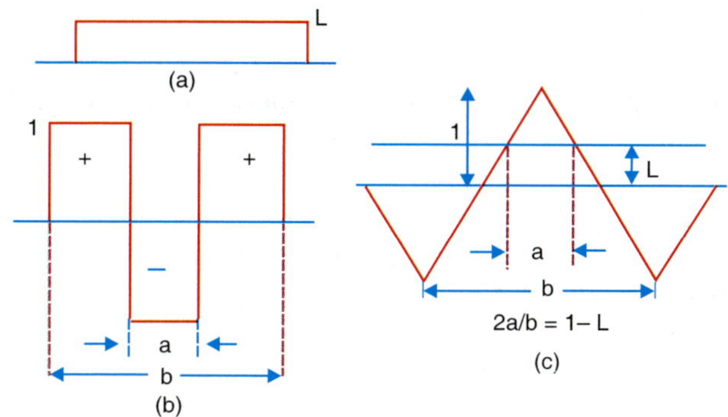

Fig. 2.23: Equivalence of a stair and a notched waveform (a) A stair (step); (b) Notched waveform; (c) Triangular waveform

The maximum value of a fundamental voltage (E_1) with sine PWM is $0.5V_{dc}$ (for $m = 1$). The maximum value of the fundamental voltage for square wave (six-step) converter is $\frac{2}{\pi}V_{dc}$. Thus, the maximum value of the normalized modulation index ($\bar{m}$) for sine PWM is given by

$$\bar{m}_{\max} = \frac{\pi}{4} = 0.785 \tag{2.49}$$

The value of $\bar{m}$ can be increased by adding triplen harmonics to the reference wave without affecting the output line voltages. This technique follows from the fact that in three phase converters, line to line voltages have no zero sequence components. In symmetrical three phase converters, the triplen harmonics present in the phase voltages are of zero sequence.

By injecting a third harmonic which is one sixth of the fundamental, it is possible to increase $\bar{m}$ to a value given by

$$\bar{m}_{\max} = \frac{\pi}{2\sqrt{3}} = 0.907 \tag{2.50}$$

REFERENCES AND BIBLIOGRAPHY

1. Kimbark E.W., *Direct Current Transmission*, Vol.1, (New York), John Wiley, 1971.
2. Uhlmann E., *Power Transmission by Direct Current*, (Berlin), Springer-Verlag, 1975.
3. Padiyar K.R., *FACTS Controllers in Power Transmission and Distribution*, (New Delhi), New Age International Publishers, 2007.
4. Schettler F., Huang H. and Christl N., "*HVDC Transmission Systems using Voltage Sourced Converters Design and Applications*", Summer Meeting, IEEE, PES, 2000.

CHAPTER 2

ANALYSIS OF HVDC CONVERTERS

3

3.1 INTRODUCTION

In the previous chapter, we looked at the converter circuits – both for Line Commutated, (current source) Converters (LCC) and Voltage Source Converters (VSC). The Graetz Bridge is the basic building block in LCC and also in VSC based HVDC converters (in multi-pulse converters). In twelve pulse LCC (which is a common configuration), two Graetz bridges are connected in series on the DC side and in parallel on the AC side. Typically, a twelve pulse LCC is fed by three single phase transformers, each with three windings (one primary and two secondary). One set of secondary windings, connected in star, feed one bridge while the other set of secondary windings, connected in delta, feed the second bridge. A UHV DC converter is made up of two twelve pulse converters (4 bridges) connected in series on the DC side.

In this chapter, we will analyze both LCC and VSC configurations and study the converter characteristics.

3.2 ANALYSIS OF LINE COMMUTATED CONVERTER

In the previous chapter, we assumed that only one valve in a commutation group of a bridge conducts. In other words, the current transfers instantaneously from one valve to the next valve when the latter is fired.

Due to the leakage inductance of the converter transformer and the impedance in the supply network, the current in a valve cannot change suddenly and thus commutation from one valve to the next cannot be instantaneous. For example, when valve 3 is fired, the current transfer from valve 1 to valve 3 takes a finite period during which both valves are conducting. This is called overlap and its duration is measured by the overlap (commutation) angle 'u'.

Each interval of the period of supply can be divided into two subintervals (see Fig. 3.1). In the first subinterval, three valves are conducting and in the second subinterval, two valves are conducting. This is based on the assumption that the overlap angle is less than 60°. As the overlap angle increases to 60°, there is no instant when only two valves are conducting. As the overlap angle increases beyond 60°, there is a finite period during an interval, when four valves conduct and the rest of the interval during which three valves conduct. Thus there are three modes of the converter as follows:

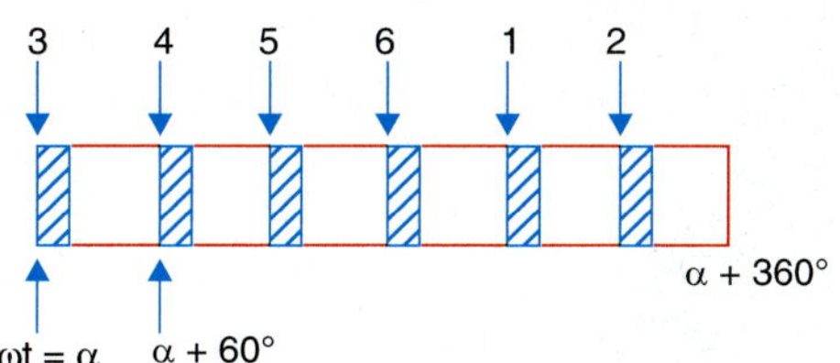

Fig. 3.1: Timing diagram

1. Mode 1—Two and three valve conduction ($u < 60°$)
2. Mode 2—Three valve conduction ($u = 60°$)
3. Mode 3—Three and four valve conduction ($u > 60°$)

Depending upon the delay angle α, the mode 2 may be just a point on the boundary of modes 1 and 3. This is explained later.

With the assumptions stated in Chapter 2 for steady state analysis, it is sufficient to consider only one interval of 60 degree duration –say the interval starting with the firing of (conduction of) valve 3.

(i) Analysis of two and three valve conduction mode

For the interval considered, the bridge circuit can be reduced to that shown in Fig. 3.2.

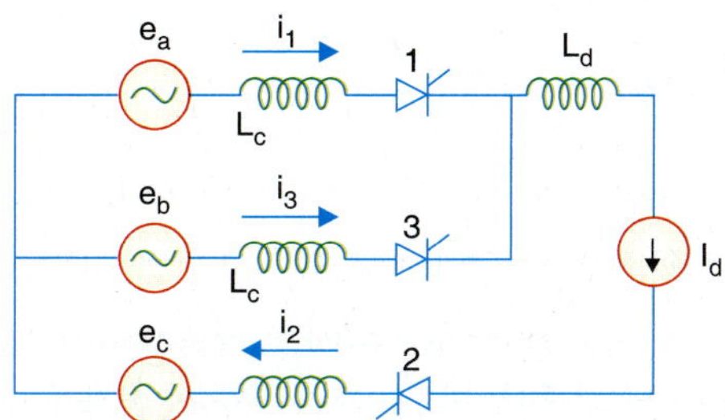

Fig. 3.2: Equivalent circuit for 3 valve conduction

For this circuit,

$$e_b - e_a = L_c\left(\frac{di_3}{dt} - \frac{di_1}{dt}\right) \tag{3.1}$$

The L.H.S. in the above equation is called the commutating emf whose value is given by

$$e_b - e_a = \sqrt{2}E_{LL}\sin\omega t \tag{3.2}$$

which is also the voltage across valve 3 before it starts conducting. Since

$$i_1 = I_d - i_3 \tag{3.3}$$

we get, $$\sqrt{2}E_{LL}\sin\omega t = 2L_c\frac{di_3}{dt} \tag{3.4}$$

Solving the above equation, we get

$$i_3(t) = I_s(\cos\alpha - \cos\omega t), \alpha \le \omega t \le \alpha + u \quad (3.5)$$

where

$$I_s = \frac{\sqrt{2}E_{LL}}{2\omega L_c} \quad (3.6)$$

Note that the solution given in 3.5 is obtained from the initial condition

$$i_3(\omega t = \alpha) = 0 \quad (3.7)$$

At $\omega t = \alpha + u, i_3 = I_d$. This gives

$$I_d = I_s\left[\cos\alpha - \cos(\alpha + u)\right] \quad (3.8)$$

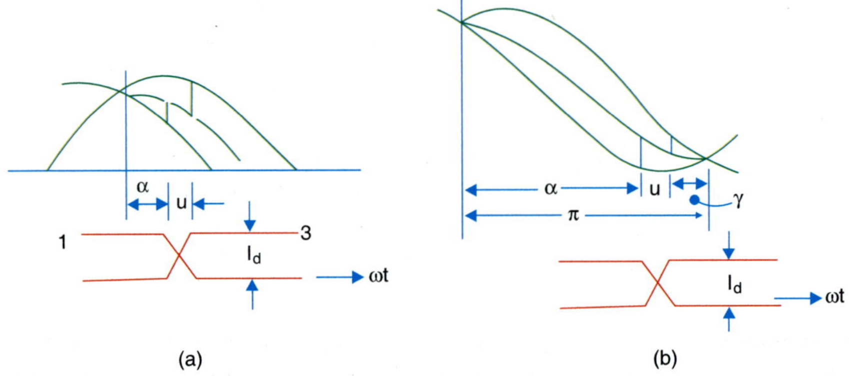

Fig. 3.3: Voltage and current waveform during firing of a valve (a) rectifier; (b) inverter

The waveforms of direct voltage and the valve currents during commutation for a rectifier and an inverter are shown in Fig. 3.3(*a*) and 3.3(*b*) respectively. It is to be noted that during commutation, the instantaneous DC voltage is $\frac{-3}{2}e_c\left(\frac{e_a + e_b}{2} - e_c\right)$ instead of $(e_b - e_c)$.

Average direct voltage

The average direct voltage can be obtained as

$$V_d = \frac{3}{\pi}\left[\int_{\alpha}^{\alpha+u} \frac{-3}{2}e_c d(\omega t) + \int_{\alpha+u}^{\alpha+60^\circ} (e_b - e_c) d(\omega t)\right]$$

$$= \frac{3}{\pi}\left[\int_{\alpha}^{\alpha+60^\circ} e_{bc} d(\omega t) - \int_{\alpha}^{\alpha+u} \frac{e_b - e_a}{2} d(\omega t)\right]$$

$$= V_{do}\cos\alpha - \frac{3}{2\pi}\int_{\alpha}^{\alpha+u} \sqrt{2}E_{LL}\sin\omega t \, d(\omega t)$$

$$= V_{do}\cos\alpha - \frac{3}{2\pi}\sqrt{2}E_{LL}\left[\cos\alpha - \cos(\alpha + u)\right] \quad (3.9)$$

Since $\frac{3}{\pi}\sqrt{2}E_{LL} = V_{do}$, we get

$$V_d = \frac{V_{do}}{2}\left[\cos\alpha + \cos(\alpha + u)\right] \tag{3.10}$$

The equation (3.8) can be substituted in (3.9) to yield,

$$V_d = V_{do}\left(\cos\alpha - \frac{I_d}{2I_s}\right) = V_{do}\cos\alpha - R_c I_d \tag{3.11}$$

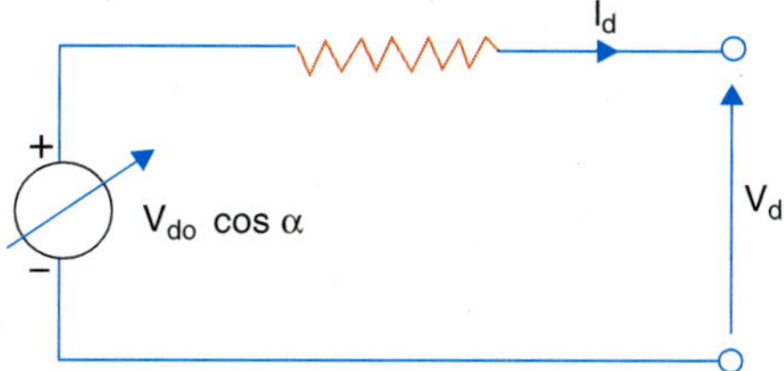

Fig. 3.4: Equivalent circuit of a bridge converter

where,

$$R_c = \frac{3}{\pi}\omega L_c = \frac{3}{\pi}X_c \tag{3.12}$$

R_c is called equivalent commutation resistance. It is something analogous to armature reaction in DC machines in the sense that it only represents a voltage drop and not a power loss. The equivalent circuit of the bridge rectifier is shown in Fig. 3.4.

DC voltage and valve voltage waveforms

Figure 3.5 shows the waveform of the voltage across the converter bridge (v_d).The valve voltage waveform (for a rectifier) is shown in Fig. 3.6. The valve voltage has now 8 jumps. The two major jumps occur at the firing and the turning off of the valve and are given by

$$V_{j1} = \sqrt{2}E_{LL}\sin\alpha$$

$$V_{j2} = \sqrt{2}E_{LL}\sin(\alpha + u)$$

The remaining 6 jumps can be divided into two groups, the first group composed of jumps that are equal in magnitude to $\frac{V_{j1}}{2}$ and the second group of jumps with magnitudes $\frac{V_{j2}}{2}$. These voltage jumps result in extra losses in the damper or snubber circuit of the valve.

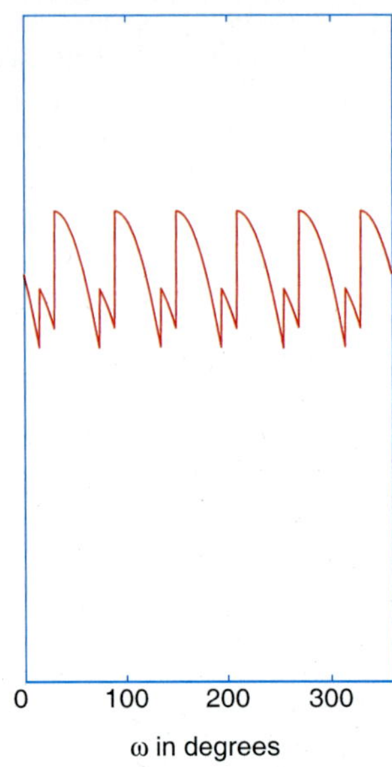

Fig. 3.5: DC voltage waveform

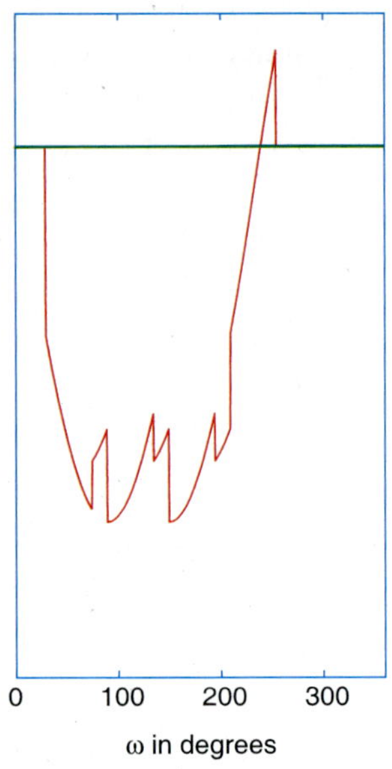

Fig. 3.6: Valve voltage waveform

Inverter equations

For an inverter, it is usual to define an angle of advance β as

$$\beta = \pi - \alpha \tag{3.13}$$

and use opposite polarity for the DC voltage with voltage rise opposite to the direction of current. Thus,

$$\begin{aligned} V_{di} &= \frac{-V_{doi}}{2}[\cos\alpha + \cos(\alpha + u)] \\ &= \frac{-V_{doi}}{2}[\cos(\pi - \beta) + \cos(\pi - \gamma)] \\ &= \frac{V_{doi}}{2}[\cos\beta + \cos\gamma] \end{aligned} \tag{3.14}$$

where the extinction angle γ is defined as

$$\gamma = \beta - u = \pi - \alpha - u \tag{3.15}$$

Similarly, it can be shown that

$$\begin{aligned} V_{di} &= V_{doi}\cos\beta + R_{ci}I_d \\ &= V_{doi}\cos\gamma - R_{ci}I_d \end{aligned} \tag{3.16}$$

The subscript '*i*' refers to the inverter.

Comparing Eqs. (3.14) and (3.16) with Eqs. (3.10) and (3.11) shows that the inverter equations are obtained by substituting γ for α in the rectifier equations. However it is to be noted that while α is directly controllable, γ is not.

AC current and DC voltage harmonics

The waveforms of the valve currents and the current in the valve winding are distorted. Thus, the expression for the fundamental component of the AC current derived for the case with no overlap is not valid. The actual expression for the current can be derived from Fourier analysis and is given by

$$I_1 = [I_{11}^2 + I_{12}^2]^{1/2} \tag{3.17}$$

$$I_{11} = I_1 \cos\phi = \frac{\sqrt{6}}{\pi} I_d \left[\frac{\cos\alpha + \cos(\alpha + u)}{2}\right] \tag{3.18}$$

$$I_{12} = I_1 \sin\phi = \frac{\sqrt{6}}{\pi} I_d \left[\frac{2u + \sin 2\alpha - \sin 2\delta}{4(\cos\alpha - \cos\delta)}\right] \tag{3.19}$$

where ϕ is the power factor angle and $\delta = \alpha + u$

From the above expressions, the power factor angle can be obtained as

$$\tan\phi = \frac{2u + \sin 2\alpha - \sin 2\delta}{\cos 2\alpha - \cos 2\delta} \tag{3.20}$$

The harmonic components in the AC current are also altered. These are reduced from the values calculated with no overlap. The reduction factor is given by

$$\frac{I_h}{I_{ho}} = \frac{1}{2x}[H^2 + K^2 - 2HK\cos(2\alpha + u)]^{1/2} \tag{3.21}$$

where

$$H = \frac{\sin\left[(h+1)\frac{u}{2}\right]}{h+1}, K = \frac{\sin\left[(h-1)\frac{u}{2}\right]}{h-1} \tag{3.22}$$

$$x = \frac{1}{2}[\cos\alpha - \cos(\alpha + u)] \tag{3.23}$$

$I_{ho} = \frac{\sqrt{6}}{\pi}\frac{I_d}{h}$ = harmonic component with no overlap. Figure 3.7 shows this reduction factor for the case $\alpha = 30°$. It can be seen that all the harmonics especially of higher order, decrease sharply with increasing values of u and the reduction factor lies in the range 0.1 to 0.2.

The harmonics in the direct voltage are also altered due to overlap. It can be shown that

$$\frac{V_h}{V_{do}} = \frac{1}{\sqrt{2}}[F^2 + G^2 - 2FG\cos(2\alpha + u)]^{1/2} \tag{3.24}$$

where

$$F = \frac{\cos\left[(h+1)\frac{u}{2}\right]}{h+1}, G = \frac{\cos\left[(h-1)\frac{u}{2}\right]}{h-1} \tag{3.25}$$

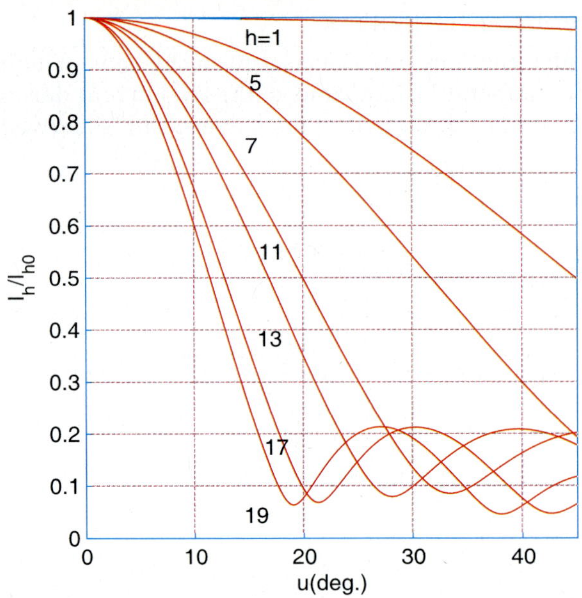

Fig. 3.7: Variation of AC current harmonics with overlap

Figure 3.8 shows $\left(\frac{V_{12}}{V_{do}}\right)$ for different values of u and α.

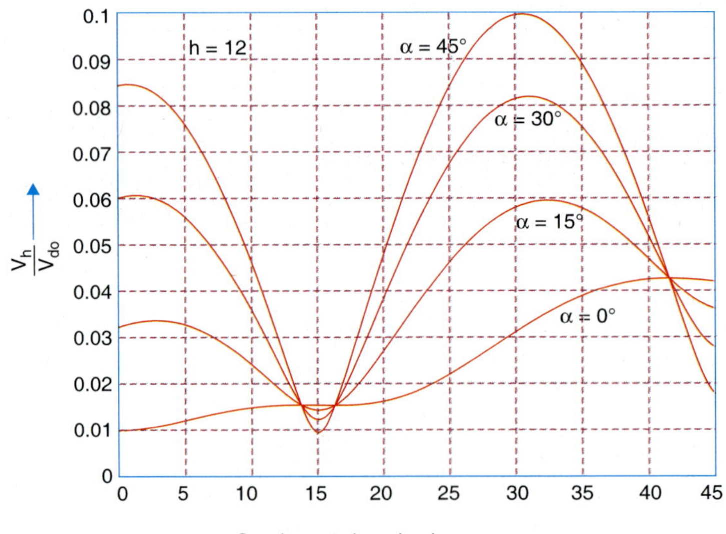

Fig. 3.8: Variation of DC voltage harmonics with overlap

Example 1

If the DC current varies with time, prove that $\frac{I_{d1}+I_{d2}}{2}=I_s\left[\cos\alpha-\cos(\alpha+u)\right]$ where I_{d1} is the DC current at the beginning of the commutation ($\omega t=\alpha$) and I_{d2} is the DC current at the end of the commutation ($\omega t=\alpha+u$). Assume two and three valve conduction mode.

Solution

Consider the circuit shown in Fig. 3.2. Substituting Eq. (3.3) in (3.1), we get

$$e_b-e_a=2L_c\frac{di_3}{dt}-L_c\frac{dI_d}{dt}$$

(Note: The above equation replaces (3.4))

Substituting Eq. (3.2) in the above equation, we get

$$2L_c i_3(t)=\frac{1}{\omega}\int_{\alpha}^{\omega t}(e_b-e_a)\,d(\omega t)+\frac{L_c}{\omega}\int_{\alpha}^{\omega t}\frac{dI_d}{dt}\,d(\omega t)$$

$$i_3(t)=\frac{\sqrt{2}E_{LL}}{2X_c}\left[\cos\alpha-\cos\omega t\right]+\frac{1}{2}\left[I_d(\omega t)-I_{d1}\right]$$

Since, $i_3(\omega t=\alpha+u)=I_{d2}$, we get

$$I_{d2}=I_s\left[\cos\alpha-\cos(\alpha+u)\right]+\frac{1}{2}\left[I_{d2}-I_{d1}\right]$$

From the above, we get the desired result

$$\frac{I_{d1}+I_{d2}}{2}=I_s\left[\cos\alpha-\cos(\alpha+u)\right]$$

Example 2

Prove that $\frac{V_d}{V_{don}}=\bar{E}\cos\alpha-\frac{\bar{X}_c}{2}\frac{I_d}{I_{dn}}$ where, the subscript n indicates rated values and $\bar{X}_c$ is the leakage reactance of the converter transformer expressed on its own base. $\bar{E}$ is the AC voltage expressed in per unit of the rated voltage.

Solution

From Eqs. (3.11) and (3.12), we get

$$\frac{V_d}{V_{don}}=\frac{V_{do}}{V_{don}}\cos\alpha-\frac{3}{\pi}\frac{X_cI_{dn}}{V_{don}}\left(\frac{I_d}{I_{dn}}\right)$$

Since, $V_{don}=\frac{3\sqrt{2}}{\pi}E_{an}$ where E_{an} is the rated line to AC voltage (on the valve side), we get

$$\frac{V_d}{V_{don}}=\frac{E_a}{E_{an}}\cos\alpha-\frac{X_cI_{dn}}{\sqrt{2}E_{an}}\left(\frac{I_d}{I_{dn}}\right)$$

where E_a is the actual line to AC voltage (Note that $V_{do}=\frac{3\sqrt{2}}{\pi}E_a$).

Noting that the rated AC current (including the harmonic current components), denoted by I_{an} is given by

$$I_{an} = \sqrt{\frac{2}{3}} I_{dn}$$

Since
$$Z_b = \frac{E_{an}}{\sqrt{3} I_{an}} = \frac{E_{an}}{\sqrt{2} I_{dn}}$$

where, Z_b is the base impedance (on the transformer base), we get finally,

$$\frac{V_d}{V_{don}} = \bar{E}\cos\alpha - \frac{\bar{X}_c}{2}\left(\frac{I_d}{I_{dn}}\right)$$

Example 3

Prove that $\dfrac{I_d}{I_{dn}} = \dfrac{\bar{E}}{\bar{X}_c}\left[\cos\alpha - \cos(\alpha + u)\right]$

Solution

This follows from Eq. (3.5) and expressing the DC current and the AC voltage in per unit of their rated values (as done in Example 2). Noting that $\dfrac{\sqrt{2} X_c I_{dn}}{E_{an}} = \dfrac{X_c}{Z_b} = \bar{X}_c$, we can derive the above expression.

(ii) Three and four valve conduction mode

Although two and three valve conduction mode is the normal mode of operation, during DC line faults or a dip in the AC voltage, it is likely that the overlap angle u would be larger and may exceed 60°.

When the overlap angle exceeds 60°, the minimum number of valves conducting are three and there are intervals when four valves are conducting. This is because when a commutation process is started, the previous commutation process is not yet completed. For example, when valve 3 is fired, the valves 1, 6, and 2 are still conducting. The equivalent circuit for this condition is shown in Fig. 3.9. It can be observed that $v_d = 0$.

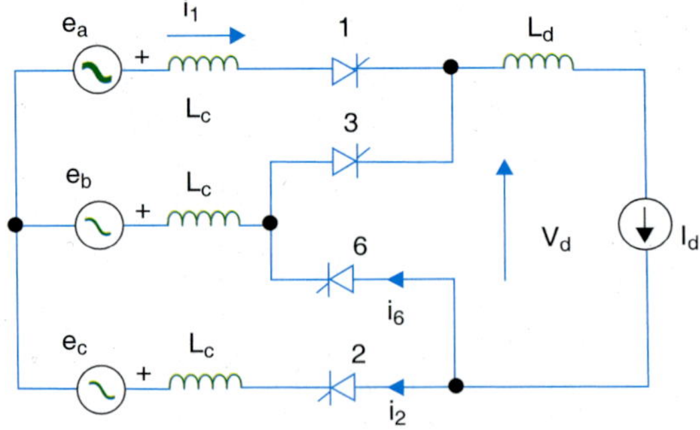

Fig. 3.9: Equivalent circuit for four valve conduction

(*a*) For $\alpha \le \omega t \le \alpha + u - 60°$

$$L_c \frac{di_1}{dt} = e_a = \sqrt{\frac{2}{3}} E_{LL} \sin(\omega t + 150°)$$

$$L_c \frac{di_2}{dt} = -e_c = \sqrt{\frac{2}{3}} E_{LL} \cos \omega t$$

Solving for i_1 and i_6

$$i_1 = I_s^1 \sin(\omega t + 60°) + A \quad (3.26)$$

$$i_6 = I_d - i_2 = I_d - I_s^1 \sin \omega t + C \quad (3.27)$$

where

$$I_s^1 = \sqrt{\frac{2}{3}} \frac{E_{LL}}{\omega L_c} = \frac{2}{\sqrt{3}} I_s$$

The constant A can be determined from initial condition

$$i_1(\omega t = \alpha) = I_d = I_s^1 \sin(\alpha + 60°) + A \quad (3.28)$$

The constant C can be determined from the final condition

$$i_6(\omega t = \alpha + u - 60°) = 0 = I_d - I_s^1 \sin(\alpha + u + 60°) + C = 0 \quad (3.29)$$

(*b*) For, $\alpha + u - 60° \le \omega t \le \alpha + 60°$

Here, with 3 valves conducting

$$2L_c \frac{di_1}{dt} = -\sqrt{2} E_{LL} \sin \omega t$$

Solving for i_1 we get

$$i_1 = I_s \cos \omega t + B \quad (3.30)$$

I_s is defined by equation (3.6)

The constant B can be determined from the continuity equation

$$i_1(\omega t = \alpha + u - 60°) = I_s^1 \sin(\alpha + u) + A$$

$$= I_s \cos(\alpha + u - 60°) + B \quad (3.31)$$

From symmetry, we have

$$i_1(\omega t = \alpha + 60°) = i_6(\omega t = \alpha)$$

$$I_s \cos(\alpha + 60°) + B = I_d - I_s^1 \sin(\alpha + u) + C \quad (3.32)$$

After some manipulations we can obtain

$$I_d = \frac{I_s^1}{2} [\cos(\alpha - 30°) - \cos(\alpha + u + 30°)] \quad (3.33)$$

Average Direct Voltage

The expression for the average direct voltage is obtained as

$$V_d = \frac{3}{\pi}\int_{\alpha+u-60^\circ}^{\alpha+60^\circ} -\frac{3}{2}e_c d(\omega t)$$

$$e_c = -E_m \cos\omega t,\ E_m = \sqrt{\frac{2}{3}}E_{LL}$$

(Note, that $V_d = 0$ when $\alpha \le \omega t \le \alpha + u - 60^\circ$)

$$V_d = \frac{3}{\pi}\cdot\frac{3}{2}E_m[\sin(\alpha+60^\circ) - \sin(\alpha+u-60^\circ)]$$

$$V_d = \frac{\sqrt{3}}{2}V_{do}[\cos(\alpha-30^\circ)+\cos(\alpha+u+30^\circ)] \quad (3.34)$$

From Eqs. (3.33) and (3.34), we get

$$V_d = V_{do}\left[\sqrt{3}\cos(\alpha-30^\circ) - \frac{3}{2}\frac{I_d}{I_s}\right]$$

$$= \sqrt{3}V_{do}\cos(\alpha-30^\circ) - 3R_c I_d \quad (3.35)$$

Comparing equation (3.35) with (3.11) shows that the equivalent commutating resistance for this case is three times that for the case with overlap angle less than 60°. When the delay angle is zero, the overlap angle u reaches a value of 60° when the current I_d increases to

$$I_d = I_s[1 - \cos 60^\circ] = \frac{1}{2}I_s$$

At $u = 60^\circ$, three valves are conducting all the time. If the current is increased further, u would tend to increase further, however, this is not possible unless $\alpha \ge 30^\circ$. For values of $\alpha < 30^\circ$, an increase in direct current is possible only when α is increased and the overlap angle is maintained at 60°. The reason why the overlap cannot be increased beyond 60° for $\alpha < 30^\circ$ is evident from Fig. 3.10. When valves 2 and 6 are still conducting, the anode of valve 3 is at a potential of

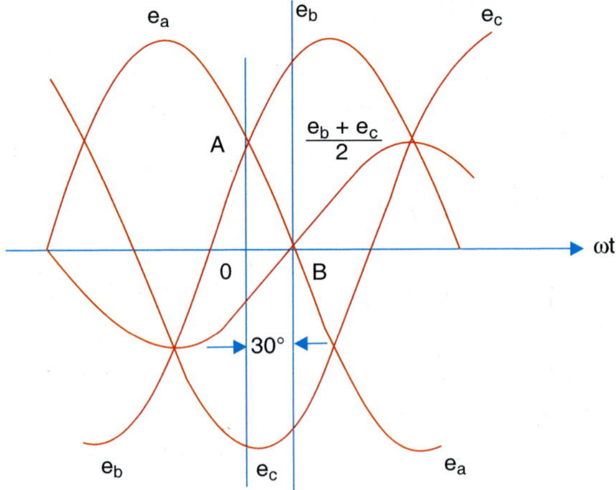

Fig. 3.10: Figure explaining minimum delay angle = 30° if overlap angle = 60° or greater

$\left(\frac{e_b + e_c}{2}\right) = \left(\frac{-e_a}{2}\right)$ and the cathode is at a potential of e_a. Hence unless e_a is negative, the valve 3 cannot conduct (even after raising the gate pulse) and this is possible only for $\alpha \geq 30°$. Thus, there is an intermediate mode when the value of overlap angle remains constant at 60°. In this mode, the increase in direct current is accompanied by automatic increase in the firing angle.

DC and valve voltage wave forms

The instantaneous voltage across the converter bridge(v_d) and the valve voltage for the 3 and 4 valve conduction modes are shown in Figs. 3.11 and 3.12 respectively (for $\alpha = 45°$ and u = 75°). The instantaneous voltage across the bridge can have both positive and negative excursions followed by periods of zero magnitude. The valve voltage has 6 jumps with three of them having the magnitude

$$V_{j1} = \sqrt{\frac{3}{2}} E_{LL} \sin(\alpha - 30°)$$

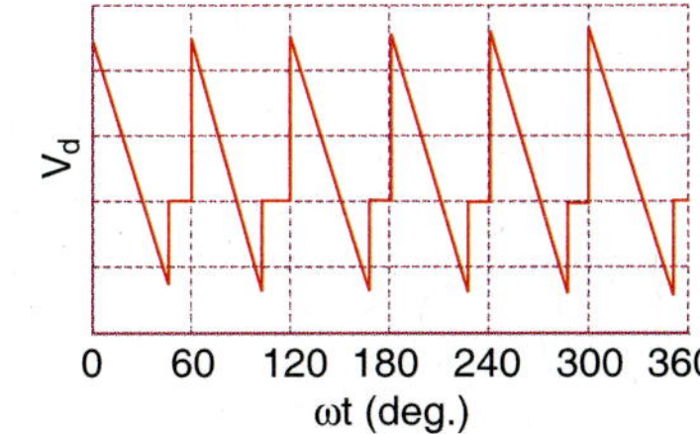

Fig. 3.11: DC voltage waveform for α = 45° u = 75°

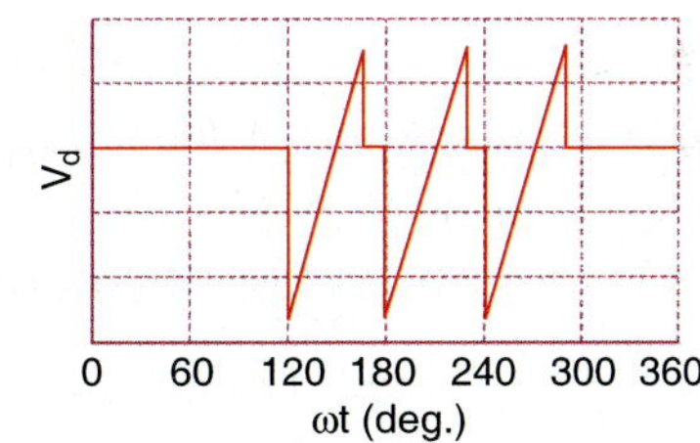

Fig. 3.12: Valve voltage waveform for α = 45°, u = 75°

The other three jumps have the magnitudes given by

$$V_{j2} = \sqrt{\frac{3}{2}} E_{LL} \sin(\alpha + u + 30°)$$

3.3 LCC BRIDGE CHARACTERISTICS

[A] Rectifier

The rectifier in general has three modes.

1. First mode: Two and three valve conduction mode($u < 60°$)
2. Second mode: Three valve conduction mode $0 \leq \alpha \leq 30°$; ($u = 60°$)
3. Third mode: Three and four valve conduction mode $\alpha > 30°$; ($60° < u \leq 120°$)

As the DC current continues to increase, the converter operation changes over from mode 1 to 2 and finally to mode 3.

The DC voltage continues to decrease until it reaches zero.

For, $\alpha \geq 30°$ mode 2 is bypassed.

for modes (1) and (3) we have respectively

$$\bar{V}_d = \frac{V_d}{V_{do}} = \cos\alpha - \frac{I_d}{2I_s} \tag{3.36}$$

$$\frac{V_d}{V_{do}} = \sqrt{3}\cos(\alpha - 30^\circ) - \frac{3I_d}{2I_s} \tag{3.37}$$

The voltage and current characteristics are linear (with different slopes) in these cases.

For mode (2), $u = 60°$. For $u = constant$, the characteristics are elliptical and the equation is given by

$$\left(\frac{\bar{V}_d}{\cos\frac{u}{2}}\right)^2 + \left(\frac{\bar{I}_d}{\sin\frac{u}{2}}\right)^2 = 1 \tag{3.38}$$

where $\bar{I}_d = \dfrac{I_d}{2I_s}$

Equation (3.38) follows from the following relations

$$\bar{V}_d = \frac{1}{2}[\cos\alpha + \cos(\alpha + u)] = \cos(\alpha + \frac{u}{2})\cos\frac{u}{2} \tag{3.39}$$

$$\bar{I}_d = \frac{1}{2}[\cos\alpha - \cos(\alpha + u)] = \sin(\alpha + \frac{u}{2})\sin\frac{u}{2} \tag{3.40}$$

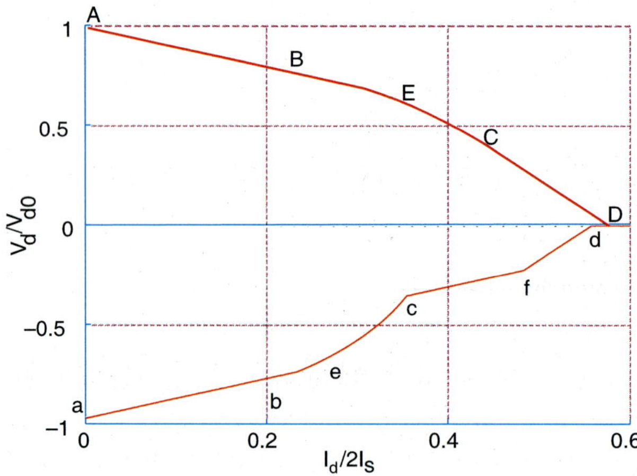

Fig. 3.13: Converter V_d–I_d characteristics

The boundary for the rectifier operation is shown in the Fig. 3.13. The coordinates of points A, B, C, D and E on the boundary are given in Table 3.1. The point E corresponds to the maximum power output of the converter.

Table 3.1: Boundary of Rectifier Characteristics

Point	α	u	$\bar{I}_d$	$\bar{V}_d$
A	0	0	0	1.0
B	0	60°	0.25	0.75
C	30°	60°	√3/4	√3/4
D	30°	120°	1/√3	0
E	15°	60°	1/(2√2)	√3/(2√2)

[B] Inverter

The inverter characteristics are similar to the rectifier characteristics. However, the operation as an inverter requires a minimum commutation margin angle during which the voltage across the valve is negative. Hence the operating region of an inverter is different from that for a rectifier.

The commutation margin angle (ξ) is equal to the extinction angle (γ) only for values of $\beta \leq 60°$. This is clear from Fig. 3.14(*a*). The voltage across the valve has a positive dent D because of the succeeding commutation. This dent normally occurs (for $\beta \leq 60°$) after the sinusoidal voltage (on which it is superposed) has become positive.

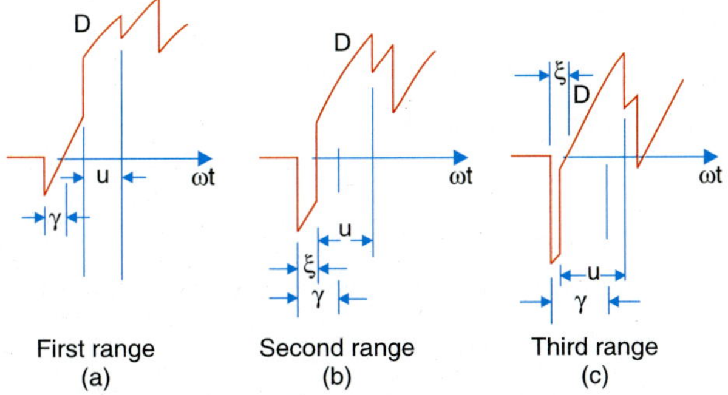

Fig. 3.14: Relationship between inverter angles

Hence under this condition $\xi = \gamma$. With increased overlap and consequently earlier ignition (firing) of the valve, the dent encroaches on the period in which the valve voltage would otherwise be negative [see Fig. 3.14(*b*) and this makes $\xi < \gamma$. After the front of the dent becomes entirely negative, further advance of the dent (Fig. 3.14(*c*)] does not decrease commutation margin further. Thus, the margin angle (ξ) has different relationship to γ depending on the range of operation, which is summarized below:

First range: $\beta \le 60°$

$$\xi = \gamma \tag{3.41}$$

Second range: $60° < \beta < 90°$

$$\xi = 60° - u = \gamma - (\beta - 60°) \tag{3.42}$$

Third range:

$$\beta > 90°$$

$$\xi = \gamma - 30° \tag{3.43}$$

In the inverter operation, it is necessary to maintain a certain minimum margin angle ξ_0. This results in 3 sub-modes of the first mode. The overall inverter characteristics are described below:

Mode 1

1(a) $\beta < 60°$ for values of $u < (60° - \xi_0)$

$$\gamma = \gamma_0 = \xi_0$$

The characteristics are linear defined by

$$\overline{V_d} = \cos\gamma_0 - \overline{I_d} \tag{3.44}$$

1(b) $60° < \beta < 90°$ for

$$u = 60° - \xi_0 = 60° - \gamma_0 = \text{constant}$$

As discussed earlier, the characteristics are elliptical defined by equation (3.38)

1(c) $90° < \beta < 90° + \xi_0$ for the values of u in the range

$$60° - \xi_0 \le u \le 60°$$

The characteristics in this case are again linear defined by

$$\overline{V_d} = \cos(\gamma_0 + 30°) - \overline{I_d} \tag{3.45}$$

Mode 2

For $u > 60°$ corresponding to $\beta > 90° + \gamma_0$

The characteristics are again linear but with a different slope and defined by

$$\overline{V_d} = \sqrt{3}\cos\gamma_0 - 3\overline{I_d} \tag{3.46}$$

The boundary of the inverter operation for a value of $\xi_0 = 15°$ is shown in Fig. 3.13. The coordinates of points a, b, c, d and f on the boundary are given in Table 3.2.

Table 3.2: Boundary of Inverter Characteristics

Point	γ	u	$\bar{I}_d$	$\bar{V}_d$
a	15°	0	0	–0.966
b	15°	45°	0.233	–0.733
c	45°	45°	0.354	–0.354
f	45°	60°	0.483	–0.224
d	45°	90°	0.558	0
e	22.5°	45°	0.271	–0.654

The maximum power supplied by the inverter occurs at $\gamma = 22.5°$ and $u = 45°$ (point e).

In the normal operation of the converter $\bar{I}_d$ is in the range of 0.08 to 0.1. Hence, only the first linear portion of the characteristics (both for rectifier and inverter) is of importance in the converter control.

Example 4

A Graetz bridge operates with a delay angle of 15°. The leakage reactance of the transformer is 10 ohms. The line to line AC voltage is 85 kV. Compute the overlap angle and DC voltage for (*i*) I_d = 2000 A, (*ii*) I_d = 4500 A.

Solution

A rectifier can operate with (*i*) $u < 60°$, (*ii*) $u = 60°$, or (*iii*) $u > 60°$. To determine the operating mode, we need to know the boundaries of the first two modes. The boundary (upper limit for I_d) for mode 1 is given by $\bar{I}_{d1} = \frac{1}{2}[\cos\alpha - \cos(\alpha + 60°)]$

$$I_{d1} = I_s[\cos 15° - \cos 75°]$$

$$I_s = \frac{\sqrt{2} \times 85}{2 \times 10} = 6 \text{ kA}$$

$$I_{d1} = 4.25 \text{ kA}$$

Similarly the upper limit on I_d, when $u = 60°$ is given by

$$I_{d2} = I_s[\cos 30° - \cos 90°] = 5.2 \text{ kA}$$

(*i*) When $I_d = 2000\text{A}$, we have mode 1 with $u < 60°$

$$\bar{V}_d = \cos\alpha - \bar{I}_d = \cos 15° - \frac{2000}{12000} = 0.8$$

$$V_{do} = \frac{3\sqrt{2}}{\pi} E_{LL} = \frac{3\sqrt{2} \times 85}{\pi} = 114.8 \text{ kV}$$

$$V_d = 0.8\, V_{do} = 91.8 \text{ kV}$$

The overlap angle u is obtained from

$$\frac{2}{12} = \frac{1}{2}[\cos 15° - \cos(15 + u)]$$

The solution for $u = 35.8°$

(*ii*) When I_d = 4500 A, the mode is $u = 60°$

$\bar{I}_d = \frac{4.5}{12} = 0.375$. $\bar{V}_d$ can be found from

$$\left(\frac{\bar{V}_d}{\cos\frac{u}{2}}\right)^2 + \left(\frac{\bar{I}_d}{\sin\frac{u}{2}}\right)^2 = 1$$

CHAPTER 3

$$\bar{V}_d = \left[1 - \left(\frac{0.375}{\sin 30^\circ}\right)^2\right]^{\frac{1}{2}} \cos 30^\circ = 0.573$$

$$V_d = 0.573 \times V_{do} = 0.573 \times 114.8 = 65.8 \text{ kV}$$

The firing angle α is obtained from

$$\bar{V}_d = \cos\alpha - \bar{I}_d$$

$$\cos\alpha = \bar{V}_d + \bar{I}_d = 0.573 + 0.375 = 0.948$$

$$\alpha = 18.6^\circ$$

Example 5

A six pulse inverter is operating at a constant margin angle of 18°. The valve side voltage is 70.7 kV (line to line) and the leakage reactance of the converter transformer is 10 ohms. Compute the extinction angle, overlap angle and DC voltage when (*i*) I_d = 2500 A and (*ii*) I_d = 4200 A .

Solution

An inverter can operate in either Mode 1 ($u < 60^\circ$, two and three valve conduction mode) or Mode 2 ($u > 60^\circ$, three and four valve conduction mode). Mode 1 can have three sub-modes.

(*a*) $\beta < 60^\circ$

(*b*) $60^\circ \le \beta < 90^\circ$

(*c*) $90^\circ < \beta \le 90^\circ + \xi_0$.

To identify the operating mode, it is necessary to know the boundaries of the various modes or sub-modes.

Mode 1 (*a*): $0 \le I_d < I_{d1}$

$$I_{d1} = I_s(\cos\gamma - \cos\beta),\ I_s = \frac{\sqrt{2}E_{LL}}{2X_c} = \frac{\sqrt{2} \times 70.7}{2 \times 10} = 5 \text{ kA}$$

$$\gamma = \xi = 18^\circ, \beta = 60^\circ, \text{ Hence } I_{d1} = 2.255 \text{ kA}$$

Mode 1 (*b*): $I_{d1} \le I_d \le I_{d2}$

When $I_d = I_{d2},\ \beta = 90^\circ,\ \gamma = \xi + 30^\circ$

Hence, $I_{d2} = I_s(\cos\gamma - \cos\beta) = 3.346$ kA

Mode 1 (*c*): $I_{d2} \le I_d \le I_{d3}$

When $I_d = I_{d3},\ \beta = 90^\circ + \xi = 108^\circ,\ \gamma = \xi + 30^\circ$

$$I_{d3} = 5 \times (\cos 48^\circ - \cos 108^\circ) = 4.891 \text{ kA}$$

(*i*) when $I_d = 2500A$, the inverter is operating in Mode 1 (b). Here, $u = 60 - \zeta = 42^\circ$ = constant

$$\left(\frac{\bar{I}_d}{\sin\frac{u}{2}}\right)^2 + \left(\frac{\bar{V}_d}{\cos\frac{u}{2}}\right)^2 = 1,\ \bar{I}_d = \frac{2.5}{10} = 0.25$$

$$\bar{V}_d^2 = \left[1-\left(\frac{0.25}{\sin 21^\circ}\right)^2\right]\cos^2 21^\circ$$

$$\bar{V}_d = 0.6689, V_d = V_{do}\bar{V}_d = 0.6689 \times \frac{3}{\pi} \times 100 = 63.88 \text{ kV}$$

Since $\bar{V}_d = \cos\gamma - \bar{I}_d, \cos\gamma = \bar{V}_d + \bar{I}_d = 0.6689 + 0.25$

$$\gamma = \cos^{-1}(0.9189) = 23.2^\circ$$

(ii) $I_d = 4200\text{A}$. The inverter is operating in Mode I (c). $\bar{I}_d = 0.42, \gamma = 48^\circ$

$$\bar{V}_d = \cos\gamma - \bar{I}_d = \cos 48^\circ - 0.42 = 0.2491$$

$$V_d = \bar{V}_d \times V_{do} = 23.79 \text{ kV}$$

To compute overlap angle

$$\bar{I}_d = \frac{1}{2}[\cos\gamma - \cos(\gamma + u)]$$

$$\cos(48^\circ + u) = \cos 48^\circ - 2\bar{I}_d$$

$$= \cos 48^\circ - 0.84$$

$$u = \beta - \gamma = 99.8^\circ - 48^\circ = 51.8^\circ$$

3.4 CHARACTERISTICS OF A TWELVE PULSE CONVERTER

As long as the AC voltages at the converter bus remain sinusoidal (with effective filtering), the operation of one bridge is unaffected by the operation of the other bridge connected in series. In this case, the converter characteristics are as shown in Fig. 3.13 with the assumption that the AC voltages at the converter bus remain constant. The region of rectifier operation can be divided into five modes as follows:

Mode 1 : 4 and 5 valve conduction
$0 < u < 30^\circ$
Mode 2 : 5 and 6 valve conduction
$30^\circ < u < 60^\circ$
Mode 3 : 6 valve conduction
$0 < \alpha < 30^\circ$; $u = 60^\circ$
Mode 4 : 6 and 7 valve conduction
$60^\circ < u < 90^\circ$
Mode 5 : 7 and 8 valve conduction
$90^\circ < u < 120^\circ$

It is to be noted that the second mode is a continuation of the first and similarly fifth is a continuation of the fourth. Five valve and seven valve conduction modes are just points on the boundaries corresponding to their previous and following modes. The region corresponding to mode 3 shrinks to a point when α exceeds 30°.

The characteristics of the inverter are similar except that a sub-mode can exist for mode 1 or 2 depending on the value of margin angle chosen (usually for mode 2). The characteristics of this sub-mode are elliptical as explained in the previous section.

CHAPTER 3

When no AC filters are provided and the source reactance is not zero the operation of either bridge is affected by the commutation process taking place in the other bridge. In this case, the operation of the twelve pulse converter is quite complex and there could be additional modes—(*i*) 5 valve conduction and (*ii*) 6-7 – 8-7 valve conduction [1]. Also, there could be new mode of 5-6–7-6 valve conduction (instead of 6 valve conduction), depending on the value of coupling factor K, defined by

$$K = \frac{X_S}{X_S + X_T}$$

where X_S is the source reactance and X_T is the converter transformer leakage reactance.

It is to be noted that the interaction between adjacent bridges can be neglected if the converter bus voltages are sinusoidal (when AC filters are provided). However, the presence of the source reactance results in the variation of the magnitude of the bus voltage (unless voltage control is provided at the converter station). This can affect the shape of the converter characteristics.

The analysis of a converter with source impedance, filters and finite smoothing reactor is described in the next section.

3.5 DETAILED ANALYSIS OF CONVERTERS

Here, the assumptions of constant DC current and constant, sinusoidal voltages at the AC bus are relaxed. The analysis takes into account impedances in both the AC and DC networks.

The steady-state solution of HVDC converter equation is periodic. With linear network elements, the equations are also piecewise linear. A method for obtaining fast steady-state solution of the system equations for the general case is described next.

Outline of the method

The method is based on the following assumptions:

1. The system is described by sets of linear differential equations and each set is applicable for particular conduction pattern of the valves in bridge.
2. AC system is symmetrical and source voltages are balanced.
3. Firing pulses are generated at equal intervals of time.

Given the initial conditions and forcing functions, the solution can be obtained analytically over each interval or sub-interval. The solution is periodic in steady-state, each period can be divided into p intervals, where p is the pulse number of the converter. Each interval, in general, can be divided into two sub-intervals as follows:

1. $0 < t < t_1$, corresponding to the conduction of $(m + 1)$ valves.
2. $t_1 < t < T_1$ corresponding to the conduction of m valves. ($T_1 = \frac{T}{p}$, T is the period of the AC voltage).

For example, in a 6-pulse converter, the normal mode consists of 3 and 2 valve conduction with $m = 2$. The derivation of the steady-state solution proceeds from first computing t_1 from a non-linear equation of the form

$$f(t_1) = 0$$

Once t_1 is obtained, the initial conditions can be calculated from sets of linear algebraic equations. These equations are derived from boundary conditions and symmetry considerations. In general, the following constraints apply in determining the boundary conditions.

1. Magnetic fluxes and electric charges must be continuous functions of time.
2. The current in the outgoing valve is zero at $t = t_1$.

Development of the method for fast steady-state analysis

It is already mentioned that there are two sub-intervals to be considered. Let the system be described by the following state equations.

$$\dot{x}_1(t) = [A_1]x_1(t) + [B_1]u_1(t), \quad 0 < t < t_1 \tag{3.47}$$

$$\dot{x}_2(t) = [A_2]x_2(t) + [B_2]u_2(t), \quad t_1 < t < T_1 \tag{3.48}$$

The orders of the state vectors x_1 and x_2 are $n + 1$ and n respectively. Since the outgoing valve current becomes zero at $t = t_1$, one state variable is eliminated in the second sub-interval. The solutions of the above equations are:

$$x_1(t) = xs_1(t) + e^{A_1 t}[x_1(0) - xs_1(0)] \tag{3.49}$$

$$x_2(t) = xs_2(t) + e^{A_2(t-t_1)}[x_2(t_1) - xs_2(t_1)] \tag{3.50}$$

where $xs_i(t)$ is the forced response in sub-interval i.

There is no loss of generality in assuming the current in the outgoing valve as x_1^{n+1}, superscript indicates the $(n + 1)^{\text{th}}$ variable in the vector x_1. Thus, we have

$$x_1^{n+1}(t_1) = 0 \Rightarrow Cx_1(t_1) = 0 \tag{3.51}$$

where $C = [0 \ 1]$ is a row vector.

The vector $x_2\,(t)$ is the subset of the vector $x_1\,(t)$ with the last variable removed from $x_1(t)$. From the consideration of continuity in the state variables, we get

$$x_2(t_1) = [I_n : 0]x_1(t_1) = [K]x_1(t_1) \tag{3.52}$$

where I_n is the identity matrix of nth order. From considerations of symmetry, we have

$$x_1(0) = [S]x_2(T_1) \tag{3.53}$$

where S is a constant matrix of dimension $(n + 1) \times n$.

The element of the matrix [S] can be obtained from the following considerations.

1. The DC quantities are periodic with period T_1. For example, the DC current i_d, satisfies the relation

$$i_d(0) = i_d(T_1)$$

2. The AC quantities satisfy the following relations (assuming $T_1 = \dfrac{\pi}{3\omega}$, ω is the system angular frequency)

$$x_a(0) = -x_c(T_1)$$
$$x_b(0) = -x_a(T_1)$$
$$x_c(0) = -x_b(T_1)$$

where x_a, x_b and x_c refer to the state variables corresponding to the phases a, b and c respectively. $x_2(T_1)$ can be obtained by applying equations (3.49) and (3.50) and is given by

$$x_2(T_1) = xs_2(T_1) + e^{A_2(T_1-t_1)}\left\{K[xs_1(t_1) + e^{A_1t_1}(x_1(0) - xs_1(0))] - xs_2(t_1)\right\} \quad (3.54)$$

Substituting for $x_1(0)$ from equation (3.53), we get

$$[Q]x_2(T_1) = g \quad (3.55)$$

where

$$[Q] = \left[I_n - e^{A_2(T_1-t_1)}Ke^{A_1t_1}S\right]$$

$$g = xs_2(T_1) + e^{A_2(T_1-t_1)}\left\{K\left[xs_1(t_1) - e^{A_1t_1}xs_1(0)\right] - xs_2(t_1)\right\}$$

The equations (3.51) and (3.55) can be used to solve for t_1 and $x_2(T_1)$. The initial condition $x_1(0)$ can be calculated from equation (3.53) and the steady-state solution can be obtained using the equations (3.49) and (3.50). This is best achieved through a digital computer program in which the state transition matrix e^{At} and the forced response $xs(t)$ are evaluated.

Remarks

1. If (Q) is non-singular matrix (of the order n) the solution for $x_2(T_1)$ is obtained from

$$x_2(T_1) = [Q]^{-1}g$$

The overlap time t_1 is determined from the non-linear equation

$$C\left\{xs_1(t_1) + e^{A_1t}[SQ^{-1}g - xs_1(0)]\right\} = 0 \quad (3.56)$$

An iterative method is used to solve the above equation.

Example

Consider the circuit shown in Fig. 3.15 where a six pulse converter is connected to a DC voltage source through an impedance. It is assumed that the operating mode consists of three and two valve conduction, with overlap angle less than 60°. Figure 3.15 shows the conduction of the three valves, 1, 2 and 3.

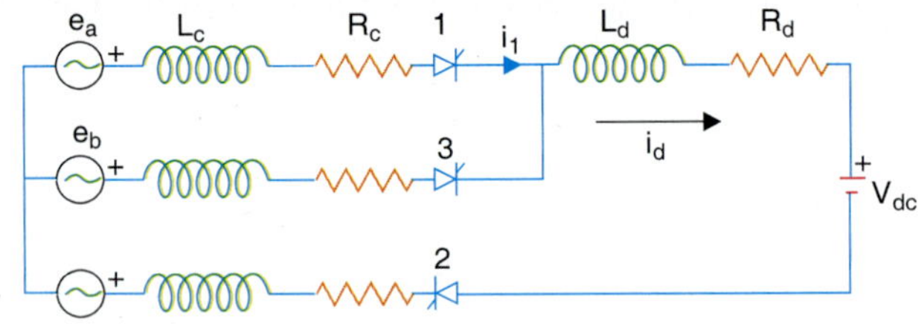

Fig. 3.15: Equivalent circuit for of 3 valve conduction (Example)

Solution:

The state vector for the circuit shown in Fig. 3.15 is chosen as

$$x_1^t = [i_d \quad i_1]$$

The matrices A_1 and B_1 are given by

$$A_1 = -[L]^{-1}[R],\ B_1 = [L]^{-1}$$

where,

$$[L] = \begin{pmatrix} 2L_c + L_d & -L_c \\ -L_c & 2L_c \end{pmatrix},\ [R] = \begin{pmatrix} 2R_c + R_d & -R_c \\ -R_c & 2R_c \end{pmatrix}$$

The vector u_1 is given by

$$u_1 = \begin{pmatrix} e_b - e_c - V_{dc} \\ e_a - e_b \end{pmatrix}$$

When valve 1 turns off, the state vector reduces to the scalar and the differential equation for $t_1 \le t \le T_1$ is given by

$$\frac{di_d}{dt} = -\frac{(2R_c + R_d)}{(2L_c + L_d)} i_d + \frac{1}{(2L_c + L_d)}(e_b - e_c - V_{dc})$$

The matrix S is a column vector of dimension 2 given by

$$[S] = \begin{pmatrix} 1 \\ 1 \end{pmatrix}$$

3.6 CAPACITOR COMMUTATED CONVERTER

The configuration of a capacitor commutated converter (CCC) is shown in Fig. 3.16. Here, series capacitors are inserted between the converter transformer and the valves to aid commutation, particularly for inverter operation, CCC permits the operation of α close to 180° and reduces the risk of commutation failures particularly with long DC cables. The requirements of the reactive power at the converter are also reduced which enable economic filters to be designed.

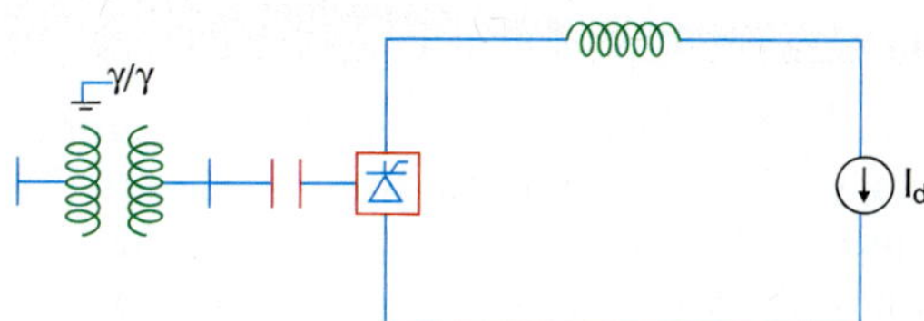

Fig. 3.16: A capacitor commutated converter

The application of series capacitors to assist the commutation process has been investigated since the days of mercury are valves. It was termed 'artificial' or 'forced' commutation [6, 7].

The first installation of CCC was at Garabi in South America. A major advantage of CCC is the improvement of the extinction angle characteristics. The apparent extinction angle viewed from the converter AC bus can approach small or even negative values depending on the size of the series capacitor. The apparent extinction angle (γ_{app}) is the margin angle (angle between the instant at which the valve turns off to the instant of the positive zero crossing of the sinusoidal commutation voltage). The actual margin angle is higher because the actual commutation voltage is the sum of the commutation voltages due to the AC bus voltage and the series capacitor voltages. For example, if γ_{app} = 2°, the actual value of the margin angle could be 22°. The reduction in (γ_{app}) improves power factor and thus reduces the need for the shunt reactive power source (normally supplied by capacitors and AC filters). Essentially, the firing angle α can approach 180° without worrying about commutation failures (described in Chapter 5).

The voltages across the series capacitors are related to the DC current and thus the actual margin angle increases with the DC current. This is opposite of what happens in a conventional HVDC converter, where the margin angle reduces with the DC current.

This characteristic of CCC is very favourable for reducing the incidences of commutation failures. A sudden decrease in the commutation voltage at the inverter (say due to a remote AC fault) results in increase of the DC current. If the DC line is a cable, the overcurrent is also due to the sudden discharge of the cable which cannot be controlled by the current controller at the rectifier. A typical variation of the actual margin angle (γ) with I_d for a constant value of γ_{app} = 2° is shown in Fig. 3.17.

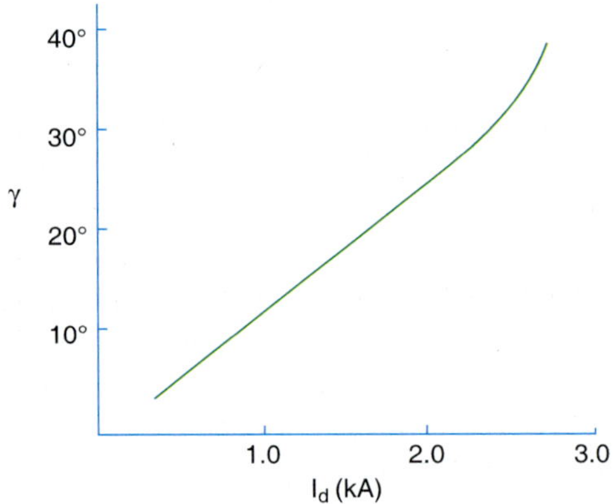

Fig. 3.17: Variation of actual margin angle (γ) with I_d

With weak AC systems connected to an inverter, the maximum power that can be supplied by the inverter could be lower than the rated power as the SCR is reduced. The maximum available power can be increased due to the series capacitors. Thus, a CCC improves the performance of a HVDC link connected to weak AC systems. The load rejection overvoltages are also reduced with CCC.

The drawbacks of a CCC are:

1. Increase in the converter valve voltage stress due to the voltages across the series capacitors.
2. Increase in the magnitude of AC harmonics as the overlap angle is reduced for a specified DC current. However, the size of the AC filters can be reduced as they don't have to supply reactive power (at the fundamental frequency). It may be possible to use active filters to filter the harmonics instead of using passive filters which can be problematic (see chapter 8).

A simplified analysis of CCC

The analysis is aimed at obtaining waveforms and relationships in steady-state operation of CCC. The following assumptions are made to simplify the analysis.

1. The DC current (I_d) is constant.
2. The source voltages (at the converter bus) are sinusoidal and balanced.
3. The firing pulses are equidistant and the firing angle α is thus, constant.
4. The overlap angle u, is less than 60° and thus, the operation of the six pulse converter is 3-2 valve conduction.
5. The losses are neglected.

There is no loss of generality in considering the time $t = 0$ at the instant when valve 3 is fired. Since the DC current is already flowing in valve 1, the capacitor in series with valve 1 is charged with voltage polarity as shown in Fig. 3.18. This voltage continues to increase as the valve current continues to flow during the overlap period until i_1 becomes zero. On the other hand, the voltage across the capacitor in series with valve 3 is negative maximum before $t = 0$ and starts to decrease as the valve current continues to increase. At $\omega t = u$, the current in valve 1, i_1 is zero and the current in valve 3, i_3 is I_d. The current in valve 3, remains constant during the interval, $u \le \omega t \le \frac{2\pi}{3}$. The capacitor voltage continues to increase (with positive polarity) when valve 5 is fired and until the current i_3 becomes zero. From symmetry considerations, the negative and positive peaks of the capacitor voltages are equal and identical in each phase. In general, it can be said that the capacitor voltages are constants when the currents through them are zero and increases from negative peak to positive peak when the valves in the upper group conduct. During the conduction of the valves in the lower group, the capacitor voltages decrease from positive peak to negative peak.

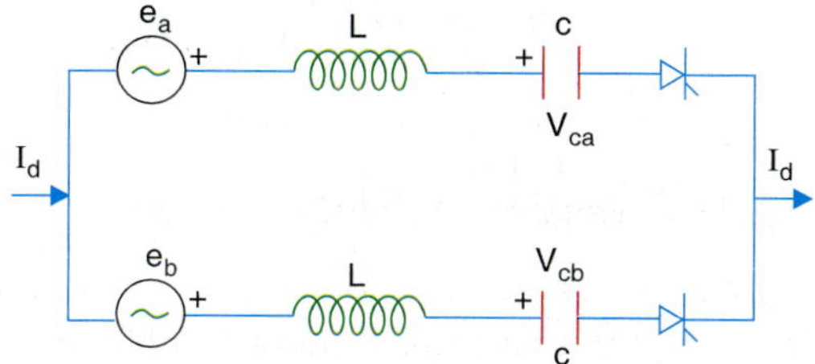

Fig. 3.18: Commutation equivalent circuit

The expressions for the valve currents in the circuit shown in Fig. 3.18 can be obtained by superposition of the two responses; one due to I_d and the other due to the AC voltages and capacitor charges. The expressions for the valve currents are

$$i_1 = i_{11} + i_{12} = -i + \frac{I_d}{2}$$

$$i_3 = i_{31} + i_{32} = i + \frac{I_d}{2} \qquad (3.57)$$

The current i is due to the AC voltage $e_{ba} = \sqrt{2}E_{LL}\sin(\omega t+\alpha)$ and the capacitor charges at $t = 0$. The capacitor voltage at $t = 0$ are

$$V_{Ca} = V_m - \Delta V_1,\ V_{Cb} = -V_m \tag{3.58}$$

where V_m is the peak voltage (magnitude) across the capacitors.

The expression for the current i be obtained as

$$i(t) = \frac{\sqrt{2}E_{LL}}{2(X_C - X_L)}\cos(\omega t+\alpha) + A\cos\omega_n t + B\sin\omega_n t \tag{3.59}$$

where $$X_C = \frac{1}{\omega C},\ X_L = \omega L,\ \omega_n = \frac{1}{\sqrt{LC}}$$

The constants A and B can be obtained from the boundary conditions on i_3 and V_{Cb}, namely $i_3 = 0$ at $\omega t = 0$, $i_3 = I_d$ at $\omega t = u$, $V_{Cb} = -V_m$ at $\omega t = 0$. However, it is to be noted that both V_m and u are unknown. But it can be shown that V_m is independent of u and is linearly proportional to I_d.

The voltages across the capacitor connected in the phase 'b', V_{Cb} at $\omega t = u$, is given by

$$V_{Cb}(\omega t = u) = \Delta V_2 - V_m \tag{3.60}$$

where ΔV_2 is defined by

$$\Delta V_2 = \frac{1}{\omega C}\int_0^u \left(i + \frac{I_d}{2}\right) d\omega t$$

The value of V_{Cb} at $\omega t = \frac{2\pi}{3}$, is given by

$$V_{Cb}\left(\omega t = \frac{2\pi}{3}\right) = \Delta V_2 - V_m + \frac{1}{\omega C}\int_u^{2\pi/3} I_d \omega t$$

$$= \Delta V_2 - V_m + \frac{I_d}{\omega c}\left(\frac{2\pi}{3} - u\right) \tag{3.61}$$

At $\omega t = \frac{2\pi}{3} + u$, V_{Cb} is given by

$$V_{Cb}\left(\omega t = \frac{2\pi}{3} + u\right) = \Delta V_2 - V_m + \frac{I_d}{\omega C}\left(\frac{2\pi}{3} - u\right) + \int_{2\pi/3}^{2\pi/3+u}\left(-i + \frac{I_d}{2}\right) d\omega t \tag{3.62}$$

Note that (by symmetry), the current in valve 3 when valve 5 is fired, is identical to the current in valve 1 when valve 3 is fired (during the overlap period). Thus, the last term in the RHS of the above equation is same as ΔV_1 and we obtain

$$V_{Cb}\left(\omega t = \frac{2\pi}{3} + u\right) = \Delta V_2 - V_m + \frac{I_d}{\omega C}\left(\frac{2\pi}{3} - u\right) + \Delta V_1 \tag{3.63}$$

Again from symmetry, the LHS = V_m

It can be observed that

$$\Delta V_1 + \Delta V_2 = \frac{1}{\omega C}\int_0^u I_d d\omega t = \frac{I_d u}{\omega C} \quad (3.64)$$

Substituting Eq. (3.64) in (3.63), we can derive

$$V_m = \frac{\pi}{3}\frac{I_d}{\omega C} \quad (3.65)$$

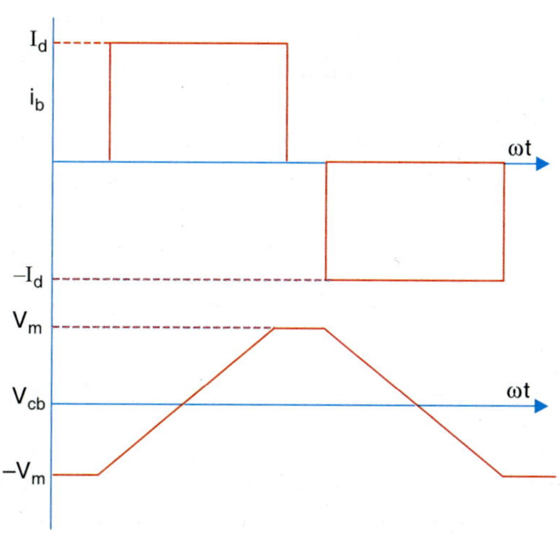

Fig. 3.19: Waveforms of i_b and V_{cb}

The waveform of V_{Cb} over a cycle is shown in Fig. 3.19 which also shows the waveform of i_b (current in phase b). Here, the overlap angles are neglected.

Neglecting overlap angles, the capacitor voltage waveform is a trapezoidal waveform which increases from zero to V_m in 60°, remains constant for 60° and reduces to zero in another 60° in a half cycle. The peak value of the fundamental component of the capacitor voltage V_{C1} is obtained as

$$V_{C1} = \frac{4}{\pi}\int_0^{\frac{\pi}{2}} V_C(\omega t)\sin\omega t \; d(\omega t)$$

$$= \frac{4}{\pi}\left[\frac{3}{\pi}V_m\int_0^{\frac{\pi}{3}}\theta\sin\theta \; d\theta + V_m\int_{\frac{\pi}{3}}^{\frac{\pi}{2}}\sin\theta \; d\theta\right]$$

$$= \frac{6\sqrt{3}}{\pi^2}V_m = 1.1V_m \quad (3.66)$$

It can be observed from Fig. 3.19 that the capacitor voltage lags the current by 90°. (Note that this is true for all the three phases). For phase '*b*',

$$i_{b1} = \frac{2\sqrt{3}}{\pi} I_d \sin(\omega t + 30^\circ) \tag{3.67}$$

$$V_{Cb1} = 1.1 V_m \sin(\omega t - 60^\circ) \tag{3.68}$$

V_{Ca1} leads V_{Cb1} by 120° and is given by

$$V_{Ca1} = 1.1 V_m \sin(\omega t + 60^\circ) \tag{3.69}$$

In addition to the commutation voltage of $\sqrt{2}E_{LL}\sin(\omega t + \alpha)$ due to the supply voltages, the capacitors in the two phases (*a* and *b*) introduce the additional commutation voltage of $(V_{Ca} - V_{Cb})$, which is equal to $\sqrt{3} \times 1.1 V_m \cos\omega t$. Note that this voltage is at the highest value when a valve is fired and is independent of the firing angle. However, the need for assisting commutation is more in the inverter operation as α tends to 180°.

It can be shown that the average DC voltage of CCC is given by [6]

$$V_d = \frac{3\sqrt{2}E_{LL}}{2\pi}\left[\cos\alpha + \cos(\alpha + u)\right] + \left(\frac{\pi}{3} - \frac{3u}{4}\right)(\Delta V_2 - \Delta V_1) \tag{3.70}$$

3.7 ANALYSIS OF A VOLTAGE SOURCE CONVERTER

A VSC based HVDC converter is similar to a STATic COMpensator (STATCOM), a shunt connected FACTS controller. Here, we consider three phase VSC using six switches made up of six GTO or IGBTs with antiparallel diodes connected as a six pulse, Graetz bridge (see Fig. 3.20). Following assumptions are made in the analysis.

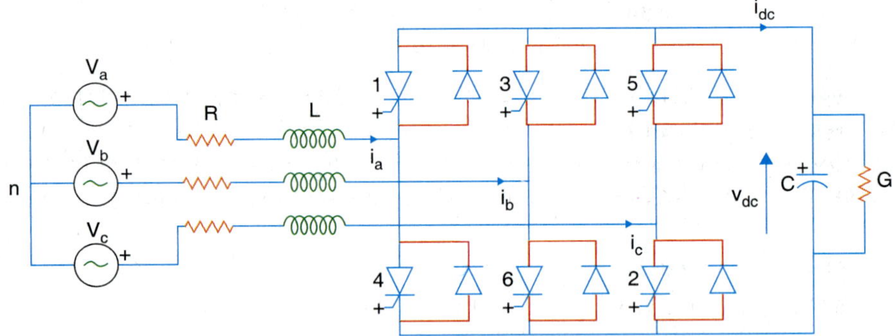

Fig. 3.20: A six pulse VSC circuit

1. The supply voltages are sinusoidal and balanced (contain only fundamental frequency, positive sequences components).
2. The losses in the switches are ignored and the switches are assumed to be ideal.
3. The inductance *L* and resistance *R* are the parameters of the series reactor connected between the supply and the VSC or represent the leakage reactance and the ohmic losses

of the interfacing transformer. G represents the losses in the DC capacitor. The VSC shown in Fig. 3.20 is described by the following differential equations.

$$L\frac{di_a}{dt}+Ri_a=v_a-e_a \tag{3.71}$$

$$L\frac{di_b}{dt}+Ri_b=v_b-e_b \tag{3.72}$$

$$L\frac{di_c}{dt}+Ri_c=v_c-e_c \tag{3.73}$$

$$C\frac{dv_{dc}}{dt}+GV_{dc}=i_{dc} \tag{3.74}$$

where e_a, e_b and e_c are the output voltages of the VSC. All the voltages are with reference to the source neutral 'n'.

We can express the voltages injected by the VSC as

$$e_a(t)=S_a(t)v_{dc}(t),\ e_b(t)=S_b(t)v_{dc}(t),\ e_c(t)=S_c(t)v_{dc}(t) \tag{3.75}$$

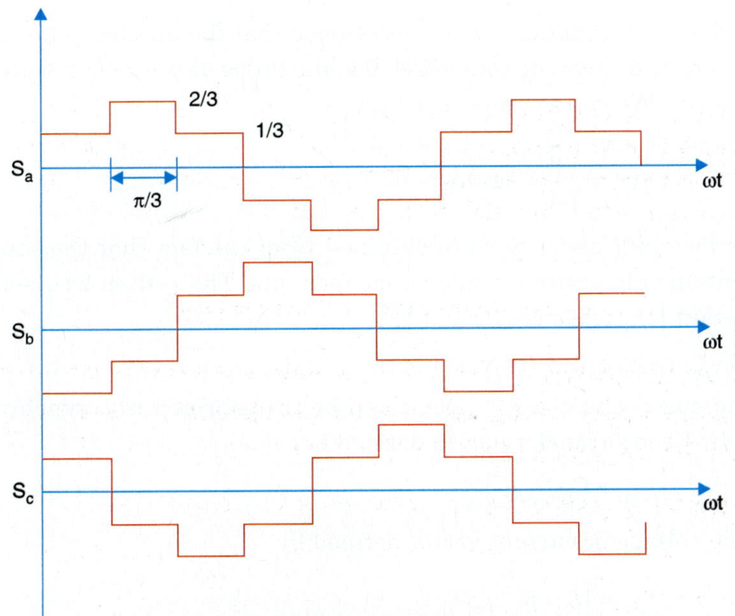

Fig. 3.21: Switching functions for six pulse VSC

where S_a, S_b and S_c are the switching functions. Without any modulation, (for square wave operation), they are periodic in steady-state and are shown in Fig. 3.21. From the principle of energy conservation, we get

$$v_{dc}(t)i_{dc}(t)=e_a(t)i_a(t)+e_b(t)i_b(t)+e_c(t)i_c(t) \tag{3.76}$$

From Eqs. (3.75) and (3.76), we can derive

$$i_{dc}(t) = S_a(t)i_a(t) + S_b(t)i_b(t) + S_c(t)i_c(t) \tag{3.77}$$

The switching functions in steady-state are related by

$$S_a(\omega t) = S_b(\omega t + 120°) = S_c(\omega t + 240°) \tag{3.78}$$

Substitution Eqs. (3.75) and (3.77) in (3.71) to (3.74) makes the system time varying. If switching functions are approximated by their fundamental frequency components only, we obtain for a two level converter.

$$e_a \simeq mv_{dc}\frac{2}{\pi}\sin(\omega_0 t + \theta + \alpha) \tag{3.79}$$

$$e_b \simeq mv_{dc}\frac{2}{\pi}\sin\left(\omega_0 t + \theta + \alpha - \frac{2\pi}{3}\right) \tag{3.80}$$

$$e_b \simeq mv_{dc}\frac{2}{\pi}\sin\left(\omega_0 t + \theta + \alpha - \frac{4\pi}{3}\right) \tag{3.81}$$

where m is the modulation index (normalized).

In deriving the above equations we have assumed that the injected voltages lead the source voltages by angle α, which can be controlled. θ is the angle of the source voltage whose line to line magnitude is V_s. We can express $v_a(t)$ as

$$v_a(t) = \sqrt{\frac{2}{3}}V_s\sin(\omega_0 t + \theta) \tag{3.82}$$

The expressions for $v_b(t)$ and $v_c(t)$ are obtained from the fact that the source voltages are balanced and contain only positive sequence components. The system frequency is assumed to be ω_0 and a constant (or varies very slowly).

Equations in D-Q reference frame

The voltages and current in the AC circuit can be transformed to a synchronously rotating reference frame by Kron's transformation defined by

$$f_{abc} = C_K f_{DQO} \tag{3.83}$$

where f_{abc} can be voltage or current vector defined by

$$f_{abc} = [f_a \quad f_b \quad f_c]^t$$

similarly,

$$f_{DQO} = [f_D \quad f_Q \quad f_O]^t$$

and C_K is the transformation matrix defined by

$$C_K = \sqrt{\frac{2}{3}} \begin{pmatrix} \cos\omega_0 t & \sin\omega_0 t & \frac{1}{\sqrt{2}} \\ \cos\left(\omega_0 t - \frac{2\pi}{3}\right) & \sin\left(\omega_0 t - \frac{2\pi}{3}\right) & \frac{1}{\sqrt{2}} \\ \cos\left(\omega_0 t + \frac{2\pi}{3}\right) & \sin\left(\omega_0 t + \frac{2\pi}{3}\right) & \frac{1}{\sqrt{2}} \end{pmatrix} \tag{3.84}$$

The subscript 'O' indicates zero sequence variable. C_K is an orthogonal matrix with the property,

$$[C_K]^{-1} = [C_K]^t \tag{3.85}$$

The advantage of the above is that the transformation is power invariant. We can derive

$$v_a i_a + v_b i_b + v_c i_c = v_D i_D + v_Q i_Q + v_O i_O \tag{3.86}$$

By applying Kron's transformation, we can derive the circuit equations in D-Q variables. These are given below:

$$L\frac{di_D}{dt} + Ri_D + \omega_0 L i_Q = v_D - mk\sin(\alpha+\theta)v_{dc} \tag{3.87}$$

$$L\frac{di_Q}{dt} + Ri_Q - \omega_0 L i_D = v_Q - mk\cos(\alpha+\theta)v_{dc} \tag{3.88}$$

$$C\frac{dv_{dc}}{dt} + Gv_{dc} = mk\left[\sin(\alpha+\theta)i_D + \cos(\alpha+\theta)i_Q\right] \tag{3.89}$$

where $k = \frac{\sqrt{6}}{\pi}$ for a two level converter. m is the (normalized) modulation index that can vary from 0 to 1. The zero sequence current is zero as $i_a + i_b + i_c = 0$.

CHAPTER 3

Remarks

1. In a high power VSC with GTO thyristor switches, the switch is turned on and off only once in a cycle to minimize switching losses. Here, m is a constant = 1.
   The magnitude of the injected voltage can be controlled by Pulse Width Modulation (PWM). There are many variations in PWM. PWM converters require several switchings in a cycle.
2. In a STATCOM with an energy source on the DC side, it is advisable to control both the magnitude and phase angle of the injected voltage by the VSC in order to control the power and reactive power output.
3. It is possible to express Eqs. (3.87) and (3.88) in terms of current and voltage phasors. By multiplying Eq. (3.87) by j on both sides and adding to Eq. (3.88), we get

$$L\frac{d\hat{I}}{dt} + (R + j\omega_0 L)\hat{I} = \hat{V} - \hat{E} \tag{3.90}$$

where

$$\hat{I} = I_Q + jI_D,\ \hat{V} = V_Q + jV_D$$

and

$$\hat{E} = mkV_{dc}\angle(\theta + \alpha) \tag{3.91}$$

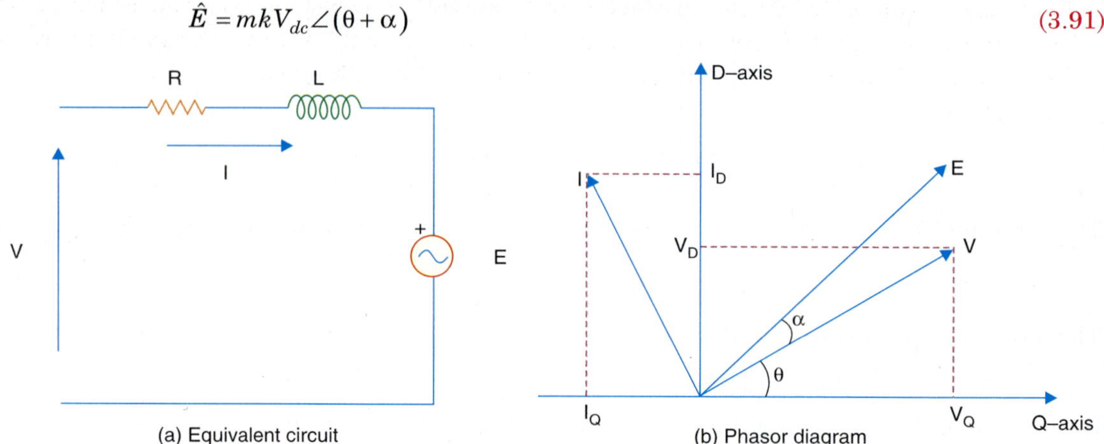

Fig. 3.22: Equivalent circuit and phasor diagram of a STATCOM

Equation (3.90) represents the simple circuit shown in Fig. 3.22. The phasor diagram is also shown here. Note that, since v_D and v_Q are the components of the (dynamic) phasor, they become constants in steady-state (corresponding to an operating point). Generally, we use upper case to denote phasors. On the other hand, the D-Q components are also instantaneous values. Thus, to simplify the notation, we will use both lower and upper case symbols interchangeably.

4. For type II converters (with control only over the phase angle of the injected voltage), the elimination of voltage harmonics requires multi-pulse configuration of the VSC. Twelve pulse operation can be arranged with two six pulse converter bridges connected in series on the AC side and in parallel on the DC side. The output voltages of the two converters are summed up by a 'summing transformer' or an 'interphase magnetics'. For connection to high voltage AC lines, it is necessary to also use a coupling transformer as in the case of a SVC.

5. For a VSC based HVDC converter, PWM is invariably used with 2 or 3 level VSC to control the magnitude of the injected voltage, while regulating the DC bus voltage. It is convenient to express the injected voltage phasor as

$$\hat{E} = mE_{ao}\angle\theta + \alpha, \quad E_{ao} = kV_{dc} \tag{3.92}$$

where m is the normalized modulation index and $m = 1$ indicates square wave operation. For a 3 level or 12 pulse converter, $k = \dfrac{2\sqrt{6}}{\pi}$.

The VSC based HVDC converters are invariably type I converters with control over both the magnitude and the phase angle of the injected voltage. This implies that both the power (P) and reactive power (Q) supplied by the converter can be controlled within limits. The typical operating region ABCD in the P-Q plane is shown in Fig. 3.23. Here the lines AB and CD are due to the maximum DC power that is carried by the HVDC cable due to the upper limit on the DC current. BC is part of a circle due to the maximum current limit through the IGBT valves. DA represents limitation on the injected reactive power due to the maximum DC voltage across capacitor.

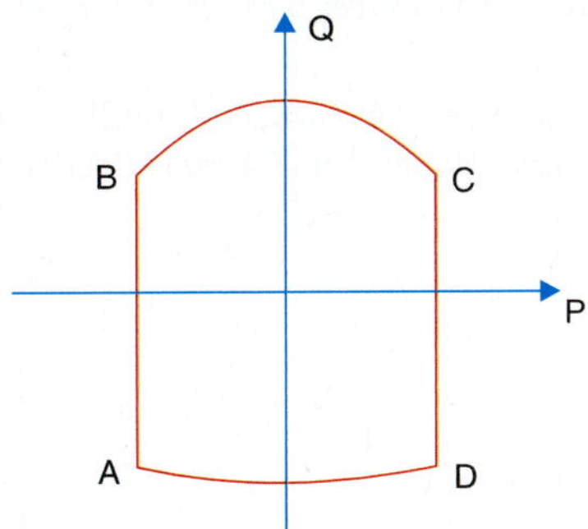

Fig. 3.23: Operating region of VSC in P-Q plane

The reactive power capability is a function of the AC bus voltage and increases with decreasing AC voltage. This is a desirable characteristic from the stability point of view. The capability curves of a VSC are similar to that of a synchronous generator. Maximum DC voltage limitation in a VSC corresponds to the maximum field current limitation in a generator while IGBT current corresponds to the armature current. However, unlike in a generator, a VSC can shift its operating point, practically instantaneously.

The HVDC cable based on XLPE technology can handle fast current reversal required to reverse the power flow in the cable. (Note that in a VSC based HVDC system, the power reversal does not require the reversal of the DC voltage which requires some time delay to accomplish).

REFERENCES AND BIBLIOGRAPHY

1. Freris L.L., '*Effect of Interaction among Groups in a Multi group Converter*', Proc. IEE, Vol. 114, No.7, July 1967, pp. 965–973.
2. Ooi B.T., Menemlis N. and Nakra H.L. '*Fast Steady-state Solution for HVDC Converters*', IEEE Trans., PAS-99, 1980, 2453–59.
3. Padiyar K.R. and Kalra P.K. '*Analysis of HVDC Converter with Finite Smoothing Reactor*, Part I–*Analysis of Six-pulse Converter*, Part II–*Analysis of a Twelve-pulse Converter*', Electric Power Systems Research 11, 1986, pp. 171–193.
4. Kimbark E.W., '*Direct Current Transmission*', Vol. I, (New York), John Wiley, 1971.
5. Uhlmann E., '*Power Transmission by Direct Current*', (Berlin), Springer-Verlag, 1975.
6. Reeve J., Baron J.A. and Hanley G.A., '*A Technical Assessment of Artificial Commutation of HVDC Converters with Series Capacitors*', IEEE Trans., Vol. PAS-87, No.10, 1968, pp. 1830–1840.

7. Gole A.M. and Menzies R.W., '*Analysis of Certain Aspects of Forced Commutated HVDC Inverters*', IEEE Trans., Vol. PAS-100, No.5, 1981, pp. 2258–2262.
8. Sadek K., Pereira M., Brandt D.P., Gole A.M. and Daneshpooy A., '*Capacitor Commutated Converter Circuit Configurations for DC Transmission*', IEEE Trans. On Power Delivery, Vol. 13, No.4, 1998, pp. 1257–1264.
9. Padiyar K.R., '*FACTS Controllers in Power Transmission and Distribution,* (New Delhi), New Age, 2007.
10. Johansson S.G., Asplund G., Jansson E. and Rudervall R., '*Power System Stability Benefits with VSC DC-transmission Systems*', Paper No. B4-204, CIGRE Conference, Paris, 2004.

4 CONVERTER AND HVDC SYSTEM CONTROL

4.1 GENERAL

One of the major advantages of a HVDC link is the rapid controllability of transmitted power through the control of firing angles of the converters. Modern converter controls are not only fast, but also very reliable and they are used for the protection against line and converter faults.

The system control in a HVDC link tends to be quite complex with a hierarchy of controllers. High speed microprocessors are being used for many of the control functions including monitoring and supervisory control.

In this chapter, the basic principles and functions of the system and converter control are presented.

LCC based HVDC transmission systems using thyristor valves is considered first. At the end, the control in a VSC based HVDC link is described.

4.2 PRINCIPLES OF DC LINK CONTROL

In what follows, a two terminal DC link is assumed. The control in multiterminal DC systems is similar and is described in a separate chapter.

The control of power in a DC link can be achieved through the control of current or voltage. From minimization of loss considerations, it is important to maintain constant voltage in the link and adjust the current to meet the required power. This strategy is also helpful for voltage regulation in the system from the considerations of the optimal utilization of the insulation. It is to be noted that the voltage drop along a DC line is small compared to the AC line, mainly because of the absence of the reactive voltage drop.

Consider the steady state equivalent circuit of a two terminal DC link shown in Fig. 4.1(*a*). This is based on the assumption that all the series connected bridges in both poles of a converter station are identical and have same delay angles. Also, the numbers of series connected bridges (n_b) in both stations (rectifier and inverter) are the same. The voltage sources E_{dr} and E_{di} are defined by

$$E_{dr} = (3\sqrt{2}/\pi)n_b E_{vr} \cos\alpha_r \tag{4.1}$$

$$E_{di} = (3\sqrt{2}/\pi)n_b E_{vi} \cos\gamma_i \tag{4.2}$$

where E_{vr} and E_{vi} are the line to line voltages in the valve side windings of the rectifier and inverter transformer respectively. From Fig. 4.1(b), these voltages can be obtained as

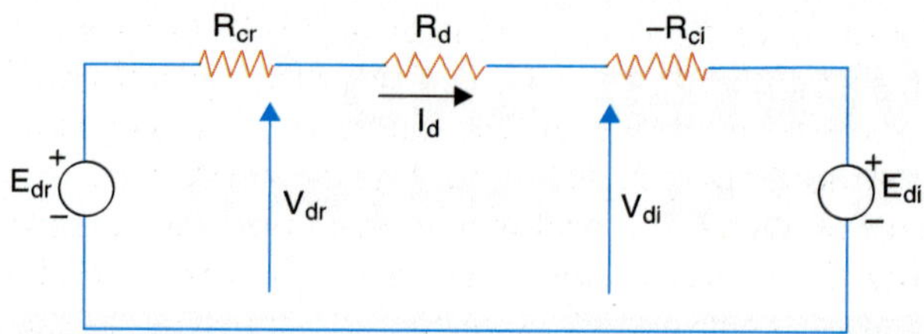

Fig. 4.1: (a) Steady-state equivalent circuit of a 2-terminal DC link

Fig. 4.1: (b) Schematic of a DC link showing transformer ratios

$$E_{vr} = \frac{N_{sr}E_r}{N_{pr}T_r}, E_{vi} = \frac{N_{si}E_i}{N_{pi}T_i} \tag{4.3}$$

where E_r and E_i are the AC (line to line) voltages of the converter buses on the rectifier and the inverter side. T_r and T_i are the off-nominal tap ratios on the rectifier and inverter side.

Combining (4.1), (4.2) and (4.3), we can write

$$E_{dr} = (A_r E_r / T_r)\cos\alpha_r \tag{4.4}$$

$$E_{di} = (A_i E_i / T_i)\cos\gamma_i \tag{4.5}$$

where A_r and A_i are constants.

It is to be noted E_{di} is defined in terms of the extinction angle γ_i rather than β_i (the angle of advance in the inverter). E_{di} can also be written as

$$E_{di} = (A_i E_i / T_i)\cos\beta_i + 2R_{ci}I_d \tag{4.6}$$

where

$$R_{ci} = (3n_b/\pi)X_{ci}, R_{cr} = (3n_b/\pi)X_{cr} \tag{4.7}$$

X_{cr} and X_{ci} are the leakage reactances of the converter transformers in the rectifier and inverter station respectively.

The steady-state current I_d in the DC link is obtained as

$$I_d = \frac{(E_{dr} - E_{di})}{R_{cr} + R_d - R_{ci}} \tag{4.8}$$

Substituting (4.4) and (4.5) in (4.8), we get,

$$I_d = \frac{(A_r E_r/T_r)\cos\alpha_r - (A_i E_i/T_i)\cos\gamma_i}{R_{cr} + R_d - R_{ci}} \tag{4.9}$$

It is to be noted that the control variables are T_r, T_i and α_r, β_i. However, for maintaining safe commutation margin it is convenient to consider γ_i as control variable instead of β_i.

As the denominator in equation (4.9) is small, even small changes in the voltage magnitudes E_r or E_i can result in large changes in the DC current, if the control variables are held constant. As the voltage changes can be sudden it is obvious that manual control of converter angles is not feasible. Hence direct and fast control of current by varying α_r, and/or γ_i in response to a feedback signal is essential. The direct and fast control over the current is also desirable from the viewpoint of limiting the overcurrents in thyristor valves which have limited short term overload capability. It is to be noted that although current and power can be controlled by changing transformer taps T_r and T_i, this can be achieved only by slow control of mechanical switches.

As mentioned in the beginning, it is desirable to control the current and regulate the voltage simultaneously in the link. The question that can be posed now is, which terminal should be assigned the task of current control, while the other terminal regulates the DC voltage? From considerations given below, it is desirable to have current control at the rectifier station under normal conditions.

1. The increase of power in the link is achieved by reducing α_r which improves the power factor at the rectifier, for higher loadings and minimizes the reactive power consumption.
2. The inverter can now be operated at minimum γ thereby minimizing the reactive power consumption at the inverter also.

It is to be noted that the current control at the inverter worsens power factor at the higher loadings as γ has to be increased. Increased γ also implies higher losses in the valve snubber circuits.

3. The operation at minimum extinction angle at the inverter and current control at the rectifier results in better voltage regulation than the operation with minimum delay angle at the rectifier and current control at the inverter (This is valid for similar values of R_{cr} and R_{ci}).
4. The currents during line faults are automatically limited with rectifier station in current control.

While there is a need to maintain a minimum extinction angle of the inverter to avoid commutation failure, it is economical to operate the inverter at constant extinction angle (CEA) which is slightly above the absolute minimum required for the commutation margin. This results in reduced costs of the inverter stations, reduced converter losses and reactive power consumption. However, the main drawback of CEA control is the negative resistance characteristic of the converter which makes it difficult to operate stably when the AC system is weak (low short circuit ratios). Constant DC voltage (CDCV) control or constant AC voltage (CACV) control are the alternatives that could be used at the inverter.

Under normal conditions, the rectifier operates at constant current (CC) control and inverter at the CEA control. Under conditions of reduced AC voltage at the rectifier it is necessary to shift the current control to the inverter to avoid run down of the DC link when the rectifier control hits the minimum α limit. This implies that current controller must also be provided at the inverter in addition to the CEA controller. A smooth transition from CEA to CC takes place whenever the link current starts falling. To avoid the clash of two current controllers, the current reference at the inverter is kept below that at the rectifier by an amount called the 'current margin'. This is typically about 10% of the rated current.

The power reversal in the link can take place by the reversal of the DC voltage. This is done easily by increasing the delay angle at the station initially operating as the rectifier, while reducing the delay angle at the station initially operating as the inverter. Thus, it is necessary to provide both CEA and CC controllers at both terminals. The on-load tap changer control at the inverter is used mainly to maintain a constant DC voltage (when the inverter is in CEA control). The tap changer control at the rectifier is designed to maintain the delay angle within certain limits (say 10° to 20°) in order to maintain certain voltage margin for the purposes of current control. A voltage margin of 3% is generally considered to be adequate to meet any sudden demand for the increase in the link current.

The feedback control of power in a DC link is not desirable for the following reasons:

1. At low DC voltages, the current required is excessive to maintain the required level of power. This can be counter productive because of the excessive requirements on the reactive power, which depresses the voltage further.
2. The constant power characteristic contributes to negative damping and degrades dynamic stability.

4.3 CONVERTER CONTROL CHARACTERISTICS

4.3.1 Basic Characteristics

The basic principles of the control of DC link have been stated in the previous section. The control characteristics of both stations are illustrated in Fig. 4.2 which shows the DC voltage at the station II versus DC current. Each station characteristic has three parts as given below

Station I	Station II	Type
ab	*hg*	minimum (α)
bc	*gf*	constant current
cd	*fe*	minimum (γ)

The intersection of the two characteristics (point A) determines the mode of operation—Station I operating as rectifier with constant current control and station II operating at constant (minimum) extinction angle.

There can be three modes of operation of the link (for the same direction of power flow) depending on the ceiling voltage of the rectifier which determines the point of intersection of the two characteristics. These are defined below:

1. CC at rectifier and CEA at inverter (operating point A) which is the normal mode of operation.
2. With slight dip in the AC voltage, the point of intersection drifts to C which implies minimum α at rectifier and minimum γ at the inverter.
3. With lower AC voltage at the rectifier, the mode of operation shifts to point B which implies CC at the inverter with minimum α at the rectifier.

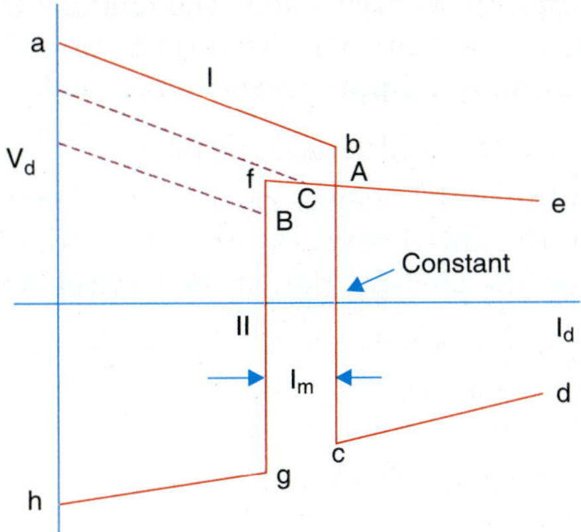

Fig. 4.2: Converter controller characteristic

The characteristic *ab* has generally more negative slope than characteristic *fe* for similar values of R_{cr} and R_{ci}. This is because of the fact that the slope of *ab* is due to the combined resistance $(R_d + R_{cr})$ while the slope of *fe* is due to R_{ci}. However, for low SCR at the inverter, the slope of *fe* could be more negative than that of *ab*.

Figure 4.3 shows the control characteristics for negative current margin I_m (or where the current reference of station II is larger than that at station I). The operating point shifts now to 'D' which implies power reversal with station I (now acting as inverter) operating with minimum CEA control while station II operating with CC control.

This shows the importance of maintaining the correct sign of the current margin to avoid inadvertent power reversal. The maintenance of proper current margin requires adequate telecommunication channel for rapid transmission of the current or power order from one station (Master) to the other (Slave) station.

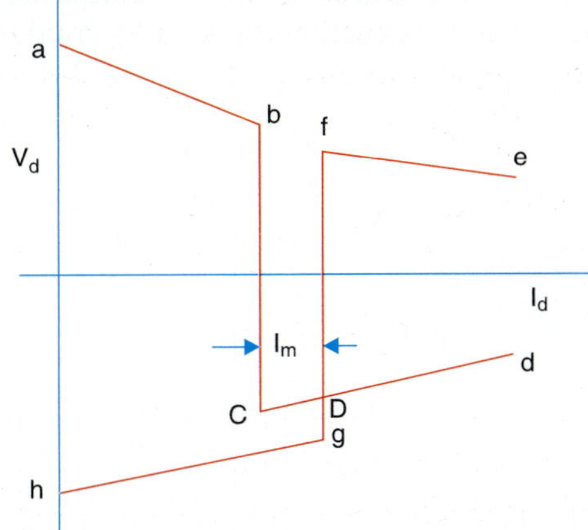

Fig. 4.3: Control characteristic for negative current margin

CHAPTER 4

In order to prevent inadvertent power reversal in the link due to failure of telecommunication channels, it is necessary to prevent the inverter from transition to the rectifier operation. This can be easily done by putting minimum limits on the delay angle of the inverter (100° to 110°).

4.3.2 Modification of the Control Characteristics

The previous discussion has outlined the need to restrict the control region to the first quadrant of the $V_d - I_d$ plane to avoid unwanted reversal of power. In addition, there are two other requirements which necessitate the modification of the control characteristics.

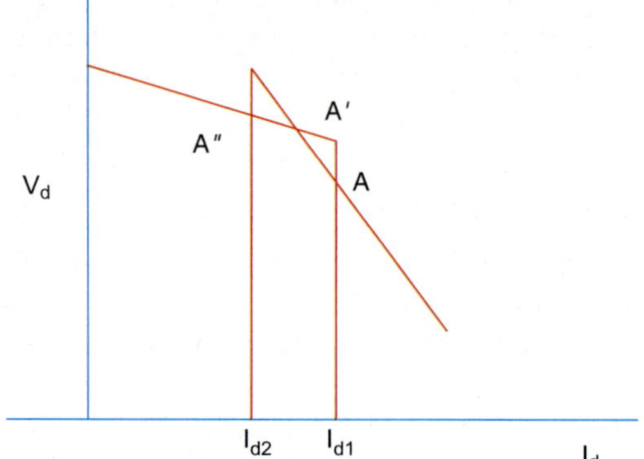

Fig. 4.4: Illustration of 3-point instability

Mode stabilization

The slope of *ab* and *fe* are nearly equal which can lead to poor definition of the intersection of point '*C*'. Further, if the slope of *fe* exceeds that of *ab* (see Fig. 4.4), there will be three possible operating points *A*, *A'* and *A"*. This implies instability of the control which will result in hunting between different modes of operation. To eliminate this problem, the inverter characteristics are modified and given a positive slope when the current is between I_{d1} and I_{d2} [see Fig. 4.5 (*a*)]. This is achieved by current error dependent γ control. Alternate solution is to modify the inverter control to maintain a constant DC voltage with back-up control of minimum CEA [see Fig. 4.5 (*b*)]. This requires the normal operating value of extinction angle to be greater than the minimum value.

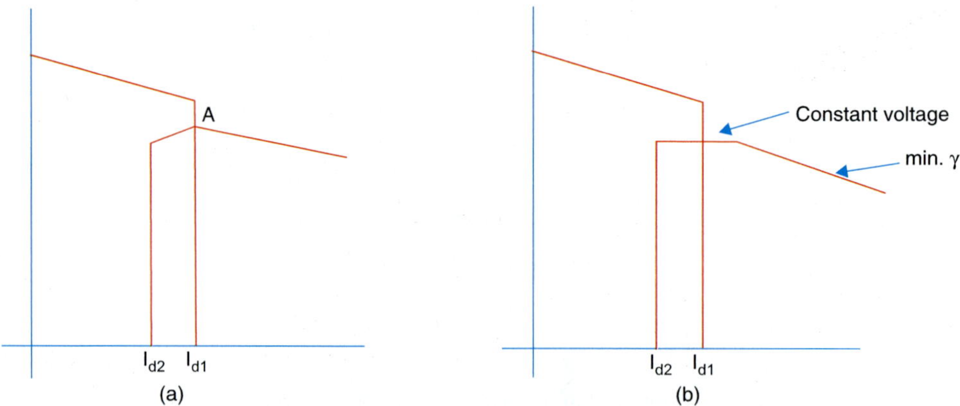

Fig. 4.5: Modification of inverter characteristics

Voltage dependent current limit

The low DC voltage in the link is mainly due to the faults in the AC system on the rectifier or inverter side. The low AC voltage due to faults on the inverter side can result in persistent commutation failure because of the increase of the overlap angle. In such cases, it is necessary to reduce the DC current in the link until the condition that led to the reduced DC voltage are relieved. Also the reduction of current relieves those valves in the inverter which are overstressed due to continuous current flow in them.

If the low voltage is due to faults on the rectifier side AC system, the inverter has to operate at very low power factor causing excessive consumption of reactive power, which is also undesirable. Thus, it becomes useful to modify the control characteristics to include voltage dependent current limits. This is shown in Fig. 4.6 which also shows current error dependent` characteristics to stabilize the mode when operating with DC current between I_{d1} and I_{d2}.

The characteristic cc' and $c'c''$ show the limitation of current due to the reduction in voltage. The DC current is reduced from I_{d1} to I'_{d1} linearly and maintained at I'_{d1} below the voltage V_{d2}. The inverter characteristic also follows the rectifier characteristic to maintain the current margin except for hd'' which is due to the lower limit imposed on the delay angle of the inverter.

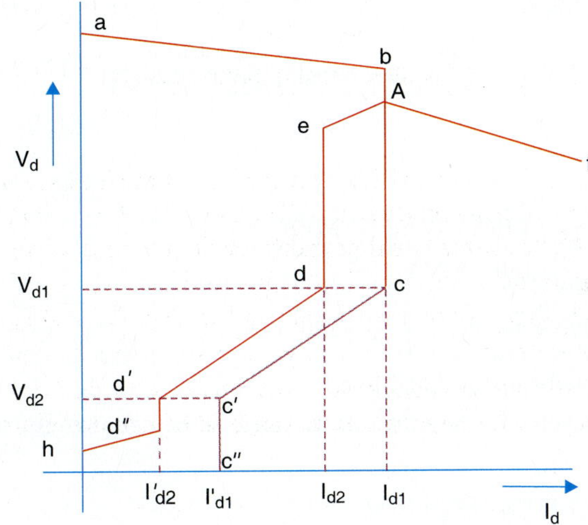

Fig. 4.6: Control characteristics including VDCOL

Example 1

A bipolar HVDC link operates with ±300 kV, 600 MW (rated voltage and power) at the rectifier end. The resistance of each line conductor is 20 ohms. The converter transformers have a leakage reactance of 0.15 p.u. on their own bases. There are two series connected converter bridges per pole. The AC bus voltage at both rectifier and inverter buses are regulated at 230 kV (rated).

(*a*) Compute the ratings and turns ratios of the converter transformers at the rectifier and inverter.

CHAPTER 4

(*b*) Find the reactive power supplied at the rectifier and the inverter stations.

Assume operating values of $\alpha_r = \gamma_i = 18°$ where α_r is the delay angle at the rectifier and γ_i is the extinction angle at the inverter.

Solution

(*a*) *At the rectifier end:*

The DC voltage at each converter bridges is $\frac{300}{2} = 150$ kV. The DC current

$$I_d = \frac{P_{dr}}{V_{dr}} = \frac{150 \times 10^3}{150} = 1000\text{A} .$$

(Note : P_{dr} is the power output of each individual bridge and V_{dr} is the voltage across the bridge). From the results given in chapters 2 and 3, we have,

$$S_t = \frac{\pi}{3} V_{dor} I_d , \quad \frac{V_d}{V_{dor}} = \cos\alpha - \frac{\bar{X}_c}{2} \frac{I_d}{I_{dn}}$$

where V_{do} is the no-load DC voltage, $\bar{X}_c$ is the pu leakage reactance of the converter transformer (on its own base). Thus, we have,

$$V_{dor} = \frac{2V_{dr}}{\left(2\cos\alpha_r - \bar{X}_c \bar{I}_d\right)} = \frac{2 \times 150}{\left(2\cos 18° - 0.15 \times 1\right)} = 171.2 \text{ kV}$$

(Note that $\bar{I}_d = \frac{I_d}{I_{dn}} = 1.0$)

$$S_{tr} = 1.047 \times 171.2 \times 1 \text{ MVA}$$
$$= 179.3 \text{ MVA}$$

The AC voltage on the valve side $E_{vr} = \frac{V_{do}}{1.35} = 126.8$ kV

The turns ratio of the transformer = 230:126.8

(Note: The turns ratio includes the effect of the on-load tap changer).

At the inverter end:

The DC voltage across each bridge (V_{di}) is given by

$$V_{di} = \frac{2V_{dr} - I_d R_d}{2} = V_{dr} - I_d \frac{R_d}{2}$$

where $R_d = 20$ ohms is the resistance of the conductor in each pole. Thus,

$$V_{di} = 150 - \frac{1 \times 20}{2} = 140 \text{ kV}$$

$$V_{doi} = \frac{2V_{di}}{\left(2\cos\gamma_i - \bar{X}_c \bar{I}_d\right)} = \frac{2 \times 140}{2\cos 18° - 0.15 \times 1} = \frac{280}{1.752} = 159.8 \text{ kV}$$

$$S_{ti} = \frac{\pi}{3} V_{doi} I_d = 167.3 \text{ MVA}$$

$$E_{vi} = \frac{V_{doi}}{1.35} = \frac{159.8}{1.35} = 118.4 \text{ kV}$$

The turns ratio = 230:118.4

A check on the results is obtained by calculating I_d from the expression,

$$I_d = \frac{E_{dr} - E_{di}}{R_{cr} + R_d - R_{ci}} = \frac{n_b\left(V_{dor}\cos\alpha_r - V_{doi}\cos\gamma_i\right)}{R_{cr} + R_d - R_{ci}}$$

where $$R_{cr} = \frac{3}{\pi}\overline{X}\left(\frac{V_b^{\,2}}{S_b}\right)\cdot n_b = \frac{3}{\pi}\times 0.15\times\frac{(126.8)^2}{179.3}\times 2 = 25.69 \text{ ohms}$$

$$R_{ci} = \frac{3}{\pi}\times 0.15\times\frac{(118.4)^2}{167.3}\times 2 = 24 \text{ ohms}$$

$$I_d = \frac{2(171.2\cos 18° - 159.8\cos 18°)}{(25.69 + 20 - 24)} = 1 \text{ kA}$$

(*b*) The power factor at the rectifier ($\cos\phi_r$)

$$= \frac{V_{dr}}{V_{dor}} = \frac{150}{171.2} = 0.876$$

The total reactive power supplied to the rectifier station = 600 × tan ϕ_r = 330 MVAR

Power factor at the inverter ($\cos\phi_i$)

$$= \frac{V_{di}}{V_{doi}} = \frac{140}{159.8} = 0.876$$

The total reactive power supplied to the inverter station = (600 – 40 × 1) tan ϕ_i = 560 × 0.55 = 308 MVAR.

Note that power factor is a function of only α (or γ) and $\overline{X}$ for a specified power.

Example 2

A monopolar HVDC link has one bridge at each terminal. The parameters of the link are: $\alpha_{min} = 5°$, $\gamma_{min} = 18°$, $R_d = 5$ ohms, $R_{cr} = 10$ ohms, $R_{ci} = 12$ ohms, $V_{dor} = 115$ kV, I_{ref} at the rectifier = 1 kA, I_{ref} at the inverter = 900 A

(*a*) If $V_{doi} = 117.5$ kV, calculate I_d, α, γ, P_i and Q_i

(*b*) Repeat (a) if $V_{doi} = 120$ kV

Solution

In order to determine I_d, we need to know the operating mode which can be any one of the three possibilities:

(*i*) Current control at the rectifier, $\gamma_i = \gamma_{min}$

(*ii*) Current control at the inverter, $\alpha_r = \alpha_{min}$

CHAPTER 4

(iii) $\alpha_r = \alpha_{\min},\ \gamma_i = \gamma_{\min}$

To identify the operating mode, we need to calculate:

$$I_M = \frac{V_{dor}\cos\alpha_{\min} - V_{doi}\cos\gamma_{\min}}{R_{cr} + R_d - R_{ci}}$$

If $I_M > I_{ref}$ at the rectifier, then the operating mode is current control at the rectifier. If $I_M < I_{ref}$ at the inverter, the operating mode is current control at the inverter. If both conditions do not apply, then the operating mode is $\alpha_r = \alpha_{\min},\ \gamma_i = \gamma_{\min}$.

(a)
$$V_{doi} = 117.5 \text{ kV}$$
$$I_M = \frac{115\times\cos 5^\circ - 117.5\times\cos 18^\circ}{5+10-12} = 937.7\text{A}$$

Since $I_{ref}(rectifier) > I_M > I_{ref}(inverter)$, the operating mode is $\alpha_r = \alpha_{\min},\ \gamma_i = \gamma_{\min}$. Hence,

$I_d = 937.7$ A, $\alpha_r = 5°$, $\gamma_i = 18°$, $P_i = V_{di}I_d$, $V_{di} = V_{doi}\cos\gamma_{\min} - R_{ci}I_d$

$$V_{di} = 117.5\cos 18^\circ - \frac{12\times 937.7}{1000} = 100.5 \text{ kV}$$
$$P_i = 100.5\times 0.9377 \text{ MW} = 94.24 \text{ MW}$$
$$Q_i = P_i\tan\phi_i,\ \cos\phi_i = \frac{V_{di}}{V_{doi}} = \frac{100.5}{117.5} = 0.855$$
$$= 57.1 \text{ MVAR}$$

(b)
$$V_{doi} = 120 \text{ kV}$$
$$I_M = \frac{115\times\cos 5^\circ - 120\cos 18^\circ}{5+10-12} = 145 \text{ A}$$

Since, $I_M < I_{ref}(inverter)$, the current controller is at the inverter. Hence, $\alpha_r = 5°$, $I_d = 0.9$ kA

$$V_{di} = V_{dr} - R_d I_d = V_{dor}\cos\alpha_{\min} - (R_{cr} + R_d)I_d$$
$$= 115\cos 5^\circ - 15\times 0.9 = 101.1 \text{ kV}$$

$$P_i = V_{di}I_d = 101.1\times 0.9 \text{ MW} = 91 \text{ MW}$$
$$\cos\phi_i = \frac{V_{di}}{V_{doi}} = \frac{101.1}{120} = 0.843$$
$$Q_i = P_i\tan\phi_i = 58.2 \text{ MVAR}$$

Since, $V_{di} = V_{doi}\cos\gamma - R_{ci}I_d$,

$$\cos\gamma = \frac{V_{di} + R_{ci}I_d}{V_{doi}} = \frac{101.1 + 12\times 0.9}{120} = 0.9325$$
$$\gamma = 21.2^\circ$$

4.4 SYSTEM CONTROL HIERARCHY

The control functions required for the HVDC link are performed using the hierarchical control structure shown in Fig. 4.7. The master controller for a bipole is located at one of the terminals and is provided with the power order (P_{ref}) from the system controller (from energy control centre). It also has other information such as AC voltage at the converter bus, DC voltage, etc. The master controller transmits the current order (I_{ref}) to the pole units which in turn provide a firing angle order to the individual valve groups (converters). The valve group or converter control also oversees valve monitoring and firing logic through the optical interface. It also includes bypass pair selection logic, commutation failure protection, tap changer control, converter start/stop sequences, margin switching and valve protection circuits.

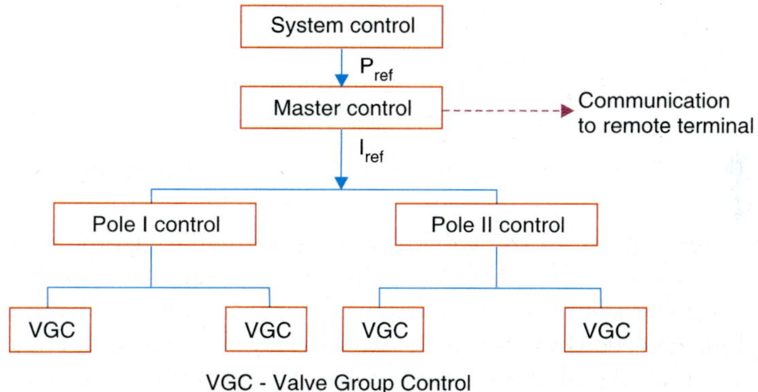

Fig. 4.7: Hierarchical control structure for a DC link

The pole control also incorporates pole protection, DC line protection and optional converter paralleling and deparalleling sequences. The master controller which oversees the complete bipole includes the functions of frequency control, power modulation, AC voltage and reactive power control and torsional frequency damping control. It also oversees pole paralleling sequences, transient pole current increases for outage of a pole, balancing of pole currents and communication of the power/current order to the remote terminal.

The block diagram of the pole and converter controls is shown in Fig. 4.8. This shows the basic control functions. The current or extinction angle controller generates a control signal V_c which is related to the firing angle required. The firing angle controller generates gate pulses in response to the control signal V_c. The selector picks the smaller of the α determined by the current and CEA controllers. In digitally based firing-pulse controller, the output of the primary regulator can be firing angle reference which is used as input to the continuously acting synchronization control with feedback from the measurement of the firing angle and the system frequency [21].

4.5 FIRING ANGLE CONTROL

The operation of CC and CEA controllers is closely linked with the method of generation of gate pulses for the valves in a converter. The following are the two basic requirements for the firing pulse generation of HVDC valves.

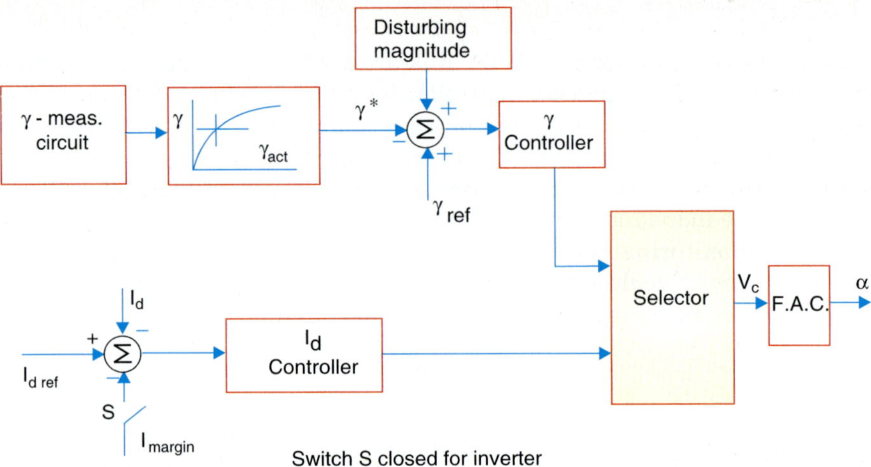

Fig. 4.8: Block diagram of pole and converter controllers

1. The firing instant for all the valves are determined at ground potential and the firing signals sent to individual thyristors by light signals through fiber-optic cables. The required gate power is made available at the potential of individual thyristor for electrically triggered thyristor (ETT) valves. However, for light triggered thyristor (LTT) valves, the light signal can be used to directly fire individual thyristors.
2. While the single pulse is adequate to turn-on a thyristor, the gate pulse generator must send a pulse whenever required, if the particular valve is to be kept in a conducting state. This is of importance when operating at low DC currents and a transient might reduce the current below the holding current.

   There are two basic firing schemes, namely:

   Individual Phase Control (IPC)

   Equidistant Pulse Control (EPC)

IPC was used in the past and has now been replaced by EPC for reasons that will be explained.

4.5.1 Individual Phase Control

This was used in early HVDC projects. The main feature of this scheme is that the firing pulse generation for each phase (or valve) is independent of each other and the firing pulses are rigidly synchronized with the commutation voltages.

There are two ways in which this can be achieved—constant α control and inverse cosine control.

Constant α control

In this scheme (see Fig. 4.9) six timing (commutation) voltages are derived from the converter AC bus via voltage transformers and the six gate pulses are generated at nominally identical delay times subsequent to the respective voltage zero crossings. The instant of zero-crossing of a particular commutation voltage corresponds to $\alpha = 0$ for that valve. The delays are produced by independent delay circuits and controlled by a common control voltage V_c derived from the current/extinction angle controllers.

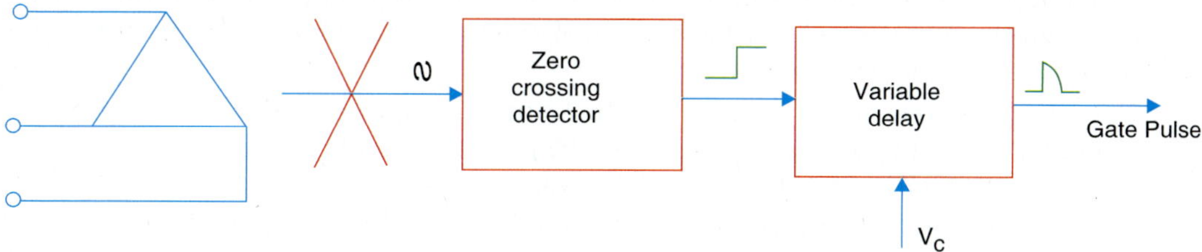

Fig. 4.9: Constant alpha control

Inverse cosine control

There are several variations of this, but in one common arrangement (see Fig. 4.10(a)), six timing voltages, (obtained as in constant α control) are each phase shifted by 90° and added separately to a common control voltage V_c. The zero crossing of the sum of the two voltages initiates the firing pulse for the particular valve considered (see Fig. 4.10(b)). The delay angle α is nominally proportional to the inverse cosine of the control voltage. It also depends on the AC system voltage amplitude and shape. The main advantage of this control scheme is that the average DC voltage across the bridge varies linearly with the control voltage V_c. It is essential in this scheme to maintain the phase shift at 90° for variations in the supply frequency.

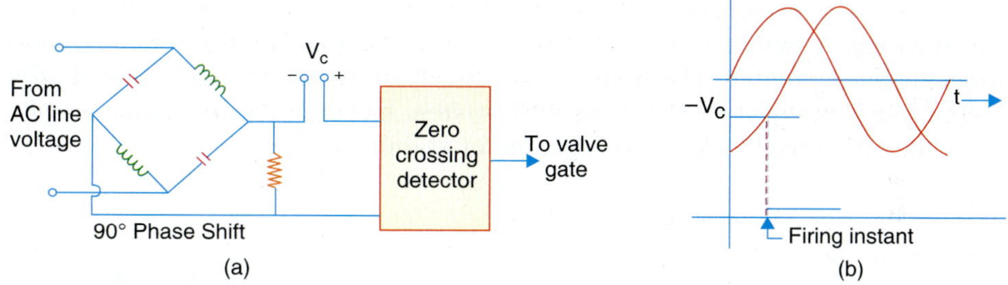

Fig. 4.10: Inverse cosine control

Drawbacks of IPC scheme

The major drawback of IPC scheme is the aggravation of the harmonic instability problem that was encountered particularly in systems with low short circuit ratios (less than 4). The harmonic instability, unlike instability in control systems, is a problem that is characterised by magnification of non-characteristic harmonics in steady-state.

This is mainly due to the fact that any distortion in the system voltage leads to perturbations in the zero crossings which affect the instants of firing pulses in IPC scheme. This implies that even when the fundamental frequency voltage components are balanced, the firing pulses are not equidistant in steady-state.

This in turn leads to the generation of non-characteristic harmonics (harmonics of order $h \neq np \pm 1$) in the AC current which can amplify the harmonic content of the AC voltage at the converter bus. The problem is aggravated at the frequencies for which the filter impedance and the system impedance are in parallel resonance.

The problem of harmonic instability can be overcome by the following measures:

1. Influencing the harmonic behaviour of AC network impedance seen by the converter (through the provision of synchronous condensers or additional filters for filtering out non-characteristic harmonics).
2. Use of filters in control circuit to filter out non-characteristic harmonics in the commutation voltages. This can, however, be problematic due to variations in the supply frequency and will add to the control delays.
3. The use of firing angle control independent of the zero crossings of the AC voltages. This is the most attractive solution and leads to the equidistant pulse firing scheme described next.

4.5.2 Equidistant Pulse Control (EPC)

In this scheme, the firing pulses are generated in steady state at equal intervals of $1/pf$, through a ring counter. This control scheme was first suggested by Ainsworth[3] using a phase locked oscillator to generate the firing pulses. There are three variations of the EPC scheme

Pulse Frequency Control (PFC)

Pulse Period Control

Pulse Phase Control (PPC)

Pulse frequency control (PFC)

In this scheme, a voltage controlled oscillator (VCO) is used, the frequency of which is determined by the control voltage V_c which is related to the error in the quantity (current, extinction angle or DC voltage) being regulated. The frequency in steady-state operation is equal of pf_0 where f_0 is the nominal frequency of the AC system. PFC system has an integral characteristic and has to be used along with a feedback control system for stabilization.

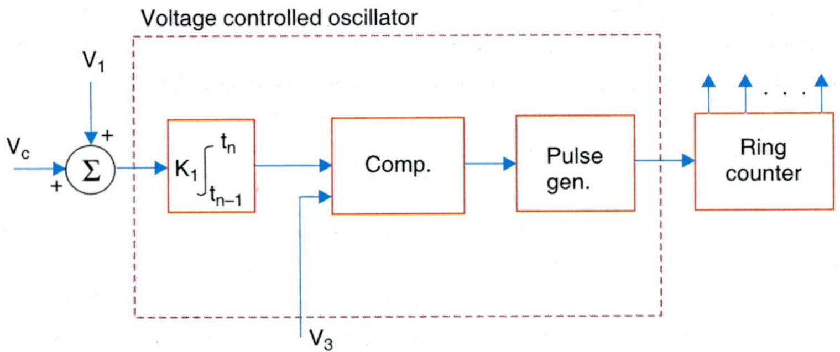

Fig. 4.11: Block diagram of PFC system

Figure 4.11 shows a simplified block diagram of the PFC system [10]. The voltage controlled oscillator (VCO) consists of an integrator, comparator and a pulse generator. The output pulses of the generator drive the ring counter and also reset the integrator. The instant (t_n) of the firing pulse is determined from the following equations

$$\int_{t_{n-1}}^{t_n} K_1(V_c + V_1)dt = V_3 \quad (4.10)$$

where V_1 is a bias (constant) voltage and V_3 is proportional to the system period. In steady state, $V_c = 0$, and from Eq. (4.10) we get

$$K_1V_1(t_n - t_{n-1}) = V_3 \quad (4.11)$$

Since

$$t_n - t_{n-1} = 1/pf_0 \quad (4.12)$$

in steady-state, the gain K_1 of the integrator is chosen as

$$K_1 = pf_0V_3/V_1 \quad (4.13)$$

The circuit shown in Fig. 4.11 does not incorporate frequency correction (when the system frequency deviates from f_o). The frequency correction according to Ainsworth is obtained by deriving V_3 as shown in Fig. 4.12. From Fig. 4.12, we have

$$V_3 = V_2/(1 + sT_1), V_2 = K_1V_1(t_{n-1} - t_{n-2})$$

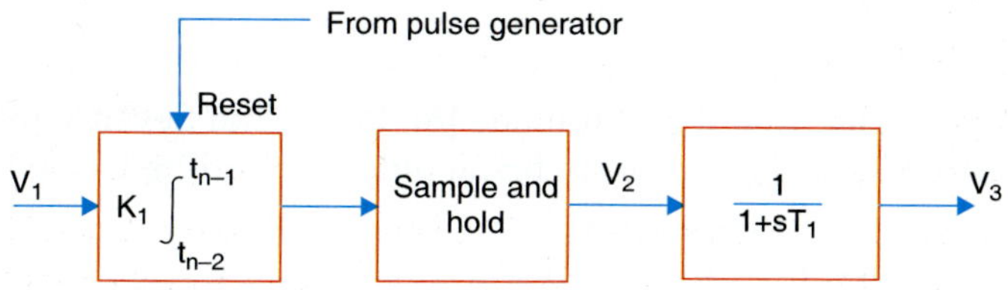

Fig. 4.12: Frequency correction for PFC

Pulse period control [4]

This is similar of PFC except for the way in which the control voltage V_c is handled. The structure of the controller is the same as that in Fig. 4.11, however, V_c is now summed with V_3 instead of V_1. Thus, the instant t_n of the pulse generation is given by

$$\int_{t_{n-1}}^{t_n} K_1V_1dt = V_3 + V_c \quad (4.14)$$

From Eq. (4.14), we get

$$K_1V_1(t_n - t_{n-1}) = V_3 + V_c \quad (4.15)$$

With $V_c = 0$, the interval between consecutive pulses, in steady state, is exactly equal to $1/pf_o$. Figure 4.13 shows how an exponentially decaying V_c which appears at $t = t_1$ shifts the delay angle by decreasing the pulse frequency between t_1 and t_2.

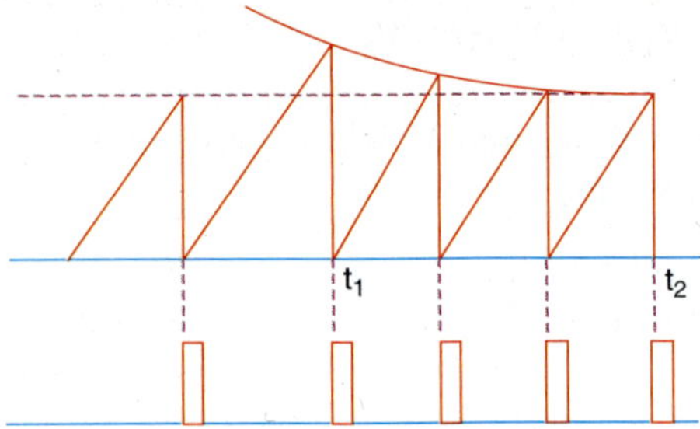

Fig. 4.13: Firing pulse train

The frequency correction in this scheme is obtained by either updating V_1 in response to the system frequency variation or including another integrator in the CC or CEA controller.

It can be seen that conceptually, the pulse period control is similar to PFC and both have integral characteristics. Ainsworth [10] claims that PFC gives better stability because the control voltage V_c is averaged in this scheme, unlike in pulse period control. However, if the ripple in control voltage V_c has harmonic components of frequency less than pf_o they are bound to affect the operation of PFC also.

Pulse phase control (PPC)

This has been suggested by Rumpf and Ranade [5], Reeve and Sevenco [6]. As suggested in [5] an analog circuit is configured to generate firing pulses according to the following equation

$$\int_{t_{n-1}}^{t_n} K_1 V_1 dt = V_{cn} - V_{c(n-1)} + V_3 \tag{4.16}$$

where V_{cn} and $V_{c(n-1)}$ are the control voltages at the instants t_n and t_{n-1} respectively. For proportional current control, the steady state can be reached when the error or V_c is constant. To reduce the error to zero, the authors of reference [5] recommend a slow α control with feedback signal taken from the measured delay angle. This signal in addition to a signal derived from frequency error (to compensate for variation in frequency) is used to control the voltage V_1.

The major advantages claimed for PPC over PFC are (*i*) easy inclusion of α limits by limiting V_c as in IPC and (ii) linearization of control characteristic by including an inverse cosine function block after the current controller. Actually, limits can also be incorporated into PFC or pulse period control system. In any case, the constraints of limits give rise to generation of nonequidistant pulses, unless special symmetrizing controls are included. This basically involves firing the subsequent $(p-1)$ valves at equal intervals once a valve is fired at, say, minimum limit, unless an increase in the firing angle is called for.

Drawbacks of EPC scheme

Although EPC scheme has replaced IPC scheme in modern HVDC projects, it has certain limitations. The first drawback is that under unbalanced voltage conditions, EPC results in less DC voltage compared to IPC. Unbalance in the voltage results from single phase to ground fault in the AC system which may persist for over 10 cycles due to stuck breakers. Under such conditions, it is desirable to maximise DC power transfer in the link which calls for IPC. However, transition from EPC to IPC under the fault conditions has to be evaluated carefully.

EPC scheme also results in higher negative damping contribution to torsional oscillations when HVDC is the major transmission link from a thermal station. However, this problem is not so serious as the problem of non-characteristic harmonics associated with IPC.

4.5.3 Control Hardware

The implementation of the controls can be achieved using analog or digital circuits. In the latter case, micro-processor based controls can be used. μP based controls were initially used where speed of control was not critical, but flexibility was an advantage. However, with the availability of high speed programmable controllers using bit-slice architecture, converter control using these devices is now feasible. μP based controllers, in general, are more reliable than analog controllers and permit redundancy. The control channel not in use operates in a hot standby mode.

4.6 CURRENT AND EXTINCTION ANGLE CONTROL

The current controller (see Fig. 4.8) is invariably of feedback type. A typical block diagram of the controller which is of PI type is shown in Fig. 4.14.

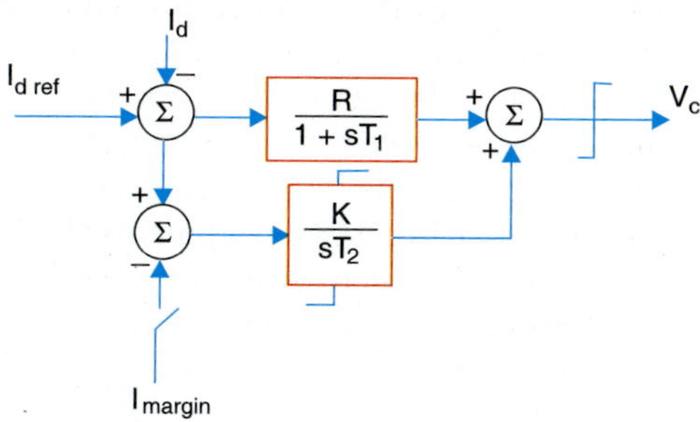

Fig. 4.14: Block diagram of current controller

The extinction angle controller can be of predictive type or feedback type with EPC control. The predictive controller is considered to be less prone to commutation failure and was invariably used in the early schemes. The feedback control with PFC type of equidistant pulse control overcomes the problems associated with IPC.

The extinction angle, as opposed to current, is a discrete variable and it was felt that feedback control of gamma is slower than the predictive type. In one of the predictive schemes proposed by Hingorani [17], the firing pulse generation is based on the following equation

$$0 = \int_{-\pi+\delta_{n-1}}^{\omega t_n} e_{cj} d(\omega t) + 2X_c I_d \tag{4.17}$$

where e_{cj} is the commutation voltage across valve j and t_n is the instant of its firing.

In general, the prediction of firing angle is based on the equation

$$\beta_j = \gamma_{ref} + u_j \tag{4.18}$$

where u_j is the overlap angle of valve j, which is to be predicted based on the current knowledge of the commutation voltage and DC current.

CHAPTER 4

In feedback control of extinction angle shown in Fig. 4.8 the measured value of γ passes through a non-linear block. The control is made faster when $\gamma < \gamma_1$ and slower when $\gamma > \gamma_2$. In the former case, the fast control reduces the incidence of commutation failure while in the latter case, the instability due to the negative resistance characteristic of fast γ control is avoided. Under large disturbances such as sudden dip in the AC voltage, signals derived from the derivative of voltage or DC current aid the advancing of delay angle for fast recovery from commutation failures.

Ekstrom and Liss [4] describe a predictive control scheme in conjunction with EPC.

4.7 STARTING AND STOPPING OF DC LINK

4.7.1 Energization and Deenergization of a Bridge

Consider N series connected bridges at a converter station. If one of the bridges is to be taken out of service, there is need to not only block, but bypass the bridge. This is because of the fact that just blocking the pulses does not extinguish the current in the pair of valves that are left conducting at the time of blocking. The continued conduction of this pair injects AC voltage into the link which can give rise to current and voltage oscillations due to lightly damped oscillatory circuit in the link formed by smoothing reactor and the line capacitance. The transformer feeding the bridge is also subjected to DC magnetization when DC current continues to flow through the secondary windings. The bypassing of the bridge can be done with the help of a separate bypass valve or by activating a bypass pair in the bridge (two valves in the same arm of the bridge). The bypass valve was used with mercury arc valves where the possibility of arc backs makes it impractical to use bypass pairs. With thyristor valves, the use of bypass pair is the practice as it saves the cost of an extra valve.

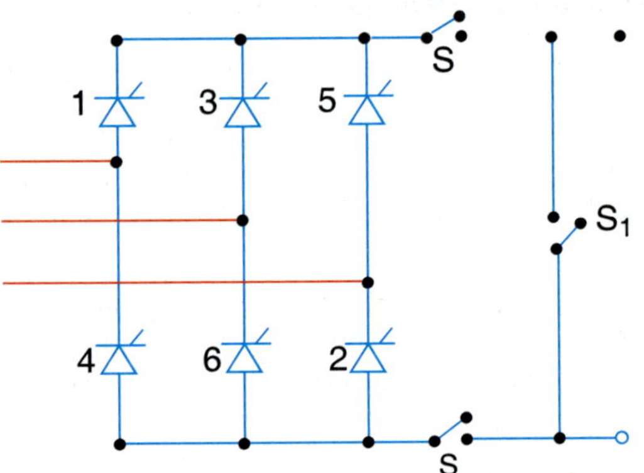

Fig. 4.15: A converter bridge with isolators

The process of deenergization of a bridge is explained with reference to Fig. 4.15. The valves 2 and 3 are assumed to be conducting initially when the blocking command is given. With the selection of bypass pair 1 and 4, the commutation from valve 2 to 4 is in the usual manner, but the commutation from valve 3 to valve 5 is prevented. In the case of a predetermined choice of the bypass path, the time lapse between the blocking command and the current transfer to bypass path can vary from 60° to 180° for a rectifier bridge. This time can be reduced (from 60°

to 120°) if the bypass pair is chosen such that the valve with the lower valve number carrying current at the instant of blocking is included in the bypass pair. In the inverter, there is no time lag involved in the activation of the bypass pair. The voltage waveforms for the rectifier and the inverter are shown in Fig. 4.16 (*a*) and (*b*) respectively. The overlap is neglected here.

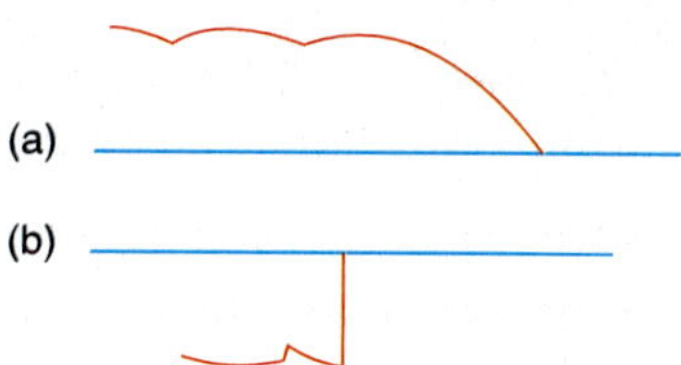

Fig. 4.16: Voltage waveforms during de-energisation for rectifier and inverter

The current from the bypass pair is shunted to a mechanical switch S_1. With the aid of the isolators S the bridge can now be isolated. The isolator pair S and switch S_1 are interlocked such that one or both are always closed.

The energization of a blocked bridge is done in two stages. The current is first diverted from S_1 to the bypass pair. For this to happen, S_1 must generate the required arc voltage and to minimize this voltage, the circuit inductance must be small. In case the bypass pair fails to take over the current, S_1 must close automatically if the current in that does not become zero after a predetermined time interval. AC breakers with sufficient arc voltage, but with reduced breaking capacity are used as switch S_1.

In the second state of energization, the current is diverted from the bypass pair. For the rectifier, this can take place instantaneously neglecting overlap. However, for the inverter, the transition requires some time lag. The voltage waveforms for this case are shown in Fig. 4.17(*a*) and (*b*).

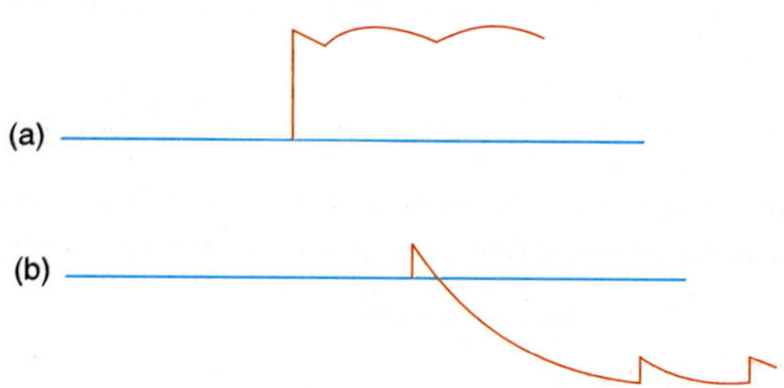

Fig. 4.17: Voltage waveforms during energisation for rectifier and inverter

It is to be noted that to avoid operation at high delay or extinction angles, the deenergization of a bridge at the rectifier (or inverter) station is accompanied by the deenergization of a bridge at the inverter (or rectifier) station.

4.7.2 Start-up of DC Link

There are two different start-up procedures depending upon whether the converter firing controller provides a short gate pulse or a long gate pulse. The long gate pulse lasts nearly 120°, the average conduction period of a valve.

Start-up with long-pulse firing

In this case, the current extinction during the start-up is not a problem. The starting sequence in this case is as follows:

1. Deblock inverter at about $\gamma = 90°$
2. Deblock rectifier at $\alpha = 85°$ to establish low direct current
3. Ramp up voltage by inverter control and the current by rectifier control

Start-up with short pulse firing

In this case, the problem of current extinction during start-up is present as the valve with forward bias is not put into conduction when the current in that falls transiently below holding current.

The starting sequence for this case is as follows:

1. Open bypass switch at one terminal
2. Deblock that terminal and load to minimum current in the rectifier mode
3. Open bypass switch at the second terminal and commutate current to the bypass pair
4. Start the second terminal also in the rectifier mode
5. The inverter terminal is put into the inversion mode
6. Ramp up voltage and current

The voltage is normally raised before raising the current. This permits the insulation of the line to be checked before raising the power. The ramping of power avoids stresses on the generator shaft. The switching surges in the line are also reduced.

The required power ramping rate depends on the strength of the AC system. Weaker systems require fast restoration of DC power for maintaining transient stability. Unfortunately, high ramping rates in such cases can give rise to large voltage drops due to the requirement of reactive power at the converter. Such voltage drops can lead to commutation failure at the inverter. The permissible ramping rates may vary from 2 to 10 p.u. per second, implying a start-up time of more than 100 to 500 msec.

4.8 POWER CONTROL

Figure 4.18 shows the block diagram of basic power and auxiliary controller used. The current order is obtained as the quantity derived from the power order by dividing it by the direct voltage.

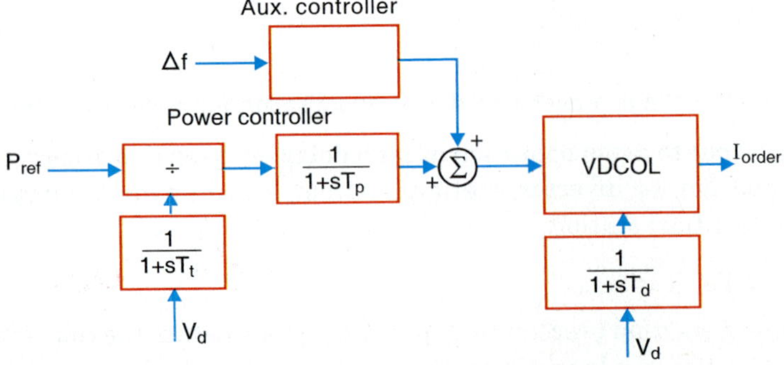

Fig. 4.18: Power and auxiliary controller block diagram

The limits on the current order are modified by the voltage dependent current order limiter (VDCOL). The objective of VDCOL is to prevent individual thyristors from carrying full current for long periods during commutation failures. However, sufficient time delay is given to prevent the action of VDCOL during normal AC system faults, which would otherwise drastically reduce DC power. The generic model of a VDCOL is shown in Fig. 4.19. G_1 (s) and G_2 (s) are simple transfer functions given by

$$G_1(s) = \frac{1}{1 + sT_{\text{down}}}$$

$$G_2(s) = \frac{1}{1 + sT_{\text{up}}}$$

By providing both converter stations with dividing circuits and transmitting the power order from the leading station in which the power order is set to the trailing station, the fastest response to DC line voltage changes is obtained without undue communication requirements. To get equal calculated current orders in the two stations, the measured DC line voltage must be referred to the same point on the DC line by compensation for DC line voltage drop. This is done by adding the term $\pm RI_d$ to the measured voltage, thus referring the voltage to a common reference point.

When the DC line resistance is large and varies considerably, e.g., when the overhead line is very long and exposed to large temperature variations, the DC line voltage drop cannot be compensated individually in the two stations. This problem can be solved by using a current order calculated in one substation only and transmitting it to the other substation.

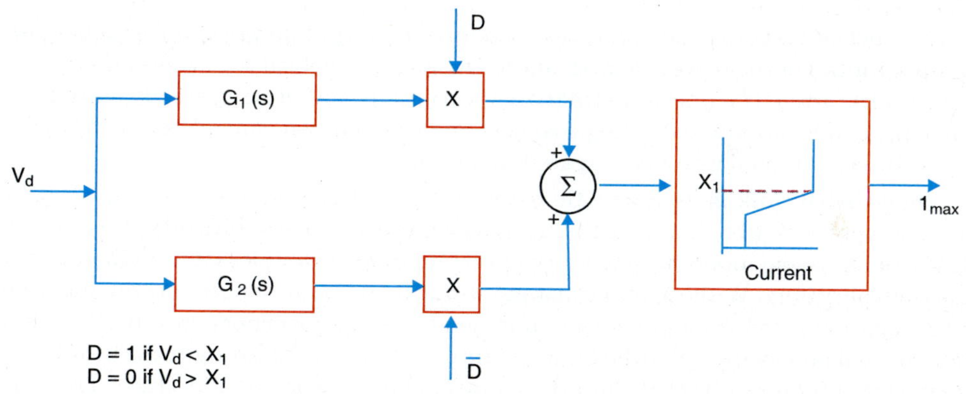

Fig. 4.19: Generic model of VDCOL (Source: Reference 18)

The power order and the corresponding rate of change or order are set with thumb wheels by the operator in one of the stations, the lead station, which may be the inverter or rectifier station. The order setting may also be transmitted there by a remote control link from the energy control centre. The order setting unit which is a digital unit based on integrated logic circuitry is intended mainly for performing power stepping and synchronization of order stepping in the two stations with the help of the communication link. The order setting unit in the trail station accepts a power order from the telecommunication system and has a supervisory function for the power order transmission. The received power is both parity checked and checked with regard to allowed change from one order message to the next. If the received order message is accepted, a receipt signal is sent back to the lead station to allow transmission of new order messages. If the receipt signal is not generated and accordingly not received in the lead station, order stepping is interrupted and the latest power order is repeatedly transmitted until accepted in the trail station at which, order stepping can start again. Thus, power transmission interruptions as a consequence of telecommunication failures are avoided.

The use of programmable logic for realization of power flow control function seems to be most appropriate solution for future applications. Standardized computer hardware equipment can be used and the functions may be adapted to individual requirement.

The auxiliary controller is designed to provide power modulation for frequency control or stabilization of AC systems. The input signal is derived from frequency or other signals.

4.9 HIGHER LEVEL CONTROLLERS

The power in a DC link can also be controlled in response to the quantities derived from the AC system (such as frequency) in order to improve the security of the overall system. This is achieved by an additional power (or current) order derived from an emergency power controller or auxiliary controller. The functions of such higher level controllers are discussed below.

4.9.1 Frequency and Power/Frequency Control

The frequency control can be used in the case of (*i*) isolated load and (*ii*) isolated generation. In the latter case, the objective is to improve the damping and reduce the wear of the generator and governor system.

The nuclear power stations are very sensitive to output power and frequency fluctuations. By proper control of the power carried on the DC link in a hybrid transmission system consisting of both HVDC and UHV lines, the frequency variations both during steady state and transient conditions, can be minimized. The prospects of a coordinated control between nuclear station and HVDC system has been investigated in Japan.

When the DC link is used as a tie between two power systems, the frequency bias can be used to adjust the power flow over the tie to assist the system in difficulty. It is to be noted that an HVDC link has no inherent sensitivity to system frequency unless it is deliberately introduced in the control system. Without it, a constant power flow can overspeed a receiving system which has become separated from part or all of its load, even if governors cut-off all primary energy inflow to local generation. Similarly, a sending system can be brought eventually to a halt, if more power is required from it than the connected generation can produce. Thus, although it would first appear that there is little benefit in the frequency control for a DC link between two systems so large that the link cannot be expected to influence the frequency of either, it is prudent to incorporate an element of frequency control which in the event of a partial break-up of a system, will prevent over speed or under speed.

4.9.2 Stabilization of AC Ties

When a DC tie is connected to a system with weak AC ties to neighbouring systems, DC link power can be varied quickly and automatically to balance the load flows and maintain stability if one the AC tie trips.

A DC tie used in parallel with an AC tie can be employed to damp the low frequency interarea oscillations in the AC tie. The control signal used can be the rate of change of AC line power (or current) or the phase angle difference across the AC tie. The DC tie can also provide frequency control for one end if the AC tie becomes disconnected and can thus permit resynchronization of the AC tie.

4.9.3 Emergency Control

In an AC tie line, the power flow is determined by conditions in the systems which it interconnects. If one of the systems suffers a disturbance there are only two options; to allow the disturbance to affect the other system also, possibly resulting in a catastrophe for both or to trip the tie in which case, the system in difficulty may lose one of the main infeeds on which it is dependent even before the disturbance.

A HVDC link on the other hand, even with simplest controls, buffers one system from disturbances on the other. Power flow can continue at worst unchanged but the option is available to vary power flow to assist the system in trouble to the extent to which the healthy system can allow without putting itself in difficulty, subject only to the rating of the link. In general, with suitable control, a disturbance originating in either system can be shared in a predetermined manner and the oscillations occurring in the two systems can be damped simultaneously. Substantial damping can be achieved with a very small amount of DC power modulation

However, if a large degree of modulation is required with the DC line already operating close to its full capacity, it is found that significant results can be achieved simply by reduction of DC power at the appropriate instants.

4.9.4 Reactive Power Control

The reactive power control is important, particularly in weak AC systems, in reducing the dynamic over voltages. Also, in inverters, the fast reactive power control can help in allowing the injection of increased power at times of need to improve the stability of the receiving end AC system. Although the reactive power requirements for a LCC station are primarily met by shunt capacitors and filters, the converter controls can be coordinated with the discrete control of the reactive power sources to provide voltage stability and eliminate voltage flicker. Modulation of the inverter extinction angle can help in HVDC power modulation by regulating the AC bus voltage at the inverter.

4.9.5 Subsynchronous Damping Control

A radial HVDC link connected to a thermal generating station can contribute to the negative damping of the torsional oscillations at subsynchronous frequency due to interactions with the current controller. This problem usually surfaces when there are no parallel AC links. The power modulation controller mentioned earlier which is designed to damp low frequency rotor oscillations, can aggravate the problem. However, a suitably designed subsynchronous damping controller (SSDC) with control signal derived from the rotor velocity can help to damp torsional frequency oscillations.

4.10 TELECOMMUNICATION REQUIREMENTS

Although manual power order setting with voice communication is possible, the utilization of inherent speed of response of the DC link requires communication channel of appropriate capacity.

For the use of stabilization control, a 1200 baud channel is required for power flow control. This gives a sampling interval of 20 ms with a resolution of 0.05 per cent of the maximum current order, which should be adequate for most cases.

The maintenance of proper current margin is essential and is the primary objective of the telecommunication channel provided. Certain precautions can help in overcoming the problems arising from the failure of communication. In raising the current order, it is preferable to carry it out first at the rectifier and then at the inverter. In reducing the current order, the inverter current is reduced before that at the rectifier.

4.11 CONTROL OF VOLTAGE SOURCE CONVERTER

A VSC used in a HVDC link injects a controllable voltage at the AC terminals given by

$$\hat{E} = mE_{ao}\angle\theta + \alpha,\ E_{ao} = kV_{dc} \tag{4.19}$$

where k is a constant depending on the converter configuration. For a two level converter, $k = \frac{\sqrt{6}}{\pi}$, m (normalized modulation index) and α (phase angle) are control variables used to control power and reactive power. θ is the angle of the AC bus of the converter. Note that $m < 1$ and α can be positive or negative depending on whether the converter is operating as an inverter or rectifier.

The reactive power injected by the converter is a function of V_s, the bus voltage magnitude. In contrast, it is to be noted that line commutated, thyristor based converter can only draw reactive power depending on the power flow through the converter.

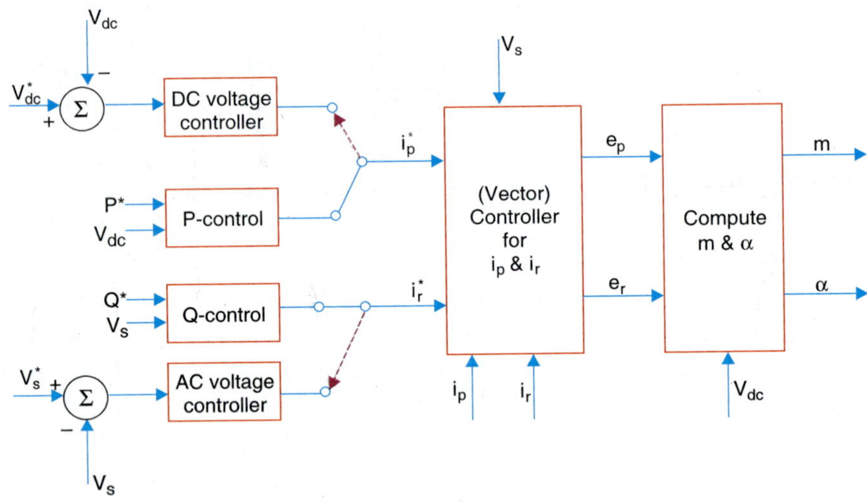

Fig. 4.20: Block diagram of VSC controller

The block diagram of the controllers for VSC is shown in Fig. 4.20. Here, i_p and i_r, the active and reactive components of the AC current drawn by the converter. That is

$i_p = \frac{P}{V_s}, i_r = \frac{Q}{V_s}$. The reference values for the active current (i_p^*) is obtained either from the power controller or DC voltage controller. The reference value for the reactive current (i_r^*) is obtained from the specified reactive power or AC voltage controller. The outputs of the current controllers (active and reactive current) are the desired converter voltage components e_p and e_r which are defined as

$$e_p = mkV_{dc} \cos\alpha, \; e_r = -mkV_{dc} \sin\alpha \tag{4.20}$$

From the above equation m and α can be calculated as

$$m = \frac{\sqrt{e_p^2 + e_r^2}}{kV_{dc}}, \; \alpha = -\tan^{-1}\frac{e_r}{e_p} \tag{4.21}$$

Design of current controllers

The variables i_p and i_r are related to the D-Q variables (i_D and i_Q) by

$$i_Q + ji_D = (i_p - ji_r)e^{j\theta} \tag{4.22}$$

Substituting this in Eq. (3.90) which is reproduced below

$$L\frac{d\hat{I}}{dt} + (R + j\omega_o L)\hat{I} = \hat{V} - \hat{E} \tag{3.90}$$

where, $\hat{I} = i_Q + ji_D, \; \hat{V} = V_Q + jV_D, \; \hat{E} = E_Q + jE_D$

we get

$$L\frac{d\hat{I}'}{dt} + j\dot{\theta}L\hat{I}' + (R + j\omega_o L)\hat{I}' = V_s\angle 0 - mkV_{dc}e^{j\alpha} \tag{4.23}$$

where $\hat{I}' = i_p - ji_r$

Note that active current i_p is in phase with the source voltage (with magnitude V_s) and the positive reactive current drawn by the converter lags the source voltage by 90°.

We can express Eq.4.23 as follows

$$\frac{d}{dt}\begin{pmatrix} i_p \\ i_r \end{pmatrix} = \begin{pmatrix} \frac{-R\omega_B}{X} & 0 \\ 0 & \frac{-R\omega_B}{X} \end{pmatrix}\begin{pmatrix} i_p \\ i_r \end{pmatrix} + \frac{\omega_B}{X}\begin{pmatrix} V_s - e_p - X\frac{\omega}{\omega_B}i_r \\ -e_r + \frac{X\omega}{\omega_B}i_p \end{pmatrix} \tag{4.24}$$

where $X = \omega_B L$, $\omega = \omega_o + \frac{d\theta}{dt}$. ω_B is the rated (nominal) frequency. (Note that we normally assume $\omega \simeq \omega_B$)

If we arrange

$$u_p = V_s - e_p - \frac{X\omega}{\omega_B}i_r \tag{4.25}$$

CHAPTER 4

$$u_r = -e_r + \frac{X\omega}{\omega_B} i_p \tag{4.26}$$

where u_p and u_r are obtained as the outputs of the active and reactive current controllers, then we get decoupled (also called as vector) control of the two current components shown in Fig. 4.21.

u_p and u_r are obtained as the outputs of controllers and defined by

$$u_p = G_c(s)(i_p^* - i_p),\ u_r = G_c(s)(i_r^* - i_r) \tag{4.27}$$

where $G_c(s) = \dfrac{K_c(1+sT_c)}{sT_c}$ for a P-I controller.

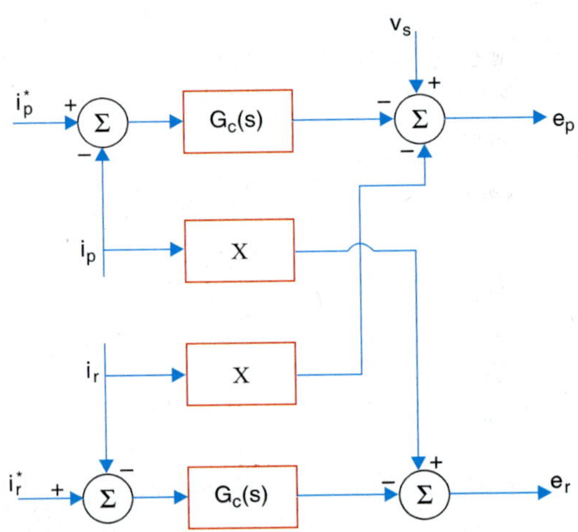

Fig. 4.21: Decoupled current contollers

It is to be noted that the decoupled structure of the current controllers is obtained by the feed forward terms involving i_p, i_r and V_s (see Eqs.4.25 and 4.26). The P-I controller time constant T_c can be chosen to cancel the poles introduced by the current flow in the inductive circuit. If $T_c = \dfrac{L}{R}$ and K_c is chosen as $K_c = \dfrac{L}{T_i}$, it can be shown that

$$i_p = \frac{1}{1+sT_i} i_p^*,\ i_r = \frac{1}{1+sT_i} i_r^* \tag{4.28}$$

The value of T_i can be chosen to provide a specified response time for the current controllers. In general, the response time is a fraction of a cycle.

The implementation of the vector control scheme involves computation of i_p and i_r which requires a phase locked loop (PLL) synchronized to the (AC) bus voltage of the converter. A typical PLL scheme is shown in Fig. 4.22.

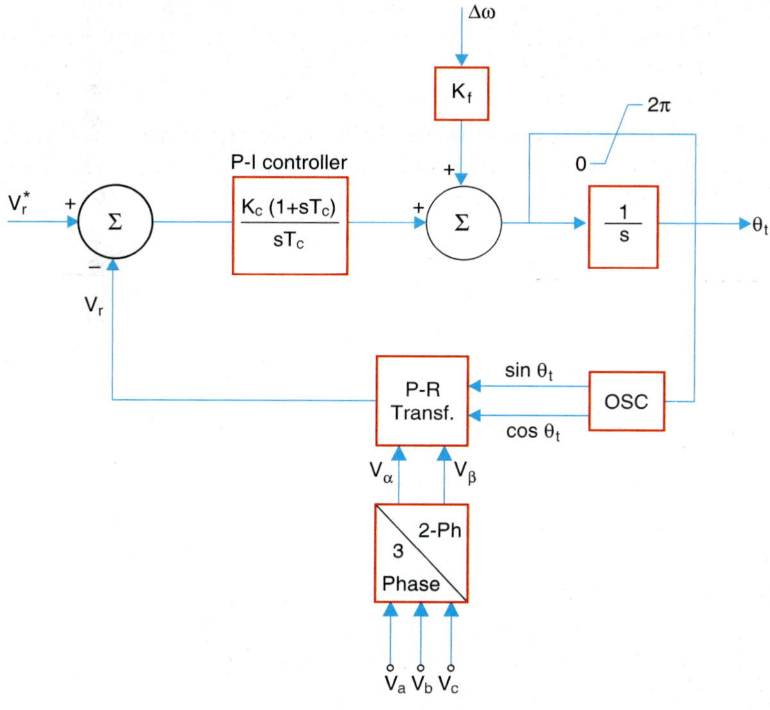

Fig. 4.22: Block diagram of PLL

Control hierarchy and system control

The control hierarchy in a VSC based HVDC link is similar to that in a LCC based HVDC link except for some major differences. For example, a VSC uses self commutated devices such as IGBTs and hence there is no problem of commutation failures caused by dips in the AC voltage. As a matter of fact, the DC voltage in a VSC based HVDC link is independent of variations in the system voltage (except under fault conditions). These characteristics have two major advantages—(1) An inverter can feed passive loads in which case the inverter controls the magnitude and the frequency of the AC voltage across the passive load. (2) The power control in a nominal HVDC link (with voltage sources at both ends) does not require fast telecommunication. In a two terminal HVDC link, one terminal, say the inverter, regulates the DC bus voltage while the other terminal regulates the power flow. Unless the limit on the DC voltage (based on the ratings) is reached, the rectifier station can supply the required power automatically. Interestingly, reduction in the AC voltage due to faults at the inverter can actually increase the DC voltage and current in the link.

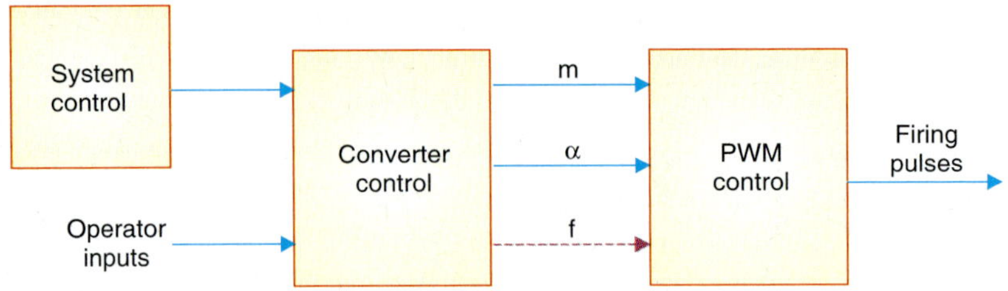

Fig. 4.23: Three levels of control in VSC based HVDC

The three levels of control in a VSC based HVDC links are shown in Fig.4.23. The valve firing sequence is regulated by the type of PWM applied. Selective Harmonic Elimination (SHE) method of PWM results in the least number of valve switchings per cycle to cancel a selected number of harmonics. The firing controller normally uses a PLL which is synchronized to an active AC system or independently locked when feeding passive loads.

The converter control sets the magnitude and phase angle of the injected voltage of the converter. Also, by adjusting the frequency of the oscillator in the firing controls, it can provide frequency control desired while supplying passive loads. It is to be noted that in this case, the power is determined by the magnitude and the frequency of the injected voltage at the inverter, while the rectifier controls the DC bus voltage and the reactive power (or AC voltage magnitude).

The converter control may also perform the following functions:

1. Balancing currents among parallel connected converters.
2. Protection of series connected converters.
3. Provide negative sequence control to minimize effects of unbalance. The introduction of negative sequence currents caused by unsymmetrical faults can inject second harmonic ripple in the DC voltage of the converter.
4. Control and limit the DC capacitor voltage.
5. Control DC voltage unbalance for multi-level converters

The system control performs the following functions:

1. Control of active and reactive power
2. Control of AC voltage magnitude subject to the limitations of the ratings
3. Enhancement of the system transient stability
4. Damping of oscillations
5. Frequency control.

Last three functions require modulation of the active (and also reactive power when useful) power based on appropriate control signals and control laws. In performing these functions, there is no significant difference between the LCC and VSC based HVDC links (except for the differences due to their individual characteristics).

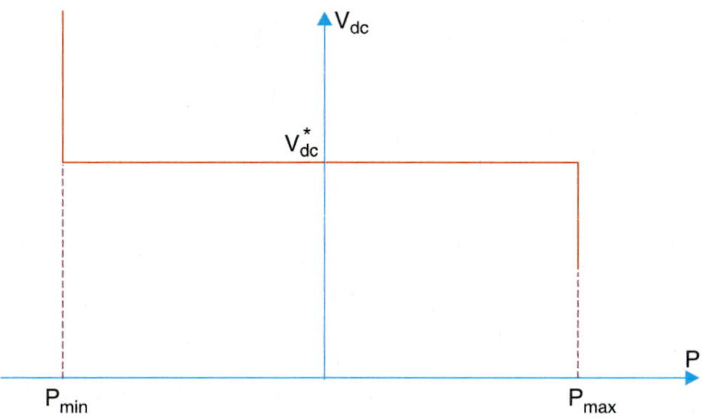

Fig. 4.24: Control characteristics for a terminal

Mode switching from DC voltage control to power control or vice versa may be required if there are limits imposed on the power flow. To prevent hunting between the DC voltage controllers at the two terminals, a voltage margin is provided. The P-V_{dc} control characteristics for a terminal are shown in Fig. 4.24. Here, positive P indicates DC power flowing away from the terminal (when the terminal is operating as a rectifier station), negative P indicates that the terminal is operating as an inverter. The rectifier voltage controller is given a reference (V^*_{dc}) which is equal to the V^*_{dc} at the inverter plus the voltage margin. The operating point is determined by the intersection of the two characteristics shown in Fig. 4.25(*a*). The inverter determines the power flow as long as it does not exceed the upper limit of power flow at the rectifier (neglecting losses). If the limit on the power flow at the rectifier is exceeded, the voltage control is shifted to the inverter automatically [see Fig. 4.25 (*b*)].

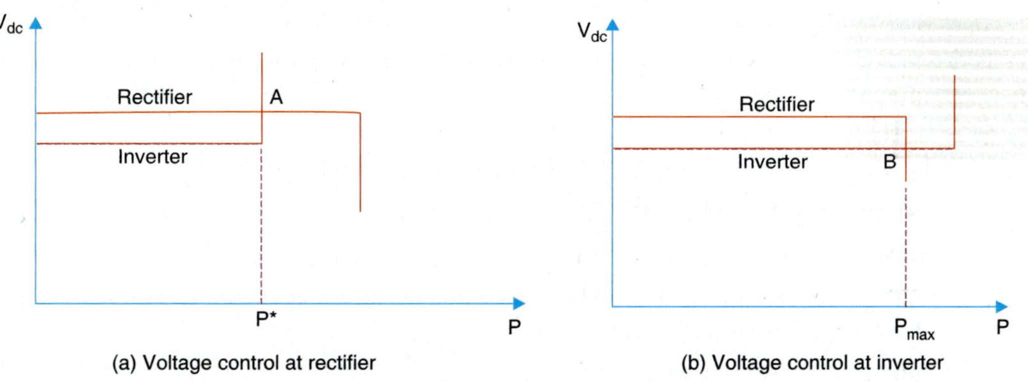

Fig. 4.25: Determination of the operating point

As mentioned earlier, VSC based HVDC link does not require fast telecommunication except under following conditions:

1. When damping of low frequency electromechanical oscillations is required at the converter not controlling the power.
2. When fast power control is required between converters in a multiterminal configuration.

REFERENCES AND BIBLIOGRAPHY

1. Ishikawa, M. et al., '*Development of a Coordinated Control System for BWR Nuclear Power Plant and HVDC Transmission System*', IEEE Trans. on Power Delivery, Vol. PWRD-1, No.3, July 1986, pp. 319–325.
2. Ainsworth, J.D. '*Harmonic Instability between Controlled Static Converters and AC Networks*', Proc. IEE, Vol. 114, pp.949–957, 1967.
3. Ainsworth, J.D. '*The Phase Locked Oscillator–A New Control System for Controlled Static Converters*', IEEE Trans., Vol.PAS.87, No.3, pp. 859–865, 1968.
4. Ekstrom, A. and Liss G., '*Theoretical Basis of a Digital Method of Grid Control for HVDC Converters*', IEEE Trans., Vol. PAS.89, No.8, pp. 2049–2055, Nov/Dec. 1970.
5. Rumpf, E. and Ranade S., '*Comparison of Suitable Control Systems for HVDC Stations Connected to Weak AC Systems*', Part I-New Control Systems, Part II-*Operational Behaviour of HVDC Transmission*, IEEE Trans., Vol. PAS.91, pp. 549–564, March/April 1972.
6. Reeve, J. and Sevenco, J.A. '*An Automatic Control Scheme for HVDC Stations Connected to Weak AC Systems*', Part I-Principles of Operation, IEEE Trans., Vol.PAS-92, pp. 2319–2324, 1973.
7. Becker, W. Ranade, S, Rumpf E. and Wess.T., '*Control Systems for Cabora Bassa HVDC Scheme: Design Criteria Simulator Studies*', CIGRE Proc., Paper 14-06, 1976.
8. McNichol, J. Ranade.S.T., Ring H. and Meyl.D, '*Parallel Operation of Nelson River HVDC Bipoles 1 and 2 Control System Simulator Studies*', Paper 81 SM 371-4, Presented at IEEE PES Summer Meeting, Portland, Oregon, July 1981.
9. Reeve, J. and Geisbrecht W.J., '*Microprocessor Control for HVDC Converters*', Proc. of Int. Conf. on 'Thyristor and variable static equipment for AC and DC transmission', (IEE 205), pp. 186–189, Dec. 1981.
10. Ainsworth, J.D. '*Development in the Phase-locked Oscillator Control System for HVDC and Other Large Converters*', IEE Conf. Publication 255, on 'AC and DC Power Transmission', Sept. 1985, pp. 98–103.
11. Jotten, R. et al., '*Control in HVDC Systems—the State-of-the Art*, Part I: Two Terminal Systems, CIGRE Conf. Paper 14-10, 1978.
12. Dewan, S.B. and Dunford W.G., '*A Microprocessor based Controller for a Three Phase Controlled Rectifier Bridge*', IEEE Trans., Vol. IA-19, Jan/Feb. 1983, pp. 113–119.
13. Alegria, C.M. and Freris L.L., '*Microcomputer Control of Power Converters*', IEEE Trans., Vol.PAS-103, No.8, Aug.1984, pp. 2011–2017.
14. Ali Mirbod and Ahmed El-Amawy, '*A General-purpose Microprocessor based Control Circuit for a Three-phase Controlled Rectifier Bridge*', IEEE Trans., Vol. IE-33, No.3, Aug.1986, pp. 310–317.
15. Stemmler, H. Guth.G. and Dahler P., '*A New High-speed Controller with Simple Program Language for the Control of HVDC and SVS*', Proc. Int. Conf. on DC Power Transmission, Montreal, June 1984, pp. 109–115.
16. Ishikawa, M. '*HVDC Transmission Control Equipment with High Reliability*', IEEE Trans., Vol. PWRD-1, No.2, April 1986, pp. 254–263.

17. Hingorani N.G. and Chadwick, P. '*A New Constant Extinction Angle Control for AC/DC Static Converters*', IEEE Trans., Vol. PAS-87, March 1968, pp. 866–872.

18. Reeve J. and Chen S.P., "*Versatile Interactive Digital Simulator based on EMTP for AC/DC Power System Transient Studies*", Paper 84-200-2 presented at IEEE PES Meeting, 1984.

19. Padiyar K.R., "*FACTS Controller in Power Transmission and Distribution*", (New Delhi), New Age Publishers, 2007.

20. Pilotto L.A.S., Roitman M. and Alves J.E.R. "*Digital Control of HVDC Converters*", IEEE Trans., Power Systems, Vol. 4, No.2, May 1989, pp. 704–711.

21. Lorden D.J., Clark K. and Larsen E.V., "*A Digitally based HVDC Firing-Pulse Synchronization Control—Description and Model Development*", IEEE Trans. Power Delivery, Vol.7, No. 3, July 1992, pp. 1405–1414.

22. Larsen E.V., Clark K. and Lorden D.J., "*Stability Analysis of a Digitally based HVDC Firing-Pulse Synchronization Control*", IEEE Trans. Power Delivery, Vol. 7, No. 3, July 1992, pp. 1415–1414.

23. Karlecik-Maier F., "*A New Closed Loop Control Method for HVDC Transmission*", IEEE Trans. Power Delivery, Vol. 11, No. 4, October 1996, pp. 1955–1960.

24. IEEE Working Group, "*HVDC Controls for System Dynamic Performance*", Paper 90SM331-9PWRD, IEEE PES Summer Meeting, Minneapolis, July, 1990.

CONVERTER FAULTS AND PROTECTION

5

5.1 INTRODUCTION

As in AC systems, the faults in a DC system are caused by (i) the malfunctioning of the equipment and controllers and (ii) the failure of insulation caused by external sources such as lightning, pollution, etc. The faults have to be detected and the system has to be protected by switching and control action such that the disruption in the power transmission is minimized.

Apart from disrupting the normal operation, the various faults that can occur also cause the stressing of the equipment due to overcurrents and overvoltages. In a converter station, the valves are the most critical equipment which need to be protected against damage caused by the rise in the junction temperature of thyristors, which is caused by excessive losses in the device and sensitivity to overvoltages.

In this chapter, the faults occurring in a converter station, their causes and effects are described. The protection against overcurrents and overvoltages is also discussed. It is seen that the converter control plays a major role in protecting the equipment.

5.2 CONVERTER FAULTS

5.2.1 General

There are three basic types of faults that can occur in converters as given below:

1. Faults due to malfunctions of valves and controllers
   - *(i)* Arc backs (or backfire) in mercury arc valves
   - *(ii)* Arc through (fire through)
   - *(iii)* Misfire
   - *(iv)* Quenching or current extinction
2. Commutation failure in inverters
3. Short circuits in a converter station.

The arc back is the failure of the valve to block in the reverse direction and results in the temporary destruction of the rectifying property of the valve due to conduction in the reverse direction. This is a major fault in mercury arc valves and is of random nature. This is a nonself clearing fault and results in severe stresses on transformer windings as the incidence of arc backs is common.

Fortunately, thyristors don't suffer from arc backs which has led to the exclusion of mercury arc valves from modern converter stations. Hence this fault will not be considered in further discussion.

Some of the converter faults such as commutation failure, arc through and misfire are self clearing if the causes that led to these faults are of transient nature. However, they can still cause a major disturbance unless the system including the controllers is properly designed.

5.2.2 Commutation Failure

Because of the turn-off time requirements of thyristors, there is need to maintain a minimum value of the extinction angle defined by

$$\gamma = 180 - \alpha - u \tag{5.1}$$

The overlap angle (u) is a function of the commutation voltage and the DC current. The reduction in the voltage or increase in the current or both can result in an increase in the overlap angle and reduction of γ below γ_{min}. This gives rise to commutation failure. The current in the incoming valve (say valve 3) will diminish to zero and the outgoing valve (valve 1) will be left carrying the full link current.

The firing of the next valve in sequence (valve 4) will result in the short circuit of the bridge as both valves in the same arm will conduct. If the commutation from valve 2 to valve 4 is successful, only valves 1 and 4 are left conducting and this state continues until the valve 6 is fired. The firing of valve 5 prior to the firing of valve 6 is unsuccessful as the valve 5 is reverse biased at the time of firing. This happens as the voltage across valve 5 (when valve 1 is conducting) is $(e_c - e_a)$ instead of the normal commutation voltage $(e_c - e_b)$. Since e_{ca} leads e_{cb} by 60°, for normal values of the firing angles for the inverter $(\beta < 60°)$, the voltage across valve 5 will be negative and hence reverse biased.

If the commutation from valve 4 to valve 6 is successful, the conduction pattern returns to normal except that the bridge voltage is negative at the instant where valve 4 ceases conduction. The voltage across valve 6, just before it starts conducting is $(e_b - e_a = -e_{ab})$. However, the voltage across the bridge after valve 4 is turned off (and when valves 1 and 6 are conducting) is $-(e_b - e_a) = e_{ab}$. If $-e_{ab} > 0$ for valve 6 to successfully turn-on, the bridge voltage when valve 4 is turned off is negative. If the causes which led to commutation failure in valve 1 in the first instance have disappeared, the bridge operation returns to normal state. Thus, a single commutation failure is said to be self clearing. The waveforms of the bridge voltage and valve voltages are shown in Fig. 5.1.

The failure of two successive commutation in the same cycle, is called 'double commutation failure'. If the commutation failure occurs when valve 4 is fired also, the valves 1 and 2 are left in the conducting state until the instant in the next cycle when valve 3 will be fired. The valve 5 is reverse biased at the instant when firing pulse is sent as e_{ca} leads e_{cb} by 60° (as mentioned earlier). Similarly, the voltage across valve 6 when firing pulse is applied to valve 6 is $-(e_c - e_b)$ instead of the normal commutation voltage $-(e_a - e_b)$. Since $-e_{cb}$ leads $-e_{ab}$ by 60°, the valve 6 is reverse biased at the instant of firing. The bridge voltage waveform for this case is shown in Fig. 5.2 and it can be seen that the double commutation failure is more severe than the single commutation failure.

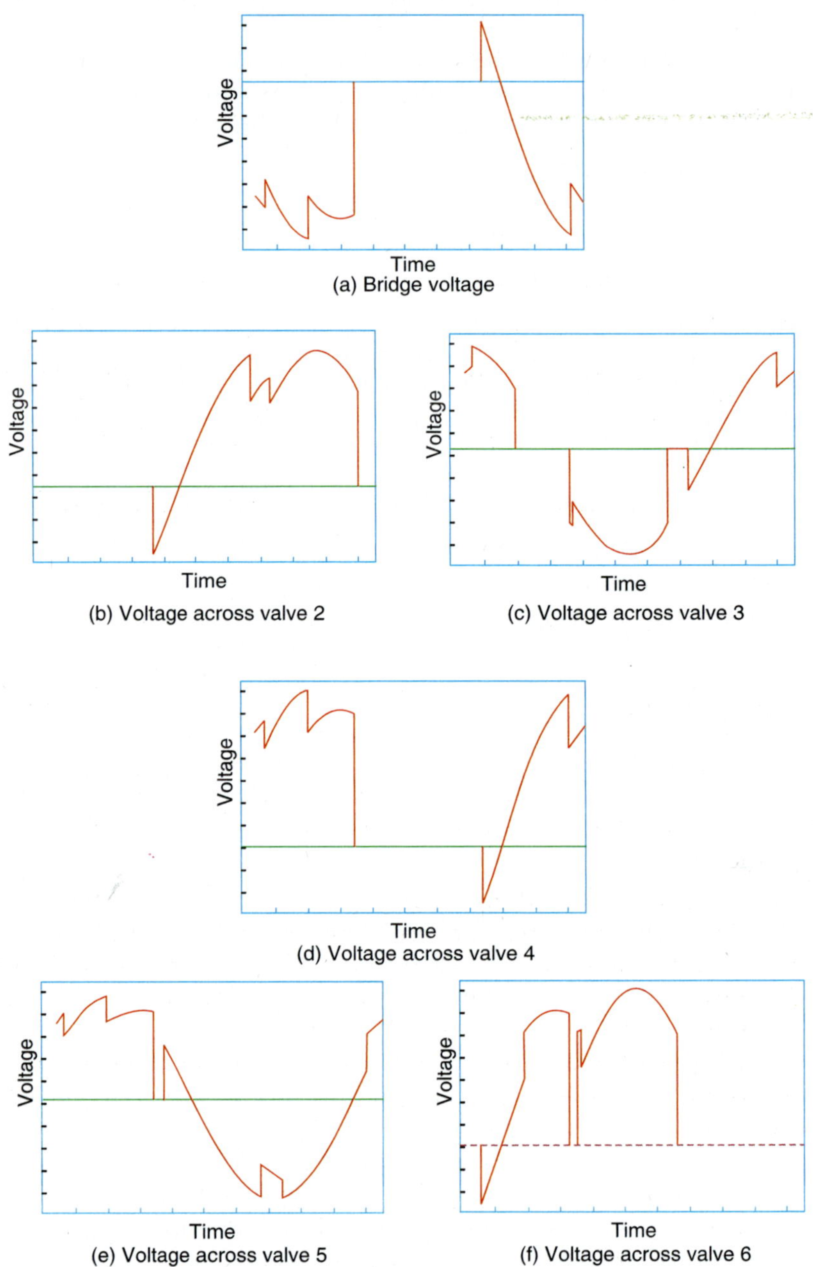

Fig. 5.1: Voltage waveforms for a single commutation failure

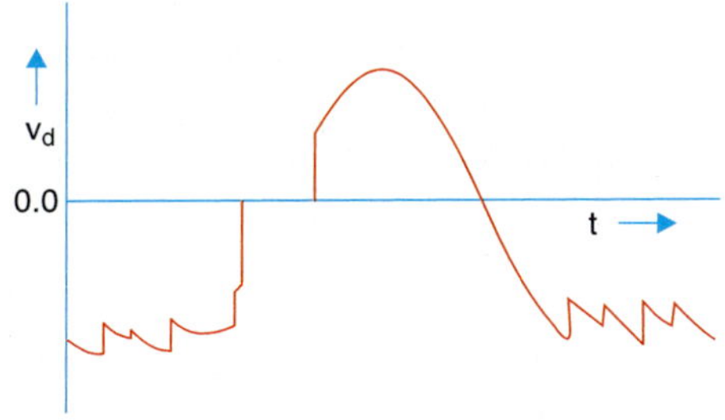

Fig. 5.2: Bridge voltage waveforms for a double commutation failure

The following are the effects of a single commutation failure:

1. The bridge voltage remains zero for a period exceeding 1/3 of a cycle, during which the DC current tends to increase
2. There is no AC current for the period in which the two valves in an arm are left conducting.

The recovery from a commutation failure depends on the following factors:

1. The response of the gamma controller at the inverter
2. The current control in the link and
3. The magnitude of the AC voltage.

If, on detection of a commutation failure, the angle of advance β is increased, there is a good chance that subsequent commutation failures may be averted. However, this also depends upon the control of DC current and the magnitude of AC voltage. The initial rate of rise of current in the inverter is limited by the smoothing reactor and the current controller at the rectifier helps to limit the current in the case of persistent commutation failure. It may be even necessary to reduce the current reference to limit the overlap angle in the case of low voltages caused by faults in the AC system.

While in most cases, commutation failures are self clearing, in the case of persistent commutation failures, the converter differential protection helps to take the converter out of service. This protection is based on the comparison of DC and the valve side AC currents. During commutation failures when the two valves in an arm of a bridge are left conducting, the AC current goes to zero while the DC current continues to flow.

The commutation failure in a bridge can lead to consequential commutation failures in the series connected bridges unless the rate of rise of current is sufficiently limited by the series connected smoothing reactors.

Example 1

Compute the maximum voltage dip at the converter bus that will not result in a commutation failure. Assume that the voltage dip occurs at the instant immediately after firing the incoming valve. Consider (*a*) symmetrical three phase voltage dip and (*b*) single phase voltage dip.

Solution (a)

It is assumed that the converter is initially operating in steady state with $I_d = I_{do}, \gamma = \gamma_0$ and $E = E_0$ (where E is the line to line AC voltage at the converter bus. As shown in chapter 3 (see example 1), the expression for the DC current can be expressed as

$$\frac{I_{do} + I'_d}{2I_{dn}} = \frac{E'}{E_{an}\bar{X}_c}[\cos\gamma_m - \cos\beta]$$

Where, E_{an} is the rated (line to line) voltage of the converter transformer at the secondary (valve) side, I_{dn} is the rated DC current, $\bar{X}_c$ is the p.u leakage reactance of the transformer (on its own base) and I'_d is the DC current at the end of commutation.

Expressing

$$I'_d = I_{do} + \Delta I_d,\ E' = E_o - \Delta E,\ \bar{E}_o = \frac{E_o}{E_{an}}$$

We can write the current equation as

$$\frac{1}{I_{dn}}\left(I_{do} + \frac{\Delta I_d}{2}\right)\cdot\frac{\bar{X}_c}{\bar{E}_0\left(1 - \frac{\Delta E}{E_0}\right)} = \cos\gamma_m - \cos\beta$$

Note that γ_m is the minimum extinction angle (typically $8°$) that ensures successful commutation. The current equation during steady state is given by

$$\frac{I_{do}}{I_{dn}}\cdot\frac{\bar{X}_c}{\bar{E}_0} = \cos\gamma_0 - \cos\beta$$

Subtracting the above equation from the previous equation and simplifying, we can solve for $\frac{\Delta E}{E_0}$ as

$$\frac{\Delta E}{E_0} = \frac{(\cos\gamma_m - \cos\gamma_0) - \frac{1}{2}\left(\frac{\Delta I_d}{I_{dn}}\right)\frac{\bar{X}_c}{\bar{E}_0}}{(\cos\gamma_m - \cos\gamma_0) + \left(\frac{I_{do}}{I_{dn}}\right)\frac{\bar{X}_c}{\bar{E}_0}}$$

Solution (b)

Let us denote e_a and e_b as

$$e_a = \sqrt{\frac{2}{3}}E_0\cos(\omega t + 60°), e_b = \sqrt{\frac{2}{3}}E_0\cos(\omega t - 60°)$$

Then the voltage across valve 1 just after it has turned off is given by

$$v_1 = e_a - e_b = -\sqrt{2}E_0\sin\omega t$$

The phasor diagram of voltages E_a , E_b and E_{ab} in steady state is shown in Fig. 5.3.

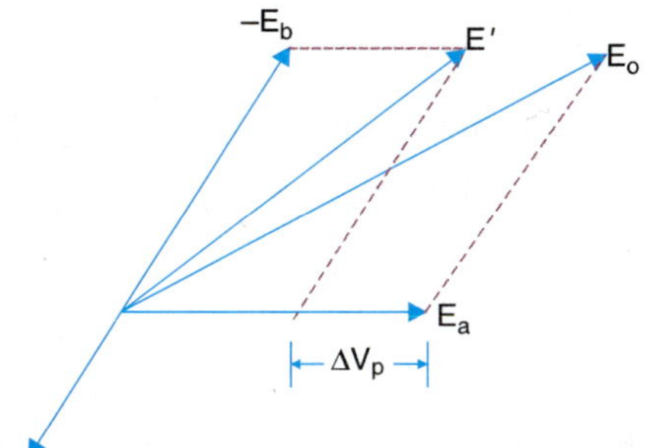

Fig. 5.3: Phasor diagram of voltages

Assuming that there is a dip ΔV_p in phase voltage (of phase 'a') at $\omega t = \beta$, immediately after the firing of the valve 3, there is a dip in the commutation voltage and the changed voltage magnitude (E') is given by (see Fig. 5.3)

$$E' = (E_0^2 + \Delta V_p^2 - \sqrt{3}E_0\Delta V_p)^{\frac{1}{2}}$$

There is also a phase shift (advance in this case) of ϕ in E_{ab} defined by

$$\sin\phi = \sin 30^\circ \cdot \frac{\Delta V_p}{E'}$$

The effect of the phase shift ϕ, in the valve voltage is equivalent to increasing the minimum extinction angle from γ_m to $\gamma_m + \phi$ [14]

To get an idea of the numbers involved, consider

$$\gamma_m = 8^\circ, \gamma_0 = 15^\circ, \Delta I_d = 0, \bar{E}_0 = 1.0 \text{ and } \bar{X}_c = 0.2$$

We get $\Delta E = 0.108$ for a symmetrical fault. If $\gamma_0 = 18^\circ$, then $\Delta E = 0.164$. This shows that operating at higher nominal extinction angles reduces the incidence of commutation failures. However, increase in γ_0 results in reduced power factor and higher voltage rating of the valves.

It is desirable to have lower values of $\bar{X}_c$ as this reduces the incidence of commutation failures while simultaneously improving the power factor and lower voltage ratings of the valves. The lower limit on $\bar{X}_c$ is determined by the surge current rating of the valves.

It is to be noted that if there is a single phase fault resulting in the voltage dip in phase *b* (instead of phase '*a*'), then ϕ is negative. In this case, the commutation failure does not occur when valve 3 is fired. However, there is likelihood of commutation failure when the valve 5 is fired. If the dip in the voltage of phase 'b' (at the instant of firing of valve 3) is detected early, the firing of valve 5 can be advanced to prevent commutation failure. In general, it can be said that if the voltage dip occurs in the phase connected to the outgoing valve, the chances of commutation failures are increased if there is no time to take corrective action. If ΔI_d is not zero, then ΔE is reduced. This shows the importance of fast current control and/or smoothing reactor. As discussed in the next chapter, the smoothing reactor value is less for back to back HVDC links.

Example 2

An inverter experiences an (symmetrical, three phase) AC voltage dip when operating at the rated current. What is the converter voltage (in pu) below which persistent commutation failures occur?

Data: $\alpha_{min} = 90°$, ξ_{min}(= margin angle) = 8°, $\bar{X}_c = 0.2$

Solution:

It can be shown (see Chapter 3) that $\bar{I}_d = \frac{I_d}{I_{dn}} = \frac{\bar{E}}{\bar{X}_c}[\cos\gamma - \cos\beta]$ while operating at mode 1 (overlap angle below 60°). For $\beta = 90°$ we have $\gamma = \xi + 30°$ and hence $\gamma_{min} = 30 + \xi_{min} = 38°$

Thus $\bar{E}$, is given by

$$\bar{E} = \frac{\bar{I}_d \bar{X}_c}{\cos\gamma_{min} - \cos\beta_{max}}$$

Substituting the values of $\bar{I}_d = 1.0$, $\bar{X}_c = 0.2$ and $\beta_{max} = 180 - \alpha_{min} = 90°$

we get $\bar{E} = 0.254$pu

Note that the converter control had advanced the firing angle to α_{min}. However, the bus voltage is too small to prevent persistent commutation failures. The VDCOL can help by reducing the operating DC current.

It should be noted that reducing α increases the reactive power requirements of the converter and can depress the AC voltage further (depending on the short circuit ratio). While the reduction of the DC current can overcome this problem, the reduction in the power flow may not be acceptable (even under transient conditions).In these situations an optimal control strategy that can reduce the incidence of commutation failures during dynamic conditions (after about one cycle following the fault initiation), is desirable [15,16]. In reference [15] the authors used predictive extinction angle control based on the detection of the d-q components of the network voltage using PLL. The required firing angle to maintain the desired commutation margin is calculated based on the prediction of the magnitude and phase angle of the commutation voltages. A similar approach is used in [16] when single phase faults are considered and control circuits based on Digital Signal Processor (DSP) are used.

It is to be noted that faults and commutation failures can introduce harmonic components in the network voltages which can complicate the analysis. A third harmonic voltage component with adverse phase in relation to the fundamental component can increase the incidence of the commutation failures.

5.2.3 Arc Through

This is a fault likely to occur mainly at the inverter station, where the valve voltages are positive most of the time (when they are not conducting). A malfunction in the gate pulse generator or the arrival of a spurious pulse can fire a valve which is not supposed to conduct, but is forward biased. For example, in a bridge, when valve 1 has successfully commutated its current to valve 3, the initial voltage across it is negative (for the duration of the margin angle) and then becomes positive. If valve 1 is fired at this time, the current will transfer back to valve 1 from valve 3.

The effects of an arc through are similar to that of a commutation failure—the voltage across the bridge falls as valve 4 is fired (with valve 1 conducting) and the AC current goes to zero when valve 2 current goes to zero. The firing of valve 5 is unsuccessful and the bridge recovers to normal operation after valve 6 is fired and the subsequent firings are according to the normal sequence. Thus a single arc through is also self -clearing if the causes that led to it are removed.

The arc throughs in thyristor valves can occur due to malfunction in the control system, the probability of which is very small. Anyway, the protection against persistent arc throughs is also through the converter differential protection scheme.

5.2.4 Misfire

While an arc through is caused by the presence of an unwanted gate pulse, misfire occurs when the required gate pulse is missing and the incoming valve is unable to fire. The probability of the occurrence of misfire is very small in modern converter stations because of duplicated converter controls, monitoring and protective firing of valves.

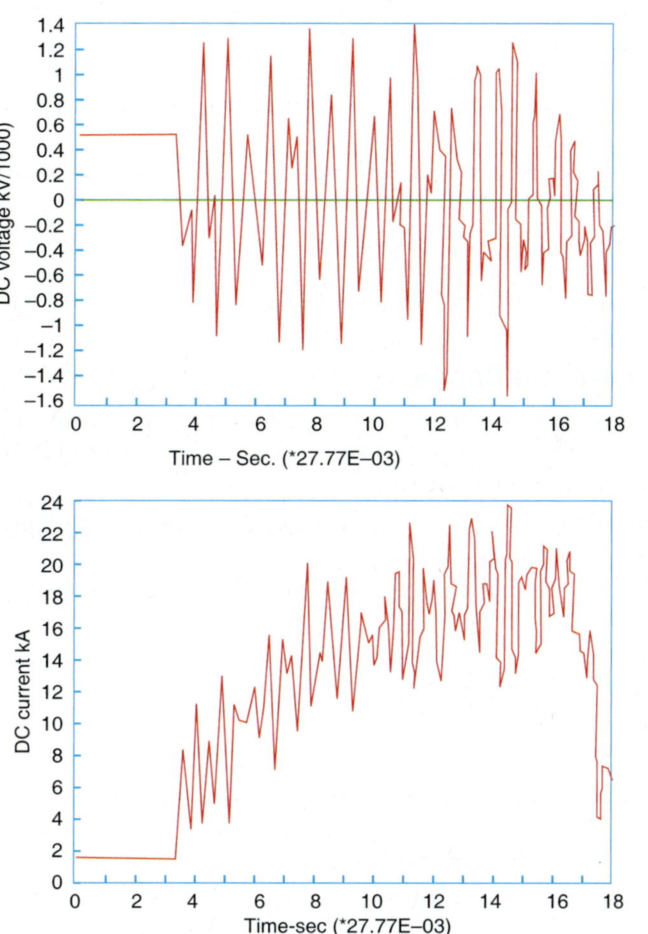

Fig. 5.4: DC voltage and current for persistent misfire

While misfire can occur in rectifier or inverter stations, the effects are more severe in the latter case. This is due to the fact that in inverters, persistent misfire leads to the average bridge voltage going to zero, while an AC voltage is injected into the link. This results in large current and voltage oscillations in the DC link as it presents a lightly damped oscillatory circuit viewed from the converter. The DC current may even extinguish and result in large overvoltages across the valves. The waveforms of the DC voltage and current in the link for persistent misfire in the inverter are shown in Fig. 5.4. The overvoltages can be controlled by controller modifications.

The effects of a single misfire are similar to those of commutation failure and arc through. When valve 3 in a bridge misfires, the valves 1 and 2 are left conducting until valve 4 is fired. However, at the end of the cycle, the normal sequence of valve firings is restored. Thus a single misfire is also self-clearing.

5.2.5 Current Extinction

The extinction of current can occur in a valve if the current through it falls below the holding current. This can arise at low values of the bridge currents when any transient can lead to current extinction. The current extinction can result in overvoltages across the valve due to current chopping in an oscillatory circuit formed by the smoothing reactor and the DC line capacitance.

The problem of current extinction is more severe in the case of short pulse firing method discussed in the last chapter. However, in modern converter stations, the return pulses coming from thyristor levels to the valve group control, indicate the build up of voltage across the thyristors and initiate fresh firing pulses when the valve is supposed to be conducting. It may happen that a number of firing pulses may be generated during a cycle when the link current is low.

5.2.6 Short Circuit in a Bridge

This fault also has very low probability as the valves are kept in a valve hall with air conditioning. However, bushing flashover can lead to a short circuit across the bridge and produce large current peaks in the valves that are conducting.

The short circuit currents are significant only in rectifier bridges.

The worst case is when the short circuit occurs at the instant of firing a valve at $\alpha = 0$. Assuming that there is no inductance in series with the bridge, the peak short circuit current (i_{peak}) is given by

$$\frac{i_{\text{peak}}}{I_s} = \frac{3}{2} + \frac{i_{do}}{2I_s} \tag{5.2}$$

where

i_{do} = the dc current at the instant of firing the valve and

$$I_s = \sqrt{2}E_{LL}/2X_c$$

The bridge current i_d waveform is shown in Fig. 5.5.

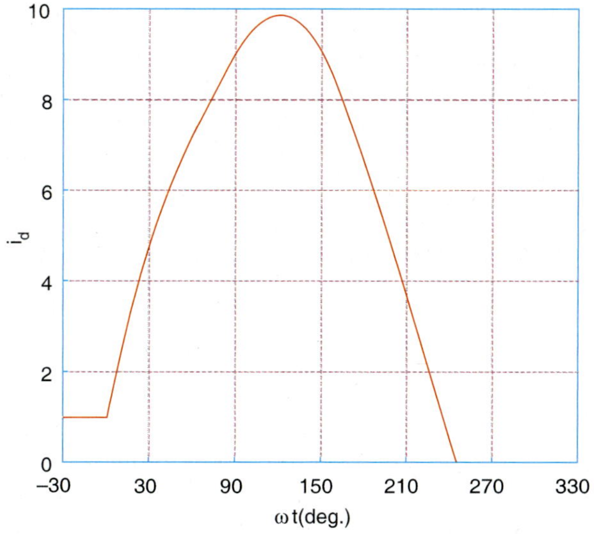

Fig. 5.5: Bridge voltage and current waveform during short circuit

Proof:

Assume that the valves 1 and 2 are conducting and valve 3 is fired at $t = 0$, corresponding to $\alpha = 0$. The supply voltages e_a,e_b and e_c in the circuit shown in Fig. 5.5 are given by

$$e_a = \sqrt{\frac{2}{3}} E_{LL} \sin(\omega t + 150°)$$

$$e_b = \sqrt{\frac{2}{3}} E_{LL} \sin(\omega t + 30°)$$

$$e_c = \sqrt{\frac{2}{3}} E_{LL} \sin(\omega t - 90°)$$

These expressions result in $e_{ba} = \sqrt{2} E_{LL} \sin \omega t$. When a bridge fault is applied at $t = 0$ (after firing of valve 3), the voltages at the terminals p and n become zero (due to symmetry). Thus the valve currents i_1 and i_3 are obtained by solving the following equations

$$L_c \frac{di_1}{dt} = e_a, \quad i_1(\omega t) = i_{do}$$

$$L_c \frac{di_3}{dt} = e_b, \quad i_3(\omega t) = 0$$

The expressions for i_1 and i_3 are obtained on

$$i_1 = \frac{2I_s}{\sqrt{3}}[\sin(\omega t + 60°) - \sin 60°] + i_{do}$$

$$i_3 = \frac{2I_s}{\sqrt{3}}[\sin(\omega t - 60°) + \sin 60°]$$

CHAPTER 5

The bridge current, i_d is given by

$$i_d = i_1 + i_3 = \frac{2}{\sqrt{3}} I_s \sin \omega t + i_{do}$$

The current in valve 1 becomes zero at $\omega t = u$ defined by

$$0 = \frac{2}{\sqrt{3}} I_s [\sin(u + 60°) - \sin 60°] + i_{do} \tag{5.3}$$

Since $i_{do} > 0$, it can be observed that $u > 60°$. For $\omega t = u$, we have the following equation as only valves 2 and 3 are conducting.

$$2L \frac{di_3}{dt} = e_b - e_c = \sqrt{2} E_{LL} \sin(\omega t + 60°)$$

The solution of this equation gives

$$i_d = i_3 = I_s [\sin(\omega t - 30°) - \sin(u - 30°)] + i_{du}$$

Where $i_{du} = i_d(\omega t = u)$ is given by

$$i_{du} = \frac{2}{\sqrt{3}} I_s \sin u + i_{do}$$

The bridge current is at its peak when $\omega t = 120°$ and i_{peak} is given by

$$i_{\text{peak}} = I_s - I_s \sin(u - 30°) + \frac{2}{\sqrt{3}} I_s \sin u + i_{do} \tag{5.4}$$

It can be shown that

$$\frac{2}{\sqrt{3}} I_s \sin u - I_s \sin(u - 30°) = \frac{I_s}{\sqrt{3}} \sin(u + 60°)$$

From Eq. (5.3), we can derive

$$\frac{I_s}{\sqrt{3}} \sin(u + 60°) = \frac{I_s}{\sqrt{3}} \sin 60° - \frac{i_{do}}{2}$$
$$= \frac{I_s}{2} - \frac{i_{do}}{2}$$

From the above, we can derive the final expression for i_{peak} given in Eq. (5.2).

In Eq. (5.2), the effect of network impedance in limiting the current is neglected. The maximum peak current in a valve results when it is conducting into a valve fault. For example, the maximum current in valve 3, when it starts conducting with short circuit across valve 1, is given by

$$\frac{i_{\text{peak}}}{I_s} = (1 + \cos \alpha) + (I_d / 2I_s) \tag{5.5}$$

The peak currents are of the order of 10 to 12 times the rated current and the thyristor valves must have surge current ratings above this value. The fault clearing is performed by blocking the pulses and when the fault current goes to zero, the valve assumes blocking state provided the voltage across it is not high. If the valve is unable to block the forward voltage, additional loops of overcurrents result and this can be avoided only by tripping the AC breaker.

The detection of bridge or valve short circuit is also performed by comparing the AC and DC currents. In this case, the DC current goes to zero while AC current tends to increase.

5.3 PROTECTION AGAINST OVERCURRENTS

The overcurrent protection in converters is based on principles similar to those used in AC systems. The factors that must be considered in designing a protection system are (*i*) selectivity, (*ii*) sensitivity, (*iii*) reliability and (*iv*) back up.

The main feature of converter protection is that it is possible to clear faults by fast controller action (in less than 20 msec) by blocking gate pulses or current regulation and control. The selectivity is also enhanced by high impedances of the smoothing reactor and the converter transformer. Further, the converters are divided into independent valve groups such that the protection system must be able to switch off only the affected valve group (or bridge).

Consider a converter station with a 12 pulse converter per pole (2 valve groups per pole). The protection system used for a pole is shown in Fig. 5.6. This does not show protection against DC line faults, undervoltage or transformer protection.

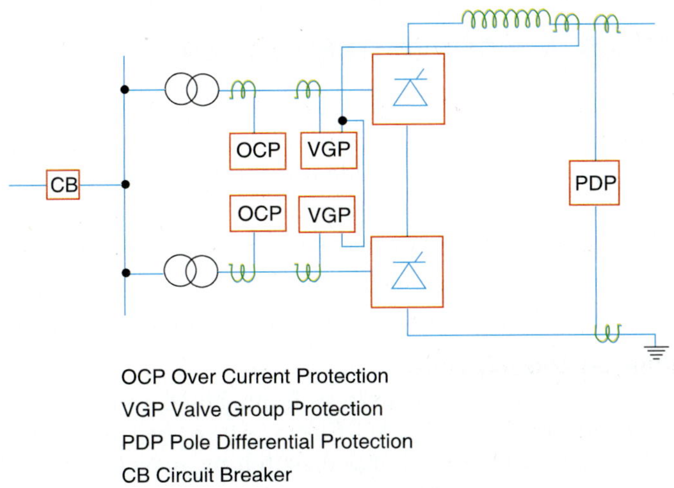

Fig. 5.6: Overcurrent protection in a pole

The basic protection against converter faults (considered in the previous section) is provided by valve group differential protection, which compares the rectified current on the valve side of converter transformer to the DC current measured on the line side of the smoothing reactor. The differential protection is employed because of its selectivity and fast detection. The overcurrent protection circuit is used as back-up. The level of overcurrent required to trip must be set higher than that of the valve group differential protection to avoid tripping with faults outside the station (that can be cleared by the control action).

The pole differential protection is used to detect ground faults which may not be otherwise detected, such as faults at the neutral bus.

The fault clearing action of these protection circuits is to block the valves and at the same time trip the AC breaker of the affected group or pole. The fast tripping sequence is used for internal faults where there is a danger of valve damage. This involves forced retard (increasing the delay angle of the rectifier to about 150°) combined with the signal to trip the AC breaker. The pulses are blocked after 20 msec. This allows the inverter action (by forced retard) at the rectifier station to try to reduce the current before the converter is blocked.

The faults producing overcurrent are classified into 3 categories:

1. Internal faults which cause high overcurrents but are very infrequent. The thyristor surge current ratings must be chosen to withstand these overcurrents.
2. Line faults which cause overcurrents in the range of 2 to 3 p.u. These are limited by current control. The protection against line faults which are frequent will be discussed in the next chapter.
3. Commutation failures at inverters may be quite frequent. However, the overcurrents are small and limited by current control. However, because of continuous conduction during commutation failures, the current reference has to be reduced.

In case 1, normal operation may commence only after the inspection of the valves to check for damages. In cases 2 and 3, normal operation can commence soon after fault clearing.

5.4 OVERVOLTAGES IN A CONVERTER STATION

5.4.1 General

The overvoltages in a converter station are caused by (*i*) disturbances originating on the AC side (*ii*) disturbances originating on the DC side (*iii*) internal faults in the converter.

The type of overvoltages, as in a AC system, can be classified into three categories:

1. The switching overvoltages (with wavefront times of more than 100 msec.)
2. Temporary overvoltages (lasting few seconds)
3. Steep front overvoltages (with wavefront times in the range of 0.3 to 3 msec).

5.4.2 Disturbances on the AC Side

The lightning strokes in the AC network cause steep-fronted high overvoltages which are, however, reduced in magnitude and steepness by AC filters. After they pass through the converter transformer, they appear only as highly damped switching surges across the converter.

The initiation and clearing (by switching action) of the faults in the AC system result in switching surges and temporary overvoltages. The energization of a converter transformer can cause high (up to 1.6 p.u) overvoltage due to the inrush magnetizing currents and last up to 100 cycles. The voltage is also distorted due to even harmonics (typically 4th harmonic). This type of temporary overvoltage can cause severe stresses on the metal oxide surge arrestors due to excessive energy dissipation. Pre-insertion resistors in circuit breakers energizing the converter transformer, are very effective in controlling these overvoltages.

The temporary overvoltages due to load rejection can be quite serious for converter stations connected to weak AC systems. The load rejection may be caused by blocking of the converters in response to the action of the protection system. During the load rejection, there is a possibility of self excitation in case of the isolated generating plant supplying the converters. This arises from the AC filters which appear as capacitive at fundamental frequency and result in self excitation of synchronous generators which cannot be controlled by AVR (automatic voltage regulator) action. The only solution is to switch off capacitor and filter banks.

5.4.3 Disturbances on the DC Side

The steep wavefront surges in DC overhead lines are produced by lightning strokes. However, when they reach the converter through the smoothing reactor they appear as switching surges.

The switching surges at the converter are also caused by ground faults on a pole of bipolar DC link. Owing to capacitive and inductive coupling between conductors, the surges also occur on the healthy pole. The magnitude and the wave shape of these surges arriving at the terminal are dependent on the type of termination inductive, capacitive or resistive. The rate of decrease of voltage at the terminal is a variable that is usually utilized in line fault detection and the setting of the threshold value is based on the knowledge of the voltage waveforms.

The overvoltages can also arise from the oscillations of current and voltage in the line caused by sudden jump in the converter voltage (due to commutation failure and other converter faults) or injection of AC voltages of fundamental frequency and second harmonic.

The switching of DC filter branches, parallel connection of poles can cause transient currents and overvoltages which will mainly stress the neutral bus and filter reactors.

5.4.4 Overvoltage Caused by Internal Converter Disturbances

The series connection of thyristors and the spread in the delay times of the thyristor turn-on result in overvoltages across the device during turn-on. However, these are repetitive and have to be taken into account in the valve design and the choice of grading circuit (snubber circuit) parameters. The spread in the reverse recovery charges and the commutation overshoot also result in repetitive overvoltages.

Transient overvoltages of very steep front may result from internal converter faults, such as a ground fault at the valve side of the smoothing reactor. The ground faults can also produce switching surge type overvoltages, for example, a fault between the valve bridge and the converter transformer. The firing of bypass pairs or closing of the bypass switch across one converter generates overvoltages across the remaining converters.

The energization of the DC line from the rectifier side with the remote terminal blocked can cause high overvoltages at the inverter which is open ended. Such events must be avoided by deblocking the inverter first and limiting the rate of decrease of the delay angle.

5.5 SURGE ARRESTERS

In the initial stages of application of DC technology, DC surge arresters were not available and the valves were protected by the spark gaps connected across them. Later, with the development of active spark gaps, it was possible to extinguish the arrester current without exceeding the protective level and DC arresters were made of nonlinear resistors in series with the active spark gaps.

With the development of metal oxide resistors with high nonlinearity, the need for a series gap has disappeared and the present DC arresters are gapless arresters. The metal oxide elements were first applied in AC arresters in 1976. Comprising primarily of zinc oxide, but containing a number of other metal oxides (such as Bi_2O_3, Sb_2O_3, MnO_2, Cr_2O_3), as additives, the material has extremely nonlinear voltage-current characteristics. A typical disc that conducts less than a milliampere of current at normal operating voltage can carry currents of thousands of amperes at twice the normal voltage. This property makes it possible to eliminate series connected spark gaps and reduce the voltage margins due to the constancy of protective levels. The voltampere curve for a typical zinc-oxide disc is shown in Fig. 5.7 for different levels of operating temperature.

It is seen that the temperature coefficient of the material is slightly negative at low currents, but becomes positive at currents above a few amperes. This makes it possible to operate the zinc oxide elements in parallel to discharge high energy surges. The long term stability of the material is satisfactory although it is influenced considerably by disc composition and processing.

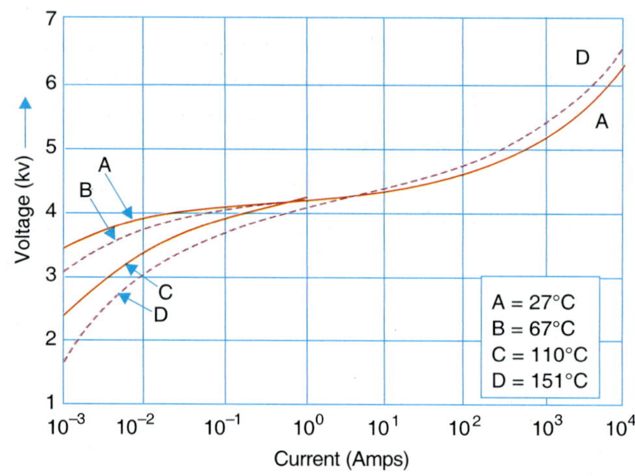

Fig. 5.7: Volt ampere curve for a typical zinc oxide disc (Source: Reference 13)

The properties of the material are such that it is possible to design arresters to control dynamic overvoltages in addition to switching surges. This results in economic insulation coordination. However, proper design of the arrester based on the evaluation of the energy losses is essential. The ultimate limit on the energy dissipation capability of a disc is imposed by the cracking of the disc under thermal shock. A single column arrester is capable of absorbing around 7 kJ per kV of the maximum continuous operating voltage (MCOV).

In many DC applications, the energy capability of a single column of discs is inadequate and multiple columns are used. A parallel column arrester is made up by selecting discs such that the voltage for each column is the same at a predetermined current. The maximum difference in the currents of parallel columns can be made less than 0.5%.

5.6 PROTECTION AGAINST OVERVOLTAGES

5.6.1 General

The basic principles of overvoltage protection is the same in DC systems as in AC systems. These are given below:

1. The overvoltage stresses in equipment with non self-restoring insulation must be limited at all times by providing surge arresters. The protection level of the arresters must be lower than the breakdown voltage of the insulation.
2. Self restoring insulation such as air may be allowed to breakdown where there is no danger to the safety of the personnel.
3. The operation of surge arresters or flashover of air insulation must not be frequent. Frequent discharges of arresters may damage them. This implies that the protective level of arresters must be higher than the maximum operating voltage in the system.
4. There must be proper coordination of the insulation and overvoltage protection in different parts of the system, taking into account the characteristics of the insulation, the nature of overvoltages, etc.

The overvoltages generated on the AC side should, as far as possible, be limited by arresters on the AC side. The overvoltages generated on the DC side must be limited by DC line, DC bus and neutral bus arresters. The critical components such as valves are directly protected by arresters connected close to the components.

5.6.2 Overvoltage Protection in a Converter Station

The typical arrangement of surge arresters in a converter station (for a pole) is shown in Fig. 5.8. For a system with two 12-pulse converters per pole, there are about 40 arresters per pole.

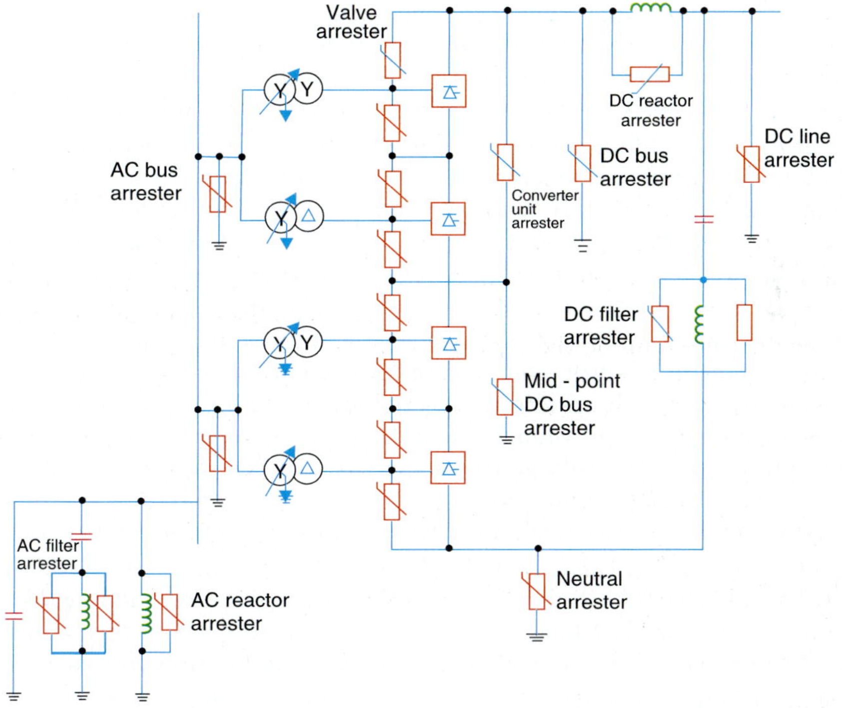

Fig. 5.8: Typical arrangement of surge arrestors for a converter pole (Source: Reference 9)

CHAPTER 5

The arresters are selected with adequate energy dissipation capabilities which vary with the location of the arresters. For example, the valve arrester protecting the commutation group at the highest potential can be subjected to higher energies than other arresters when a ground fault occurs between the valve and the converter transformer in the upper bridge. This is due to the discharge of the line and the DC filter.

The closing of a bypass switch across a converter results in increasing the DC voltage across the remaining converter. The converter unit arrester is stressed in such cases.

The protective firing of a valve is the backup protection that is available for overvoltages in the forward direction.

5.7 PROTECTION AGAINST FAULTS IN A VOLTAGE SOURCE CONVERTER

The protection requirements are similar to those of an LCC based HVDC link. The objectives of the protection scheme are:

1. Protect against damage by an external fault or overvoltage
2. Limit the damage against internal faults and
3. Feed the fault to permit the operation of fault detection relays in an isolated system.

The last requirement is unique to a VSC based HVDC station. Since VSC based links can supply passive loads and the current is normally held close to the nominal current even at low network voltages, it may be required to feed AC faults to allow the operation of overcurrent relays.

A VSC is equipped with free-wheeling diodes which prevent overvoltages in the reverse direction impacting the IGBT valves. However, it may be necessary to provide overcurrent protection (against inrush current when the DC voltage is reduced). Also, DC voltage-clamping devices are provided to protect against overvoltages in the forward direction.

For temporary external faults, the converter may be blocked temporarily to protect IGBT and free-wheeling diodes from overcurrents. After a few milliseconds, the valves are deblocked and an attempt is made to restart. If automatic deblocking fails one or more times, permanent blocking is also carried out. Permanent blocking is also carried out for internal faults along with tripping of AC breakers. After permanent blockings, deblocking is done manually. A fault across the converter requires tripping of the AC circuit breaker to prevent overcurrents and isolate the converter against voltage stresses.

A specific requirement in a VSC station is the protection against progressive failure of VSC DC capacitor elements, which results in the reduction of the capacitance value and the asymmetric distribution of the DC voltage. Monitoring the voltages across each capacitor helps in detecting the failures as the voltage ripple increases as the capacitance value decreases. However, change in the capacitance value is not so critical compared to that in a harmonic filter. If the reduction in the DC capacitance is significant, the AC breaker is tripped.

Unlike in the case of a CSC based HVDC link, moderate changes in the AC bus voltages can be handled by the control action of the converter (in regulating the injected current). If there is a failure in the control system, there can be overcurrents due to dips in the AC voltage. The overvoltages due to lightning and switching are handled by surge arresters.

The sudden shifting in the phase of the AC bus voltage results on a sudden power change on a VSC that requires a fast acting PLL to counter the phase shift. Otherwise, the overcurrent detection will result on VSC blocking and tripping.

AC voltage unbalance can lead to large ripples in the DC capacitor voltage. The control system needs to be designed to inject negative sequence voltage temporarily to reduce the impact of phase imbalance. In the worst case, the converter has to be tripped.

Interestingly, the dip in the AC voltage at the inverter can cause overvoltage across the DC capacitor due to energy imbalance (the power output of the inverter falling below the input received from the rectifier). Voltage clamping circuits can be used to regulate the DC voltage. Alternately, control action can be used to reduce the power level at rectifier (particularly in back to back DC links).

The post-fault recovery of VSC based HVDC links is very crucial when they are used to evacuate power from wind forms. If the VSC based HVDC link is stopped for too long, the wind generator will be tripped by its over-speed protection resulting in the tripping of the entire wind farm.

REFERENCES AND BIBLIOGRAPHY

1. Kauferle J. and Povh D., *'Concepts of Overvoltage and Overcurrent Protection of HVDC Converters'*, CIGRE Paper 14-08,1978.
2. Kaufhold W. and Povh D. *'Recovery of HVDC Transmission after Faults in the AC System'*,IEE Conf. Publication 205 on Thyristor and Variable Static Equipment for AC and DC Trans- mission', London, 1981, pp.171–175.
3. Bowles J.P., *'Overvoltages in HVDC Transmission Systems Caused by Transformer Magnetizing Inrush Currents'*, IEEE Trans., Vol. PAS-93, 1974, pp. 487–493.
4. Giesner D.B. and Arrillaga J. *'Behaviour of HVDC Links under Unbalanced AC Fault Conditions'*, Proc. IEE, Vol. 119, No. 2, 1972, pp. 209–215.
5. Kimbark E.W., *'Transient Overvoltages Caused by Monopolar Ground Faults on Bipolar DC Lines—Theory and Simulation'*, IEEE Trans., Vol. PAS-89, 1970, pp. 584–592.
6. Hingorani N.G., *'Transient Overvoltages on Bipolar HVDC Overhead Lines Caused by DC Line Faults'. ibid*, 1970, pp. 597–610.
7. Clerici A and Taschini A. *'Transient Overvoltages Caused by Earth Fault on Bipolar DC Lines'*, IEE Conf. Publication No. 107, on HVDC/AC Power Transmission, Nov. 1973, pp. 196–200.
8. Melvold D.I., Odam P.C. and Vithayathil,J.J., *'Transient Overvoltages on HVDC Bipolar Line during Monopolar Line Faults'*, IEEE Trans., Vol. PAS-96, No. 2, March/April, 1977, pp. 591–601.
9. CIGRE Study Committee No. 33, *'Application Guide for Insulation Coordination and Protection of HVDC Converter Stations'*, 33.83 (SC 03-21, WD).
10. Peterson H.A., Phadke A.G. and Reitan D.K., *'Transients in EHV DC Power Systems, Part I-Rectifier currents'*, IEEE Trans., Vol. PAS-88, No. 7, July 1969, pp. 981–989.
11. Reeve J. and Kapoor S., *'Analysis of Transient Short Circuit Currents in HVDC Power Systems'*, IEEE Trans., Vol. PAS-90, No. 3, 1971, pp. 1174–1182.
12. Thio C.V. et al., *'Switching Overvoltage on the Nelson River HVDC System–Studies, Experience and Field Tests'*, IEEE Conf. on 'Overvoltages and Compensation on Integrated AC- DC Systems', Winnipeg, 1980, pp. 15–26.

13. Breuer G.D., Krege J.S. and Shakshaug E.C., '*Zinc-oxide Overvoltage Control and Insulation Coordination*', Electric Forum, Vol. 9, No. 3, 1984, pp. 50–55.
14. Thio C.V., Davies J.B. and Kent K.L., '*Commutation Failures in HVDC Transmission Systems*', IEEE Trans., Power Delivery, Vol. 11, No. 2, 1996, pp. 946–957.
15. Tamai S., et al, '*Fast and Predictive HVDC Extinction Angle Control*', IEEE Trans., Power Systems, Vol. 12, No. 3, 1997, pp. 1268–1275.
16. Hansen A. and Havemann H., '*Decreasing the Commutation Failure Frequency in HVDC Transmission Systems*', IEEE Trans., Power Delivery, Vol.15, No. 3, 2000, pp.1022–1026.

SMOOTHING REACTOR AND DC LINE

6

6.1 INTRODUCTION

The analysis, control and protection of converters was presented in Chapters 3 to 5. In this chapter, some aspects of the equipment on the DC side is taken up. In particular, the smoothing reactor and the DC line will be discussed in some detail.

The smoothing reactor is connected before the DC filter and in series with the converters. Apart from smoothing the direct current, it also serves as a buffer between the converters and the DC line. The sizing of the reactor depends on various requirements including the reduction of ripple in the current.

The DC line, in the case of a bipolar system, consists of two conductors, one operating at positive polarity (with respect to the ground) and other at negative polarity. Extra high voltage DC lines have bundled conductors to minimize the effects of corona. The design of a DC line is carried out with similar considerations as an AC line. However, the performance of insulators is different with direct voltage and the problems of pollution are severe compared to AC operation. But the converter control allows reduced voltage operation under heavy polluted conditions.

The protection against faults in a DC line is invariably through the converter control. This is mainly due to the non-availability of DC breakers in the early stages of introduction of HVDC systems. However, the reliability of converter control has led to its continued use even when DC breakers have been developed. The DC breakers may be required under special operating requirements such as transfer from ground to metallic return and multiterminal system operation.

The corona effects, overvoltages, insulation and protection against faults are some of the aspects that will be discussed in this chapter. Some aspects of the metallic return and the operation of AC and DC lines in the same right of way will also be considered.

6.2 SMOOTHING REACTORS

The smoothing reactors have several functions as given below:

1. They reduce the incidence of commutation failure in inverters caused by dips in the AC voltage at the converter bus.
2. They prevent consequent commutation failures in inverters by reducing the rate of rise

of direct current in the bridge when the direct voltage of another series connected bridge collapses.

3. They smooth the ripple in the direct current in order to prevent the current becoming discontinuous at light loads.
4. They decrease harmonic voltages and currents in the DC line.
5. They limit the crest current in the rectifier due to a short circuit on the DC line.
6. They limit the current in the valves during the converter bypass pair operation, due to the discharge of shunt capacitances of the DC line.

It is to be noted that in back to back HVDC systems, the last three functions are not relevant. For systems with DC transmission lines, the inductors of value from 0.27H to 1.5H have been used. For back to back HVDC systems, the value ranges from 12 mH to 200 mH.

The sizing of the reactor is done not only from the considerations mentioned above, but also from the point of view of minimizing the effect of low order harmonic resonances in the AC/DC system. It is necessary to avoid series resonance of the DC system at fundamental frequency and also at the second harmonic. The effect of the inductor value on the resonant frequency (corresponding to the first peak of the admittance as seen by the converter) as a function of the DC filter capacitance is shown in Fig. 6.1. Here, $L_3 > L_2 > L_1$ shows that the resonant frequency is reduced by increasing the inductance.

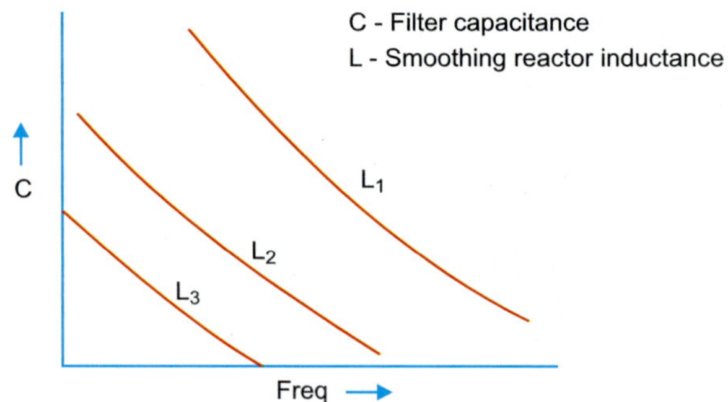

Fig. 6.1: Series resonant frequency as a function of smoothing inductor and filter capacitance

The inductance value must remain practically constant with variations in the direct current. This requires air-core construction. The smoothing reactor helps to limit the fault current in the DC line as mentioned above. This is feasible only if the reactor does not get saturated by the fault current. By the way, it is to be noted that saturable reactors used for the limitation of di/dt in thyristor valves, cannot replace the smoothing reactor as smoothing reactors unaffected by saturation reduce commutation failures.

The location of the smoothing reactor can be either at the high voltage terminal or at the ground terminal as shown in Figs. 6.2 (*a*) and (*b*) respectively. In the latter case, it is also necessary to have a small reactor of the order of 5 to 10 mH on the line side, to protect the converter station from the consequences of lightning strokes to the line. The advantage of having the reactor at the ground side is that it allows the converter ground faults to be cleared by converter control (of increasing the delay angle to its maximum limit to control the current). The insulation level of the reactor is reduced for this location of the reactor.

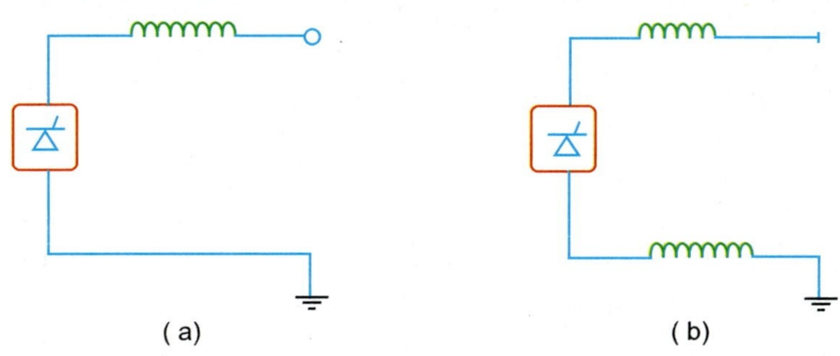

Fig. 6.2: Location of smoothing reactor

There is little information available on the choice of the optimum size of the DC smoothing reactor. One criterion used is the S_i factor [1,2] defined below

$$S_i = V_{dn} /(LI_{dn}) \tag{6.1}$$

where V_{dn} and I_{dn} are the rated direct voltage (in kV) and direct current (in kA) respectively. L is the DC circuit inductance in mH which includes the transformer leakage inductance. The S_i factor for back to back HVDC links varies from 0.24 to 1.3 ms^{-1}. Higher the factor, higher is the rate of rise of the fault current.

Alternate criterion for the sizing of the reactor is the ripple in the direct current. For a converter feeding a constant voltage source through a constant inductor (see Fig. 6.3), the peak value of the ripple is given by

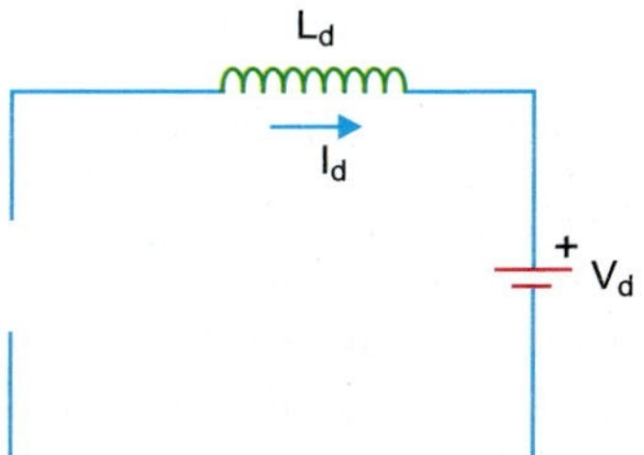

Fig. 6.3: Equivalent circuit for calculating ripple

$$i_d^{peak} = (sV_{do} / \omega_1 L_d)[1-(\pi/p)\cot(\pi/p)]\sin\alpha \tag{6.2}$$

where V_{do} is the no load DC voltage of a converter, s is the number of converters connected in series, p is the pulse number, L_d is the inductance of the reactor, ω_1 is the fundamental frequency. In deriving Eq. (6.2), the leakage reactances of the converter transformers are neglected. A major assumption used in deriving Eq. (6.2) is that the delay angle α exceeds $\bar{\alpha}$ where

$$\tan\bar{\alpha} = (p/\pi) - \cot(\pi/p) \tag{6.3}$$

for $\quad p = 6,\ \bar{\alpha} = 10°$ and for $p = 12, \bar{\alpha} \simeq 5°$

In the former case

$$i_d^{peak} = 0.0931(sV_{do})\sin\alpha/(\omega_1 L_d) \tag{6.4}$$

while for $p = 12$

$$i_d^{peak} = 0.0230(sV_{do})\sin\alpha/(\omega_1 L_d) \tag{6.5}$$

For the case where $\alpha = 0$

$$i_d^{peak} = k(sV_{do})/(\omega_1 L_d) \tag{6.6}$$

where the constant $k = 0.00947$ for $p = 6$ and $k = 0.00119$ for $p = 12$.

It is necessary to avoid discontinuous conduction of current in the converter valves during the system operation. This is because of the possibility of overvoltage across the valves that are transmitted from the line side.

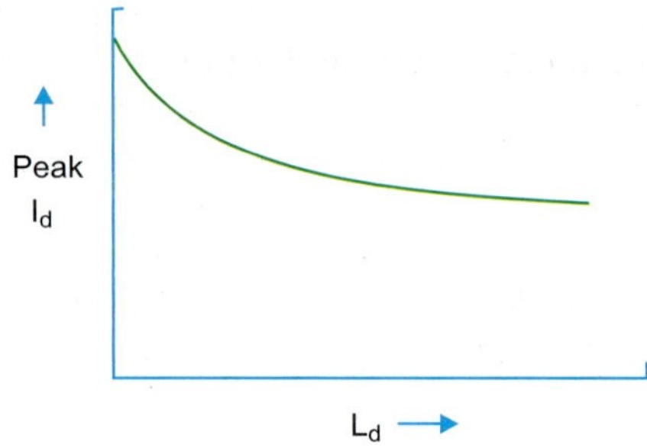

Fig. 6.4: Peak current as a function of smoothing reactor

Another factor that could affect the sizing of the smoothing reactor is the peak direct current during the bypass pair operation caused by commutation failures. Figure 6.4 shows the effect of the smoothing reactor inductance. This shows that the reduction in the surge current is not commensurate with the increase in L_d, beyond a certain point.

6.3 DC LINE

6.3.1 Corona Effects

The corona is defined as a 'luminous discharge due to ionization of air surrounding a conductor caused by a voltage gradient exceeding a certain value'. The ionization takes place in a zone which is a very thin circumferential layer (not more than 2 cm) surrounding the conductor surface. Within this zone, the high field strength causes high velocity particles to collide with the air molecules. Electrons are removed from the atoms of the air molecules and are accelerated towards the positive conductor or away from the negative conductor. These high velocity electrons collide with other air molecules releasing additional electrons in an avalanche process. The ions carrying the same charge as the adjacent conductor are repelled from the ionization zone at initial velocities of about 1.4 cm/sec for positive ions and 1.6 cm/sec for negative ions for every V/cm of the field strength.

The ions moving into the inter-electrode region (between pole to pole or between pole to the ground) recombine with oppositely charged ions or neutral molecules. To maintain the net charge in this region, a corona current flows from the conductor by the movement of electrons in the ionization zone and by ions beyond this zone.

The ion velocities at the ground level are in the range of 3 m/sec for typical voltage gradients. The ion movement in perfectly still air conditions is restricted to the electric field direction. However, with wind movement, the ions are randomly dispersed downwind from the DC line. The losses are also increased.

The effects of the corona are:

1. Corona loss
2. Radio and television interference
3. Audible noise
4. Space charge field

While the first three effects occur on AC lines also, the last one is peculiar to DC lines. It involves the effect of movement of ions which results in the increase in the voltage gradient at the ground level. The 'electrostatic field' is defined as the field resulting from the charges on or near the conductor surface. The total 'electric field' results from the superposition of the electrostatic and space charge fields.

Corona loss

The power losses due to corona can be expressed as [3, 4]

$$P_{\text{loss}} = [2V(k+1)k_c n r 2^{0.25(g-g_0)}] \times 10^{-3} \text{ kW/circuit} - \text{km} \tag{6.7}$$

where

V = pole to ground voltage in kV

n = number of subconductors

r = radius of each subconductor in cm

g = maximum conductor surface gradient at operating voltage (in kV/cm)

$g_o = 22\delta$ kV/cm where δ = relative air density

k_c = conductor surface coefficient which varies from less than 0.15 for smooth, clean conductor to more than 0.35 for conductor with imperfections

$k = (2/\pi)\tan^{-1}(2H/S)$

H = mean height of conductors

S = pole spacing

The relative air density δ is given by

$$\delta = 2.94p/(273+T) \tag{6.8}$$

where

p = barometric pressure in kilo-Pascals

T = temperature in centigrade

Morris and Maruvada [5] have applied Mangolt equation to the calculation of conductor surface gradients for bipolar lines. The maximum gradient g, is given by

$$g = \frac{[1+(n-1)(r/R)]V}{nr \cdot \ln\left[\dfrac{2H}{(nrR^{n-1})^{1/2}[(2H/S)^2+1]^{1/2}}\right]} \tag{6.9}$$

where R = radius of the circle passing through the centres of all subconductors in a bundle, in cm. The other quantities are as defined earlier. ln refers to the natural logarithm.

The power loss predicted by Eq. (6.7) gives correctly the mean fair-weather corona loss levels. With rain, the DC losses may increase by a factor of 10 to 1 while AC losses may increase by 50 to 1. However, the average ratio of rain weather loss to fair weather loss varies from 2 to 4, with higher ratios applicable to operation at lower gradients.

The bipolar corona loss per pole is higher than the monopolar corona loss by a factor of 1.5 to 2.5. However, for a given voltage, the positive and negative polarity losses are approximately equal. DC corona losses usually increase with wind velocity in the range of 0 to 10 m/sec.

Radio interference (RI)

The most predominant corona effect that may determine the conductor design is the ratio interference. This is measured at a frequency of 1 MHz and for a receiver bandwidth of 9 kHz, at a horizontal distance of 30 meters from the outermost conductor.

The RI is mainly due to the positive conductor. This is because of the fact that the corona discharges from the negative conductor are in the form of Trichel pulses which are uniformly distributed over the conductor surface. Positive corona discharges are of three types—Hermstein glow, plume discharge and streamers. Plumes and streamers are randomly distributed and the more persistent discharges are usually associated with high stress points due to surface imperfections. These are mainly responsible for the RI.

The expression for RI (in decibels above the field strength of 1 μV per meter) is empirically obtained as

$$RI = 25 + 10\log n + 20\log r + 1.5(g - g_0) \tag{6.10}$$

This is due to the positive conductor. The RI due to the negative conductor is about 20 dB lower and is not of consequence.

Interestingly, DC-RI levels are decreased by rain and wet snow which completely wet the conductor. This phenomenon is opposite to that in AC conductors.

DC-RI levels are increased by wind, with maximum increase during the wind flow from negative to positive conductor.

The bipolar lateral RI profile is symmetric about the positive pole and attenuates inversely as the square of the distance from the conductor initially (up to 50 meters) and inversely as the distance thereafter.

The television interference (TVI) with DC lines is mainly due to the ion currents and is of little consequence at distances greater than 25 meters from the right of way.

Audible noise (AN)

The corona discharges from the conductor produce compressions and rarefactions that are propagated through the medium as acoustical energy. The portion of the acoustical energy spectrum that lies within the sonic range is perceived as audible noise (AN).

The sound level is expressed in decibels and is defined as

$$\text{dB} = 20\log(P/P_R) \tag{6.11}$$

where P is the measured sound-pressure level and P_R is the reference pressure level. The standard level for P_R is 20μ Pascal which is the average threshold of audio perception at 1 kHz. Test line studies [3] indicate that ± 600 kV DC lines would produce an audible noise of 45 to 55 dB measured at 30 meters from the ROW centre line. This is not considered to be serious. In general, the annoyance produced by the audible noise varies linearly with the conductor surface voltage gradient.

The test results also show that positive polarity conductor is the primary source of AN. Rain causes a very slight reduction of AN.

Audible noise is produced in converter stations by converter transformers and smoothing reactors due to the phenomenon of magnetostriction. Although higher AN is to be expected in converter transformers (compared to transformers in AC substations) due to the presence of harmonics, this is counterbalanced by the operation at lower flux densities in the core. Audible noise has not been a serious problem in converter stations.

Space charge (Ion flow) field

As mentioned earlier, the ions produced by corona on overhead DC lines drift through space under the action of electric field and wind. The possible environmental effects due to electric fields and space charges, particularly at the ground level are matters of concern and have to be considered in transmission planning.

In AC lines, the problem of ion flows is not present as all the ions created during the one-half cycle are recaptured during the second half-cycle by the polarity reversal on the conductor. The net effect is the increase in the diameter of the ionization zone as the surface gradient increases above the critical level. Very few ions escape the ionization zone.

Although ion flow fields are absent around AC lines, the time varying field can induce currents in humans or objects due to capacitive coupling. Experimental studies have shown that the magnitudes of these induced currents can be about 100 μA for a person directly under a typical EHV or UHV line. Such capacitive coupling effects are not present with DC fields.

The current induced under DC fields is due to the flow of ions. The current density (in Amps/meter^2) can be obtained from the relationship

$$j = k\rho E \tag{6.12}$$

where

k = ion mobility (1.8×10^{-4} m/s/v/m for negative ions)

ρ = charge density in coulombs/m^3

E = electric field strength in volts per meter.

The charge density at ground level can be calculated from

$$\rho = \varepsilon \nabla E \tag{6.13}$$

where ε is the electric permittivity.

The total current intercepted by a person at ground level is the current density multiplied by the equivalent area which is related to the height. For a person of height 1.73 meters, the maximum current intercepted is around 3 μ A under a line operating at ±600 kV. Incidentally, the threshold of perception level for direct currents is 5.2 mA for men and 3.5 mA for women. These levels are higher than the threshold levels for AC currents.

The total electric field at the ground level is superposition of the space charge field and the electrostatic field. The lateral ground level profiles of the electrostatic and electric fields are shown in Fig. 6.5.

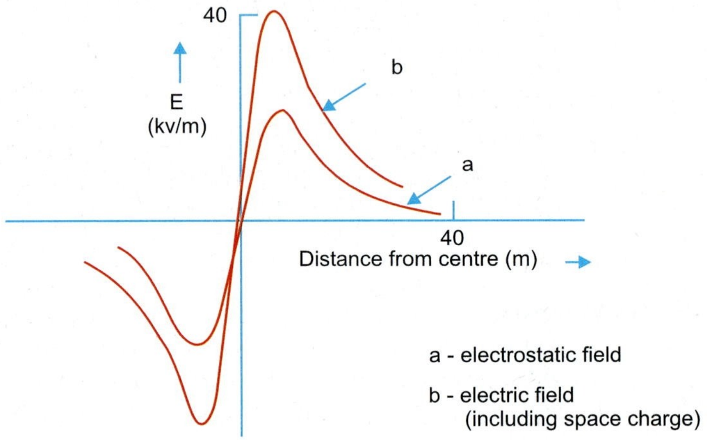

Fig. 6.5: Lateral ground level profile of electrostatic and electric fields

The computation of ion flow fields is complex as non-linear partial differential equations are involved. The computations are simplified by Deutsch's major assumption that the space charge affects only the magnitude and not the direction of the electric field. This reduces the two-dimensional ion field problem to one-dimensional problem [6]. Recent research efforts [7,8] are directed at improving the accuracy of computations without Deutsch's assumption, thus allowing the considerations of the effects of wind velocity.

6.3.2 DC Line Insulators

The insulator strings of a DC line are made up of porcelain or toughened glass as in the case of AC lines. The insulator performance is evaluated in terms of critical flashover voltage (CFO) and critical withstand voltage (CWS). The critical flashover voltage is defined as the statistical mean (with 50% probability) of a data group (usually 20 flashovers) on a specific insulator specimen. The analysis assumes Gaussian distribution.

The critical withstand voltage (CWS) is defined as

$$CWS = CFO - 3\sigma \qquad (6.14)$$

where σ is the standard deviation. The CWS represents a voltage level with 0.13% probability of flashover.

It is to be noted that the overvoltage should result only in flashover and not puncture of the insulators.

There are special problems with direct voltage stresses on an insulator. The first problem is that of ion migration that takes place in the insulating material, which is a function of the temperature. An insulating material should have very little ion migration even at high ambient temperatures. There can be thermal runaway due to the ionic conduction in a very narrow zone of the dielectric which results in ageing of the dielectric. The current that flows in the insulator body is dependent on the resistivity which is a function of the temperature.

There is also the phenomenon of electrolytic process due to the creepage currents along the surface of the insulator .These currents cause the electrodes to increase in volume, partly through the deposition of material and partly as a result of chemical processes. The insulator design must take this into account to avoid mechanical failures.

The insulator string is subjected to the direct voltage in addition to transient impulses of different waveforms. The creepage distance of an insulator is determined by the operating direct voltage, while the string length is determined by the impulse voltage level. Increasing the string length influences the height of the towers and their costs.

As transient overvoltages are smaller with DC lines (as compared to AC lines), there is an economic incentive to make the ratio of creepage distance to string lengths as great as possible.

A major problem with DC insulators is that the accumulation of contamination through direct voltage stresses is higher than with AC lines. The collection of dirt particles is also less uniform than with AC. Providing collector rings close to the ends of the string helps in making the field distribution more uniform and reducing the accumulation of contamination.

The insulators are cleaned by heavy rain while the fog or dew, which makes the dirt conductive without removing it, can result in flashover. At times of unfavourable weather conditions, it may be desirable to operate the DC line at lower voltages by converter control or by removing some of the series connected bridges.

The geometry between the conductor and the tower can be viewed as that of a rod- plane gap. The withstand strength of such gaps is higher with the combination of impulse and direct voltage than with pure impulse alone. This behaviour is shown in Fig. 6.6 where the CWS is plotted as a function of the gap length.

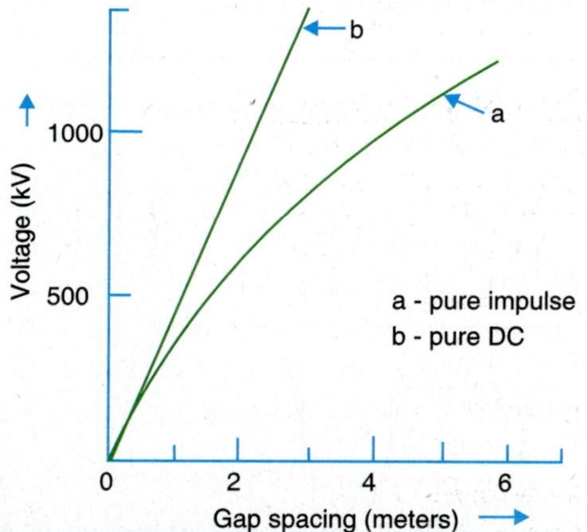

Fig. 6.6: Withstand voltage versus gap length

The flashovers are caused by lightning strokes which usually affect one pole of the bipolar line. Due to coupling between the two poles, overvoltages can occur in the healthy pole. The insulation design must take this into account and prevent consequential fault on the healthy pole.

6.4 TRANSIENT OVERVOLTAGES IN DC LINE

Pole to ground faults in bipolar DC lines can result in transient overvoltages in the healthy pole of magnitudes exceeding 2.0 p.u. [9, 10] (although typically they are 1.7 p.u.). The maximum overvoltage occurs at midpoint in a pole when the fault is also at the midpoint of the other pole. For off-centre faults, the maximum overvoltage occurs at a location which is a mirror image of the fault location (with respect to the midpoint). For example, if a fault occurs at a point of one third distance from the sending end, the overvoltage occurs at a point of one third distance from the receiving end.

There are two modes of travelling waves on a bipolar DC line. One is termed as the pole to pole mode (analogous to positive sequence) while the other is termed as the pole to ground mode (analogous to zero sequence). If the prefault voltage is V at a given location, it can be shown that immediately after the fault, the voltage at the other pole rises by an amount ΔV, given by

$$\Delta V = \left[(Z_0 - Z_1)/(Z_0 + Z_1)\right]V \tag{6.15}$$

Where Z_1 and Z_0 are the surge impedances of the first and the second mode respectively. Generally $Z_0 > Z_1$. For typical values of Z_0 and Z_1, $\Delta V \simeq 0.3V$.

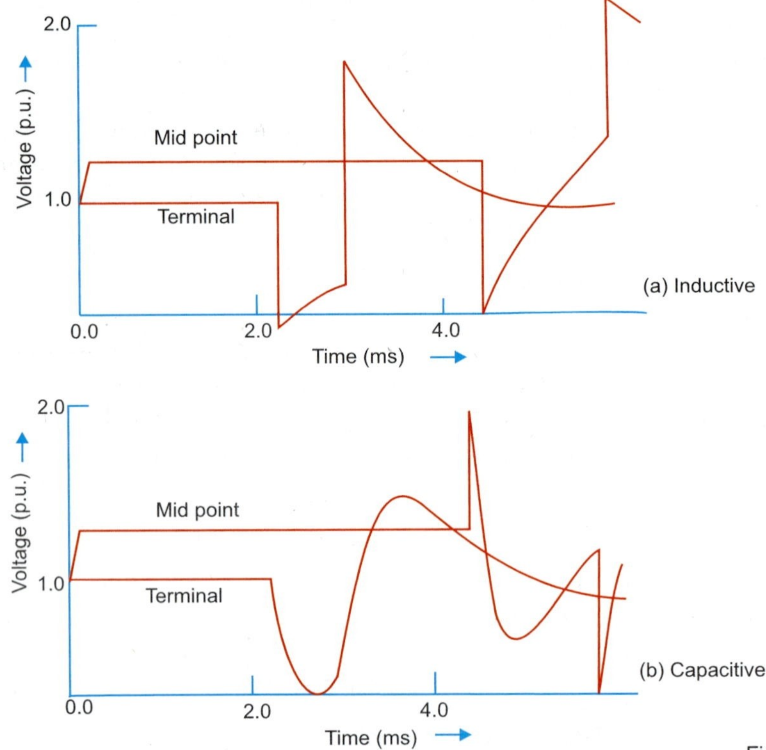

Fig. 6.7 (Contd...)

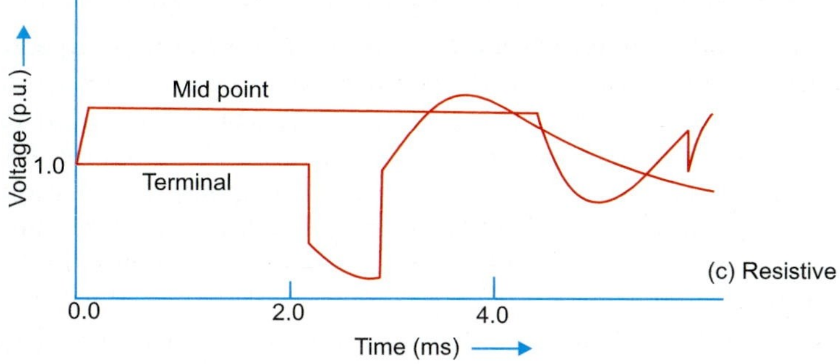

Fig. 6.7: Transient voltage waveforms (at midpoint and terminal for (a) inductive, (b) capacitive and (c) resistive termination) (Source: Reference 10)

The travelling waves originating at the fault location travel in both directions and are reflected by the terminals. The kind of termination at the converter station-inductive, capacitive or resistive has a bearing on the voltage waveform at the converter and in the line.

The typical voltage waveforms at the midpoint and the terminal are shown in Fig. 6.7 for (*i*) inductive, (*ii*) capacitive and (*iii*) resistive terminations. The parameter of the DC filter also affects the nature of the terminal voltage waveform. The following conclusions can be drawn from a digital computer study [10] of the transient overvoltages.

1. With capacitive termination to the wavefront, overvoltage in the unfaulted pole is caused by pole to pole mode. The attenuation and distortion of this mode is slight.
2. The surge capacitor, while helping in reducing the overvoltage at the terminal, does this at the expense of the DC line. It is desirable to use a surge arrester to protect the terminal against overvoltages.
3. The resistive termination to the wavefront with $R = Z_1$, is the best and can be achieved by putting $R = Z_1$ in the high pass DC filter and not having a surge capacitor.
4. With inductive termination to the wavefront, overvoltages are caused by the pole to ground mode for which the attenuation is substantial. But the overvoltage at the terminal is high.
5. It is desirable to provide extra insulation around the middle portion of the DC line to take care of the highest overvoltages that may occur in that zone.
6. With capacitive termination, the lowest limit on the dv/dt setting of the line fault protection is determined by the maximum fault resistance above which the fault will not be detected.

It is also necessary to provide a time delay in the operation of the protection (using dv/dt) to prevent the tripping of the healthy pole. Typically, a time delay of about 3 milliseconds may be adequate. The tests carried out on Pacific intertie [11] and subsequent simulation using digital computer have shown that the modeling of frequency dependent characteristics of the line for both modes is essential for accurate prediction of the overvoltages [12].

6.5 PROTECTION OF DC LINE

The fault current immediately following a fault is limited by the surge impedance and is of the order of 2.0 p.u. The current controller at the rectifier acts to limit this current to the prefault current. If the inverter current control is also active and power reversal in the inverter is permitted, the inverter current is maintained at a value which differs from the rectifier current by the current margin. This implies that the fault current in steady state is equal to the current margin (of the order of 0.1 p.u.). Although this is much less than the currents in AC line faults, the clearing of the fault requires the current to be brought to zero and sufficient time given for the deionization of the arc path.

For fast clearing of DC line faults, it is necessary that the inverter is not allowed to operate as a rectifier and the rectifier is put into the inverter mode by sudden increase in the delay angle to its maximum limit (about 135°). The converters at both terminals then help in discharging the energy stored in the DC circuit (in capacitances and inductances) and delivering it to the AC system. The current and voltage in the DC line fall to zero and help in deionizing the arc path. After some time has elapsed (say 0.2 to 0.5 sec), the line may be automatically energized by restarting the converters (in the usual manner by ramping up the direct voltage and current). If the restart is unsuccessful, due to a persistent fault, the protective action will deenergize the line again. Normally three attempts are made to restart automatically with increasing dead time. The failure to restart even after 3 attempts implies a permanent fault and requires shut down of the link until the fault is located and cleared. Alternately, reduced voltage restart may be tried out if the insulation failure is due to heavy pollution on the line insulators and in bad weather.

The automatic deenergization and restarting of the DC link is similar to the clearing of the fault and automatic reclosure in AC lines. The major difference in the two cases being that while breakers are used in AC lines, the clearing and restarting is performed in DC systems using only converter control with the help of protective relays.

6.5.1 Detection of Line Faults

As mentioned earlier, the normal converter control is not adequate for clearing the fault. The deenergization of the line by driving the rectifier into the inverter mode requires special control action which is initiated on detecting the line fault.

The detection of the line fault is based on one of the following conditions: (i) a sudden drop in the DC voltage measured on the line side of the reactor, or (ii) sustained low direct voltage. Mathematically the two conditions can be expressed as

(*i*) $V_d < k_1$ and $dV_d/dt < -A$ for a duration of $\tau > \tau_1$

(*ii*) $V_d < k_2$ for a duration of $\tau > \tau_2$

where k_1, k_2 and A are positive parameters that are chosen along with τ_1 and τ_2.

The criteria for selecting these parameters are as follows:

1. Selectivity —to ensure that the protection operates rapidly for DC line faults and not for AC faults or converter faults. The smoothing reactor prevents the rate of voltage drop in the line exceeding a limit for AC and converter faults. By choosing A higher than this limit, it is ensured that the voltage derivative unit does not operate for faults

other than DC line faults. Selectivity is desirable as the dead time required for deionization of the arc path in the case of DC line faults, is unnecessary in the case of AC or converter faults. To ensure that the DC fault protection does not operate for AC faults, it is also necessary to set k_2 below a certain limit determined by the most common AC faults (single line to ground faults). Furthermore, by setting τ_2 greater than the clearing time of AC breakers, the operation with multiphase faults is also avoided. Alternatively, low AC voltages can be used to block the operation of the voltage level unit (which operates subject to the second condition).

2. Sensitivity—the DC line protection must operate satisfactorily for normal values of the arc resistance. This puts an upper limit on A.
3. Reliability—the protection must operate reliably for all DC line faults including faults through a high resistance. This is the reason for providing the voltage level unit.

k_1 and τ_1 are selected such that the protection does not operate for the healthy pole (and only for the faulted pole). This is essential as, in bipolar DC lines, the power can be transmitted at sufficient levels even after a monopolar ground fault.

6.5.2 Protection Against DC Line Faults with VSC

Unlike in CSC based HVDC links, the DC voltage in VSC based links cannot be reversed and the DC capacitors at both rectifier and inverter stations discharge into the faults. In addition, the VSC can feed current to the fault. The blocking of the converter does not help as the free-wheeling diodes will continue to feed the fault. This implies that on detecting DC line faults, the converters need to be isolated from the AC system by opening the AC breakers. As an alternative, DC breaker can be employed to isolate the converters from the DC line, when a line fault is detected. This can be faster than the operation of the AC breakers.

The VSC based HVDC projects commissioned, so far, invariably use DC cables instead of overhead lines. Although the cable faults tend to be permanent, the probability of cable faults is much less than in a overhead line in which temporary insulation failure can occur due to lightning strikes or due to tree branches, pollution or mechanical failures.

6.6 DC BREAKERS

The development of HVDC circuit breakers has been under way for some time in late eighties, and a 500 kV breaker with current interrupting capability up to 4000A, has been reported [13]. Although several types of DC breakers have been developed by different manufacturers, the basic concepts are the same [14] in all the cases. These are reviewed below.

6.6.1 Basic Concepts of DC Circuit Interruption

The major problem in the current interruption in DC circuits is that there is no natural current zero as in the case of AC circuits. The current can be brought to zero only by applying a counter voltage higher than the system voltage. The second problem is the dissipation of large energy stored in the inductance of the circuit.

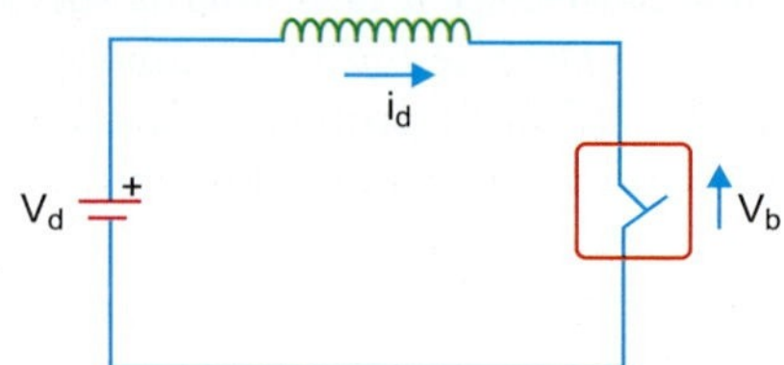

Fig. 6.8: Simple representation of DC circuit with breaker

Consider the simple representation of DC circuit shown in Fig.6.8. The breaker has a counter voltage V_b. It can be shown that the energy absorbed by the breaker is

$$W_b = \frac{1}{2} L I_d^2 \left(V_b / (V_b - V_d) \right) \tag{6.16}$$

and the time required to bring the current to zero is given by

$$T_i = L I_d / (V_b - V_d) \tag{6.17}$$

where I_d is the current in the breaker prior to interruption. It is obvious that reduction of I_d has a significant effect on the breaker costs as W_b is decreased which results in the reduced requirements of the energy capability of a breaker.

The current I_d is brought down to the rated value by the normal action of current control in converters. Although this requires certain amount of time (say 30 to 50 msec), the time required for interruption reduces. The alternate strategy is to interrupt the fault current fast but this will increase the interruption time T_i apart from increasing the cost of the breaker.

The counter voltage produced by the arc that is struck when the breaker contacts separate, is not sufficient in HVDC breakers. This requires an auxiliary circuit in which a capacitor is inserted to develop the required counter voltage. However, a capacitor is unable to dissipate the energy and the current in the capacitor has to be commutated to nonlinear resistors which then dissipate the energy without undue increase in the voltage across them.

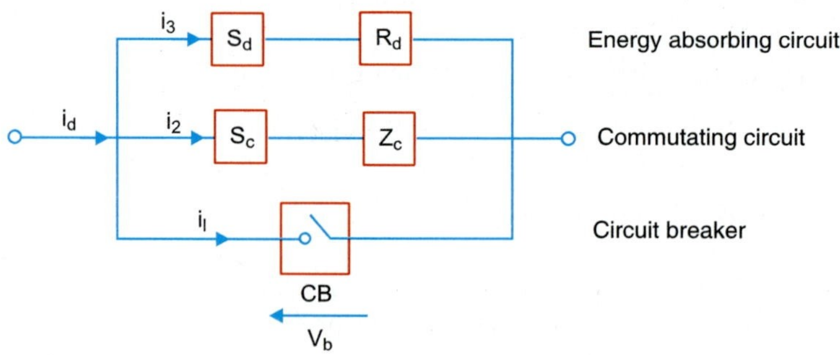

Fig. 6.9: Arrangement of a DC breaker

The general arrangement of a HVDC circuit breaker is shown in Fig. 6.9. The current in the breaker (when closed) is normally carried through CB with moving metallic contacts. This may be vacuum, oil, airblast or SF_6 device. After a trip signal is given to the breaker, the breaker contacts open to draw an arc. This is initiated at time t_1(see Fig. 6.10 which shows the

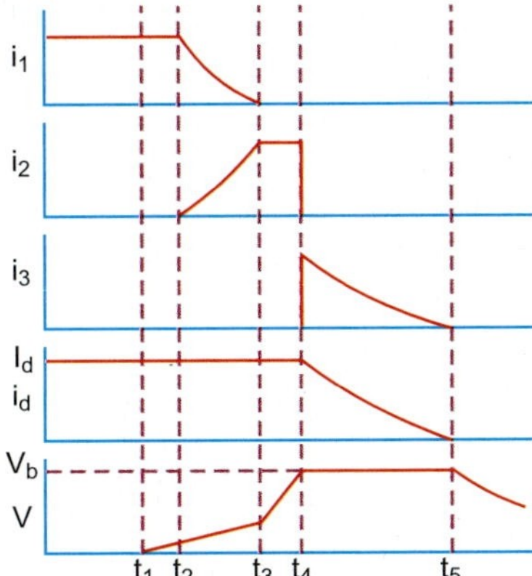

Fig. 6.10: Current and voltage waveforms

current and voltage waveforms). At a short time later at t_2, the commutation circuit is inserted through the insertion device S_c. The commutating impedance is primarily made up of a series L-C circuit which is tuned to a certain frequency. The capacitor may or may not be precharged. The insertion devices S_c may be a triggered vacuum gap or spark gap or in the so called 'passive' commutation circuit, just a solid connection (no switching required). The main purpose of inserting a commutating impedance is to create current zero in *CB* and transfer the current to Z_c with sufficient contact separation in *CB* to regain its dielectric strength. This current transfer is completed by time t_3. The DC circuit current i_d, flowing through the capacitor in Z_c rapidly builds up to a high voltage V_b across the breaker. When the voltage reaches V_b at time t_4, the energy absorber R_d is inserted through the device S_d. The nonlinearity of the resistance R_d acts as a switch which closes when its clipping voltage is reached. The direct current now decays to zero by discharging its energy to R_d after the interval of T_i (at time t_5). The breaker operation is completed by t_5 if the three parallel paths *CB*, Z_c and R_d have adequate voltage withstand capability. Otherwise, fast acting isolators that operate at zero current level may be used to prevent overstressing of the breaker until the converter control restores the voltage to a level that can be withstood by the breaker under steady state conditions.

In the passive commutation method, a series L-C circuit is connected across *CB* [15, 16]. The interrupter element *CB* is an air blast or SF_6 breaker similar to that used in AC circuits. Field tests have shown that air blast breaker is superior in performance compared to SF_6 [17, 14].

The commutation of the current to Z_c in the passive commutation method is initiated by the arc itself. The differential equation for the current i_2 (see Fig. 6.11) is given by

$$L\left(d^2i_2/dt^2\right)+(dV_{arc}/di_1)(di_2/dt)+(1/C)i_2=0 \tag{6.18}$$

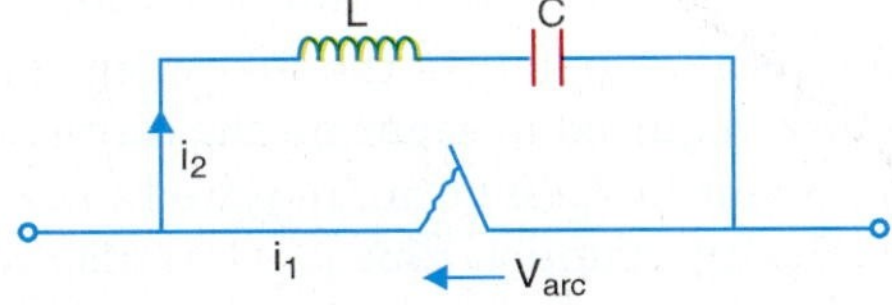

Fig. 6.11: Commutation circuit

Since the arc voltage decreases with the current, its differential resistance dV_{arc}/di_1 is negative. This causes the build up of increasing current oscillations in the circuit and the arc current goes to zero when the amplitude of the oscillations reach the value of direct current i_d (see Fig. 6.12). While this is a simple explanation of the phenomenon, the process of commutation is affected by the geometry of the breaker nozzle and the interruption chamber, the natural frequency and the stray resistance of the circuit. The increase in the oscillation magnitude may also occur in steps.

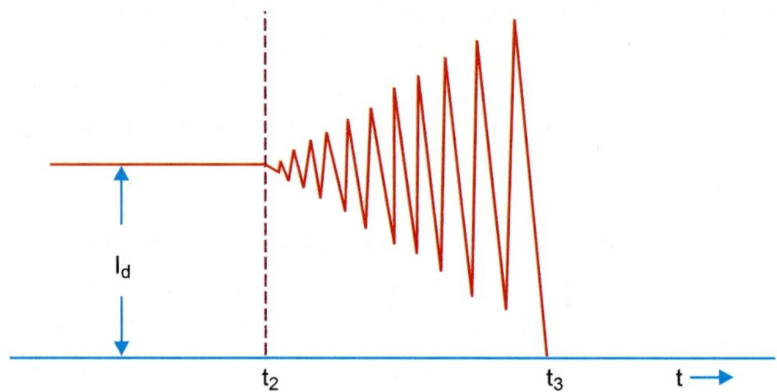

Fig. 6.12: Current oscillations in the commutating circuit.

6.6.2 Characteristics and Types of DC Breakers

The breaker is characterized by four variables of interest in its application to the system. These are:

1. Voltage capability
2. Current capability
3. Energy capability
4. Switching time

The voltage capability is related to two parameters—(*i*) the voltage during the interruption process and (*ii*) steady-state operating voltage and transients in the system. The breaker voltage requirements can be kept low by coordination with converter control and fast isolators. But this is not desirable. In general, the highest permissible voltage gives the best performance in terms of lower energy absorption and shorter switching time [refer to Eqs. (6.16) and (6.17)].

The interrupting capability of a breaker does not have to be much above the rated current in the circuit, because the converter control can be relied on to bring the fault current down to the rated levels.

The required energy capability of a breaker depends upon many factors such as the inductance, converter voltage and current, breaker voltage and the duty cycle. In HVDC breakers, series connected modules are used to give the required voltage and energy capabilities.

The switching time of a breaker includes the following components: (*i*) time required to generate the trip signal (*ii*) time required to separate the contacts in the main breaker (*iii*) time required to commutate the current to Z_c (*iv*) time required to commutate current from Z_c to R_d (energy absorbing circuit) (*v*) the interruption time (T_i) (*vi*) time required to bring the DC system back to steady-state post fault condition.

While the first component can be kept small (say 2 ms for station faults), there is no real advantage in generating the trip signal much faster than the time taken to reduce the fault current using converter control (around 30–50 ms).

The second component is around 10 ms. The third and fourth components can be reduced by using charged capacitor in the commutation circuit. Again, there is no real advantage in keeping the total time below 10 ms as there is the problem of restriking of the arc in the main breaker.

The interruption time is related to V_b and I_d. The voltage is normally around 1.7 p.u. The reduction in I_d reduces T_i in a linear fashion. The last component depends on the voltage capability of the breaker. With low voltage capability, this component of the switching time could be of the order of 100 ms due to the time required for the operation of the isolator, the converter control and the communication delay (required for coordination of the breaker operation and converter controls).

The above analysis shows that there is significant saving in switching time possible by increased voltage capability while the current capability may be limited to the rated current. The requirements on the switching time depend on the AC system requirements for maintaining DC power flow in transient conditions, taking into account the considerations of system stability.

6.6.3 Applications of DC Breakers

The application of DC breakers is required mainly for fault clearing in MTDC systems. However, even for two terminal DC systems, the DC breakers can be useful in the following situations:

1. When the converters feed two parallel DC lines
2. When parallel connected converters feed the same line
3. When current needs to be transferred from the ground return to the metallic return during the monopolar operation. The breakers for this application are termed as Metallic Return Transfer Breakers (MRTB). Although voltage capability required for this application is moderate, the energy capability required may be substantial.

6.7 MONOPOLAR OPERATION

If there is a permanent fault in one of the poles (either in the DC line or converters), the other pole can still continue to function until the fault is rectified. This is in contrast to AC systems where unbalanced operation is not feasible for more than a second.

The monopolar operation of a bipolar DC system involves current return through (i) ground, or (ii) metallic return (using the conductor in the other pole). It is to be noted that even if the insulation is damaged in a pole conductor, it can still be used as a return path as steady-state voltage stress is small. With metallic return, only one terminal is grounded and the other terminal is ungrounded.

There are advantages and drawbacks associated with each type of return. With ground return, the losses are less (losses are associated only with the ground electrodes). However, there is a danger of corrosion of buried pipelines. The ground current may also interfere with DC type rail road signals and may cause saturation in transformers that are located in the vicinity. Although these problems can be overcome, the solution may be complicated and restrictions are usually placed on extended operation with ground return.

While the metallic return does not have the problems associated with the ground return, it has the following drawbacks:

1. Increased losses due to voltage drop in the metallic return
2. The increased voltage drop requires larger tap range in converter transformers. Alternative to this is the operation of converters at large angles which results in higher valve stresses and additional losses in valve grading/damping circuits. Both options are costly and third alternative is to derate the DC link during the metallic return operation
3. Reduced reliability due to the fact that two conductors are exposed as compared to one conductor with ground return
4. Increased requirements for the protection of the ungrounded neutral.

The last factor arises from the consideration of the transient overvoltage to which the ungrounded neutral is subjected to, when valve groups are blocked/deblocked, or line faults, commutation failure takes place. According to a study carried out on Pacific Intertie system [19], it is preferable to ground the inverter terminal as the worst condition for overvoltage is the blocking of an inverter group if it is floating.

Capacitors in shunt with spark gaps were used initially for protection against transient overvoltages. But the disadvantage is the large value of capacitance required and increased telephone interference due to harmonic currents flowing through the capacitor. The present day solution to the problem is the use of zinc oxide arresters in place of the capacitors [20].

6.7.1 Ground Electrodes

The ground return is economical not only due to the elimination of a return conductor (significant saving with long submarine transmission), but also due to reduced power losses, particularly in long lines. As most of HVDC projects are staged with the first stage being monopolar, the use of ground return is attractive. Even when bipolar operation is introduced in the following stage, the use of ground return improves reliability/availability of the HVDC link. The ground electrodes are used at both ends of the link to transfer current from the converter to the ground. The ground electrodes need to be located at a distance (8 to 80 km) from the converter station to prevent DC current leaking into the ground mat of the converter station. The injection of DC currents to the transformer windings when the neutral is grounded, results in the saturation of the magnetic core in one direction, that can lead to the generation of even harmonics in the magnetizing currents. The problem is more severe in the case of single phase transformers compared to the case with three phase transformers with three limbs as the flux due to the DC current is forced to flow through the high reluctance path from the top yoke to the bottom yoke. The first indication of core saturation is the noticeable increase in audible noise when the DC current per phase is more than 1.2 times the AC exciting current. It is to be noted that the exciting current in modern transformers may be as low as 1% of the rated current.

There are three types of electrodes, namely

1. Land electrodes, located far away from the sea
2. Shore electrodes, located on a shore against saline sea water. Shore electrodes can be located either on the beach at a short distance (less than 50 m) from the waterline or in the water protected by a breakwater
3. Sea electrodes, located in the water at some distance (greater than 100 m) from the coastline.

The ground return can cause electrolytic corrosion of buried or immersed metallic structures. Also, ground currents can interfere with operation of other services such as AC power transmission, railway signaling and ship compasses. There are also safety issues which require the step and touch voltages to be kept within tolerable limits. The distribution of ground currents in the vicinity of an electrode is determined by the electric fields which are affected by the location and configuration of the electrodes and geology of the subsoil. However, at a distance (x) from the electrodes station, the voltage (against remote earth) is proportioned to the resistivity (ρ) and inversely proportional to x. The voltage gradient is inversely proportional to x^2 [30].

The performance requirements of the electrodes depend on the duty envisaged for continuous or intermittent duty. The electrolytic corrosion of the electrodes (anode) itself and of foreign metal objects is cumulative and thus depends on the ampere-hours of use. The other factor that affects the performance is the heating and drying of the soil. An electrode tends to dry the soil in two ways—first by heating and second by electro-osmosis (around an anode) which moves water in the direction of the current. To avoid electro-osmosis, it is necessary to limit the current density to less than 1 A/m^2. The resistivity of the soil decreases with increasing temperature up to the boiling point of water. The hottest points in the soil are at the surface of the electrode. When the temperature at the surface reaches the boiling point, arcing increases abruptly, thus, increasing the losses that lead to more evaporation of the moisture. This condition is called 'thermal instability' and must be avoided. The amount of moisture increases and the variation of the temperature of the soil decreases with the depth below the surface. The electrodes resistance is reduced along with the step voltage. Thus it is advisable to bury the electrode to as great a depth that is economically feasible.

The design of the three categories of electrodes (land, shore and sea) are discussed in detail in [30] and [31]. The form of the electrode can be of several types—(*i*) ring, (*ii*) straight, (*iii*) six pointed star, or (*iv*) branched star.

The line connecting the neutral to the ground electrode site is called the 'electrode line' and is usually made of two sections in parallel to facilitate detection of faults in electrode lines (by differential protection). Although steady-state voltages on electrode lines is small (less than 5 kV), the transient voltages can exceed 300 kV for line faults. The voltage is maximum at the midpoint of the electrode line and is highest when it is on the same tower as the DC line. Providing a separate right of way for the electrode line reduces these overvoltages considerably.

The design of ground electrodes are based on the requirements of (*i*) low resistance, (*ii*) adequate current carrying capacity, (*iii*) low maintenance cost, (*iv*) accessibility, (*v*) avoiding damage to other facilities and (*vi*) safety.

6.8 EFFECTS OF PROXIMITY OF AC AND DC TRANSMISSION LINES

For reasons of economy and constraints on the right of way, it is possible that AC and DC lines may be situated in the same right of way. Also, it is possible to have them on the same tower when one circuit of an AC line is converted to a DC line. Some studies [22-26] have been reported on the effects of proximity of AC and DC lines.

Nakra et al. [22] reported from a simulator study that contact faults between a DC pole and an AC phase result in the superposition of AC voltage and currents on the DC line and DC voltages and currents on the AC line. This can result in transformer saturation in some phases and overvoltages on other phases (up to 2.0 p.u.). Otherwise they did not find appreciable effects of electromagnetic coupling between AC and DC line under steady-state.

In contrast, Larsen et al. [23] report the possibility of steady-state induction of fundamental frequency voltages in DC lines due to proximity, which can cause saturation of converter transformers. To avoid problems associated with the transformer saturation, they recommend that the fundamental frequency current in the DC line be restricted to 0.1% of the rated DC current. To limit the induction of AC voltages in DC lines, they suggest the use of DC filters, control modulation and transposition of AC and DC lines. Chopra et al. [24] have observed the possibility of large zero sequence transients in AC lines caused by transients in a neighbouring ground return DC lines. A study was carried out to explain the tripping of AC lines in a common corridor due to the operation of ground current relays. This event happened at Nelson River HVDC link when the transition of metallic return to ground return took place. Interestingly, the AC line was at a distance of over 180 meters from the DC lines. From an analytic study, the authors observe that the phenomenon is due to magnetic coupling between two circuits and is affected by the ground resistivity. The zero sequence currents induced on AC lines are of low frequencies (1-30 Hz) and occur during transients. No induction is present in steady-state.

The effect of interactions of electric field and ion flow in hybrid AC/DC lines have been reported in [25] and [26]. According to a study by Maruvada and Drogi [25], the presence of alternating field component has negligible effect on the trajectories of ions emitted by DC conductors. For lines on the same tower, ionized fields resulting from DC conductors in corona, can cause direct current induction in AC lines. The maximum current induced can be around 0.4 mA per km length of the line. With AC circuit below the DC circuit, there is a significant screening effect on the electric field, current and charge distribution at the ground level. However, maximum field at the conductor surface is increased substantially resulting in higher RI and AN.

Clairmont et al. [26] show that DC conductors in adjacent towers can shield the AC electric field and vice-versa. They use the concept of sensation level to characterize field interactions and state that below a critical separation level, the effects of proximity of AC and DC lines should be considered. The critical separation for a 400 kV DC line and a 230 kV AC line, is less than 40 ft. while for a 800 kV line, it is 110 ft.

REFERENCES AND BIBLIOGRAPHY

1. Melvold D.J., et al., '*Effects of Operating Configurations on Multi-terminal HVDC Systems on DC Filter Performance*'. IEEE Trans., Vol. PWRD-2, No. 3, July 1987, pp. 924–930.
2. Melvold D.J. and Long W.F., '*Back-to-back HVDC System Performance with Different Smoothing Reactors*', Paper 88WM 113-3, Presented at IEEE PES Winter Meeting 1988.
3. Electric Power Research Institute '*Transmission Line Reference Book for ±600 kV*', EPRI, Palo Alto.
4. Knudsen N. and Iliceto F., '*The Electrical Design of EHV DC Overhead Lines*', IEEE Paper No. CP 369 PWR 1970.
5. Morris R.M. and Maruvada P.S., '*Conductor Surface Gradients on Bipolar HVDC Transmission Lines*', IEEE Trans., Vol. PAS 95, No.6, Nov./Dec. 1976, pp. 1934–1945.

6. Maruvada, P.S. and Janischewskyj W., '*Analysis of Corona Losses on DC Transmission Lines,* Part I: *Unipolar lines; Part II-Bipolar lines*', IEEE Trans., Vol. PAS-88, 1969, pp. 718–731, 1476–1491.
7. Takuma T. and Kawamoto T., '*A Very Stable Calculation Method for Ion Flow Field of HVDC Transmission Lines*', IEEE Trans., Vol. PWRD-2, No.1, January 1987, pp. 189–198.
8. Qin B.L et al., '*Accurate Calculation of Ion Flow Field under HVDC Bipolar Transmission Lines*', IEEE Trans., Vol. PWRD3, No.1, January 1988, pp. 368–376.
9. Kimbark E.W., '*Transient Overvoltages caused by Monopolar Ground Faults on Bipolar DC lines—Theory and Simulation*', IEEE Trans., Vol. PAS-89, No.4, 1970, pp. 584–592.
10. Hingorani N.G., '*Transient Overvoltages on Bipolar HVDC Overhead Line caused by DC Line Faults*', IEEE Trans., Vol. PAS-89, No.4, 1970, pp. 592–610.
11. Melvold D.J., Odam P.C. and Vithayathil J.J., '*Transient Overvoltages on HVDC Bipolar Line during Monopolar Line Faults*', IEEE Trans., Vol. PAS-96, No.2, 1977, pp. 591–601.
12. Melvold D.J. and Vithayathil J.J., '*Transient Overvoltages on Pacific HVDC Intertie during Monopolar Line Tests*', Proc. Symposium on HVDC Power Transmission, Portland, Oregon, 1978.
13. Pauli B., et al., '*Development of a High Current HVDC Circuit Breaker with Fast Fault Clearing Capability*', IEEE Trans., Vol. PWRD-3, No.4, October 1988, pp. 2072–2080.
14. Vithayathil J.J., '*HVDC Breakers*', Proc. National Workshop on HVDC Transmission, January 1987 , I.I.T., Kanpur, India.
15. Bachmann B.,et al., '*Development of a 500 kV Air Blast HVDC Circuit Breaker*', IEEE Trans., Vol. PAS-104, No.9, Sept. 1985, pp. 2460–2466.
16. Lee A., et al., '*The Development of a HVDC* SF_6 *Breaker*', IEEE Trans., Vol. PAS-104, No.10, October, 1985, pp. 2721–2729.
17. Vithayathil J.J., et al., '*HVDC Circuit Breaker Development and Field Tests*', IEEE Trans., Vol. PAS- 104, No.10, October 1985, pp. 2693–2705.
18. Tokuyama S., et al., '*Development of Interrupting Tests on* 250 *kV* 8 *kA HVDC Circuit Breaker*', IEEE Trans. on PAS-104, No.9, Sept. 1985, pp. 2453–2459.
19. Hingorani N.G., '*Monopolar Metallic Return Operation of Long Distance HVDC Transmission System*', IEEE Trans., Vol. PAS- 93, No.2, March/April 1974, pp. 554–563.
20. Lasseter R.H., Demarest D.M. and Ellen F.J., '*Transient Over-voltage on the Neutral Bus of HVDC Transmission Systems*', Paper A 78 607-4 Presented at IEEE/PES Summer Meeting, 1978.
21. Meliopoulos A.P.S. and Dunlap J., '*Investigation of Grounding Related Problems in AC/DC Converter Stations*', IEEE Trans., Vol. PWRD-3, No.2, April 1988, pp. 558–567.
22. Nakra H.L., et al., '*System Considerations in Converting One Circuit of a Double Circuit AC Line to DC*', IEEE Trans., Vol. PAS-103, No.10, Oct. 1984, pp. 3096–3103.
23. Larsen E.V., et al., '*Parallel AC/DC Transmission Lines-Steady-state Induction Issues*', Paper 88WM084-6, IEEE/PES Winter Meeting, 1988.
24. Chopra N., et al., '*Zero Sequence Currents in AC Lines caused by Transients in Adjacent DC Lines*', IEEE Trans., Vol. PWRD- 3, No.4, Oct. 1988, pp. 1873–1879.
25. Maruvada P.S. and Drogi S., '*Field and Ion Interactions of Hybrid AC/DC Transmission Lines*', IEEE Trans., Vol. PWRD-3, No.3, July 1988, pp. 1165–1172.

26. Clairmont B.A., et al., *'The Effect of HVAC-DC Line Separation in a Hybrid Corridor'*, IEEE Trans., Vo1. PWRD-4, No.2, April 1989, pp. 1338–1350.

27. Pargamin L., et a1., *'Considerations on the Choice of Insulators for HVDC Lines'*. CIGRE Conf. Proc., Paper 33–11, 1984.

28. Prabhakara F.S., et al., *'Design, Commissioning and Testing of IPP Ground Electrodes'*, IEEE Trans., Vol. PWRD-3, No.4, Oct. 1988, pp. 2037-2047.

29. Peixoto C.A.O., et al., *'DC Arc Extinction on Long Electrode Lines for HVDC Transmission'*, Int. Conf. on DC Power Transmission, Montreal, Canada, 1984.

30. Kimbark E.W., *'Direct Current Transmission'* (New York), Wiley-Interscience, 1971.

31. CIGRE Working Group, 14.21-TF2, *"General Guidelines for the Design of Ground Electrodes for HVDC Links"*, July 1998.

REACTIVE POWER CONTROL

7

7.1 INTRODUCTION

The converters in conventional HVDC power stations are line commutated, which implies that the current initiation in the valve can only be delayed with reference to the zero crossing of the converter bus AC voltage. This results in lagging power factor operation of the converters, requiring reactive power sources connected at the converter bus for better voltage control. The reactive sources are required at both the rectifier and inverter stations. While the rectifier station appears as a load in the system, the inverter station can be viewed as a generator consuming reactive power. This characteristic of the inverter is generally not desirable and requires suitable modifications by providing adequate var compensation.

The reactive power sources that are used vary from switched capacitors to static var systems. The requirements of voltage control and the costs dictate the choice of the speed of response of the reactive power control under dynamic conditions. The operations of weak AC systems can be problematic due to voltage instability and dynamic overvoltages. A better coordination of var sources is desired under such conditions. The ease of control of firing angles makes it possible to operate the converter as a static var compensator in addition to power conversion. This feature of the converter control is being increasingly used in modern HVDC converter stations.

In this chapter, the characteristics of reactive power sources employed in a converter station are described. The reactive power requirements of the converter and how they are affected by the converter control methods are presented. The var control during transients is discussed along with the description of the static var system.

7.2 REACTIVE POWER REQUIREMENTS IN STEADY STATE

7.2.1 Conventional Control Strategies

As described in Chapter 4, a DC link is operated with current control at the rectifier and the minimum extinction angle control at the inverter under normal conditions. This method of control leads to the minimum reactive power requirement at both ends.

The equations for the reactive power as a function of active power are conveniently expressed in terms of per unit quantities. The following per unit system is used for convenience.

Base converter voltage $V_{db} = (3\sqrt{2}/\pi)V_n$ where V_n is the rated (line to line) voltage at the valve side winding.

Base DC current (I_{db}) = rated DC current (I_{dn})

Base DC power $(P_{db}) = n_b V_{db} I_{db}$

where n_b is the number of bridges connected in series.

Base AC voltage (on the valve side) $(V_b) = V_n$

Base AC power = Base DC power = $(\sqrt{18}/\pi)V_n I_{db} n_b$

The average DC voltage across a converter bridge is given by

$$\bar{V}_d = \bar{V}\cos\alpha - \bar{R}_c \bar{I}_d \tag{7.1}$$

where

$$\bar{V}_d = V_d / V_{db}, \bar{I}_d = I_d / I_{db}, \bar{V} = V / V_b$$

$$\bar{R}_c = \bar{X}_c / 2$$

$\bar{X}_c$ = p.u. leakage reactance of the transformer on its own base.

The power factor is given by

$$\cos\phi \simeq (V_d / V_{do}) = (\bar{V}_d / \bar{V}) = \cos\alpha - (\bar{R}_c \bar{I}_d / \bar{V}) \tag{7.2}$$

The power and reactive power in per unit are given by the following equations

$$\bar{P}_d = \bar{V}\,\bar{I}_d \cos\phi \tag{7.3}$$

$$\bar{Q}_d = \bar{V}\,\bar{I}_d \sin\phi \tag{7.4}$$

Equation (7.1) is also valid for the inverter if α is replaced by γ.

For typical values of $\alpha = 15^\circ$, $\bar{X}_c = 0.2$, $\bar{V} = 1.0$ the variation of Q_d versus P_d (expressed in per unit) is shown in Fig. 7.1.

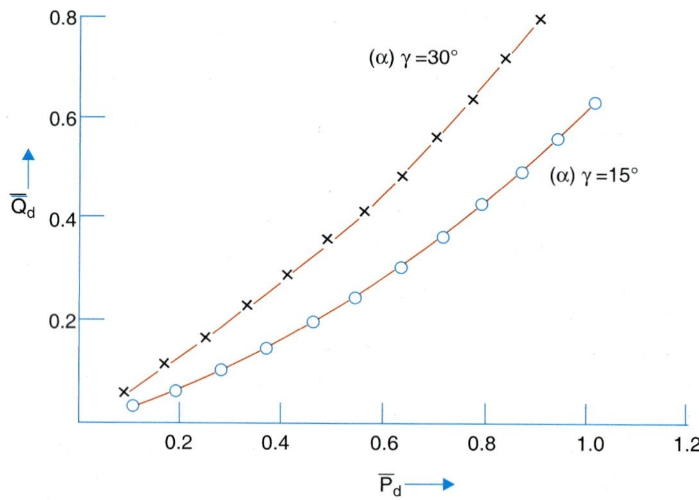

Fig. 7.1: Variation of $\bar{Q}_d$ with $\bar{P}_d$

It is to be noted that the rated DC power is less than 1 p.u. as rated voltage is less than the DC base voltage. Actually, the two are related by

$$V_{dn}/V_{don} = \bar{V}_{dn} = \cos\phi_n \tag{7.5}$$

where $\cos\phi_n$ is the rated P.F. of the converter. In the above example, the rated P.F. is 0.866. It reduces to 0.851 if the delay angle is 18°. The reactive power is around 0.6 times the real power at its rated value. The reactive power requirements can be brought down by reducing X_c. For example, at $X_c = 0.12$ p.u., Q_d is less than 0.5 P_d. The increase in the firing angle results in sharp increase in Q_d. For $\alpha = 30°$, Fig. 7.1 shows that Q_d increases by about 60% at the rated power. This shows the importance of maintaining low firing angles in steady state. However, too low values of α can result increased frequency of mode shifts (transfer of current control from rectifier to inverter) and too low values of γ can result in increased incidence of the commutation failure.

The reactive power is also affected by the magnitude of the AC voltage. The reduction in V leads to increase in Q_d. However, on-load tap changer can control V within limits. A 10% reduction of voltage from 1.0 p.u. to 0.9 p.u. requires about 15% increased current at rated power, which results in over 30% increase in the losses.

7.2.2 Alternate Control Strategies

The region of operation of a converter bridge is bounded by the limits on the DC current and the firing angle. Neglecting the minimum current limit, the operating region of a bridge in $P_d - Q_d$ plane is shown in Fig. 7.2 which is drawn for a constant (rated) AC voltage. This region is bounded by (*i*) minimum α characteristic, (*ii*) minimum γ characteristic and (*iii*) constant rated DC current.

In general, the locus for a constant DC current is part of a circle in the $P_d - Q_d$ diagram and the constant DC voltage characteristic is a straight line passing through the origin. These observations are made from Eqs. (7.2) to (7.4) and noting that V is a constant.

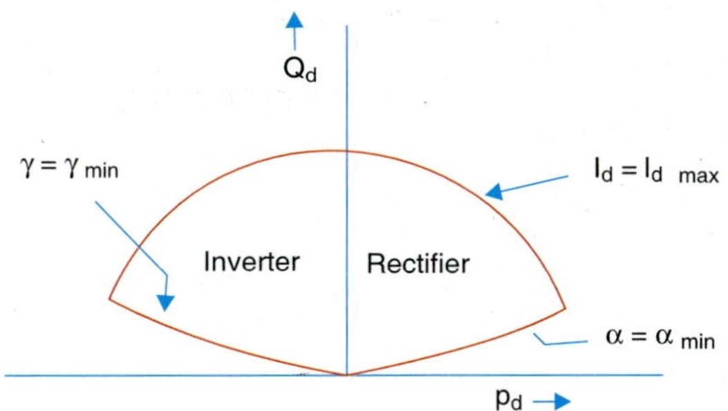

Fig. 7.2: Operating region of a bridge in P_d–Q_d plane

The operation at constant DC voltage implies constant power factor characteristic at the converter bus, (if the valve side AC voltage is kept constant through the action of the tap-changer). At the rectifier, the characteristic is that of a load with lagging power factor, while at the inverter, this can be viewed as a generator with leading power factor operation. If there is no voltage support provided at the converter bus, the stability limit is considerably reduced. This is shown from the analysis of a simplified system shown in Fig. 7.3. In Fig. 7.3 (*a*), the rectifier is shown as a constant (lagging) power factor load while Fig. 7.3 (*b*) is applicable to the inverter operation. The phasor diagrams for both cases are shown in Fig. 7.4.

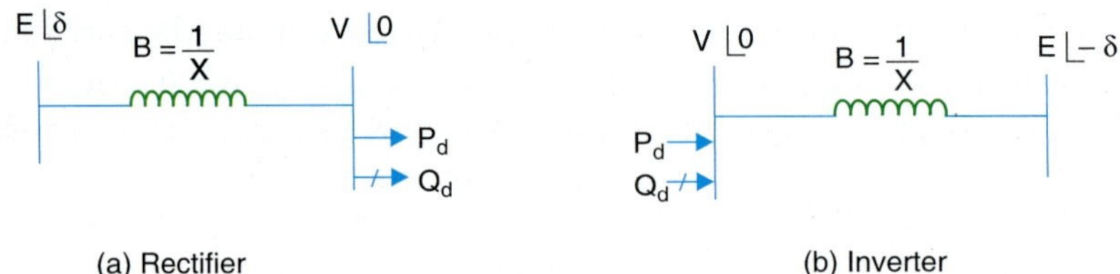

Fig. 7.3: Simplified system diagram (a) rectifier (b) inverter

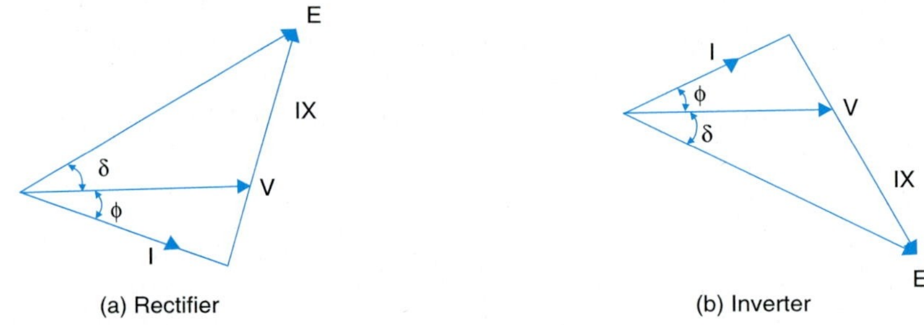

Fig. 7.4: Phasor diagram (a) rectifier (b) inverter

It can be shown from the phasor diagram that

$$V = \frac{E\cos(\delta+\phi)}{\cos\phi} \tag{7.6}$$

where ϕ is the power factor angle.

The power expression is given by

$$P = VEB\sin\delta \tag{7.7}$$

Substituting Eq. (7.6) in Eq. (7.7), we get

$$P = \frac{E^2B\cos(\delta+\phi)\sin\delta}{\cos\phi} = \frac{E^2B}{2\cos\phi}\left[\sin(2\delta+\phi)-\sin\phi\right] \tag{7.8}$$

It can be shown that the maximum power transfer is obtained when

$$\delta = \delta_m = 45^\circ - \phi/2 \tag{7.9}$$

$$P_{max} = \frac{E^2B}{2}\tan\delta_m$$

The maximum power (for $\phi = 30^\circ$) is given by

$$P_{max} = 0.2887E^2B \tag{7.10}$$

This is much less than what can be obtained in the following cases

(i) $\phi = 0,\ P_{max} = 0.5E^2B$

(ii) $\phi = -30^\circ,\ P_{max} = 0.866E^2B$

(iii) $V = E,\ P_{max} = E^2B$

It is to be noted that the provision of a shunt capacitor (of susceptance B_C) at the converter bus results in the modification of the maximum power expression from Eq. (7.10) to

$$P_{max} = \frac{0.2887E^2B}{(1 - B_C/B)} \qquad (7.11)$$

For $B = 3.0$ p.u., $B_C = 0.5$ p.u., this results in an increase of 20% in the maximum power given by Eqn. (7.10).

The above analysis shows that there is a need to modify the reactive power characteristics of the converter station by either (i) choice of the reactive power sources or (ii) adjusting the converter control characteristics. When the DC link involves long distance transmission, the minimization of power losses in the line dictates operation at constant DC voltage and flexibility of converter operation is not feasible. However, with back to back links, the operation at constant voltage is not critical and alternate converter control strategies can be adopted. These are:

(i) constant reactive power characteristic (ab, $a'b$)

(ii) constant leading power factor characteristic (ac, $a'c$)

These characteristics are shown in Fig. 7.5.

It is to be noted that by providing a constant reactive power source of q_n (in per unit) at the converter bus, the characteristic ab or $a'b$ results in unity power factor operation of the converter. Similarly, by providing a reactive source of $2q_n$ and operating the converter along the characteristic ac or $a'c$, the power factor angle is changed from ϕ to $-\phi$.

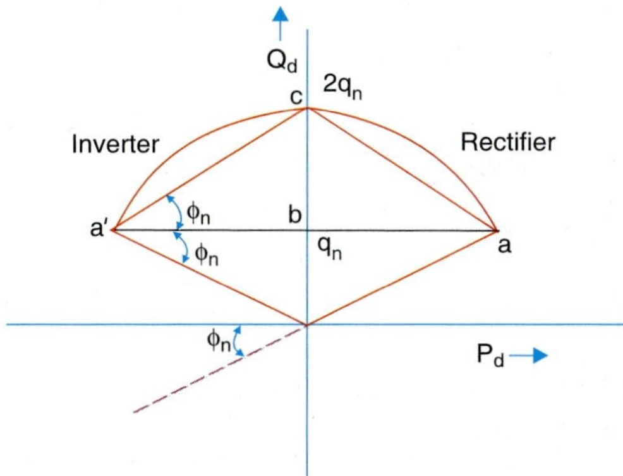

Fig. 7.5: Alternate reactive power control characteristics

The expression for the DC current and voltage for the two characteristics are given by

(i) $i_d = (p_{dn}/v)[\tan^2\phi_n + (P_d/P_{dn})^2]^{\frac{1}{2}}$ (7.12)

(ii) $$i_d = (2q_{dn}/v)\left[1-(P_d/P_{dn})+\left(P_d/(2P_{dn}\sin\phi_n)\right)^2\right]^{\frac{1}{2}} \tag{7.13}$$

$$v_d = p_d/i_d \tag{7.14}$$

In the above expressions, the lower case letters indicate the variables in per unit.

Remarks:

1. The expression (7.12) is derived by noting that

$$vi_d = \left(p_d^2 + q_d^2\right)^{\frac{1}{2}} = p_{dn}\left[\left(\frac{p_d}{p_{dn}}\right)+\left(\frac{q_d}{p_{dn}}\right)^2\right]^{\frac{1}{2}} \tag{7.15}$$

and

$q_d = q_n$ (for the constant reactive power characteristic) (7.16)

$q_n = p_{dn}\tan\phi_n$ (7.17)

Substituting Eqs. (7.16) and (7.17) in (7.15) we get (7.12)

The variations of i_d, v_d and α with variation in P_d are shown in Figs. 7.6 and 7.7 for cases (*i*) and (*ii*) respectively. These are drawn for $V = 1.0$ and $\phi_n = 30°$. These are also applicable equally for the inverter operation except that α is replaced by γ. The increase in α or γ above the minimum implies additional losses in the snubber circuits.

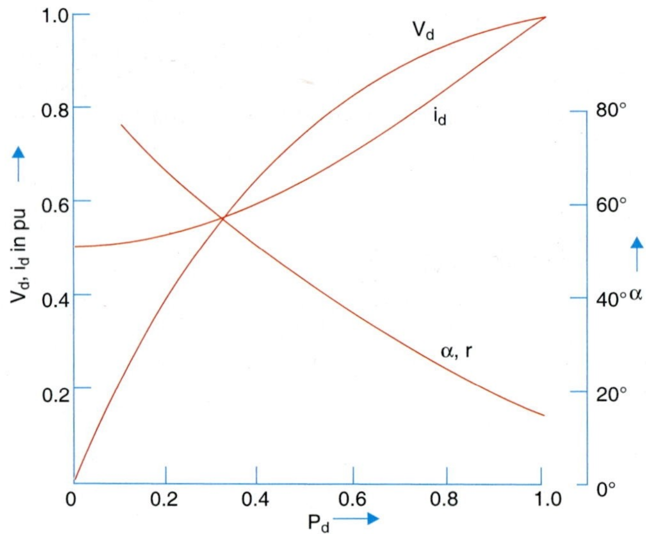

Fig. 7.6: Variation of i_d, V_d, and α with P_d for constant reactive power

2. The expression (7.13) is also derived in a similar fashion by noting that

$$q_d = 2q_n - \frac{p_d}{p_{dn}}q_n \tag{7.18}$$

For the constant leading power factor characteristic, substituting Eqs. (7.17) and (7.18) in (7.15) leads to the expression (7.13)

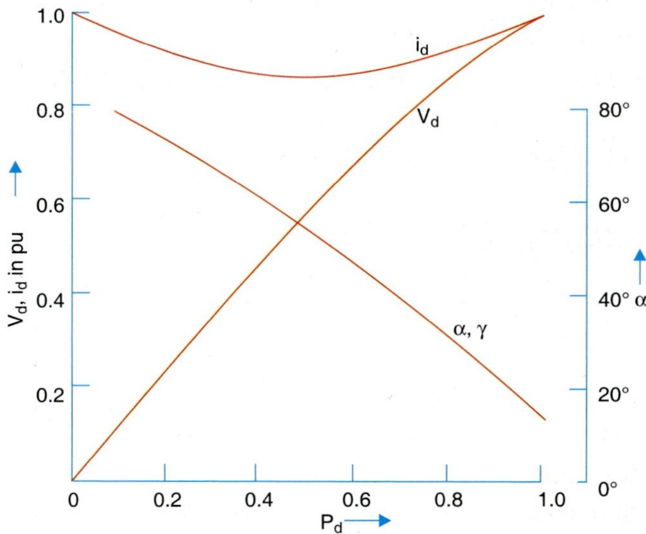

Fig. 7.7: Variation of i_d, v_d, and α with P_d for constant leading power factor

Example 1

(*a*) A back to back HVDC link with one bridge at each end is transmitting 100 MW with V_d = 100 kV. If, $\alpha = 15°$, $\gamma = 18°$, find V_{dor}, V_{doi}, Q_r and Q_i. Assume $R_{cr} = R_{ci}$ = 12 ohms.

(*b*) If the DC link is controlled such that Q_i is kept at the value calculated earlier, find V_d, I_d, Q_r, α and γ for P_d = 50 MW.

Solution

(*a*)

$$I_d = \frac{P_d}{V_d} = \frac{100}{100} = 1 \text{ kA}$$

$$V_{dor} = \frac{V_d + R_{cr} I_d}{\cos\alpha} = \frac{100 + 12 \times 1}{\cos 15°} = 115.95 \text{ kV}$$

$$V_{doi} = \frac{V_d + R_{ci} I_d}{\cos\gamma} = \frac{100 + 12 \times 1}{\cos 18°} = 117.76 \text{ kV}$$

$$\cos\phi_r = \frac{V_d}{V_{dor}} = \frac{100}{115.95} = 0.8624, \ \phi_r = 30.4°$$

$$\cos\phi_i = \frac{V_d}{V_{doi}} = \frac{100}{117.76} = 0.8492, \ \phi_i = 31.88°$$

$$Q_r = P_d \tan\phi_r = 58.69 \text{ MVAR}$$

$$Q_i = P_d \tan\phi_i = 62.19 \text{ MVAR}$$

(b)

$$Q_i = 62.19 = \text{constant} = P_d \tan\phi_i$$

$$\tan\phi_i = \frac{Q_i}{P_d} = \frac{62.19}{50} = 1.244,\ \phi_i = 51.2°$$

$$V_d = V_{doi}\cos\phi_i = 117.76 \times \cos 51.2° = 73.79 \text{ kV}$$

$$I_d = \frac{50 \times 10^3}{73.79} = 677.6A$$

$$\cos\phi_r = \frac{V_d}{V_{dor}} = \frac{73.79}{115.95} = 0.6364$$

$$Q_r = P_d \tan\phi_r = 60.60 \text{ MVAR}$$

$$\cos\alpha = \frac{V_d + R_{cr} I_d}{V_{dor}} = \frac{73.79 + 12 \times 0.6776}{115.95} = 0.7065,\ \alpha = 45.05°$$

$$\cos\gamma_i = \frac{V_d + R_{ci} \grave{I}_d}{V_{doi}} = \frac{73.79 + 12 \times 0.6776}{117.76} = 0.6957,\ \gamma_i = 45.92°$$

7.2.3 Forced Commutation

The reactive power requirements of a converter can be reduced to zero or even reversed if forced commutation is used. This also helps in avoiding commutation failure in inverters.

Forced commutation involves addition of a voltage component to the normal commutation voltage to shift the zero-crossing. One simple method of implementing is by providing series capacitors. This method is adopted in capacitor commutated converter (CCC).

7.3 SOURCES OF REACTIVE POWER

The reactive power requirements of the converter are met by one or more of the following sources:

1. AC system, 2. AC filters, 3. Shunt capacitors, 4. Synchronous condensers.
5. Static Var Compensator (SVC) or STATCOM. These are shunt connected FACTS controllers.

These are shown schematically in Fig. 7.8.

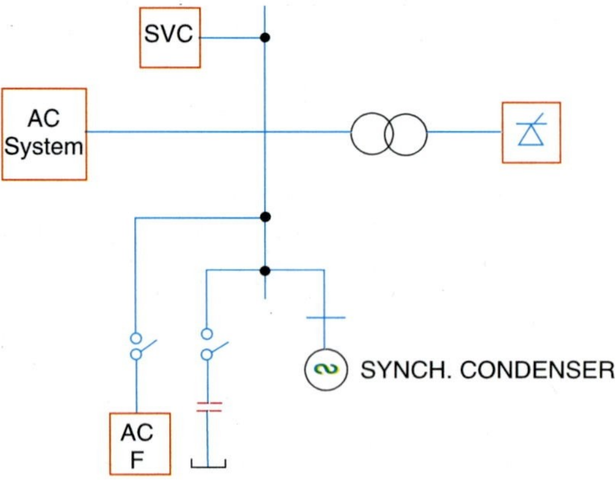

Fig. 7.8: Reactive power sources at a converter bus

From voltage regulation, losses and stability considerations, it is not desirable to draw reactive power from the system except at low loads. The voltage regulation at the converter bus is desirable not only from the voltage control view point but also from the minimization of loss and stability considerations. This requires adjustable reactive power source which can provide variable reactive power as demanded. For slow variations in the load, switched capacitors or filters can provide some control. However, this is discrete type of control and can result in voltage flicker unless the size of the unit, which is switched, is made sufficiently small. In contrast, the synchronous condensers, SVC and STATCOM provide continuous control of the reactive power and can follow fast load changes.

The passive AC filters that are provided at the converter bus for filtering out AC current harmonics appear as capacitors at the fundamental frequency and thus provide reactive power. These filters and shunt capacitors are mechanically switched. Although these devices are less expensive than SVC or synchronous condensers, they suffer from the inability of continuous control. Also, they can cause low order harmonic resonances with the network impedance, resulting in harmonic overvoltages.

The synchronous condensers are essentially synchronous motors operating at no load, with excitation control to maintain the terminal voltage. Their advantages are as follows:

1. The availability of voltage source for commutation at the inverter even if the connection to the AC system is temporarily interrupted. This also implies an increase in SCR as the fault level is increased. When the load supplied by the inverter is passive, the synchronous condenser is essential for providing voltage sources for the line commutation at the inverter.
2. Better voltage regulation during a transient due to the maintenance of flux linkages in the rotor windings. The effect of the armature reaction is counteracted during a transient by induced currents in the field and amortisseur circuits.

There are also disadvantages of synchronous condensers. These are (*i*) high maintenance and cost—the former necessitated by slip rings and brushes on the rotor and (*ii*) possibility of instability due to the machine going out of synchronism.

The Static Var Compensator (SVC) provides fast response following a disturbance. The configurations normally used are (*i*) Fixed Capacitor (FC), Thyristor Controlled Reactor (TCR) or (*ii*) Thyristor Switched Capacitors (TSC) -TCR combination.

The Static Synchronous Compensator (STATCOM) based on Voltage Source Converter (VSC) is an advanced type of SVC that is comparable to a synchronous condenser in terms of its control characteristics. Actually STATCOM was originally named as Static Condenser (STATCON) with very fast response, reduced maintenance requirements and free from the problems of loss of synchronism. The VSC based HVDC transmission does not require a separate reactive power source as the voltage source converter also functions as a STATCOM. As a matter of fact, a BTB VSC based HVDC link can function as two STATCOMs when the DC link opened.

7.4 SVC AND STATCOM

The Static Var Compensators were initially used for load compensation where the objective is to dynamically control the reactive power demand of large fluctuating loads such as rolling mills. They were subsequently used for voltage control applications in transmission systems, where, by maintaining voltage support at specified locations, it is possible to provide increased power transfer capability, control of dynamic overvoltages and improve voltage stability. By using auxiliary control signals, it is also possible to damp low frequency and subsynchronous frequency oscillations.

In HVDC converter stations, the provision of SVC mainly helps to have fast control of reactive power flow, thereby controlling voltage fluctuations and also to overcome the problem of voltage instability. The first example of the installation of SVC is at Chateaguay HVDC link in Canada in 1984.

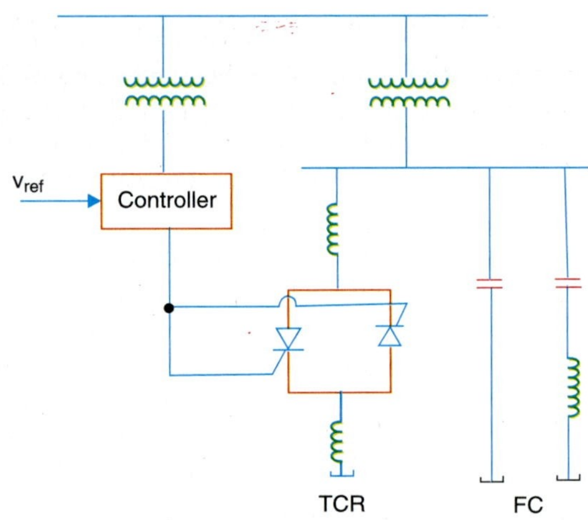

Fig. 7.9: The schematic diagram of FC-TCR

SVC is a variable impedance (or susceptance) device made up of FC (or TSC) and TCR in parallel (see Fig. 7.9) while TSC (Thyristor Switched Capacitor) provides discrete control over the capacitor susceptance. TCR (Thyristor Controlled Reactor) provides continuous control of inductive susceptance, while FC or TSC does not inject harmonics, TCR injects harmonics as discussed below.

7.4.1 Thyristor Controlled Reactor (TCR)

The single phase thyristor controlled reactor is shown in Fig. 7.10. By controlling the firing angle of the back to back connected thyristors, the current in the reactor can be controlled. This is shown in Fig. 7.11. For $\alpha = 90^\circ$, the current is maximum, while for $\alpha = 180^\circ$, the current is zero.

The fundamental component of the inductor current is given by

$$I_1 = \frac{\sigma - \sin\sigma}{\pi X_L} \cdot V \tag{7.19}$$

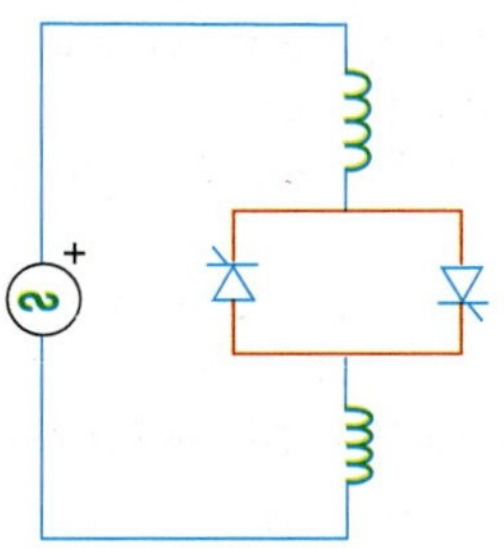

Fig. 7.10: Single phase thyristor controlled reactor

where V is the rms voltage across the TCR, X_L is the fundamental frequency reactance and σ is the conduction angle related to α by the following equation.

$$\sigma = 2(\pi - \alpha) \tag{7.20}$$

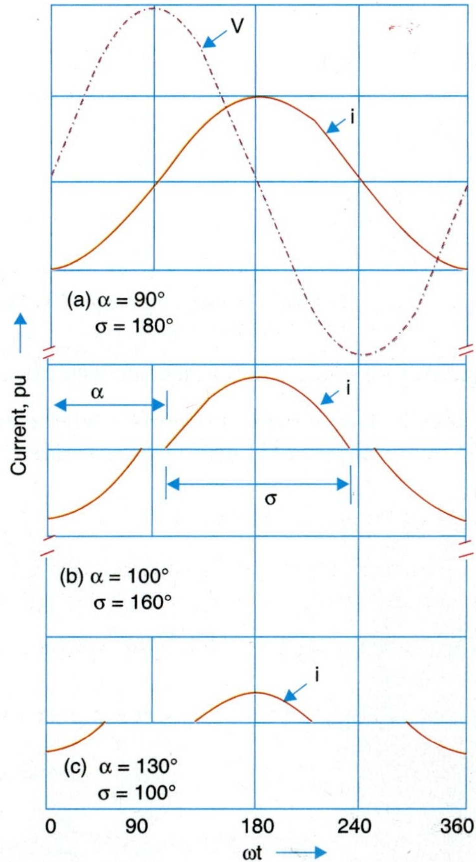

Fig. 7.11: Current control in TCR

Equation (7.19) can be written as

$$I_1 = B(\sigma)V \tag{7.21}$$

where

$$B(\sigma) = \frac{\sigma - \sin\sigma}{\pi X_L} \tag{7.22}$$

is the variable susceptance of the reactor resulting from the control action.

The harmonic component of the current corresponding to harmonic of order 'h' is given by

$$I_h = (4/\pi)(V/X_L)\left[\frac{\sin[(h+1)\alpha]}{2(h+1)} + \frac{\sin[(h-1)\alpha]}{2(h-1)} - \frac{\cos\alpha\sin[h\alpha]}{h}\right] \tag{7.23}$$

$$h = 3,5,7,......$$

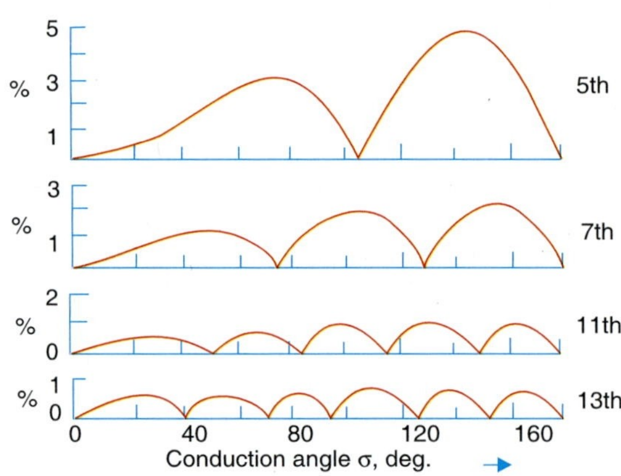

Fig. 7.12: Variation of lower order harmonics with the conduction angle

The variation of the lower order of the harmonics with conduction angle σ is shown in Fig. 7.12. The variation of the fundamental and the total harmonic components is shown in Fig. 7.13.

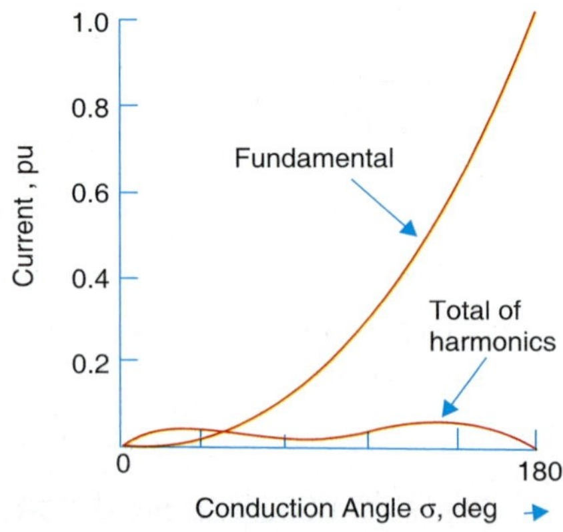

Fig. 7.13: Variation of the fundamental and total harmonics

The triplen harmonics in the lines are eliminated by the delta connection of the three single phase TCRs.

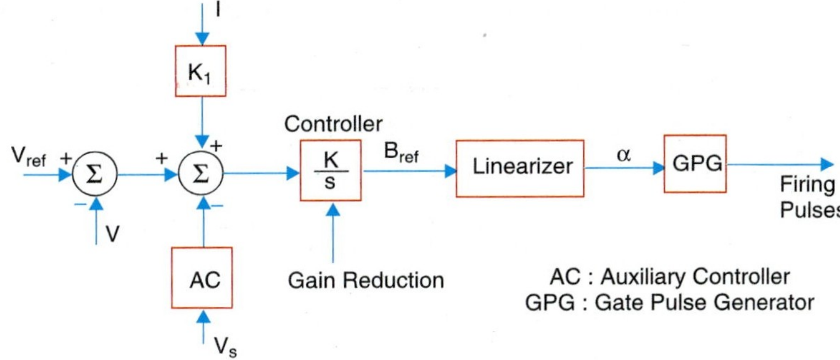

Fig. 7.14: A typical control system for a TCR

The typical control system for a TCR is shown in Fig. 7.14 where the control signals are obtained from the voltage and the reactor current. The controller is usually an integral controller with variable gain to avoid the problems of control instability. The auxiliary signal V_s may be derived from the bus frequency, line reactive power or other locally measured quantities.

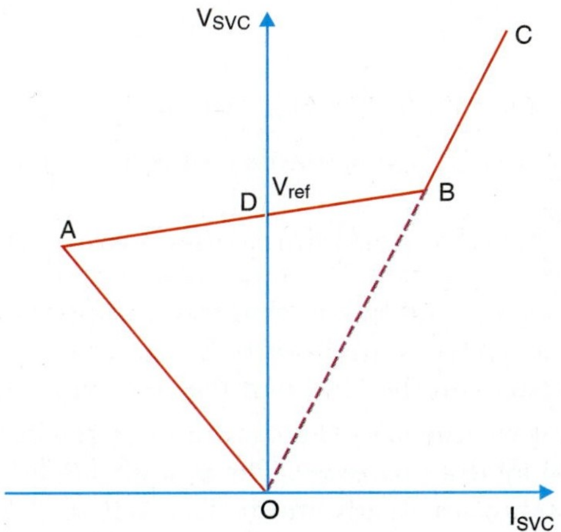

Fig. 7.15: Control characteristics of a SVC

The steady-state control characteristics of a SVC in the V-I plane is shown in Fig. 7.15. The control range is AB which shows a positive slope which can be adjusted from the gain in the current feedback path.

The harmonics injected by TCR into the system can be considerably reduced either with twelve pulse arrangement or with additional filters tuned to 5th and 7th harmonics.

7.4.2 STATCOM

The schematic diagram of a static compensator, also termed on static condenser is shown in Fig. 7.16. The VSC consists of either GTO thyristor switches (which are turned on and off only once in a cycle) or IGBT switches which have a much higher speed and operate up to 2 kHz.

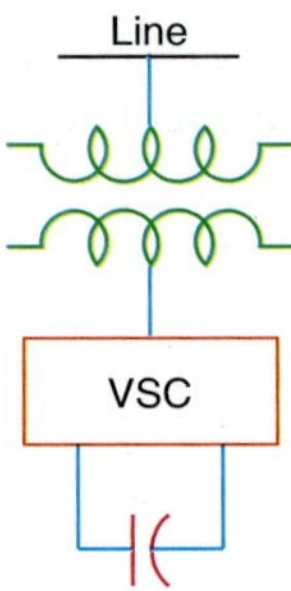

Fig. 7.16: Shunt connected STATCOM

The equivalent circuit and the phasor diagram of STATCOM are shown in Fig. 3.22. Using dynamic phasors, the equation for the circuit is given by Eq. (3.90) and is reproduced here.

$$L\frac{d\hat{I}}{dt}+(R+j\omega_0 L)\hat{I}=\hat{V}-\hat{E} \tag{7.24}$$

where

$\hat{I}=I_Q+jI_D,\ \hat{V}=V_Q+jV_D=V\angle\theta,\ \hat{E}=kV_{dc}\angle(\theta+\alpha)$

V_{dc} is the voltage across the DC capacitor described by the following equation

$$c\frac{dV_{dc}}{dt}+GV_{dc}=mk\left[\sin(\alpha+\theta)I_D+\cos(\alpha+\theta)I_Q\right] \tag{7.25}$$

where G is the conductance across the capacitor (representing the losses in the capacitor). The losses in the converter are neglected. L and R are the inductance and resistance of the reactor (or the leakage impedance) between the VSC and the converter bus.

m and α are the control variables of the converter that affect the magnitude and phase angle of the voltage injected by the converter. The magnitude is normally controlled by PWM (keeping the DC voltage constant in steady state). The control of α affects the power flow that supply the losses in the STATCOM; whereas the control of m affects the reactive power output. In converters with GTO switches, m is normally kept constant (at the maximum possible value using square wave operation) and only α is varied to control the reactive power by varying the magnitude of the DC voltage. By increasing α, the reactive current drawn is increased (note that the reactive current drawn is assumed to be positive when it is inductive).

7.4.3 Comparison between SVC and STATCOM

While SVC is a variable impedance type compensator, STATCOM is a variable voltage type compensator. A STATCOM has several technical advantages over a SVC as follows:

(*a*) Faster response (fraction of a cycle) which is independent of the network impedance.

(*b*) Requires lesser space as bulky passive components (such as reactors) are eliminated

(*c*) Inherently modular and relocatable

(*d*) It can be interfaced with real power sources such as battery, fuel cell or SMES (Superconducting Magnetic Energy Storage)

(*e*) A STATCOM has superior performance during low voltage condition as the reactive current can be maintained constant. It is even possible to increase the reactive current in a STATCOM under transient conditions if the devices are rated for the transient overload. In a SVC, the maximum reactive current is determined by the rating of the passive components—reactors and capacitors. Also, the reactive current in a SVC drops linearly with the voltage drop at the limit.

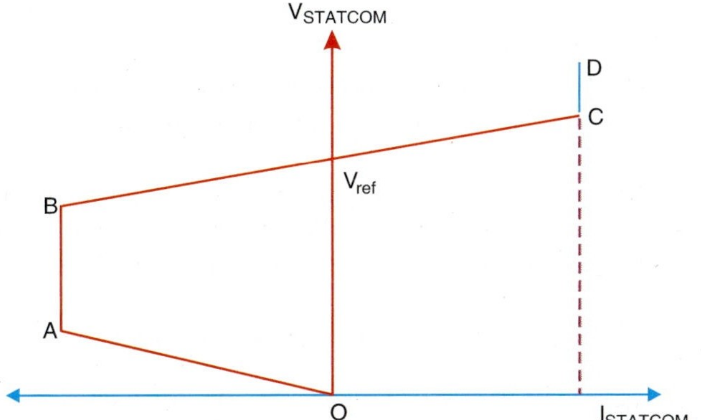

Fig. 7.17: Control characteristics of a STATCOM

The last point can be explained with the help of steady state control characteristics of a STATCOM (shown in Fig. 7.17) and comparing it with the SVC characteristic shown in Fig. 7.15. The reactive current limits (AB and CD) shown in Fig. 7.17 can be increased by selecting devices (GTO or IGBT) with a higher transient rating.

In reference [8], the dynamic performance of a STACOM at an HVDC inverter feeding a very weak AC system (SCR < 2) is investigated. It was observed from a case study, that the SVC exhibited poor dynamic performance during recovery from inverter close-in faults, caused by increased frequency of commutation failures. A simple explanation for this behaviour is that the SVC causes TCR to back off during the recovery as it senses low voltage immediately after the fault is cleared. Such action reduces ESCR even further, resulting in repeated commutation failures. In contrast, the STATCOM performed satisfactorily as its equivalent circuit is similar to that of a synchronous condenser (a voltage sources behind a small inductance) and the reactive current can be held constant even at lower voltages.

7.5 REACTIVE POWER CONTROL DURING TRANSIENTS

The control characteristics of the reactive power sources has a bearing on the system behaviour during a transient. In this context, it is to be noted that the converter control characteristics can be modified to control the reactive power demand. This has already been discussed in Section 7.2. However, this is feasible mainly for back to back links.

Suitable control of reactive power is required during a transient for the following reasons:

1. Control of dynamic overvoltages caused by load rejection
2. Speedy recovery of power following a fault in the inverter system
3. Control of instability

The dynamic overvoltages are mainly due to the excess of reactive power released by the sudden blocking of the converters. This requires a fast control of the reactive power generation from capacitive to inductive. Static Var Systems (including SVC or STATCOM) can achieve the speed, however the initial control of overvoltage may not be feasible unless the rating of the SVS is increased, which may be uneconomic.

The distortion of the voltage waveform produced during the recovery period when the power is increased can cause commutation failures unless the rate of change of power is limited. This is crucial, particularly for low SCR. In this case, the voltage support is also critical as increase in power can result in the reduction of voltage magnitude unless fast control of reactive power is implemented.

The voltage instability can also be a problem at low SCR and can be tackled by suitable converter control strategy or the provision of SVS at the converter bus. The modulation of extinction angle (γ) at the inverter has also been applied.

REFERENCES AND BIBLIOGRAPHY

1. Kangiesser, K.W. and Lips H.P., '*Control Methods for Improving the Reactive Power Characteristics of HVDC Links*', IEEE Trans., Vol. PAS-89, No. 6, July/Aug. 1970, pp. 1120–1125.
2. Kauferle, J., Sadek K. and Povh D., '*System Aspects of the Application of Static var Systems at AC Bus of HVDC System*', IEEE Int. Conf. on Overvoltages and Compensation or Integrated AC-DC Systems', Winnipeg, 1980.
3. Barthold L.O., et al., '*Static Shunt Devices for Reactive Power Control*', CIGRE Conf. Paper 30–08, 1.
4. Romegialli G. and Beeler H., '*Problems and Concepts of Static Compensator Control*', IEE Proc., Vol. 128, Pt. C, No. 6, November 1981, pp. 382–388.
5. Gyugyi L., '*Reactive Power Generators and Control by Thyristor Circuits*', IEEE Trans., Vol. IA-15, No. 5, Sept./Oct. 1979, pp. 521–532.
6. Brewer G.L. et al., '*The Application of Static var Compensators to the English Terminal of the 2000 MW HVDC Cross Channel Link*', CIGRE Conf. Paper 14-07, 1986.
7. Salgado E., et al., '*Technical and Economic Aspects of the use of HVDC Converters as Reactive Power Controllers*', CIGRE Conf. Paper 14-06, 1986.
8. Zhuang Y., Menzies R.W., Nayak O.B. and Turanli H.M., "*Dynamic Performance of a STATCON at an HVDC Inverter Feeding a Very Weak AC System*", IEEE Trans on Power Delivery, Vol.11, No. 2, April 1996.
9. Padiyar K.R., '*FACTS Controllers in Power Transmission and Distribution*', (New Delhi) New Age Publishers, 2007.

HARMONICS AND FILTERS 8

8.1 INTRODUCTION

HVDC converters introduce both AC and DC harmonics which are injected into the AC system and DC line respectively. There are several problems associated with the injection of harmonics and these are listed below:

1. Telephone interference
2. Extra power losses and consequent heating in machines and capacitors connected in the system
3. Overvoltages due to resonances
4. Instability of converter controls, primarily with Individual Phase Control (IPC) scheme of firing pulse generation
5. Interference with ripple control systems used in load management.

AC filters are invariably used to filter out AC current harmonics which are critical. These filters are of band pass (tuned) or high pass type and also supply reactive power. DC smoothing reactor along with DC filters perform the function of filtering DC harmonics. While most of the filters used in converter stations are passive filters using inductors and capacitors, active filters based on VSC has been introduced since 1993 in conjunction with passive filters. At present, there are 14 active filters on the DC side and one active filter on the AC side.

In addition to the harmonics which cause telephone interference, the harmonics at the carrier and radio frequencies are also generated by the converters and may require suitable filters. The harmonic generation and the various types of filters used in converter station are described in this chapter. The criteria and some of the design aspects are also presented.

8.2 GENERATION OF HARMONICS

8.2.1 Characteristic Harmonics

The characteristic harmonics are harmonics which are always present even under ideal operation —balanced AC voltages, symmetric three phase network and equidistant pulses. In the converter analysis given in Chapter 3, the DC current is assumed to be constant. In this case, there are harmonics in AC current of the order

$$h = np \pm 1 \tag{8.1}$$

where p is the pulse number, n is any integer. The harmonics in the converter DC voltage are of the order

$$h = np \tag{8.2}$$

When the smoothing reactor is of finite impedance, there are harmonics in the DC current also of the order given by Eq. (8.2). The AC current harmonics, in this case, are still of the order defined in Eq. (8.1), although their magnitudes will be different with the presence of the ripple in DC current.

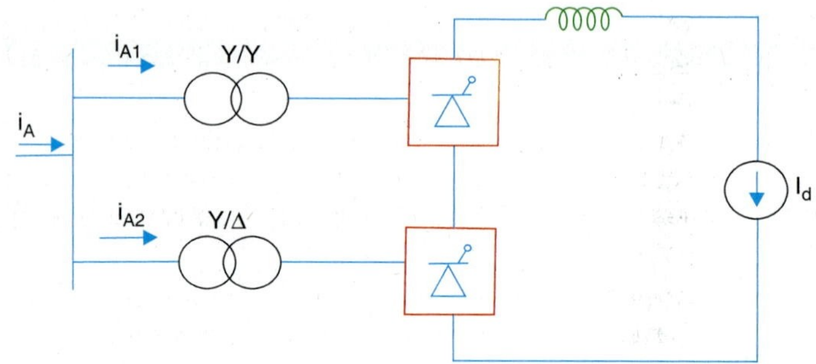

Fig. 8.1: A 12 pulse converter unit

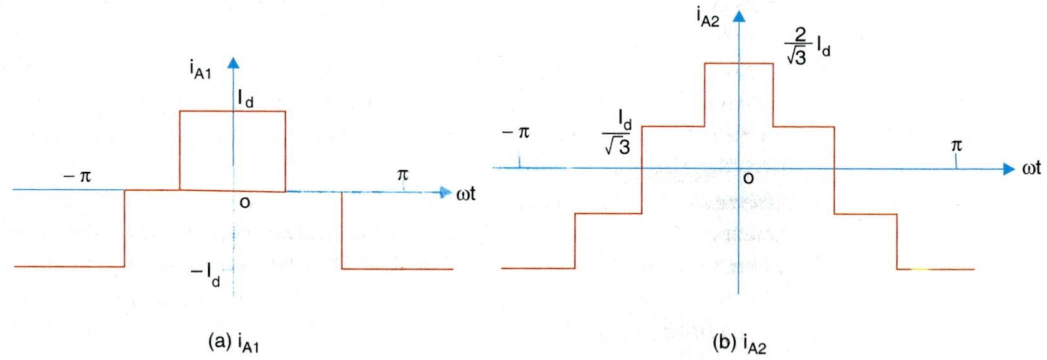

(a) i_{A1} (b) i_{A2}

Fig. 8.2: AC current waveforms

Calculation of characteristic AC harmonics

Consider a 12 pulse converter unit shown in Fig. 8.1. Neglecting overlap, the waveforms of currents i_{A1} and i_{A2} are shown in Fig. 8.2. For convenience, the ordinate axis (corresponding to $\omega t = 0$) is chosen such that the waveform have even symmetry. The waveforms also have half-wave symmetry so that even harmonics are zero. Hence, we can express i_{A1} as

$$i_{A1} = \sum_n a_{n1} \cos n\omega t \tag{8.3}$$

where

$$a_{n1} = \frac{4}{\pi} I_d \int_0^{\frac{\pi}{3}} \cos n\omega t \, d(\omega t)$$

$$= \frac{4I_d}{\pi}\left[\frac{\sin n\omega t}{n}\right]_0^{\frac{\pi}{3}} = \frac{4I_d}{n\pi}\sin\left(n\frac{\pi}{3}\right)$$

It is obvious $a_{n1} = 0$ for triplen harmonics. Also $a_{n1} = -\frac{2\sqrt{3}}{n\pi}I_d$ for $n = 6k - 1$ and $a_{n1} = \frac{2\sqrt{3}}{n\pi}I_d$ for $n = 6k + 1$ where k is a positive integer.

The current i_{A2} can be expressed as

$$i_{A2} = \sum_n a_{n2} \cos n\omega t \tag{8.4}$$

where

$$a_{n2} = \frac{4}{\pi}I_d\left[\int_0^{\frac{\pi}{6}} \frac{2}{\sqrt{3}}\cos n\omega t\, d(\omega t) + \int_{\frac{\pi}{6}}^{\frac{\pi}{2}} \frac{\cos n\omega t\, d(\omega t)}{\sqrt{3}}\right]$$

$$= \frac{4I_d}{\sqrt{3}\pi n}\left[\sin\frac{n\pi}{2} + \sin\frac{n\pi}{6}\right]$$

$$= 0 \text{ for } n = 3\,k$$

$$= \pm\frac{2\sqrt{3}}{n\pi}I_d \quad \text{for } n = 6\,k - 1$$

The positive sign applied for odd values for k and negative sign applies for even values of k. Similarly, it can be shown that

$$a_{n2} = \mp\frac{2\sqrt{3}}{n\pi}I_d \quad \text{for } n = 6k + 1$$

The negative sign applies to odd values for k and positive sign applies for even values of k.

The current I_A is given by

$$i_A = i_{A1} + i_{A2} = \left(4\sqrt{3}/\pi\right)I_d[\cos\omega t - \left(\frac{1}{11}\right)\cos 11\omega t + \left(\frac{1}{13}\right)\cos 13\omega t - \left(\frac{1}{23}\right)\cos 23\omega t + \left(\frac{1}{25}\right)\cos 25\omega t - ...] \tag{8.5}$$

From the above expression, it can be observed that

$$I_{10} = \left(2\sqrt{6}/\pi\right)I_d \tag{8.6}$$

$$I_{ho} = \left(I_{10}/h\right) \tag{8.7}$$

where I_{10} and I_{ho} are rms values of the fundamental component and harmonic of order 'h'. The second subscript 'o' indicates that the overlap angle u is assumed to be zero.

CHAPTER 8

When overlap angle is non-zero, the expression for I_h is given by

$$I_h = \frac{I_{ho}\left[A^2 + B^2 - 2AB\cos(2\alpha + u)^{1/2}\right]}{\cos\alpha - \cos\delta} \tag{8.8}$$

where

$$A = \frac{\sin(h+1)\frac{u}{2}}{h+1},\quad B = \frac{\sin(h-1)\frac{u}{2}}{h-1},\quad \delta = (\alpha + u)$$

The above expression is valid for $u \leq 60°$. For higher values of the overlap angle, the expression given by (8.8) can still be used if, α, u and δ are replaced by α', u' and δ' where

$$\alpha' = \alpha - 30°,\ u' = u + 60°,\ \delta' = \delta + 30° \tag{8.9}$$

The effect of the overlap angle is to prevent steep changes in the AC current. This implies that I_h is smaller than I_{ho}. For example, for $\alpha = 15°$, $u = 25°$, the 11th and 13th harmonic currents are reduced to about 0.25 I_{ho}.

The magnitudes of the characteristic harmonics are also functions of the load current. This is shown in Fig. 8.3.

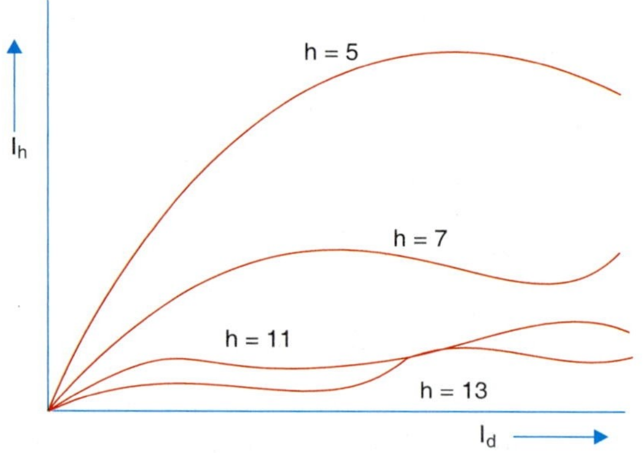

Fig. 8.3: Harmonic magnitudes with variation in DC current

DC voltage harmonics

From the Fourier analysis of DC voltage waveform, we can obtain

$$V_h = \frac{V_{do}\left[C^2 + D^2 - 2CD\cos(2\alpha + u)\right]^{1/2}}{\sqrt{2}} \tag{8.10}$$

where

$$C = \frac{\cos\left[(h+1)\frac{u}{2}\right]}{h+1},\ D = \frac{\cos\left[(h-1)\frac{u}{2}\right]}{h-1}$$

8.2.2 Non-characteristic Harmonics

The harmonics of the order other than the characteristic harmonics are termed as non-characteristic. These are due to (*i*) imbalance in the operation of two bridges forming a 12 pulse converter (*ii*) firing angle errors (*iii*) unbalance and distortion in AC voltages and (*iv*) unequal transformer leakage impedances.

The harmonics produced due to the first cause are termed as residual harmonics. These are mainly due to the difference in the firing angles in the two bridges which lead to unequal cancellation of the harmonics of order 5, 7, 17, 19, etc. The unequal leakage impedances of the two converter transformers feeding the two bridges also lead to residual harmonics.

The last three causes can lead to the generation of triplen and even harmonics and their analysis is complex.

Effect of Firing Angle Errors

It is convenient to neglect overlap in the analysis. The errors in the firing angles can be due to the jitter or the nature of the control system. The equidistant pulse control scheme ideally, has no inherent errors except due to the jitter.

To study the effect of firing angle errors, we will simplify the analysis by considering a single Graetz bridge fed from a star/star connected transformer. We consider the error ε_j as the delay in the firing of valve *j* from the instant corresponding to the desired value of the firing angle α.

There is no loss of generality in assuming the firing error for valve 1 is zero as the analysis applies to steady state conditions, where it is assumed that the waveform is periodic with the fundamental frequency of ω. Thus, there are five independent parameters ε_2, ε_3, ε_4, ε_5, and ε_6.

To illustrate the analysis, we will consider an example with following data:

$$\varepsilon_3 = \varepsilon_5 = 0, \varepsilon_2 = \varepsilon_4 = \varepsilon_6 = \varepsilon$$

The waveform of the phase currents is similar to the waveform of i_{A1} shown in Fig. 8.2. To analyze the waveforms, we can consider them as the sum of two waveforms—one ideal with no firing errors (as shown in Fig. 8.2) and the other that represent the effect of firing errors. Thus, we define

$$i_a(t) = i_{a0}(t) + \Delta i_a(t)$$
$$i_b(t) = i_{b0}(t) + \Delta i_b(t)$$
$$i_c(t) = i_{c0}(t) + \Delta i_c(t)$$

where i_{a0}, i_{b0}, and i_{c0} represent ideal waveforms neglecting firing angle errors. For the data given, the waveforms of Δi_a, Δi_b and Δi_c are shown in Fig. 8.4.

Note that $\Delta i_a = 0$ except when $-\frac{2\pi}{3} \le \theta \le -\frac{2\pi}{3} + \varepsilon$ and $\frac{2\pi}{3} \le \theta \le \frac{2\pi}{3} + \varepsilon$, $\theta = \omega t$

CHAPTER 8

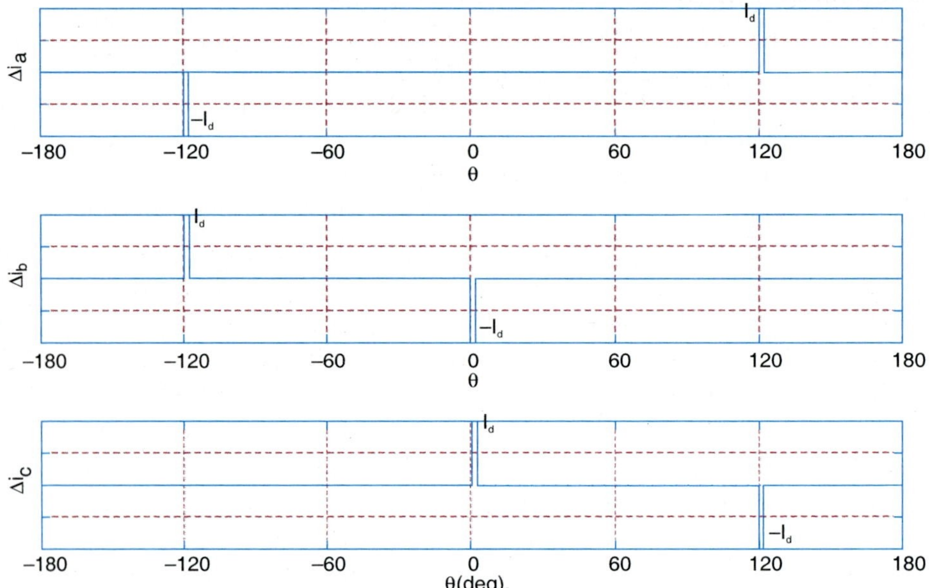

Fig. 8.4: Waveforms of phase current deviations

Similarly $\Delta i_b = 0$ except when $-\dfrac{2\pi}{3} \le \theta \le -\dfrac{2\pi}{3} + \varepsilon$ and $0 \le \theta \le \varepsilon$ and $\Delta i_c = 0$ except when $0 \le \theta \le \varepsilon$ and $\dfrac{2\pi}{3} \le \theta \le \dfrac{2\pi}{3} + \varepsilon$.

To analyze these waveforms, let us consider the complex Fourier Series for a function $f(t)$, defined by

$$f(t) = \sum_{n=-\infty}^{n=\infty} C_n e^{jn\omega t}$$

where

$$C_n = \frac{1}{2\pi}\int_{-\pi}^{\pi} f(t)e^{-jn\omega t}d\theta, \quad \theta = \omega t$$

Thus,

$$C_{na} = -\frac{I_d}{jn2\pi}\left[-e^{-jn\theta}\Big|_{\theta=\frac{2\pi}{3}}^{\theta=\frac{2\pi}{3}+\varepsilon} + e^{-jn\theta}\Big|_{\theta=\frac{2\pi}{3}}^{\theta=\frac{2\pi}{3}+\varepsilon}\right]$$

$$= \frac{-I_d}{jn2\pi}\left(e^{-jn\varepsilon} - 1\right)\left[-e^{j\frac{2\pi}{3}n} + e^{-j\frac{2\pi}{3}n}\right]$$

$$= \frac{I_d}{n\pi}\left(e^{-jn\varepsilon} - 1\right)\sin\left(n.\frac{2\pi}{3}\right)$$

Similarly, it can be shown that

$$C_{nb} = \frac{-I_d}{n\pi}\left(e^{-jn\varepsilon} - 1\right)\sin\left(\frac{2\pi n}{6}\right)e^{j\frac{n\pi}{3}}$$

$$C_{nc} = \frac{-I_d}{n\pi}\left(e^{-jn\varepsilon} - 1\right)\sin\left(\frac{2\pi n}{6}\right)e^{-j\frac{n\pi}{3}}$$

It is obvious that for $C_{na} = C_{nb} = C_{nc} = 0$ for $n = 3k$, where k is a positive integer. It can be shown that for $n = 2$, the current components form a negative sequence and for $n = 4$, the current components form a positive sequence. In general, the sequence is negative when $n = 3k - 1$ and positive $n = 3k + 1$, where k is an integer. It can also be shown that the firing angle errors also affect the characteristic harmonics and the fundamental (positive sequence) current components. The fundamental component (peak value), for small values of ε is given by

$$I_{1peak} = \frac{2\sqrt{3}}{\pi} I_d \cos(\varepsilon/2) \tag{8.11}$$

The magnitude (peak value) of the non-characteristic harmonic is given by

$$I_n = \frac{\sqrt{3}I_d \sin\left(\frac{n\varepsilon}{2}\right)}{n\pi} \simeq \frac{\sqrt{3}I_d}{2\pi}\varepsilon = I_{10}\,\varepsilon/2 \tag{8.12}$$

The approximation applies for $\frac{n\varepsilon}{2} < 30°$. This shows that the magnitudes of all the non-characteristic harmonics are identical for all values of n for which the approximation applies. For $n < 30°$, the harmonic distortion in the current due to non-characteristic harmonics alone with $\varepsilon = 2°$, is more than the distortion introduced by the characteristic harmonics for a 12 pulse converter. Thus, there is no significant advantage in increasing the pulse number beyond 12.

The major source of firing angle errors in modern converter stations is due to the ripple in the current feedback signal and not due to the equipment. The tolerance in the firing pulses does not exceed ± 0.2°.

In twelve pulse converters, it is necessary to operate at equidistant pulse control in steady-state to eliminate residual harmonics. This implies that the delay angle for each bridge is identical (when the voltages are balanced). This involves increased reactive power consumption which can be reduced by having independent control of bridges. However, this is not desirable as the harmonic generation is increased while trying to reduce the reactive power consumption.

Effect of unbalanced voltages

The presence of the negative sequence component in the AC voltage shifts the zero crossing of the commutation voltages. With individual phase control (IPC) system, this introduces firing angle dissymmetry and results in non-characteristic harmonics. With a 5% negative sequence voltage, the third harmonic current generated can be as large as 5% of the fundamental component.

With EPC firing scheme, this is avoided. However, if the smoothing reactor is finite, there will be DC current harmonic of second order which will generate AC harmonic of third order.

8.3 DESIGN OF AC FILTERS

8.3.1 Criteria of Design

The major design objective of AC filters is to reduce the telephone interference. This can be measured by any of the following performance indices.

Harmonic distortion

This can be measured in two ways:

$$D = \left(\sum_{n=2}^{m} I_n Z_n / E_1 \right) \times 100$$

in percentage where I_n, Z_n and E_1 are the harmonic current injected, the harmonic impedance of the system and the fundamental component of the line to neutral voltage respectively. m is the highest harmonic considered (49 or 50).

The second definition is

$$D_{RSS} = \left(\left[\sum_{n=2}^{m} (I_n Z_n)^2 \right]^{1/2} / E_1 \right) \times 100 \tag{8.13}$$

In some cases, the harmonic distortion can be defined individually for a single harmonic as

$$D_n = (I_n Z_n / E_1) \times 100 \tag{8.14}$$

Telephone influence factor

This is an index of possible telephone interference and is defined as

$$TIF = \left(\left[\sum_{n=1}^{m} (I_n Z_n F_n)^2 \right]^{1/2} / E \right) \tag{8.15}$$

where

$$F_n = 5 n f_1 p_n$$

p_n is the C message weighting used by Bell Telephone Systems (BTS) and Edison Electric Institute (EEI) in USA. This weighting reflects the frequency dependent sensitivity of the human ear and has a maximum value 1.0 at the frequency of 1000 Hz.

Telephone harmonic form factor (THFF)

This is analogous to TIF except that

$$F_n = (n f_1 / 800) W_n \tag{8.16}$$

where W_n is the psophometric weight at the harmonic order n, as defined by the Consultative Commission on Telephone and Telegraph Systems (CCITT). While TIF is used in USA, THFF is popular in Europe. The maximum value of $W_n = 1.0$, at the frequency of 800 Hz.

IT Product

In BTS-EEI system, there is another index called IT product and is defined by

$$IT = \left[\sum_{n=1}^{m}(I_n F_n)^2\right]^{1/2} \tag{8.17}$$

KIT product is defined as

$$KIT = IT/1000 \tag{8.18}$$

Although there are no specific standards on the performance requirements, the suggested values of the above mentioned indices are given in Table 8.1.

Table 8.1: Performance Indices

Index	Range	Suggested Value
TIF	25–50	35
KIT	25–100	–
D_{RSS}	2–5	2
D_n	–	1

8.3.2 Types of Filters

There are basically two types of filters that can be used. These are (1) Passive filters and (2) Active filters. The passive filters can be further divided into (*a*) tuned (high Q) filters and (*b*) damped (low Q) filters.

The passive filters have been used in HVDC converter station right from the beginning and continue to be used. Tuned filters are used to filter out individual harmonics – both characteristic and non-characteristic. For example, in present day twelve pulse converter stations, tuned AC filters are used for 11th and 13th harmonics in addition to say 3rd harmonic which may be problematic (particularly if the DC circuit has a resonant frequency corresponding to the second harmonic).

The requirement of AC and DC filters with overhead DC lines adds to the cost of converter stations. However, in the case of AC filters, passive filters act as sources of reactive power required for the Line Commutated Converters (LCC) with thyristor valves. The smoothing reactors on the DC side contribute to the filtering of DC current harmonics. Back to back (BTB) HVDC links are not provided with DC filters as they are not required. DC filters are designed to minimize primarily telephone interference in the wire-line communication circuits in the vicinity and costs associated with these filters can be avoided if there are no interference problems. However, with overhead lines, the interference from the DC harmonics can be critical and to save costs associated with filtering, active filters (based on VSC) in series with the passive filters have been suggested. Although active filters are costly, there are several advantages over the passive filters and a hybrid solution is generally preferred if the performance of filters is to be improved. With AC filters, the active filters are not yet cost advantageous unless the reactive power requirements at the converter station are minimal (for example, in the case of capacitor commutated converters).

8.4 PASSIVE AC FILTERS

A single tuned filter and a damped filter are shown in Fig. 8.5 (*a*) and (*b*) respectively. For both these filters, we define $\omega_r = \frac{1}{\sqrt{LC}}$ as the resonance frequency and $X_0 = \omega_r L = \frac{1}{\omega_r C} = \sqrt{\frac{L}{C}}$ as

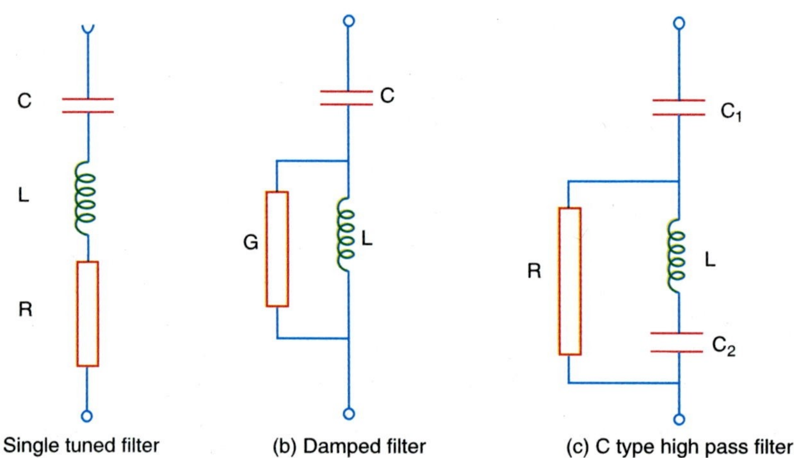

Fig. 8.5: Passive filters

the reactance of the inductor or capacitor at the resonance frequency. Note that both filters become identical when $R = G = 0$. In this case the impedance of the filter is purely reactive and becomes zero at $h = h_r$, where h_r is the order of the harmonic for which the filter is designed. We express the sharpness of tuning in terms of the quality factor (Q) defined as

$$Q = \frac{X_0}{R} \text{ for the tuned filter} \tag{8.19}$$

$$Q = \frac{1}{GX_0} \text{ for the damped filter} \tag{8.20}$$

Note that G is the conductance of the resistor in parallel with the inductor for the damped filter.

Impedance of single tuned filter

The impedance (Z_{Fh}) of the single tuned filter at the harmonic order 'h' is given by

$$Z_{Fh} = R + j\left(h\omega L - \frac{1}{h\omega C}\right) \tag{8.21}$$

where ω is the fundamental frequency which can vary with the power system operating conditions.

A tuned filter is designed to filter a single harmonic (of order h_r). If $h_r\omega = \omega_r$, then obviously $Z_{Fh} = R = \frac{X_0}{Q}$ and is minimum. The variation of $\frac{Z_{Fh}}{X_0}$ as a function of $\frac{h}{h_r}$ is shown in Fig. 8.6.

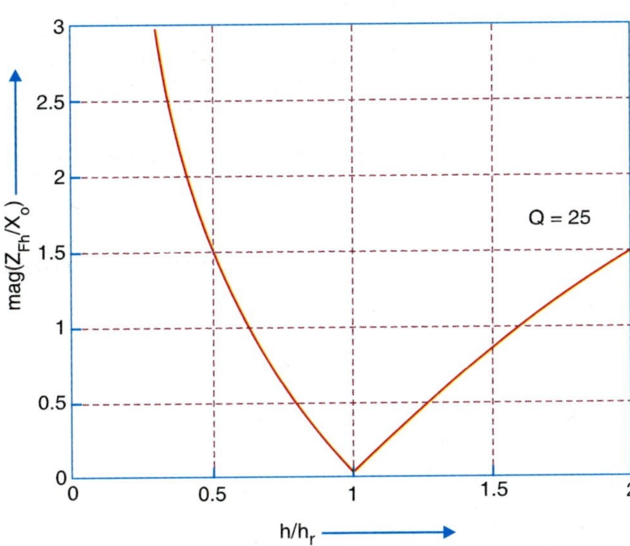

Fig. 8.6: Variation of impedance (tuned filter)

Here, h is viewed as continuous variable. Since ω is variable and there could be errors in the tuning ($\omega_r \neq h_r\omega_n$ where ω_n is the nominal (rated) system frequency), it is necessary to compute the impedance of the tuned filter as a function of the detuning parameter (δ) defined by

$$\delta = \frac{h_r\omega - \omega_r}{h_r\omega_n} = \frac{\omega}{\omega_n} - \frac{\omega_r}{h_r\omega_n} \tag{8.22}$$

Considering variations in the frequency (f), inductance (L) and capacitance (C), we can express δ as

$$\delta = 1 + \frac{\Delta f}{f_n} - \left[\left(1+\frac{\Delta L}{L_n}\right)\left(1+\frac{\Delta C}{C_n}\right)\right]^{-\frac{1}{2}}$$

$$\delta \simeq \frac{\Delta f}{f_n} + \frac{1}{2}\frac{\Delta L}{L_n} + \frac{1}{2}\frac{\Delta C}{C_n} \tag{8.23}$$

where L_n and C_n are the nominal values of L and C such that $h_r\omega_n = (L_nC_n)^{-\frac{1}{2}}$.

It is to be noted that ΔL can be treated as the error in setting the value of L. The variation in C can be due to (*i*) error in the initial setting of C and (*ii*) the variation in C due to the temperature dependence of the dielectric constant. We can express Z_{Fh} as

$$Z_{Fh} = R + jX_0\left[\frac{\omega}{\omega_n}\frac{L}{L_n} - \frac{\omega_n}{\omega}\frac{C_n}{C}\right] \tag{8.24}$$

where $$X_0 = h_r\omega_n L_n = \frac{1}{h_r\omega_n C_n}$$

The quantity inside the brackets in the RHS of Eq. (8.24) can be expressed as

$$\left[\frac{\omega}{\omega_n}\frac{L}{L_n} - \frac{\omega_n C_n}{\omega C}\right] = \left[\left(1+\frac{\Delta f}{f_n}\right)\left(1+\frac{\Delta L}{L_n}\right) - \left(1+\frac{\Delta f}{f_n}\right)^{-1}\left(1+\frac{\Delta C}{C_n}\right)^{-1}\right]$$

$$= \left[\left(1+\frac{\Delta f}{f_n}+\frac{\Delta L}{L_n}\right) - \left(1-\frac{\Delta f}{f_n}\right)\left(1-\frac{\Delta C}{C_n}\right)\right]$$

$$= \left[2\frac{\Delta f}{f_n} + \frac{\Delta L}{L_n} + \frac{\Delta C}{C_n}\right] \tag{8.25}$$

From Eq. (8.23), the RHS of Eq. (8.25) is equal to 2δ. Thus, we can finally derive,

$$Z_{Fh} = R + jX_0\cdot 2\delta$$

$$= X_0\left(Q^{-1} + j2\delta\right) \tag{8.26}$$

where Q is defined in Eq. (8.19)

Impedance of damped filter

The tuning in a damped filter is not critical, hence we will ignore the effects of variations in the system frequency (ω) impedance (L) and capacitance (C).

Defining $\omega_r = \frac{1}{\sqrt{LC}} = h_0\omega$, $Q = \frac{R}{X_0}$, $X_0 = h_0\omega L = \frac{1}{h_0\omega C}$ we can express the impedance of the filter as

$$Z_{Fh}(h) = \frac{-j}{h\omega C} + \frac{jR\omega hL}{R + j\omega hL}$$

$$= X_0\left[\frac{-jh_0}{h} + \frac{jQ}{\left(\frac{Qh_0}{h} + j1\right)}\right]$$

$$= \frac{X_0\left[Q + j\frac{h_o}{h}\left(Q^2 - 1 - \left(\frac{Qh_0}{h}\right)^2\right)\right]}{1 + \left(\frac{Qh_0}{h}\right)^2} \tag{8.27}$$

The variation in the (normalized) magnitude of the damped filter $\left(\frac{Z_{Fh}}{X_0}\right)$ as a function of h/h_0 is shown in Fig. 8.7 for three different values of the quality factor Q. It is observed that the impedance remains practically constant at higher frequencies.

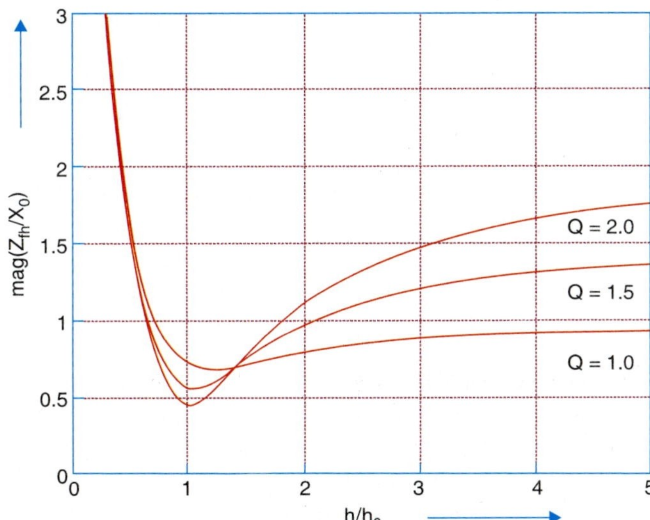

Fig. 8.7: Variation of impedance (Damped filter)

A damped filter is used in converter stations as a second order high pass filter to filter out higher order harmonics. The tuning of these filters is not critical. The losses at the fundamental frequency can be reduced by using a C type filter where the capacitor C_2 in series with L, provides a low impedance path to the fundamental component of current [see Fig. 8.5(c)].

A typical converter station with 12 pulse converters has double tuned (or two single tuned) filter banks to filter out 11th and 13th harmonics and a high pass filter bank to filter out the rest of harmonics. Sometimes a third harmonic filter may be included to filter out the non-characteristic harmonic of the third order (particularly with weak AC systems where some voltage unbalance is expected).

All the filters branches appear capacitive at fundamental frequency and supply reactive power.

Design of a single tuned filter

The single tuned filters are designed to filter out (usually) characteristic harmonics of single frequency. The equivalent circuit seen by the harmonic current I_h generated by the converter is shown in Fig. 8.8. Z_{Fh} and Z_{Sh} are the filter and system impedances at the harmonic frequency (hf).

The harmonic current in the filter is given by

$$I_{Fh} = \frac{I_h |Z_{Sh}|}{|Z_{Sh} + Z_{Fh}|} \tag{8.28}$$

CHAPTER 8

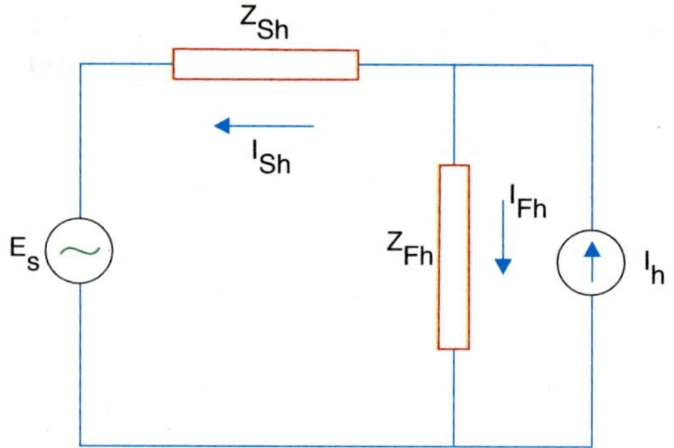

Fig. 8.8: Equivalent circuit for harmonic current

The harmonic voltage at the converter bus is

$$V_h = I_{Fh}|Z_{Fh}| = \frac{I_h}{|Y_{Fh} + Y_{Sh}|} = \frac{I_h}{|Y_h|} \tag{8.29}$$

The basic objective in designing the filter is to select the filter admittance Y_{Fh} in order to minimize V_h or satisfy the constraints on V_h. The problem of designing a filter is complicated by the uncertainty about the network admittance (Y_{Sh}). There are two possible representations of system impedance in the complex plane. These are (*a*) impedance angle is limited [see Fig. 8.9 (*a*) & (*b*)] the impedance is limited both in angle and impedance. The impedance is assumed to lie in the region shown in Fig. 8.9 (*b*) in which R_1, R_2 and θ_m obtained from the system characteristics.

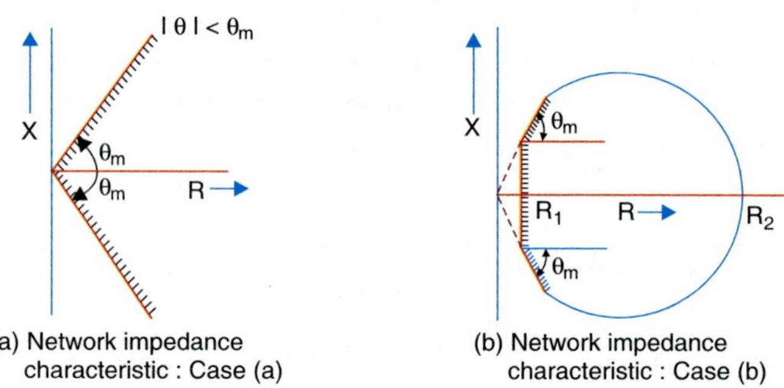

(a) Network impedance characteristic : Case (a)

(b) Network impedance characteristic : Case (b)

Fig. 8.9

It is to be noted that (*a*) or (*b*) reflect the broad assumptions about the nature of system impedance. The assumption (*b*) requires more knowledge of the system than (*a*). Having complete information on the network impedance characteristic will result in economic design of filters and also a realistic assessment of the harmonic distortion of the converter bus voltage.

Although assumption (*a*) is rather simplistic, it allows a simplified computation of the optimum value of Q. In computing the optimum value of Q, we need to minimize the maximum value of V_h. Since we do not have the exact knowledge of the network impedance (except that its angle is limited) we assume the worst case scenario for the network admittance that results in maximizing V_h. The optimum value of Q corresponds to the lowest value of the upper limit on V_h.

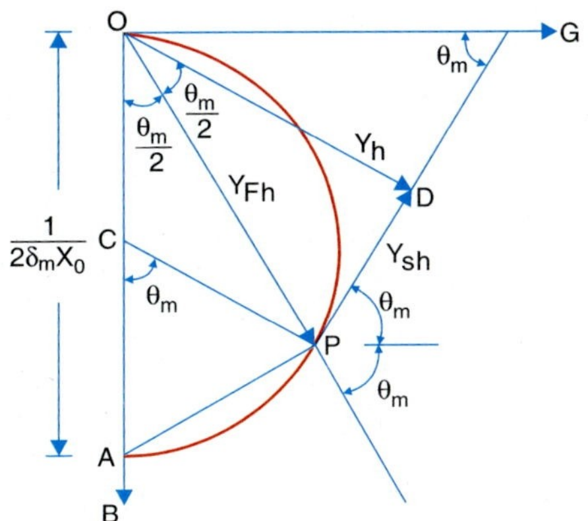

Fig. 8.10: Loci of filter and network admittance

The value of Y_h (the net admittance seen by the converter) is reduced if the detuning parameter δ is maximum = δ_m. For a specified δ_m and X_0, the locus of the filter impedance(as Q is varied) is a semicircle in the 4th quadrant of the G-B plane as shown in Fig. 8.10. The diameter of the circle is $1/(2\delta_m X_0)$. It can be seen, from the geometry of the figure, that minimum Y_h occurs when $\theta = \theta_m$. The optimum value of Q can be obtained from game-theoretic analysis. If one selects Y_{Fh} (OP) arbitrarily (the tip of Y_{Fh} lying along the semicircle), the network can select Y_{Sh} such that the vector Y_h (OD) is perpendicular to PD (the vector Y_{Sh})and ensure Y_h is a minimum. To maximize the minimum magnitude of Y_h , it is necessary to have PD tangential to the circle at point P. Thus, we select Y_{Fh} to maximize Y_h when the network (viewed as an adversary not under our control) tries to minimize it.

To compute optimum Q, we note that triangle OPC is isosceles with angle COP = angle $CPO = \theta_m/2$. The angle of the impedance Z_{Fh} is $(90 - \theta_m/2)$ and we have,

$$\tan(90 - \theta_m/2) = Q_{opt} \times (2\delta_m) = \cot(\theta_m/2)$$

Hence,

$$Q_{opt} = \frac{\cot(\theta_m/2)}{2\delta_m} = \frac{1+\cos\theta_m}{2\delta_m \sin\theta_m} \tag{8.30}$$

Since $|Y_h| = Y_{Fh}\cos(\theta_m/2)$, $|Y_{Fh}| = \cos(\theta_m/2)/(2\delta_m X_0)$, we can derive the expression for V_h using Eq. (8.29). We finally get,

$$V_h = \frac{I_h}{|Y_{Fh} + Y_{Sh}|} = \frac{4\delta_m X_0 I_h}{1+\cos\theta_m} \tag{8.31}$$

Minimum cost tuned filters

The costs of the reactor and the capacitor which make up the tuned filter are dependent on their respective ratings. The rating of the capacitor is given by

$$S_C = I_{F1}^2 X_{C1} + I_{Fh}^2 X_0 \tag{8.32}$$

where I_{F1} is the fundamental frequency current in the tuned filter given by

$$I_{F1} = \frac{V_1}{(X_{C1} - X_{L1})} = \frac{V_1}{X_0\left(h - \frac{1}{h}\right)} = \frac{hV_1}{(h^2 - 1)X_0} \tag{8.33}$$

In deriving the above equation, the effect of the resistance, R is neglected. V_1 is the fundamental frequency component of the voltage at the filter bus. (Actually, the harmonic components in the bus voltage in steady state, are neglected as filtering of harmonics in the source current supplied to the converter is assumed to be effective). $X_{C1} = X_0 h$ is the reactance of the capacitor and $X_{L1} = \frac{X_0}{h}$ is the reactance of the inductor at the fundamental frequency.

The rating of the reactor (S_L) is given by

$$S_L = I_{F1}^2 X_{L1} + I_{Fh}^2 X_0 \tag{8.34}$$

The size of the filter (S) is defined as the reactive power supplied by the capacitor (at the fundamental frequency). Thus, we get

$$S_C = S + I_{Fh}^2 X_0 \tag{8.35}$$

$$S_L = \frac{S}{h^2} + I_{Fh}^2 X_0 \tag{8.36}$$

The total cost of the filter (K) is given by

$$K = S_C U_C + S_L U_L \tag{8.37}$$

where U_C and U_L are the unit costs of the capacitor and inductor, respectively. Substituting Eqs. (8.35) and (8.36) in (8.37) we get an expression for the total cost as a function of the size of the filter. Since our objective is to determine the optimum size of the filter that will minimize the total cost, we need to express X_0 in terms of the size S. It can be easily shown that

$$X_0 = \frac{N}{S}, \quad N = \frac{V_1^2 h^3}{(h^2 - 1)^2} \tag{8.38}$$

Substituting Eq. (8.38) in (8.37), we get,

$$K = AS + BS^{-1} \tag{8.39}$$

where $A = U_C + \frac{U_L}{h^2}$, $B = I_{Fh}^2 N (U_C + U_L)$.

It is to be noted that I_{Fh} is expressed in terms of I_h (the harmonic current injected by the converter) and is defined by Eq. (8.28).

Since, A and B are constants, the cost of the filter is minimum when the size of the filter is minimum given by

$$S_{opt} = (B/A)^{\frac{1}{2}} \tag{8.40}$$

The filtering would improve with increased size of the filter. Hence the choice of the filter size will not be below that given by the above expression.

Design of a High Pass Filter

This is a second order (damped) filter described earlier. For harmonic frequencies equal to or higher than 17, second order high pass filter is usually provided. By defining the following parameters,

$$h_0\omega = 1/\sqrt{LC}\,, X_0 = \sqrt{L/C},\; Q = R/X_0 \tag{8.41}$$

the following values can be chosen

$$0.5 < Q < 2$$

$$h_0 \leq \sqrt{2}h_{\min} \tag{8.42}$$

where $h_{\min}$ is the smallest value of h to be handled by the filter. The choice of h_0 given above implies that the filter impedance at the $h_{\min}$ has decreased approximately to the value of R.

The filter impedance is given by Eq. (8.27). The reactive power supplied by the filter Q_F is

$$Q_F = \left(h_0/\left(h_0^2 - 1\right)\right)\left(V_1^2/X_0\right) \tag{8.43}$$

In deriving the above, the voltage across the inductor at the fundamental frequency is neglected. It can be shown that the filtering is improved if Q_F is increased and higher value of h_0 can be chosen. Hence, it is advantageous in designing high pass filter to exclude six pulse operation.

There is no optimum quality factor, Q in the case of high pass filters. Q is chosen for providing the best performance of the filter over a band of frequencies higher than those covered by tuned filters. The performance of these filters is insensitive to the deviations in the frequency or filter parameters.

Double and Triple Tuned Filters

A single tuned filter can be used to filter one characteristic harmonic (or a non-characteristic harmonic when required). Multiple single tuned filters are required to filter several (discrete) harmonics. For example, at least two single tuned filters are required for twelve pulse converters (to filter 11th and 13th harmonics). Since AC filters also provide the required reactive power which is a function of the power flow is the HVDC link, there is need to arrange the filters into switchable sections (filter banks) that adds to the cost. An alternative is to provide switchable shunt reactors that are switched in at low DC power levels. This also is a costly option with increased space requirements. Weak AC systems (with low SCR) also impose an upper limit on the size of the filter bank to limit the voltage fluctuations (flicker) caused by the switching of a filter bank. Thus, coordination of the filtering and supplying reactive power considerations dictate the adoption of double and triple tuned filters shown in Fig. 8.11 and 8.12 respectively. The advantages also include (*i*) only one inductor (L_1) is subject to full line impulse voltage and (*ii*) power loss at the fundamental frequency is considerably reduced.

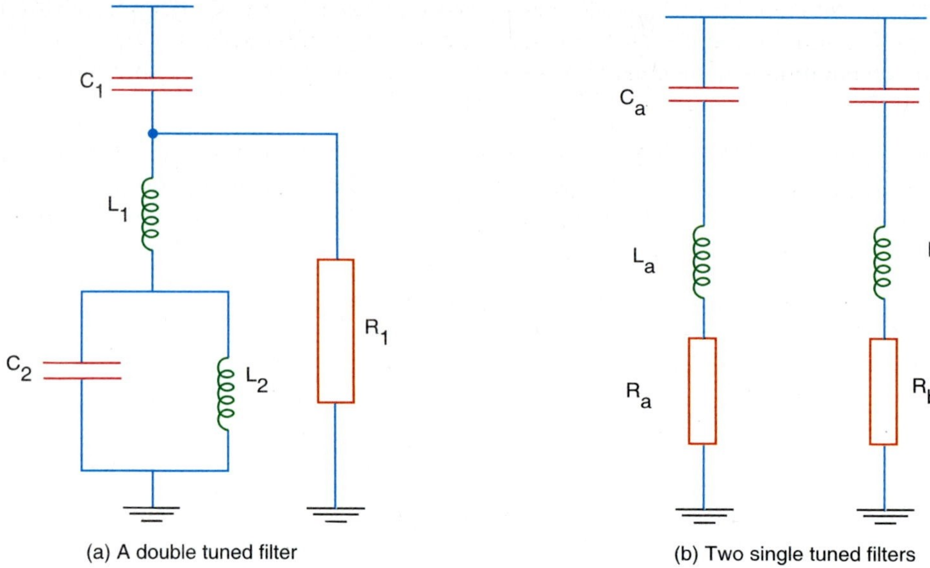

(a) A double tuned filter

(b) Two single tuned filters

Fig. 8.11: Configuration of a double tuned filter and its equivalent

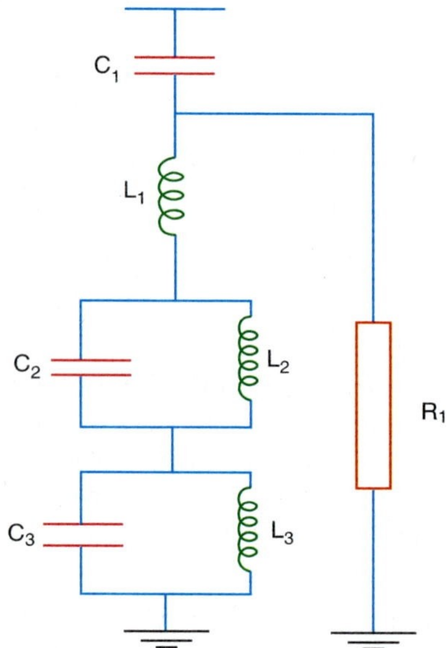

Fig. 8.12: A triple tuned filter

Electrically, neglecting losses, a double tuned filter is identical to two single tuned filter branches shown in Fig. 8.11 (*b*). Similarly a lossless triple tuned filter is identical to three single tuned filter branches.

The double and triple tuned filters shown in Figs. 8.11 and 8.12 have a single resister R_1 each. Additional damping resistors can be applied in parallel with reactor L_1 and/or L_2 and L_3 to produce a damped characteristic.

Analysis of a double tuned filter

The impedance of the filter shown in Fig. 8.11 (a) (neglecting the effect of R_1) as a function of the complex frequency (s) is given by

$$Z_F(s) = L_1 s + \frac{1}{C_1 s} + \frac{1}{C_2 s + \dfrac{1}{L_2 s}}$$

$$= \frac{L_1\left(s^4 + as^2 + b\right)}{s\left(s^2 + \dfrac{1}{L_2 C_2}\right)}$$

$$= \frac{L_1\left(s^2 + \omega_1^2\right)\left(s^2 + \omega_2^2\right)}{s\left(s^2 + \Omega_1^2\right)} \tag{8.44}$$

where $$a = \frac{L_1 C_1 + L_2 C_2 + L_2 C_1}{L_1 L_2 C_1 C_2},\quad b = \frac{1}{L_1 L_2 C_1 C_2},\quad \Omega_1^2 = \frac{1}{L_2 C_2}$$

The zeros of $Z_F(s)$ lie on the imaginary axis corresponding to the two frequencies ω_1 and ω_2. The poles are also on imaginary axis corresponding to $\omega = 0$ and $\omega = \Omega$. From the properties of the impedance of one-port L-C networks, we can state

$$0 < \omega_1 < \Omega_1 < \omega_2 \tag{8.45}$$

Since the double tuned filter is equivalent to the two single tuned filter shown in Fig. 8.12 (a), we can compare the impedances of the two circuits and obtain the following results

$$L_1 = \frac{L_a L_b}{(L_a + L_b)},\ L_2 = L_a + L_b,\ C_1 = C_a + C_b,\ C_2 = \frac{C_a C_b}{(C_a + C_b)} \tag{8.46}$$

From the above, it follows that

$$L_1 L_2 = L_a L_b,\ C_1 C_2 = C_a C_b \tag{8.47}$$

The impedance of the double tuned filter as a function of the frequency is obtained by substituting $s = j\omega h$ in the expression given in Eq. (8.44). Note that ω is the fundamental (system) frequency and h is viewed as a continuous variable.

Defining $\omega_r = \dfrac{1}{\sqrt{L_1 C_1}},\ X_1 = \omega_r L_1 = \dfrac{1}{\omega_r C_1},\ h_r = \dfrac{\omega_r}{\omega},\ H_1 = \dfrac{\Omega_1}{\omega},\ X_2 = \Omega_1 L_2 = \dfrac{1}{\Omega_1 C_2}$

We can express the impedance as a function of h as

$$Z_F(j\omega h) = jX_1\left(\frac{h}{h_r} - \frac{h_r}{h}\right) + \frac{X_2}{j\left(\frac{h}{H_1} - \frac{H_1}{h}\right)}$$

$$= j\left[X_1\frac{h^2 - h_r^2}{hh_r} - \frac{X_2H_1h}{h^2 - H_1^2}\right] \tag{8.48}$$

To compute ω_1 and ω_2, we equate the RHS of Eq. (8.48) to zero and get the following quadratic equation in h^2 given below

$$h^4 - \left(h_r^2 + H_1^2 + \frac{X_2}{X_1}H_1h_r\right)h^2 + h_r^2H_1^2 = 0 \tag{8.49}$$

The reactive power (Q_F) supplied by the filter is given by

$$Q_F = \frac{V_1^2}{Z_{F1}}$$

where $Z_{F1} = \left(h_r^2 - 1\right)\frac{X_1}{h_r} - \frac{X_2H_1}{\left(H_1^2 - 1\right)}$ (8.50)

Example

A double tuned AC filter at Rihand-Delhi HVDC converter station has the following parameters.

$$C_1 = 0.77\ \mu\text{F}, L_1 = 94.43\ \text{mH}, C_2 = 31.69\ \mu\text{F}, L_2 = 2.29\ \text{mH}, f = 50\ \text{Hz}, V_1 = 400\ \text{kV}.$$

Compute (*i*) ω_1 and ω_2 (*ii*) the reactive power Q_F.

Solution

$$\omega_r = \frac{1}{\sqrt{L_1C_1}} = 3708.5\ \text{rad/sec}, f_r = 590.2\ \text{Hz}$$

$$\Omega_1 = \frac{1}{\sqrt{L_2C_2}} = 3712\ \text{rad/sec},\ F_1 = 590.8\ \text{Hz}$$

$$h_r = \frac{f_r}{f} = \frac{590.2}{50} = 11.8, H_1 = \frac{F_1}{50} = 11.8$$

$$X_1 = 350\ \text{ohms}, X_2 = 8.5\ \text{ohms}$$

We can solve the quadratic equation (8.49) to obtain the solution for h = 10.9 and 12.76.

Note that this filter is designed to filter 11th and 13th harmonics.

The reactive power supplied by the filter is given by

$$Q_F = \frac{V_1^2}{Z_{F1}} = \frac{400^2}{4100} = 39\ \text{MVAR}$$

It is interesting to note that there are two more double tuned AC filters at Rihand-Delhi project. These are (1) Filters for 3rd and 36th harmonic and 2) filters for 5th and 27th harmonic. Data for 3rd and 36th filters:

$$C_1 = 0.77\ \mu F,\ L_1 = 14.13\ \text{mH},\ C_2 = 1.99\ \mu\text{F},\ L_2 = 416.6\ \text{mH}$$

Data for 5th and 27th filters:

$$C_1 = 1.39\ \mu\text{F},\ L_1 = 12.38\ \text{mH},\ C_2 = 6.11\ \mu\text{F},\ L_2 = 54.78\ \text{mH}$$

In both these filters, instead of R_1, damping resistors R_H and R_L are connected across L_1 and L_2 respectively. For the first double tuned filter (3, 36) R_H = 480 ohms, R_L = 13700 ohms. This results in quality factors at 3rd and 36th harmonics as 27 and 3 respectively. Essentially, the filters combine a tuned and a damped filter. The quality factors for the second filter are 53 (for 5th harmonic) and 3.3 (for 27th harmonic).

Analysis of triple tuned filters

The impedance of a lossless triple tuned filter as a function of complex frequency s is given by

$$Z_F(s) = \frac{L_1(s^2+\omega_1^2)(s^2+\omega_2^2)(s^2+\omega_3^2)}{s(s^2+\Omega_1^2)(s^2+\Omega_2^2)} \tag{8.51}$$

where ω_1, ω_2 and ω_3 are frequencies which have to be filtered. Ω_1 and Ω_2 are defined by

$$\Omega_1 = \frac{1}{\sqrt{L_2C_2}}, \Omega_2 = \frac{1}{\sqrt{L_3C_3}}$$

The impedances of the filter as a function of $h\omega$ (h is treaded as a continuous variable) is given by

$$Z_F(jh\omega) = jX_1\left(\frac{h}{h_r} - \frac{h_r}{h}\right) + \frac{X_2}{j\left(\frac{h}{H_1} - \frac{H_1}{h}\right)} + \frac{X_3}{j\left(\frac{h}{H_2} - \frac{H_2}{h}\right)} \tag{8.52}$$

where

$$H_1 = \frac{\Omega_1}{\omega},\ H_2 = \frac{\Omega_2}{\omega},\ h_r = \frac{\omega_r}{\omega},\ \omega_r = \frac{1}{\sqrt{L_1C_1}}$$

$$X_1 = \omega_r L_1 = \frac{1}{\omega_r C_1},\ X_2 = \Omega_1 L_2 = \frac{1}{\Omega_1 C_2},\ X_3 = \Omega_2 L_3 = \frac{1}{\Omega_2 C_3}$$

To compute ω_1, ω_2 and ω_3 (the zeroes of Z_F) we equate the RHS of Eq. (8.52) to zero. We finally get a cubic equation in h^2 given by

$$h^6 - a_1 h^4 + a_2 h^2 - a_3 = 0 \tag{8.53}$$

where

$$a_1 = h_r^2 + H_1^2 + H_2^2 + h_r\left(\frac{X_2}{X_1}H_1 + \frac{X_3}{X_1}H_2\right)$$

$$a_2 = H_1^2H_2^2 + h_r^2(H_1^2 + H_2^2) + h_r H_1 H_2\left(\frac{X_2}{X_1}H_2 + \frac{X_3}{X_1}H_1\right)$$

CHAPTER 8

$$a_3 = H_1^2 H_2^2 h_r^2$$

$h_1 = \frac{\omega_1}{\omega}$, $h_2 = \frac{\omega_2}{\omega}$, $h_3 = \frac{\omega_3}{\omega}$ are the three real roots of Eq. (8.53). It is easily observed that the product of the real roots is $\sqrt{a_3}$. This implies that

$$\omega_1 \omega_2 \omega_3 = \omega_r \Omega_1 \Omega_2 \tag{8.54}$$

we have to select Ω_1 and Ω_2 such that

$$0 < \omega_1 < \Omega_1 < \omega_2 < \Omega_2 < \omega_3 \tag{8.55}$$

Automatic tuning of filters

The optimum value of the quality factor of a tuned filters is inversely proportional to δ_m (which primarily depends on the frequency deviations in the system). The low Q (quality factor) results in increased losses and increased harmonic distortion at the converter bus. To maintain the same level of filtering, the value of the capacitance has to be raised increasing the cost. Only advantage of low Q is the reduction in the harmonic currents in the filters caused by ambient harmonics (sources other than the HVDC converters).

The automatic tuning was tried earlier by varying the inductance of the filter by the use of a tapped coil with tap-changing mechanism or by a variometer (a fixed coil in series with a movable coil so that the coupling between the two coils is variable). The controller works to reduce the phase angle between the harmonic voltage and the current.

Recently ABB has developed Continuously Tuned (ConTune) AC filters (for single tuned filters) by variation of the inductance of the filter reactor without moving parts. The variable inductance is achieved with an iron core placed inside the reactor. Around the iron core is wound a control winding in which a variable DC current can flow. The principle of this method is based on the variation of permeability of the magnetic material by applying a transverse magnetic field (perpendicular to the main flux direction). As the current increases from zero, the inductance decreases continuously. The controller uses a P-I regulator and a 6-pulse controlled rectifies (as an amplifier).

A test installation of an 11th harmonic ConTune filter was made in the Lindome Station of the 300 MW Konti-Skan 2 HVDC link in 1993. The filter generates reactive power of 11.6 MVAR at 132 kV. The filters was designed to accommodate frequency and component variations that represent detuning of –3 to 2 Hz. The ConTune filter had Q factor of 200 and resulted in substantial reduction in the 11th harmonic distortion. A commercial installation of ConTune filter was subsequently done at Celilo terminal of the Pacific Intertie HVDC link in western USA.

Design of combined AC filters

While Eq. (8.30) can be used for finding the optimum Q factor of each tuned branch, the harmonic voltages have to be computed by considering the entire filter as a unit. Accurate computation of the harmonic voltages requires a good knowledge of the harmonic characteristics of the AC system. This is best done by using digital computer programs.

The size of individual filter branches has to be determined by the considerations of (i) demands on the reactive power at various loading conditions and how they are to be met and (ii) the filtering considerations of satisfying constraints on telephone interference. The optimization of filter design involves several trials by checking various combinations (of filter branches and shunt capacitors/reactors) and operating conditions that also affect the AC system harmonic characteristics.

Ratings of filter components

The voltage and current stresses of AC filters consist of the fundamental frequency and harmonic frequency components. Their magnitudes in steady-state depend on the AC system voltage, harmonic currents generated by the converters and AC system impedances. It is necessary to determine the highest steady-state current and voltages for each filter component, based on calculations performed over the entire operating range.

The study of transient voltage and current stresses enable the computation of energy duty of filter resistors and arresters in addition to the insulation levels for all the filter components. To calculate the highest stresses, the following cases are investigated.

1. Single phase ground fault: The fault is applied on the converter bus next to the AC filter. It is assumed that the filter capacitor is charged to a voltage level corresponding to switching impulse protective level of the AC bus arrester.
2. Switching surge: For the calculation of the switching surge stresses, a standard wave of 250/2500 μs with a switching impulse protection level of the AC bus arrester is applied at the converter bus.
3. Filter energization: To determine the maximum inrush currents for AC filters, they are energized at the instant of the (maximum) AC peak voltage at the filter bus.
4. Fault recovery after three-phase ground fault: To determine the maximum energy stresses for the filter arresters and resistors, various fault-clearing scenarios are considered. Worst case stresses occur if the HVDC converter are blocked after fault initiation, while AC filters remain connected to the AC bus after fault clearing and recovery of the AC System voltage. In this case, load rejection by blocking of the HVDC converters result in temporary overvoltages with high content of non-characteristics harmonics due to transformer saturation and low harmonic resonances between the filter and AC network.

Protection of filters

The filter is exposed to an overvoltage during switching in and the magnitude of this overvoltage is a function of the short circuit ratio (higher with low values of SCR) and the saturation characteristics of the converter transformer.

During switching in, the filter currents (at filter frequencies) can have magnitudes ranging from 20 to 100 times the harmonic currents in normal (steady-state) operation. The lower values are for tuned filters and higher values are applicable to high pass filters. These overcurrents must be taken into consideration in the mechanical design of reactor coils.

When filters are disconnected, their capacitors remain charged to the voltage at the instant of switching. The residual direct voltages can also occur on bus bars. To avoid this, the capacitors may be discharged by short-circuiting devices or through converter transformers or by voltage transformers loaded with resistors.

If the network frequency deviates from the nominal valve, higher currents and losses will result in AC filters. If they exceed the limits, the filters have to be disconnected.

8.5 DC FILTERS

8.5.1 Criteria of Design

The harmonics in the DC voltage across the converter contain both characteristic and non-characteristic orders. These harmonics result in current harmonics in DC lines and cause noise in telephone circuits.

The harmonic current generated in the line can be computed from the knowledge of harmonic voltage sources at the converters, smoothing reactor, DC filter and line parameters. The harmonic current varies with the distance (from the converter station) along the line.

The effectiveness of the DC filter is judged by one of the following criteria:

1. maximum voltage TIF on DC high voltage bus
2. maximum induced noise voltage (INV) in millivolts/km in a parallel test line one kilometer away from the HVDC line
3. maximum permissible noise to ground in dBrnc in telephone lines close to HVDC lines.

The use of TIF criterion does not take into account the effects of coupling between the telephone line and the HVDC line. It also does not consider the standing wave phenomenon of the various harmonic currents along the DC line and characteristics of the telephone line, such as balance and shielding.

The test line method has been used for designing HVDC filters for the Pacific Intertie Project in Western USA. The Induced Noise Voltage (INV) computed is the longitudinal voltage along the test line and establishes a norm for a specific system. According to Bell Telephone Standards (BTS), the permissible total longitudinal noise voltage is 54.8 mV. This permissible voltage assumes a 50 dB balance in the telephone line. This implies that the transverse voltage induced in the telephone line is $1/10^{2.5} = 1/316$ times the longitudinal voltage. The permissible induced noise voltage is 0.173 mV. This is based on the criterion, the acceptable noise (according to BTS) must not exceed 20 dBrnc that is 20 dB above the reference level of 10^{-12} W (1 pico watt). Thus, the permissible noise power is 10^{-10} W. Not more than half of this (0.5×10^{-10} W) should be from any one source. Since telephone lines are usually terminated in impedance of 600 ohms, transverse noise voltage (C-message weighted) corresponding to the noise power of (0.5×10^{-10} W) is

$$\sqrt{PR} = 0.173 \text{ mV}$$

To use criterion independent of a particular system, the equivalent disturbing current method is now widely used. This is described in [18] and is outlined below.

The induced noise on a communication circuit near a HVDC line is computed from

$$V_g = \left[\sum_{n=1}^{N} \sum_{j=1}^{k} (I_{jn} Z_{mjn} K_{sn} P_n)^2 \right]^{\frac{1}{2}}$$

where n is the order of the harmonic, N is determined by the system frequency. An upper limit of 5000 Hz corresponds to $N = 83$ in a 60 Hz network and $N = 100$ in a 50 Hz network. k is the number of conductors in the line. I_{jn} is the nth harmonic current in the jth conductor. Z_{mjn} is the mutual impedances between the telephone line with the jth conductor at n^{th} harmonic. This impedance depends on the earth resistivity. K_{sn} is the shielding factor and P_n is the weighting factor at n^{th} harmonic.

The harmonic current I_{jn} is a function of the distance from the sending or receiving end. Typically, the current is obtained as superposition of two components—one due to the source voltages at the sending end and the second , due to the voltages at the receiving end. If the phase angles of the harmonic voltages at the two ends of the line are not known exactly, a root sum of the squares of the currents at the two ends is used. Thus we have,

$$V_g = [\sum_{n=1}^{N}\sum_{j=1}^{k}(I_{jns}^2 + I_{jnr}^2)(Z_{mjn}K_{sn}P_n)^2]^{\frac{1}{2}} \tag{8.56}$$

We assume that the disturbing effect (on the telephone line) can be approximated by a current on a single conductor geometrically located between the conductors on the DC line. Neglecting variations in the mutual impedance as function of j, we can obtain the expression for the induced noise voltage on nearby a telephone line as

$$V_g = [\sum_{n=1}^{N}(I_nP_nK_{sn}Z_{mn})^2]^{\frac{1}{2}} \tag{8.57}$$

where $$I_n^2 = I_{ns}^2 + I_{nr}^2$$

Definition of equivalent disturbing current

Assuming that the majority of the wire-line circuits susceptible to induced noise will be shielded cable, either crossing or close to the DC line, we can assume $K_{sn}Z_{mn}$ as independent of the frequency and we can express Eq. (8.57) as

$$V_g = I_e \cdot Z_e \tag{8.58}$$

where Z_e is the effective average value of $K_{sn}Z_{mn}$ typically calculated at 1000 Hz, with C-message weighted noise calculations. I_e is defined as

$$I_e = [\sum_{n=1}^{N}(I_nP_n)^2]^{\frac{1}{2}} \tag{8.59}$$

I_e is termed as the equivalent disturbing current and represents current at single frequency (typically, 1000 Hz) which has the same disturbing effect. I_e can be determined by the DC system design. For the chosen DC line routing, Z_e can be determined by analysis of the routing of the communication circuits. The basic advantage of using the concept of I_e is to decouple the studies required to design the DC filters from studies involved in inductive coordination.

The specification on I_e determines the DC filter design . A value of I_e = 270 mA (C-message weighted) has been used in USA. This specification is equivalent to the specification of 10 mV/km longitudinal INV used at Pacific Intertie ($I_e = \frac{V_g}{Z_m}$, where Z_m is calculated at 0.0370 ohm/km).

A sample profile of I_e with and without DC filters is shown in Fig. 8.13 for a specific HVDC link.

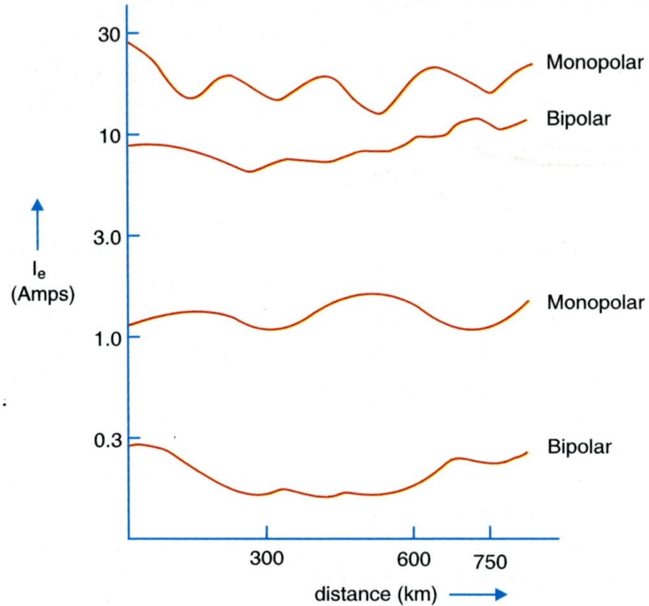

Fig. 8.13: Equivalent disturbing current profile

8.5.2 Passive DC Filters

Both tuned and damped passive filters have been used for DC filters. Back to back(BTB) HVDC links and those using cable systems do not have DC filters as induced noise is not an issue. The first HVDC scheme to use DC filters is the Sardinia link which has both cable and overhead line sections. Unlike in AC filters, the size of the filters does not depend on the reactive power requirements as this is not an issue with DC lines. The design is aimed (primarily) at reducing the filtering costs while meeting the specifications on the I_e (equivalent disturbing current).

The factors that affect the design of passive DC filters are given below:

1. Mixed cable/overhead line systems—These usually have more than one natural resonance frequency. DC filters, if used, should consider the possibility of overvoltages due to line resonance. Sardinia, Hokkaido-Honshu and Skagarrek schemes are examples of mixed cable/overhead line systems.
2. Modes of operation—Although bipolar mode of operation is used most of the times, monopolar ground return operation under extreme detuned conditions is the worst scenario to be considered. The monopolar metallic return mode results in – 3 dB noise

level compared to the worst case whereas the bipolar (unbalanced) mode has – 9 dB noise level (compared to the worst case).

3. DC line resonance overvoltages—HVDC line with smoothing reactors and DC filters has a natural first resonance at a low frequency (45–75Hz). If the line is resonant at the fundamental frequency, any continuous malfunction such as misfires can cause serious overvoltages. Smoothing reactor and/or DC filter parameters need to be changed to overcome this problem. While this may not always be feasible, it has been claimed that HVDC control systems can reduce such overvoltages.

   Overhead line or mixed cable and overhead line systems may have resonance frequencies corresponding to second and fourth harmonic. The excitation at these frequencies due to AC system faults can cause overvoltages. Although DC filter cannot significantly change the resonance frequencies, they can be designed to reduce the magnification level.

4. Overvoltages due to DC line faults—The overvoltages on the healthy pole during monopolar ground fault are affected by the line termination. High pass damped filters are most effective in reducing the overvoltages by presenting resistive termination. Even line surge capacitors reduce the terminal overvoltages compared to those at the midpoint of the line. Limiting line and terminal overvoltages is an important consideration in the selection of type and size of DC filters.

Presence of triplen harmonics in DC voltage

Although the classical theory does not predict the presence of triplen harmonics in the DC voltage (smallest characteristic harmonic with a Graetz bridge is of order six), measurements of DC harmonics have disclosed the occurrence of triplen harmonics of magnitudes which cannot be explained by factors such as unbalance, firing angle errors, tolerances in the leakage impedances etc. This has led to the development of a three-pulse model to explain the phenomenon [16, 17]. The cause for triplen harmonics were predicted to be the presence of stray capacitances inherent in converter transformers, bushings etc [14].

It has been shown in Chapter 2 that a Graetz bridge is a series connection of two valve groups. Each valve group can be modeled as a three-pulse converter. When the neutral point is not grounded there is no path for the flow of triplen harmonic currents. It can be easily seen that the neutral point voltage is not zero, but a third harmonic voltage (see Fig. 8.14 (a) for the equivalent circuit when 2 valves in a bridge, say 1 and 2 are conducting) which has a magnitude of half the line to line voltage ($e_a - e_c$). If there are stray capacitances to the ground, triplen harmonic currents can flow in the ground path. The DC side equivalent circuit for a Graetz bridge, is shown in Fig. 8.14 (b). The analysis for a 12 pulse converter is also presented in [16]. The implications of the presence of triplen harmonics are the following [16].

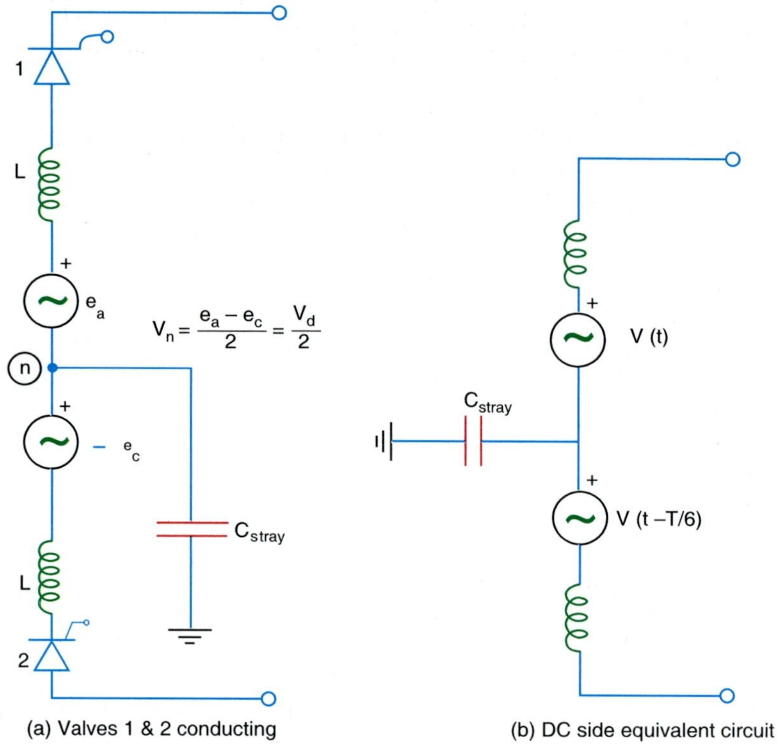

Fig. 8.14: A Graetz bridge represented as a series connection of two 3 pulse converters

1. Without a neutral bus capacitor, the telephone interference levels are several times (10–30) those calculated ignoring stray capacitances.
2. While neutral bus capacitor can decrease the level of ground currents of all triplen harmonics, this may not be sufficient. DC filters with sufficiently low impedance at the troublesome triplen harmonics will be required. The filter impedances at the characteristic harmonics must also be considerably lower than what is predicted by the classical analysis. The resulting DC filters can become both large and complex with multiple tuned frequencies.

The authors of [16, 19] claim that the utilities are specifying extremely low disturbing current criteria which may not be required for the problem of telephone interference.

8.6 ACTIVE FILTERS

Active filters were considered for DC filtering to meet the stringent requirements from power utilities in limiting telephone interference. The use of passive filters alone can increase the costs substantially. In 1991 a test installation was established at the Lindome station of Konti-Skan HVDC link. The first commercial active DC filter was installed in 1993 at the Skagerrak-3 HVDC intertie followed by Baltic Cable HVDC link in 1994. In India, active filters were installed at Chandrapur-Padghe link in 1998.

The active filter used in DC filtering is actually a hybrid active filter – that is an active filter in series with a passive shunt filter. The configuration of the DC active filter is shown in Fig. 8.15. Here, a double tuned filter (for 12th and 24th harmonic) is connected in series with a VSC-based active filter. IGBT devices are used as switches in the VSC with Pulse Width Modulation (PWM). The transformer provides galvanic separation between the VSC and the HVDC line. It also helps in raising the voltage injected to the desired level.

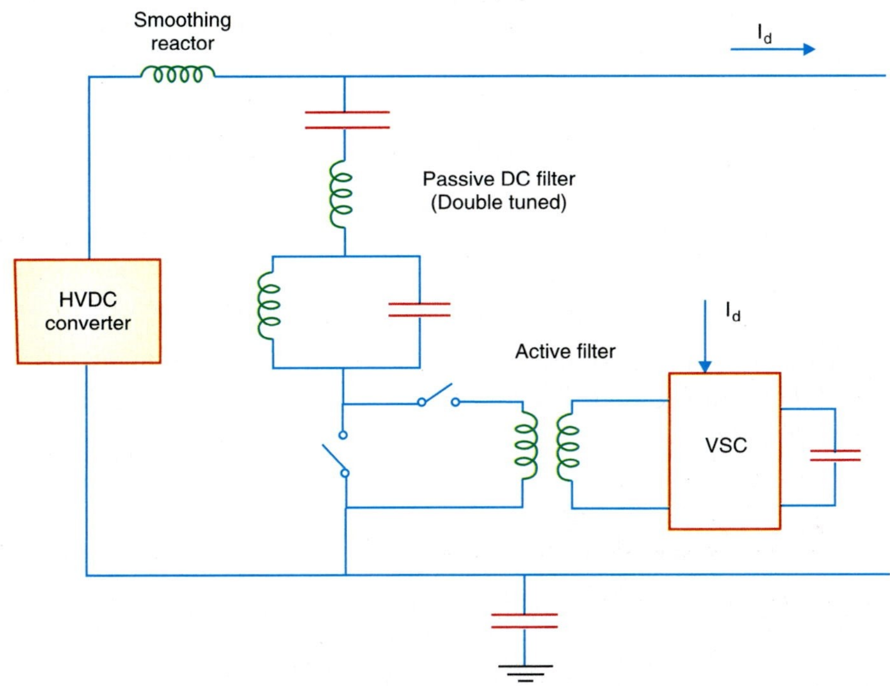

Fig. 8.15: Hybrid active filter for DC filtering

The control strategy of the active filter is to inject the required harmonic voltages of appropriate magnitudes and phase angles to cancel the harmonic currents flowing in the line.

At Chandrapur-Padghe line, active filters are installed at both stations, each filter connected in series with the passive filter (12/24). The active filter is designed to operate in the range of 350-2500 Hz. There are four active filters in all. The use of active filter replaced an additional passive filter branch in each pole compared to a purely passive solution. The disturbing current was found to be 170 mA in bipolar mode of operation, carrying full load of 1500 MW.

8.7 CARRIER FREQUENCY AND RI NOISE

HVDC converter stations can produce high levels of electrical noise in the carrier frequency band from 20 kHz to 490 kHz. They also generates radio interference (RI) noise in the mega Hertz range of frequencies. However, converters are usually located in buildings which are effectively shielded against electromagnetic radiation. Hence the contribution due to direct radiation from valves can be neglected. The radiation from the switchyard can predominate over that produced by DC lines. Effective PLC-RI filters are necessary to minimize the impact of the noise and elimination of interference with power line carrier communication.

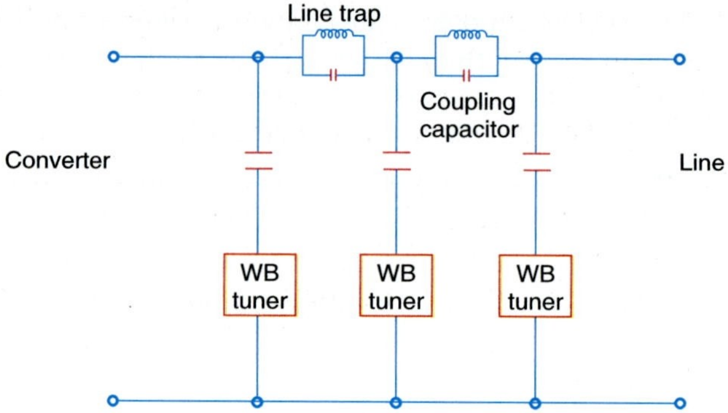

Fig. 8.16: Configuration of PLC/RI filter

The configuration of a PLC/RI filter is shown in Fig. 8.16. The attenuation requirements of the filter are shown in Fig. 8.17. The actual attenuation of the noise by the filter must be above the curve shown in Fig. 8.17.

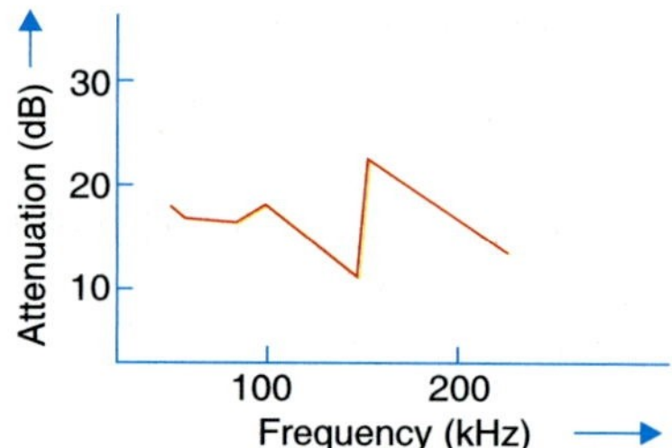

Fig. 8.17: Attenuation requirements of PLC/RI filters

The problem of interference to carrier systems is primarily from the turn-on of the valves in a converter which results in discharge of stray and other capacitances in the converter station. The analysis of interference problems involve (*i*) determination of the magnitude of the noise produced by the converter valves, (*ii*) coupling between the noise source and the carrier systems (*iii*) methods of reducing the coupling to avoid harmful interference.

Measurements and study of the carrier frequency noise result in the following conclusions

1. The magnitude of the conducted converter valve noise is generally higher than the ambient corona noise in the AC switchyard. It is also higher than the corona noise on the AC and DC transmission lines.
2. The differences in these levels are more at the lower frequencies.

REFERENCES AND BIBLIOGRAPHY

1. Ainsworth J.D., '*Filters, Damping Circuits and Reactive Volt-Amperes in HVDC Converters*', Chap. 7 in High Voltage Direct Current Converters and Systems, edited by B.J. Cory, MacDonald (London), 1965.
2. Phadke A.G. and Harlow J.H., '*Generation of Abnormal Harmonics in High Voltage AC-DC Power Systems*', IEEE Trans., Vol. PAS-87, No. 3, March, 1968, pp. 873–883.
3. Reeve J. and Krishnayya P.C.S., '*Unusual Current Harmonics Arising from High-voltage DC Transmission*', IEEE Trans., Vol. PAS-87, No. 3, March 1968, pp. 883–893.
4. Reeve J. and Baron J.A., '*Harmonic Interactions between HVDC Converters and AC Power Systems*', IEEE Trans., Vol. PAS-90, No. 6, Nov./Dec. 1971, pp. 2785–2791.
5. Hingorani N.G. and Burbery M.F., '*Simulation of AC System Impedance in HVDC System Studies*', IEEE Trans., Vol. PAS-89, No. 5/6, May/June 1970, pp. 820–826.
6. Reeve J. and Subbarao T., '*Dynamic Analysis of Harmonic Interaction between AC and DC Power Systems*', IEEE Trans., Vol. PAS-93, No. 2, March/April 1973, pp. 640–646.
7. Subbarao T. and Reeve J., '*Harmonics caused by Imbalanced Transformer Impedances and Imperfect Twelve Pulse Operation in HVDC Conversion*', IEEE Trans., Vol. PAS-95, No. 5, Sept./ Oct. 1976, pp. 1732–1737.
8. Harrison R.E. and Krishnayya P.C.S., '*System Considerations in the Application of DC Filters for HVDC Transmission*', CI-GRE Paper 14-09, 1978.
9. Mathur R.M. and Sharaf A.M., '*Harmonics on the DC Side in HVDC Conversion*', IEEE Trans., Vol. PAS-96, No. 5, Sept./Oct. 1977, pp. 1631–1638.
10. Lasseter R.H., Kalley P.W. and Lindh C.B., '*DC Filter Design Methods for HVDC Systems*', IEEE Trans., Vol. PAS-96, March/April 1977, pp. 571–578.
11. Breuer G.D., et al., '*HVDC-AC Harmonic Interaction,* Parts I & II', IEEE Trans., Vol. PAS-101, March 1982, pp. 701–718.
12. Melvold D.J., Haglof L. and Degn P., '*Carrier Frequency Harmonics Generated by HVDC Conversion*', IEEE Trans., Vol. PAS- 99, No. 2, March/April 1980, pp. 565–572.
13. Sarma M.P. and Gilsig T., '*A Method of Calculating the RI from HVDC Converter Stations*', IEEE Trans., Vol. PAS-92, No. 3, May/June 1973, pp. 1009–1018.
14. Larson E.V., Sublich M. and Kapoor S.C., '*Impact of Stray Capacitance on HVDC Harmonics*', IEEE Trans., Vol. PWRD- 4, No. 1, January 1989, pp. 637–645.
15. Prabhakara F.S. and Vancers I., '*Telephone Noise Induction Analysis for Square Butte HVDC Transmission Line*', Paper A78 584-5 Presented at IEEE PES Summer Meeting, Los Angeles, July 1978.
16. Shore N.L., Andersson G., Canelhas A.P. and Asplund G., '*A Three-pulse Model of DC Side Harmonic flow in HVDC Systems*', IEEE Trans., Power Delivery, Vol. 4, No.3, July 1989, pp. 1945–1954.
17. Dickmander D.D. and Peterson K.T., '*Analysis of DC Harmonics using Three-pulse Model for the Intermountain Power Project HVDC Transmission*', IEEE Trans., Power Delivery, Vol. 4, No. 2, April 1989, pp. 1195–1204.

18. Patterson N.A. and Fletcher D.E., "*The Equivalent Disturbing Current Method for DC Transmission Line Inductive Coordination Studies and DC Filter Performance Specification*", Int. Conf. on DC Power Transmission, June 1984, Montreal, Canada.

19. Johansson A.V. and Ekstrom A., "*Telephone Interference Criteria for HVDC Transmission Lines*", IEEE Trans., Power Delivery, Vol. 4, No.2, April 1989, pp. 1408–1421.

20. Zhang W., Asplund G., A. Aberg. U. Jonsson and O. Loof, "*Active DC filter for HVDC System–A Test Installation in the Konti-Skan DC Link at Lindome Converter Station*", Paper 92 SM–469–7 PWRD, IEEE/PES 1992 Summer Meeting, Seattle USA.

21. Patterson N.A., "*Carrier Frequency Interference from HVDC Systems*", IEEE Trans., Vol.PAS-104, No. 11, Nov. 1985, pp. 3255–3261.

22. Einarsson O., Peterson K.J. and Shockley P.R., "*Intermountain Power Project Carrier Interference Study*", IEEE Trans., Power Delivery, Vol. 4, No.2, April 1989, pp. 1205–1216.

23. Adhikari T., Isacsson G. and Ambekar V.D., "*The Chandrapur–Padghe HVDC Bipole Transmission*", CIGRE Symposium, Kaula Lampur, Malaysia, Sept. 1999.

24. C. Bartzsch, H. Huang, R. Roessel and K. Sadek, "*Triple Tuned Harmonic Filters–Design Principle and Operating Experiences*", IEEE Power Con. 2002, Kunming, China, 2002.

25. Juhlin L.E. et al., "*Measurement of Radio Frequency Interference from High Voltage Substations – Backgrounds and Considerations for Future Emission Requirements*", Paper C4-102, CIGRE Conference, Paris, 2008.

26. LaForest J.J., Lindh C.B., Lasseter R.H. and Reeve J., "*Modelling of the DC Converter Equipment in Carrier Band*", IEEE Trans., Vol.PAS-101, No.7, July 1982, pp. 2137–2145.

MULTITERMINAL AND MULTI-INFEED DC SYSTEMS

9

9.1 INTRODUCTION

HVDC transmission systems designed and operated so far are point to point systems with two terminals (converter stations). A multiterminal DC (MTDC) system has more than two converter stations, some of them operating as rectifiers and others as inverters. The simplest way of building a MTDC system from an existing two terminal system is to introduce tappings. Parallel operation of converters and bipoles can also be viewed as multi terminal operation.

Unlike in AC systems, the task of extending two terminal systems to multiterminal systems is not trivial. The complexities of control and protection increase considerably, and the use of HVDC breakers is generally required in the MTDC systems.

In recent times, with growth in HVDC transmission, there could be more than one converter in proximity feeding same load area. The operational issues of such Multi-Infeed DC (MIDC) systems are similar to those of MTDC systems. Hence, MIDC systems are also discussed in this chapter. With recent advances in the emerging technology of VSC-HVDC transmission, the application of MTDC systems is becoming more attractive than before.

9.2 POTENTIAL APPLICATIONS OF MTDC SYSTEMS

Apart from tapping of power from existing two terminal systems, there are three specific areas of applications for MTDC systems. These are listed below:

1. Bulk power transmission from several remote generating stations to several load centres. Here, each generating plant (or unit) is connected directly to a rectifier station thereby dispensing with the AC collector system. Similarly, a converter station at each load centre eliminates the need to build additional AC (or DC) lines for flexible energy exchange.

An MTDC system has several advantages over the alternative of point to point systems. For example, consider a system of two generating stations and two loads as shown in Fig. 9.1. This is a radial system with two rectifiers and two inverters. For ensuring the same level of flexibility in energy exchange, three two terminal DC links will be required in addition to a link connecting the two receiving systems, which could be AC or DC (see Fig. 9.2). This would result in extra costs for the converter stations, lines and additional power losses in increased number of conversions.

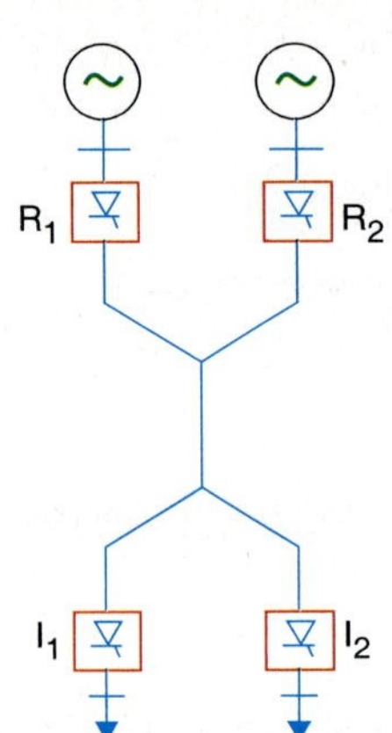

Fig. 9.1: MTDC system configuration for bulk power transmisstion

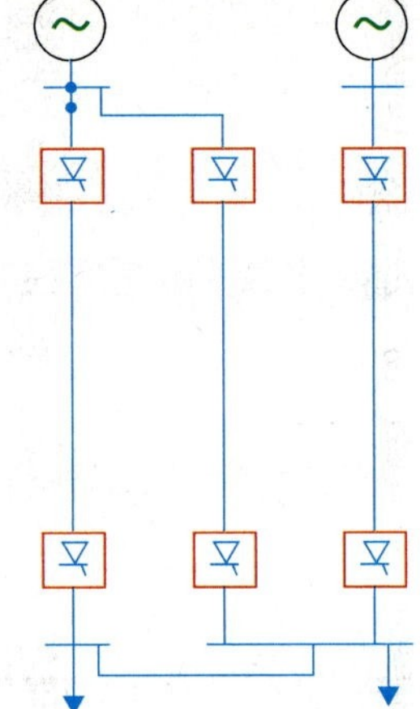

Fig. 9.2: Bulk power transmission using two terminal links

The elimination of AC collector system at the remote hydro generating stations can result in better efficiency in the operation of hydraulic turbines which are free to run at speeds independent of the system frequency.

2. Asynchronous interconnection between adjacent power systems. The advantages of asynchronous interconnection have already been described in Chapter 1. When more than two systems are involved, a MTDC system for interconnection is more flexible and economical than employing several two terminal DC links.
3. Reinforcing of an AC network which is heavily loaded. Consider an urban power system which is fed by a distant power station. It would be advantageous to arrange the power injection at more than one point so that the underlying AC network is not overloaded. This is easily achieved using a MTDC system with one rectifier station and several inverter stations (see Fig. 9.3).

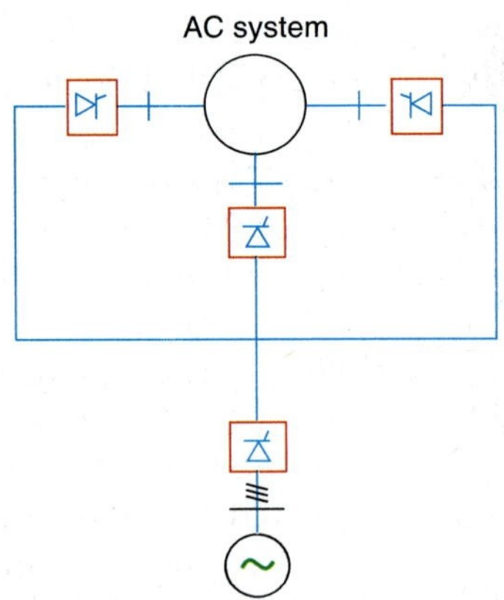

Fig. 9.3: Reinforcing of AC network using MTDC link

9.3 TYPES OF MTDC SYSTEMS

There are two possible types of MTDC systems

(*i*) Series

(*ii*) Parallel

The parallel MTDC systems can be further subdivided into the following categories:

(*a*) Radial

(*b*) Mesh

9.3.1 Series MTDC System

This is a natural extension of the two terminal system which is a series connected system. A three-terminal MTDC system is shown in Fig. 9.4. This shows a monopolar arrangement; however, a homopolar arrangement with two conductors is also possible. The system is grounded at only one point which may be conveniently chosen. If the line insulation is adequate, the grounding point can be shifted, based on changes in the operating conditions. Grounding capacitors may also be used to improve insulation coordination and system performance during transients.

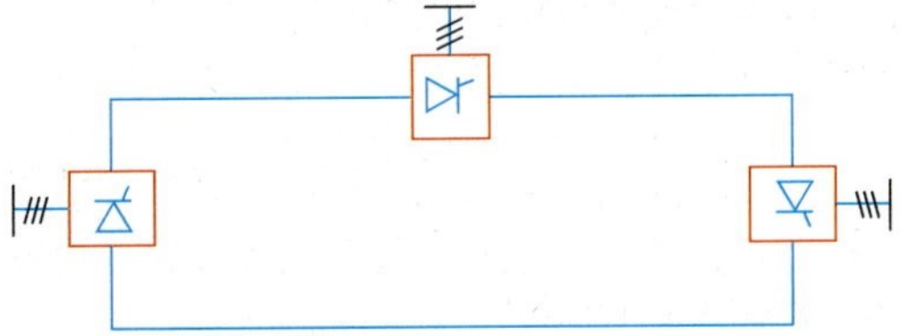

Fig. 9.4: A series connected MTDC system

In a series connected system, the current is set by one converter station and is common for all the stations. The remaining stations operate at constant angle (extinction or delay) or voltage control. In order to minimize the reactive power requirements and the losses in valve damper circuits, the normal operating values of firing angles may be adjusted using tap changer control. At all times, the sum of the voltages across the rectifier stations must be larger than the sum of the voltages across the inverter station. In case of a drop in the voltage at the current controlling rectifier station, the inverter with the larger current reference takes over the current control.

The switching in or out of a bridge is accomplished by deblocking/block and bypass in a manner similar to that in a two terminal system. The clearing of a fault in the DC line is also similar. The power reversal at a station is also done as in a two terminal system, by reversing the DC voltage by converter control.

The power control in a two terminal system is accomplished by adjusting the current while trying to maintain a constant voltage in the system. This is done to minimize the losses. However, in a MTDC series system, central control would be required to adjust the current in response to changing loading conditions. The local control of power would imply adjusting voltage at the converter station using angle and tap controls. Using only one bridge or a 12 pulse unit for the voltage control and operating the remaining bridges at minimum (or maximum) delay angle can reduce the reactive power requirements.

9.3.2 Parallel MTDC System

Here, the operating philosophy of constant voltage AC systems is extended to DC systems. The currents in all the converter stations except one are adjusted according to the power requirement. One of the terminals operates as a voltage setting terminal at constant angle or voltage. An example of 3 terminal radial system is shown in Fig. 9.5. This shows a monopolar system but bipolar arrangement would be normally used.

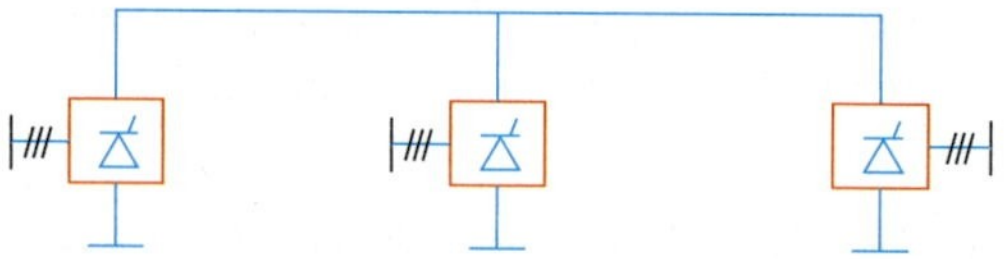

Fig. 9.5: A parallel connected radial MTDC system

A radial system is one in which the disconnection of one segment of transmission would result in interruption of power from one or more converter stations. In a mesh system, the removal of one link would not result in a disruption, provided the remaining links are capable of carrying the required power (with increased losses). Evidently, a mesh system can be more reliable than a radial system. An example of a 4 terminal mesh system is shown in Fig. 9.6.

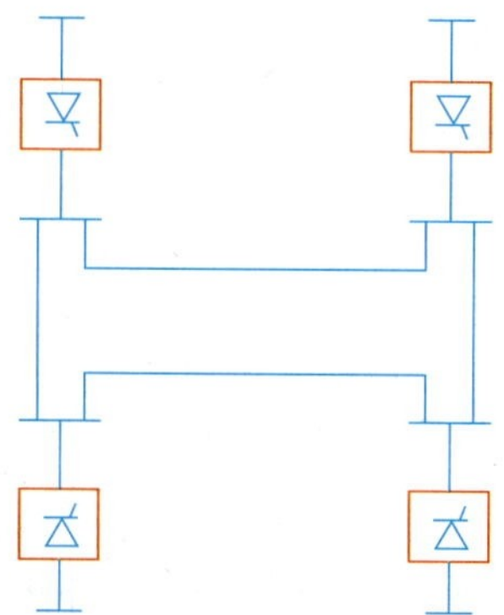

Fig. 9.6: A parallel connected mesh type MTDC system

The power reversal in a parallel MTDC system would involve mechanical switching as the voltage cannot be reversed. Also, loss of a bridge in one converter station would require either the disconnection of a bridge in all the stations or disconnection of the affected station.

9.3.3 Comparison of Series and Parallel MTDC Systems

The advantages and disadvantages of series and parallel MTDC systems are given below:

1. High speed reversal of power is possible in series systems without mechanical switching. This is not possible in parallel systems.
2. The valve voltage rating in a series system is related to the power rating, while the current rating in a parallel connected system is related to power. This would imply that for small power ratings of the tap, series connection may be cheaper even though valves have to be insulated for full voltage to ground. The parallel connection has the advantage of staged development in the converter stations by adding parallel converters as the power requirements increase.

3. There are increased losses in the line and valves in series systems, in comparison to parallel systems. The system operation in series systems can be optimized by operating the largest inverter at rated voltage.
4. Insulation coordination is a problem in series systems as the voltage along the line varies.
5. The permanent fault in a line section would lead to complete shutdown in a series connected system, while it would lead to only the shutdown of a converter station connected to the line section in a radial MTDC system. With provisions for fast identification and clearing of faults in mesh connected system, there is no disruption of power transfer.
6. The reduction in AC voltages and commutation failures in an inverter can lead to overloading of converters as current is transferred from other terminals in a parallel system. The problem is severe if the rating of the inverter is relatively small. Increased values of smoothing reactor and voltage dependent current limits can reduce the severity. However, the valve ratings would increase, resulting in increased unit costs. A recent study shows that the cost of a 500 MW DC equipment (at ±500 kV) would be 74% of the cost of a 1000 MW DC equipment [11]. It is concluded that a practical limit to unequal inverter ratings may be 75% : 25%.
7. The control and protection philosophy in a series MTDC system is a natural extension of that in a two terminal system. However, extension to parallel systems is not straightforward. Increased communication requirements and problems in recovery from commutation failures are associated with parallel systems. HVDC breakers of appropriate rating may be required for clearing faults in the DC line or converter stations.

From the relative merits and demerits of series and parallel MTDC systems described above, it may be concluded that series connection is appropriate for taps of rating less than 20% of the major inverter terminal. Parallel connection is more versatile and is expected to be widely used as in AC systems. The first application of a MTDC system is the Sardinia-Corsica-Italy link where an existing link between Sardinia and Italy is tapped at Corsica. This is a 50 MW parallel connected tap with two 100 kV six pulse thyristor bridges connected in series. A series tap was rejected for two reasons (*i*) the operating current due to frequency control can be as low as 10% of the rated current. This increases the voltage rating of the series tap and (*ii*) A series tap in inverter operation reduces voltage at the main inverter terminal, requiring increased extinction angles. This is harmful to mercury arc valves as the probability of arc-through increases.

Commutation failure at Corsica can result in overcurrents of 7 p.u. Smoothing reactors of 2.5H are chosen to limit the overcurrents due to disturbances in the AC system.

Quebec-New England link from Radisson in Quebec to Sandy Pond in Massachusetts with a converter at Nicolet became operational in 1992 with two inverters and one rectifier of capacity 2250 MW. While it was planned to integrate an existing two terminal HVDC link (from Des Canton to Comerford) with this system to form a 5 terminal MTDC system, this plan was dropped. Other plans to introduce MTDC system [13] was also shelved. The setback to planning for MTDC was primarily due to the inadequate preparation in the development of control and protection concepts required for reliable system operation.

9.4 CONTROL AND PROTECTION OF MTDC SYSTEMS

There are several methods of control in MTDC systems. Only parallel MTDC systems are considered as these involve complexities in extending the existing control methods. The various methods suggested are reviewed below.

9.4.1 Current Margin Method

This is most widely considered and the natural extension of control philosophy in a two terminal system. One of the converter station which is operating at the angle limit (minimum α or minimum γ) determines the DC voltage (dependent on the AC voltage and the tap ratio). The remaining terminals operate as current controlling terminals. The current through the voltage setting terminal (say n) is given by

$$I_n = -\sum_{j=1}^{n-1} I_j \tag{9.1}$$

where n is the number of terminals. In the above equation, the inverter currents are treated as negative, while the rectifier currents are treated as positive. The current controlling terminals operate with a voltage margin which may become zero or negative during disturbances in the AC system. As even small disturbances can affect the voltage margin, it is necessary to maintain the current and power distribution in the system with minor changes, during the disturbances. This is possible if current control is also provided at the voltage setting terminal (or slack terminal) such that it tries to maintain the same current as before. Because of measurement errors and the requirements of a smooth transition from angle (or voltage) control to current control, the current reference at the voltage setting terminal (VST) is chosen to satisfy the following equation:

$$\sum_{j-1}^{n} I_{jref} = I_{\text{margin}} \tag{9.2}$$

where I_{margin} is positive quantity.

The converter with the lowest voltage ceiling always acts as a voltage setting terminal. The changes in the voltage setting terminal due to disturbances in the AC system are called mode shifts. Uncontrolled mode shifts can be minimized by selecting a terminal with highest short circuit ratio as the voltage setting terminal. Due to the negative resistance characteristics of the constant extinction angle control, it would be advisable to choose a rectifier terminal as VST. The magnitude of the current margin is critical as converters of lower ratings can be overloaded when operating at angle limit.

The central controller that regulates the current orders at all the converter stations is termed as Current Reference Balancer (CRB) and is shown in the analog version in Fig. 9.7. Here, the current orders calculated from local power controllers are adjusted in order to satisfy Eq. (9.2). The limits on the current orders are taken into account in balancing current references. The actual implementation of CRB can be performed by using microprocessors.

Satisfactory operation of MTDC systems requires a reliable central CRB that operates at all times. This requires reliable two way communication between a central station and each converter station. If there is loss of a station and this information is not communicated, the system operation is adversely affected. In case of loss of a rectifier station, the power transfer is interrupted by voltage collapse. In case of loss of an inverter station, other stations will be overloaded.

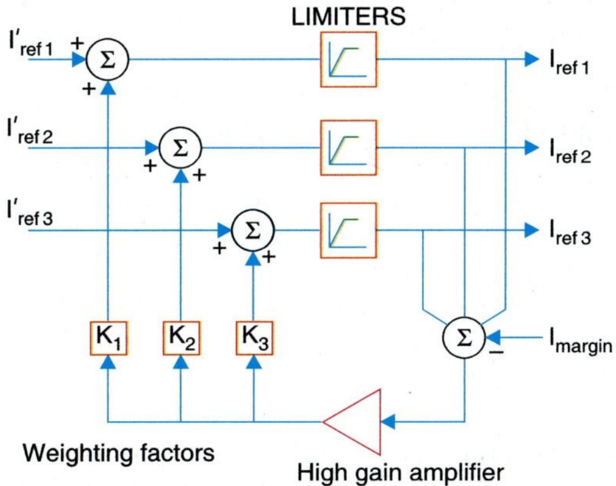

Fig. 9.7: Current reference balancer (*Source*: Reference 4)

In the current margin method, the change in the voltage setting terminal requires the operation of the tap changer in converter transformer to modify the voltage margin. This can be slow and results in less flexible control to deal with mode shifts. An improvement has been suggested by using a modified control scheme termed as voltage margin control method. In this method, all converter stations are provided with automatic voltage regulators (AVR) along with automatic current regulators (ACR). In the voltage setting terminal, AVR reference voltage is set to the rated voltage and in other stations, AVR reference voltage is set higher by an amount ΔE.

The operation of this control technique is illustrated by Fig. 9.8 which represents the control characteristics for a MTDC system with two rectifiers and two inverters. When the voltage setting terminal is shifted from INV2 to REC1, the voltage margin ΔE is added to the reference value at INV2 and subtracted from REC1. The voltage margin control method is also not free from the requirements of the centralized control and fast communication.

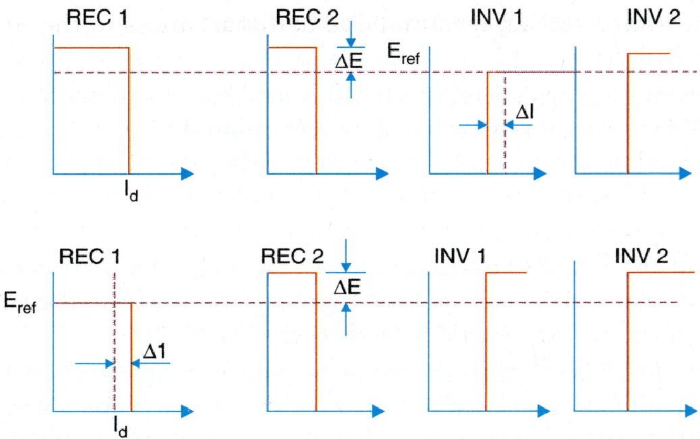

Fig. 9.8: Operation with voltage margin control method (a) VST = INV1 (b) VST = REC1

In order to facilitate the operation of MTDC systems even when there is failure of communication system, the following modifications to the basic current margin method of control have been suggested:

1. Voltage Limiting Control [1]
2. Use of Decentralized Current Reference Balancer (DCRB) [9]
3. Two ACR (Automatic Current Regulator) method [6].

These are discussed below.

9.4.2 Voltage Limiting Control

In the voltage limiting control method, the rectifier and inverter characteristics are arranged as shown in Fig. 9.9. The loss of a rectifier station does not have any significant effect on the system operation even in the absence of communication failure. The inverter currents are reduced in order to prevent voltage collapse. Also, the loss of an inverter station (operating on current control) would not overload the voltage controlling inverter because of its voltage reserve.

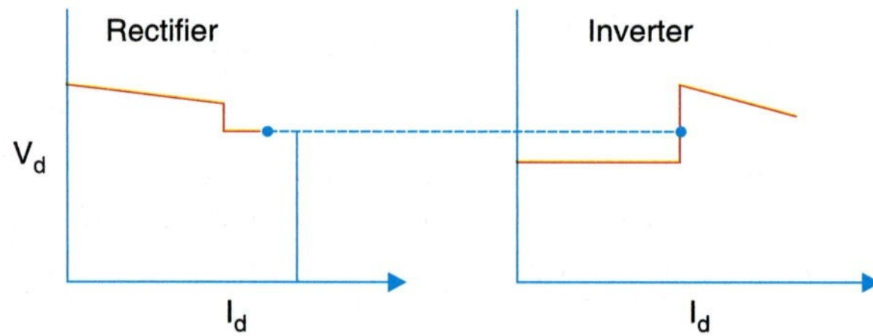

Fig. 9.9: Operation with voltage limiting method

The drawbacks of this scheme are (*i*) the disturbance in the AC system connected to an inverter station can result in other inverters getting unloaded (due to drop in the DC voltage). This may cause adverse effect on the AC systems supplied by healthy inverter stations, (*ii*) Two or more terminals can operate in the voltage controlling mode forcibly, in the case of loss of a terminal resulting in indeterminate distribution of currents in those terminals, (*iii*) Currents during DC line fault or commutation failure are likely to be higher without additional measures.

9.4.3 Decentralized Current Reference Balancing

This method is similar to the one described by Foerst et al. [1] without CRB and which operates on the current and voltage characteristics. DCRB is designed to allow rapid recovery of MTDC system operation following severe disturbances which require a momentary rundown of the DC system. The DCRB is supposed to permit the system recovery without coordinated centralized control of converters and even if one or more of the terminals is removed from service due to the disturbance.

The operation of the DCRB is illustrated with the help of a 3 terminal system with two rectifiers and one inverter. In the case of loss of one rectifier station, the characteristics of the remaining two converter stations do not intersect (see Fig. 9.10).

The object of the DCRB is to adjust the converter control characteristics around a preselected voltage level V_o in the manner indicated in Fig. 9.10. The slopes of the balancing characteristics σ_r and σ_i determine the relative adjustments that must be made to the current orders. In DCRB, the minimum and maximum current limits are preselected. The intersection of the balancing characteristics determines the steady state operation of the converters following a disturbance. The normal post-disturbance characteristics are also shown in Fig. 9.10 by solid lines. The current order required for this is locally determined by the balanced current and the specified current margin. In case of a rectifier, a fraction of the current margin would be added to the balanced current, while in the case of an inverter, a fraction of the margin would be subtracted.

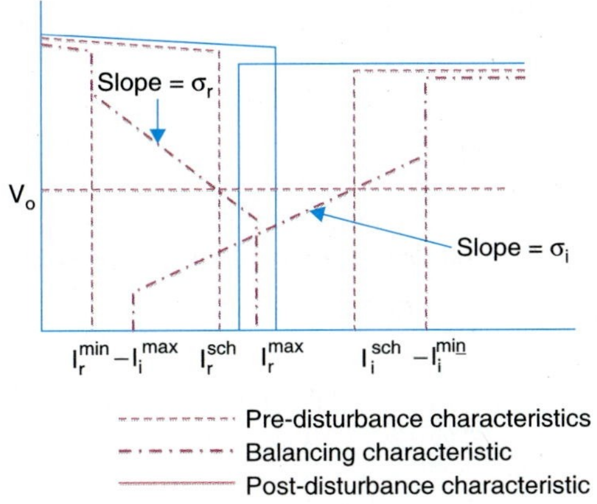

Fig. 9.10: Decentralized current reference balancing method

The slopes of the balancing characteristics along with the voltage level V_o are chosen, based on the considerations of voltage limits and magnitude of the variations in the DC network voltages during the balancing period.

Simulator studies have been carried out to demonstrate the functioning of DCRB and it is claimed that this technique is quite satisfactory in rapid restarting of the system with communication failure.

9.4.4 Two ACR Method

In this method, each-converter is provided with two automatic current regulators (ACR) and one voltage regulator. The rectifier and inverter control characteristics are shown in Fig. 9.11.

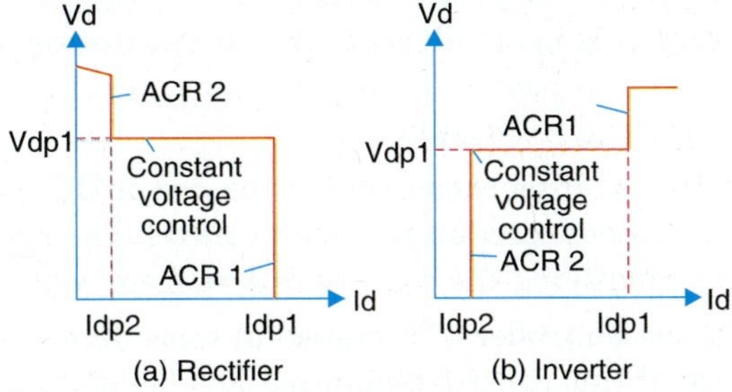

Fig. 9.11: Static $V_d - I_d$ characteristics of a converter in two ACR method (a) Rectifier, (b) Inverter

During the normal mode of operation, the rectifier station with the lowest voltage reference or the inverter station with highest voltage reference acts as voltage setting terminal (see Fig. 9.12). The remaining converter stations operate under the control of ACR1. I_{ref1} of the voltage setting terminal is set larger than the operating current (determined by other converters) by the current margin. The ACR2 operates during a contingency and its order can be fixed at the minimum operating current of the inverter.

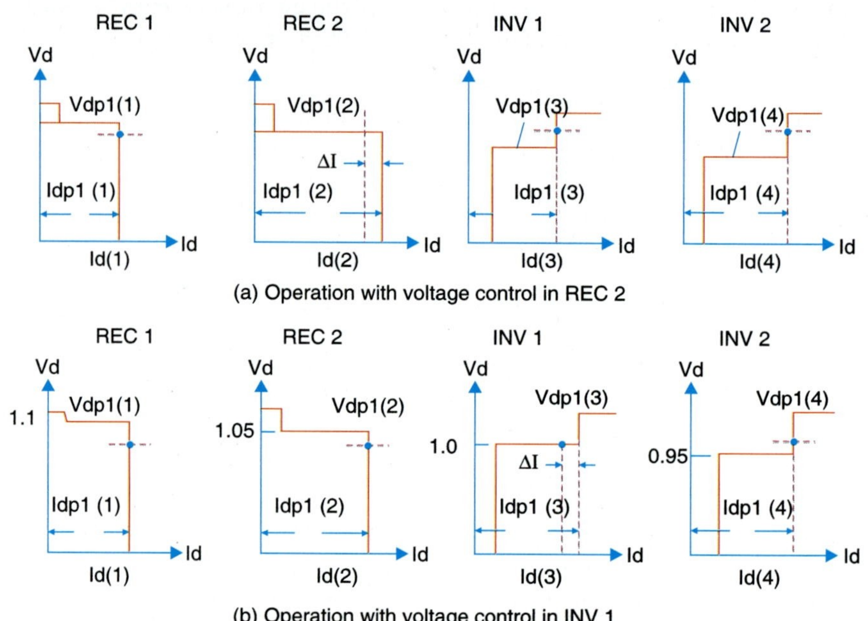

Fig. 9.12: Static $V_d - I_d$ characteristics of a 4 terminal system (a) Operation with voltage control in REC2 (b) Operation with voltage control in INV 1

Without additional measures, the inverter may not be able to recover from a commutation failure. Voltage dependent current order limit (VDCOL) is introduced to overcome this problem. The operation of a 4 terminal system both during normal conditions and when one of the converter stations is shutdown, is shown in Fig. 9.13.

The rectifier with the highest voltage order and inverter with lowest voltage order are always controlled by ACR1, thus resulting in small power deviations during a disturbance. This can be utilized to prevent shutdown of a base station (such as a nuclear power station) or prevent interruption of supply to a critical AC system.

It is claimed that the two ACR method permits the use of DC circuit breakers to isolate a line fault or converter fault without coordination with control. The system operation following a disturbance is optimized by modifying the current orders from central controller.

The implementation of the controller is discussed in some detail in [6,12] using control block diagrams. The performance of this control technique has been studied using digital dynamic simulation and HVDC simulator.

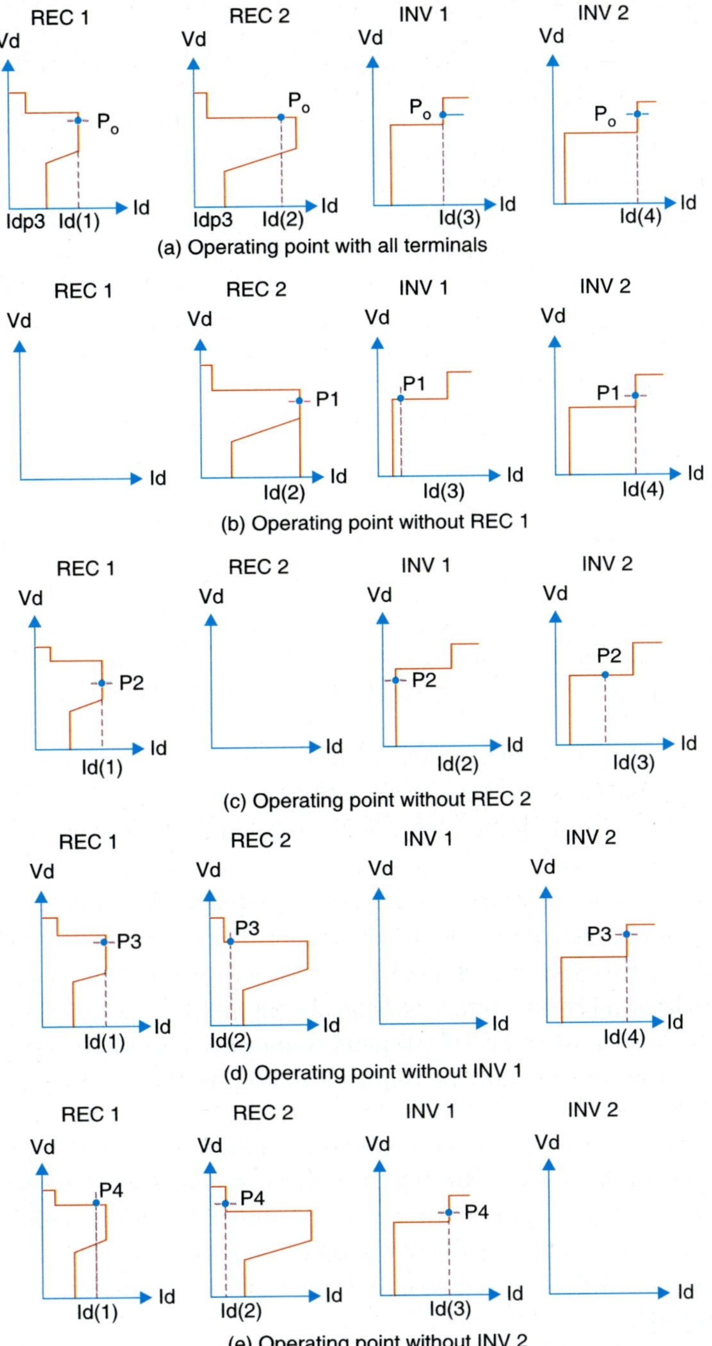

Fig. 9.13: (a) Operating point under normal conditions (b) Operating point without REC 1 (c) Operating point without REC2 (d) Operating point without INV 1 (e) Operating point without INV 2

9.4.5 Protection of MTDC Systems

One of the issues under discussion regarding MTDC system operation is the evaluation of protection requirements and coordination with control.

Conceptually, the system can be shut down following a fault in DC line or converter station and the faulted component isolated using high speed disconnect switches. The system can be restarted after adequate time for deionization of the arc path (in case of short circuits).

System reliability considerations dictate the need for fast clearing of fault with minimum disturbance to the healthy parts of the system. The DC breaker ratings can be minimized by utilizing the intervention of fast current controls to reduce the magnitude of the fault currents.

The detection of DC line faults gets complicated in a mesh system. A major problem is the large drop in the DC voltage even for distance faults. This requires fast detection and clearing of faults to maintain power transfers. Differential type of protection or directional sensitive measurements of the currents can be used to locate the fault. Communication would be required in both methods.

DC switches would be required primarily for parallel MTDC systems and particularly in mesh systems to utilize the improved security provided by such configurations.

9.5 STUDY OF MTDC SYSTEMS

The planning, design and operation of MTDC systems require detailed study using simulation tools. The new control and protection concepts have been formulated using HVDC simulators. The performance of the controllers are judged not only under normal conditions with changing load conditions, but also under disturbances caused by faults in the AC system and DC lines.

Some of the typical problems that have been considered for study are as follows:

1. Operation of small inverter taps connected to weak AC systems
2. Integration of existing HVDC converter stations in MTDC systems without major modifications in control
3. Evaluation of communication, reactive power and filtering requirements
4. Security of power transfer without fast communication links
5. Power and reactive power modulation strategies in MTDC systems.

Some of the conclusions from studies carried out [8] indicate the type of problems that can be anticipated in the operation of MTDC systems connected to weak AC systems.

1. Large smoothing reactor may be required to help in the recovery of a small inverter from AC faults.
2. The speed of recovery of the entire MTDC system depends upon the recovery of the small tap, if it is still connected to the system following clearance of a fault.
3. It is necessary to consider the effect of mode shifts and develop additional protection sequences, particularly for faults in DC line.
4. The provision of VDCOL at each rectifier is beneficial and its characteristics have to be adjusted suitably.

With a good knowledge of HVDC controls and their accurate modeling, HVDC simulators (both physical and digital) can be used to analyze problems and search for solutions. For example, during the tests for parallel operation of the two bipoles in Nelson River HVDC transmission system in Manitoba, Canada, in 1985, it was observed that there was a current oscillation of 6 Hz in the two inverters although the current in the two rectifiers was steady. The V_d–I_d characteristics for the rectifiers and the inverters are shown in Fig. 9.14 (*a*) and (*b*) respectively. The operating voltage is also shown here, which indicates that the inverter 2 operates in the mode of current error dependent gama (γ) control mode. In simulating the system using EMTDC program with a generic modeling of the controls, it was not possible to identify the source of the (oscillation) problem. However, from the knowledge of the actual valve group controller (provided by the manufacturer) it was possible to identify the source of the problem to a particular $K/(1 + sT)$ type delay block in the actual controller. By decreasing the time constant T from 22 ms to 3 ms, it was possible to eliminate the current oscillation problem in the real system. Incidentally, the digital program also allows one to experiment and determine how the oscillations are initiated from a stable operating point. In this case, changing the tap ratios from the set point initiated the low frequency current oscillations [15].

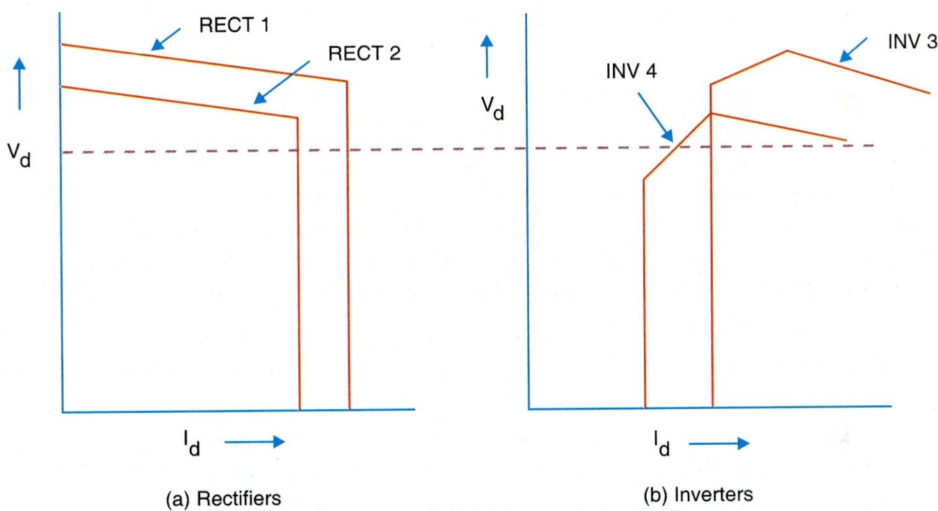

Fig. 9.14: Multiterminal characteristics

A HVDC simulator study of a four terminal MTDC system indicated the possibility of adverse interaction between the control and DC network dynamics. This can result in multiple mode shifts in rectifiers following a remote three phase fault at a rectifier [23]. A commutation failure at the small inverter results in the total DC current (from the rectifiers) diverted into the faulted inverter due to voltage drop in the inverter caused by the commutation failure. The transient current is even higher due to the discharge of line capacitances. The peak current can be limited partly by the use of higher smoothing reactor. However, it is difficult for the inverter to recover even when the AC system is strong. The system recovery is made possible by initiating Force-Retard (FR) on the rectifiers (increasing α to values above 90°, say 145°) as done in the case of DC line faults in a two terminal system. FR can be initiated without tele-communications since the collapse of the DC voltages at the rectifiers can be used as signals for the protective action. After a pre-determined time, the converter firing angles are returned to their normal values at a controlled speed. However, there is still a danger of a subsequent commutation failure at the small inverter due to current overshoots. The characteristics of a VDCOL at the inverter have to be carefully chosen to avoid this problem.

9.6 MULTI-INFEED DC SYSTEMS

Although multiterminal DC systems are not yet common, it is possible that more than one HVDC link feeds into the same load area [16-18]. In reference [16], an attempt was made to answer the question of possible limits of HVDC power supplied to an AC system. The author identified two major issues that can affect the limits–(*i*) current harmonics injected by the converter and (*ii*) compensation of reactive power. The first issue is closely connected with the generation of non-characteristic harmonics and the possibility of harmonic instability (which is essentially the phenomenon that results in the sustained non-characteristic harmonics due to AC-DC system interactions). According to [16], this is mainly dependent on the frequency dependence of the AC system (including the AC filters) impedance viewed from the converter terminals and not necessarily dependant on the Short Circuit Ratio (SCR). Harmonic instability can occur even at higher SCR. Incidentally, there was also a belief earlier that the problem of harmonic instability was completely solved after the development of Equidistant Pulse Control (EPC) scheme of firing pulse generation. However, this assumption is also not valid. There could be adverse harmonic interactions even with EPC.

The second issue of reactive power compensation is related to the problem of voltage instability and voltage flicker (caused by switching a filter or capacitor bank). This is primarily dependent on the SCR and the problem can be aggravated at low SCR. There is also the problem of overvoltages caused by load rejection which can be tackled by applying SVC or STATCOM for reactive power compensation.

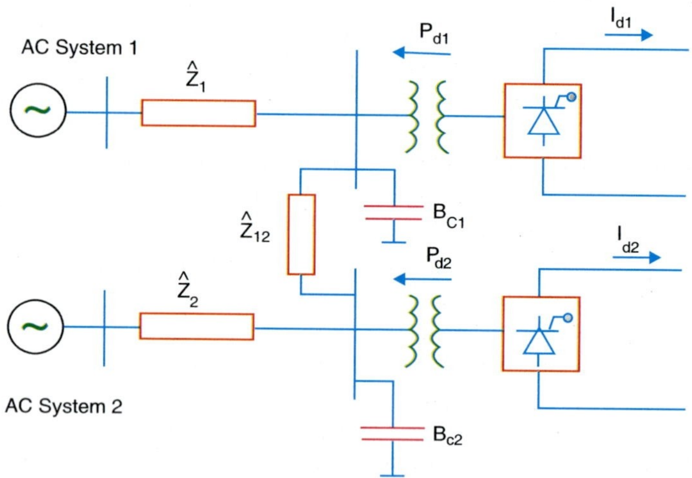

Fig. 9.15: A multi-infeed system

References [17] and [18] formulate the problem of MIDC systems by considering the system shown in Fig. 9.15. The AC system shows an equivalent circuit based on Thevenin impedances.

The interaction between the two HVDC converters that may belong to two different HVDC links can be ignored if the impedance Z_{12} is large compared to Z_1 or Z_2. Thus, in a large network with several embedded DC links, the DC control design can be made based exclusively on SCR at the AC connection point, without considering the other DC links, if the transfer impedances between the commutation busbars are high [18].

The control strategies at the rectifier and the inverter have a bearing on the voltage stability of the system. Current control rather than power control at the rectifier improves matters while constant DC voltage control at the inverter is much better than constant extinction angle control.

The interactions among inverters in close proximity requires coordinated design of controllers. The voltage distortions caused by persistent commutation failures in one inverter, also affect the other inverter. The VDCOL characteristics of one inverter that decide the speed of recovery following DC faults can affect the voltage at the inverter.

The control and protection in MIDC systems must be carefully designed to avoid adverse interactions. Interestingly, a BTB link interconnecting two asynchronous AC systems can introduce low frequency current oscillations in a nearby rectifier [23]. For example, even if the nominal frequencies of the AC systems interconnected by a BTB link are same, there can be variations in the frequencies. If one frequency is 50 Hz while the other is 49 Hz, the DC link with 12 pulse converters will have voltages of frequencies 600 Hz and 588 Hz. This can result in the AC bus voltages having a 12 Hz frequency component and the current controller in the rectifier in proximity can react to this. Such oscillations have been observed in HVDC simulator studies [23].

9.7 MTDC SYSTEMS USING VOLTAGE SOURCE CONVERTERS

The VSC based HVDC transmission is ideally suited for multiterminal operation due to the following advantages:

1. The power reversal in a converter is achieved by current reversal which is easily arranged by control action without having to use mechanical switches to reverse the polarity of the converter connections to the conductor in a parallel MTDC system.

   In line commutated converters, the power reversal requires voltage reversal and this implies that the converter has to be connected to the conductor of the opposite polarity.
2. There is no problem of commutation failures in an inverter based on VSC. A VSC based inverter can even supply passive loads.
3. The use of Pulse Width Modulation (PWM) eliminates low frequency harmonics and simple AC filters can be supplied.
4. There is no need for reactive power compensation of VSC. Actually, a VSC can supply reactive power and can help in the control of the AC voltage.

However, it is to be noted that harmonic interactions can occur in VSC also and the choice of the circuit parameters and control design must consider this problem. However, the problem is likely to be less complicated compared to LCC.

One of the drawbacks with VSC is the problem of handling DC faults. Since DC capacitors can discharge into the fault, the protection must disconnect the converter from the line or block the firing pulses to the converter and trip the converter. In the case of LCC rectifier, the action of forced retard (FR) can help in clearing the fault by deionization of the fault arc. This difference in the action of LCC and VSC rectifiers has prompted the application of hybrid MTDC schemes [20], with LCC rectifiers and VSC inverters. Of course, when a terminal has to operate in either of the two (rectifier or inverter), obviously VSC is the choice.

Note that in the case of cable transmission, the probability of DC faults is less (although any cable fault can be non-self healing).

In urban areas, the increasing demand of power is generally met by importing from distant generators. The application of DC transmission has the advantage of not contributing to short circuit levels (the increase in short circuit levels would require upgrading the ratings of switchgear). Thus, introduction of MTDC systems based on VSC in urban areas is quite attractive.

9.8 SUMMARY

Although the concept of MTDC systems has been discussed for a long time, the implementation has been very limited. The complexities in control and protection have deterred many utilities from experimenting with MTDC system operation. Introduction of MTDC systems requires extensive R & D activities to solve the problems associated with MTDC systems. Meanwhile, it has been recognized that MTDC systems have similarities with MIDC systems in terms of operational issues. It is anticipated that emerging technology of VSC-HVDC transmission can help solve many of the problems and encourage utilities to introduce MTDC systems.

REFERENCES AND BIBLIOGRAPHY

1. Foerst R., Heyner, G. Kangiesser K.W. and Waldmann, H. '*Multiterminal Operation of HVDC Converter Stations*', IEEE Trans., Vol. PAS-88, No.7, pp. 1042–50, July 1969.
2. Kangiesser K.W., Bowles J.P. Ekstrom A., Reeve J. and Rumpf E. '*HVDC Multiterminal Systems*', CIGRE 14-08, 1974.
3. J. Reeve, '*Multiterminal HVDC Power Systems*', IEEE Trans., Vol. PAS-99, No.2, pp. 729–37, March/April 1980.
4. Joetten R., Bowles J.P., Liss, G. Martin C.J.B. and Rumpf E. '*Control in HVDC Systems–The State of the Art*', Part II: Multiterminal Systems, CIGRE 14-07, 1980.
5. Ishikawa I., Machida T. Watanabe A., Konishi H. and Murai K. '*Development of Centralized Control System for Multiterminal HVDC Transmission Systems*', IEEE PES Winter meeting, A78-122-4, 1978.
6. Sakurai T., Goto K., Irokawa S. , Imai K. and Sakai T., '*A New Control Method for Multiterminal HVDC Transmission without Fast Communication Systems*', IEEE Trans., Vol. PAS-102, pp. 1140–50, May 1983.
7. Reeve J. and Chen S.P., '*Digital Simulation of a Multiterminal HVDC Transmission System*', IEEE PES Winter Meeting, 84 WM 199-6, 1984.
8. Krishnayya P.C.S., Lefebvre, S. Sood V.K. and Balu N.J. '*Simulator Study of Multiterminal HVDC System with Small Parallel Lap and Weak AC System*', IEEE Trans., Vol. PAS-103, pp. 3125–32, Oct. 1984.
9. Nozari F., Grund C.E. and Hauth R.L. '*Current Order Co-ordination in Multiterminal DC System*', IEEE Trans., Vol. PAS-100, pp. 4628–35, Nov. 1981.

10. Bergstrom L., Juhlin, L.E., Liss G. and Svensson S., '*Simulator Study of Multiterminal HVDC System Performance*', IEEE Trans., Vol. PAS 97, No. 6, pp. 2057–66, Nov./Dec. 1978.
11. Long W.F., Reeve K., McNichol J.R., Harrison R.E. and Bowles J.P., '*Considerations for Implementing Multiterminal DC Systems*', IEEE Trans., Vol. PAS 104, No. 9, pp. 2521–30, Sept. 1985.
12. Ishikawa M., Horiuchi S., Irokawa S., Imai K., Hirose S. and Sekiya K., '*Simulator Study of Multiterminal HVDC Transmission Systems without Fast Communication*', IEEE Trans., Vol. PWRD-1, No. 3, pp. 218–227, July 1986.
13. Long W.F. et al., '*Application Aspects of Multiterminal DC Power Transmission*", IEEE Trans., Power Delivery, Vol. 5, No.4, pp. 2084–2098, November 1990.
14. Blahouse L., '*Extension of the Pacific Intertie by Two Parallel Bipolar Stations–Concept and Factory Simulator Tests*', Paper 14–205, CIGRE, Paris, 1990.
15. Shome T., Gole A.M., Brandt D.P. and Hamlin R.J., '*Adjusting Converter Controls for Paralleled DC Converters using a Digital Transient Simulation Program*', IEEE Trans., Power Systems, Vol. 5, No.1, pp.12–19, Feb. 1990.
16. Lips H.P., '*Aspects of Multiple Infeed of HVDC Inverter Stations into a Common AC System*', IEEE Trans., Vol. PAS-92, No.2, pp. 775–779, 1973.
17. Bui L.K., Sood V.K. and Laurin S., '*Dynamic Interactions between HVDC Systems Connected to AC Buses in Close Proximity*', IEEE Trans., Power Delivery, Vol. 6, No.1, pp. 223–230, Jan. 1991.
18. Szechtman M. et al., '*The Behaviour of Several HVDC Links Terminating in the Same Load Area*', paper 14-201, CIGRE, Paris, 1992.
19. Sadek K., Periera M., Christl N. and Lutzelberger P., '*Control Concept and Fault Performance for Multiterminal HVDC Systems*', Paper-204, CIGRE, Paris, 1990.
20. Hayashi T., Seki N., Sampei M. and Sekine Y., '*Development and Field Test of a Multiterminal DC Link using Voltage Sourced Converters and New Control and Protection Schemes*', Paper 14–204, CIGRE, Paris, 2000.
21. Jiang H. and Ekstrom A., '*Multiterminal HVDC Systems in Urban Areas of Large Cities*', IEEE Trans., Power Delivery, Vol. 13, No. 4, pp. 1278–1284, October 1998.
22. Tang L. and Ooi B.T., '*Protection of Multiterminal HVDC against DC Faults*', IEEE 33rd Power Electronics Specialists Conference (PESC), Queensland, Australia, Vol.2, pp. 719–724, 2004.
23. Sood V.K., '*HVDC and Facts Controllers*', (Boston), Kluwer Academic Publishers, 2004.

POWER FLOW ANALYSIS IN AC/DC SYSTEMS

10

10.1 GENERAL

Power flow analysis is an essential component of system studies carried out for planning, design and operation of power systems. It is basically simulation of the system in steady-state and determines the operating point which is later used for initializing variables in transient and dynamic system simulation. The power flow or load flow analysis of AC systems has been thoroughly investigated in terms of numerical algorithms for obtaining the solution to the nonlinear algebraic equations. Although, there is little work reported on analytical aspects such as the existence and number of solutions, the convergence characteristics of different algorithms etc., there are several production grade computer programs that can handle large systems with thousands of buses and lines. For most of the cases, the use of flat start results in convergence to a solution that is in most of the cases, the unique and stable operating point. The Gauss-Seidel method has given way to the use of Newton's method which results in fast convergence. The computations are further simplified using Fast Decoupled Load Flow (FDLF) method in which the corrections to the bus voltage estimates are found from solving the following equations [1]:

$$\Delta P/V = [B']\Delta\theta \tag{10.1}$$

$$\Delta Q/V = [B'']\Delta V \tag{10.2}$$

where ΔP_i, ΔQ_i are mismatches of real and reactive powers at bus i, $\Delta\theta$ and ΔV are the correction vectors to bus angles and voltage magnitudes. B' and B'' are constant matrices of appropriate sizes and consist of elements that are related to the reactances of the elements of the network. The sparse nature of B' and B'' matrices permits the solution of very large scale networks with moderate computational costs. The existence of DC links in the power system requires the modeling of DC systems for power flow analysis. While the modeling of DC systems for power flow is fairly standard, the solution methodology varies. The sequential or alternating method which does not require major changes in the software available for the power flow analysis of AC systems, is widely used. The mathematical models and the solution methods are discussed in this chapter, and multiterminal DC systems are considered.

The analysis is general and is directly applicable to MTDC systems. Two terminal and multi-infeed DC systems are actually special cases of MTDC systems. The analysis is first given for

DC systems with LCC (Line Commutated Converters) using thyristor valves. Since, several VSC based HVDC links have been commissioned over the last decade and this is an emerging technology, the modeling of VSC based HVDC systems is also described.

10.2 POWER FLOW ANALYSIS–AN OVERVIEW

The solution of nonlinear algebraic equations describing the AC/DC system can be obtained using Newton, Gauss-Seidel or other iterative techniques. There are several publications [2–7] which deal with the method of modeling the DC systems in the power flow analysis. The approach to the solution of power flow problem in AC/DC system is broadly divided into two categories (1) sequential method (2) simultaneous method. The sequential method, involves the iterative solution of the AC and DC system equations separately and alternately until convergence is obtained. Here, the advantage is that any well developed AC power flow program can be used. In the simultaneous or unified method, both the AC and DC system equations are combined and the resulting set of equations are solved. However, there is a need to examine the problem (of power flow analysis in AC/DC systems) afresh and evolve an appropriate methodology for the solution.

The power flow analysis in AC or AC/DC systems involves the solution of nonlinear equations subject to certain constraints. In general, the equations can be written as

$$g(x,w) = 0 \tag{10.3}$$

$$h_{\min} < h(x,w) < h_{\max} \tag{10.4}$$

$$u_{\min} < u(x,w) < u_{\max} \tag{10.5}$$

where, x = vector of state variables

w = vector of specified variables

h = vector of variables which are constrained (for example in AC systems, it can represent line flows or generator reactive power)

u = vector of control variables.

If 'w' is specified in advance (subject to the limits) and if inequalities (10.4) and (10.5) are satisfied, then the solution of Eq. (10.3) corresponds to ordinary power flow analysis. If 'w' is not specified in advance, it can be chosen by minimizing a cost function subject to the constraints (10.3) to (10.5) and this becomes the solution to Optimal Power Flow (OPF).

In an AC system, the control variables 'u' and the specified variables 'w' are identical (for example generator real power (P_g) , and its terminal voltage (V_g). In the case of DC systems, the control variables are (*i*) the delay angle α or extinction angle γ (θ in general) and (*ii*) the off-nominal transformer tap ratio 'a', but the specified variables can be power (P_d) or current (I_d) or voltage (V_d), reactive power (Q_d) or power (factor ($\cos\phi$) or converter bus voltage (V). Sometimes the control variables θ or 'a' themselves are specified, which however reduces the degree of freedom on the choice of control. Normally the control specification in DC system is based on some criteria like minimizing losses, cost or control effort or maintaining constant voltage profile. For example, the real and/or reactive power at the converter buses may be specified based on an optimal real and/or optimal reactive power flow of the AC system. The control variables of the DC system have to be adjusted to meet these specifications without violating the constraints on the control variables 'a', θ and also the dependent variables such as (I_d), V etc. Thus, the distinction between the scheduled and the control variables in the DC system is important. The use of control variables 'a' and θ as specified variables by Mahseredjian et al. [6] reduces the degrees of freedom on adjustments to be made to satisfy the limits on different variables.

Since there are two control variables per terminal, we can have two specified variables for each terminal. For a 'n' terminal system there will be '$2n$' specifications which can be divided into two subsets. The first set of specification $\left(W_{dc1}^{spec}\right)$ corresponds to specifying (P_d) or (I_d) at all terminals except at VST (Voltage Setting Terminal) where (V_d) is specified. VST is analogous to slack bus in the AC system and one such terminal is essential in the operation of DC systems as the losses in the DC system cannot be fixed apriori. The second set of specification $\left(W_{dc2}^{spec}\right)$ has 'n' specified variables which could be θ or (Q_d) or (cosϕ) or V for each terminal. It is shown later that the real power at the AC/DC interface are functions of first set of specification $\left(W_{dc1}^{spec}\right)$ whereas the reactive power is a function of both the first $\left(W_{dc1}^{spec}\right)$ and the second set of specifications $\left(W_{dc2}^{spec}\right)$. Thus, only $\left(W_{dc1}^{spec}\right)$ is needed to perform DC power flow, while both $\left(W_{dc1}^{spec}\right)$ and $\left(W_{dc2}^{spec}\right)$ are needed to perform AC power flow. In other words, out of '$2n$' scheduled variables in a 'n' terminal MTDC system, only 'n' variables are required to solve the DC network. Also, it is not always necessary that θ be specified as has been reported in the literature. It could be any one of (θ, Q_d, cosϕ or V). It will be shown that specifying any one of these, is equivalent to specifying θ for a given $\left(W_{dc1}^{spec}\right)$. Once θ is known, reactive power (Q_d) at the converter buses can be calculated. It is also shown later that once the first set of specification $\left(W_{dc1}^{spec}\right)$ is given, the value of ($aV\cos\theta$) is fixed. As θ is known from the second set of specifications, the product 'aV' is calculated.

When both the real and reactive powers are known at a converter bus, they can be treated as loads at that bus and AC power flow can be performed. From the voltages obtained from AC power flow, the taps can be calculated as the product of 'aV' is already determined from the knowledge of $\left(W_{dc1}^{spec}\right)$ and $\left(W_{dc2}^{spec}\right)$.

10.3 DC SYSTEM MODEL

The following basic assumptions are made in the derivation of the equations representing the AC/DC converter.

(*i*) The AC bus voltages are sinusoidal and are of fixed frequency. The three phase voltages and currents are balanced.

(*ii*) All the harmonics of voltage and current produced by the converter are filtered out and do not appear in the AC system.

(*iii*) The converter transformers have zero resistances and magnetizing currents. Only the leakage reactances are considered.

(*iv*) The valves of the converter are ideal and have no voltage drops while conducting.

(*v*) Only average DC voltage and currents are considered and the harmonics are ignored.

(*vi*) In a bipolar DC link, the two poles are balanced.

(*vii*) The magnitude of the AC current injected by the converter is directly proportional to the DC current and independent of the firing angle or the overlap angle.

The assumptions (*iii*), (*iv*) and (*vi*) are made only for convenience and can be relaxed if required.

10.3.1 Basic Model of the Converter

The DC system can be represented by the DC network with converters modeled by their equivalent circuits. Figure 10.1 shows the basic converter used in this work. For multibridge converter, assuming equal transformer reactances for all the bridges, they may be replaced by a single equivalent converter bridge with the DC voltage and AC current multiplied by the number of bridges (n_b).

10.3.2 Converter Equations

The average DC voltage in volts (V_d) (pole to ground) at a converter terminal is given by

$$V_d = \frac{3\sqrt{2}atn_bV\cos\theta}{\pi} \pm \frac{3X_cI_dn_b}{\pi} = n_b\left(V_{do}\cos\theta - R_cI_d\right) \tag{10.6}$$

where a = off-nominal tap ratio of converter transformer on the DC side (see Fig. 10.1)

t = nominal turns ratio

n_b = number of bridges connected in series per pole

V = converter bus (line-line rms) voltage (on the primary side in volts)

θ = converter control angle (α for rectifier and γ for inverter)

X_c = leakage reactance referred to the secondary side of converter transformer (in ohms)

I_d = average DC current (in amperes), $R_c = \pm\frac{3}{\pi}X_c$

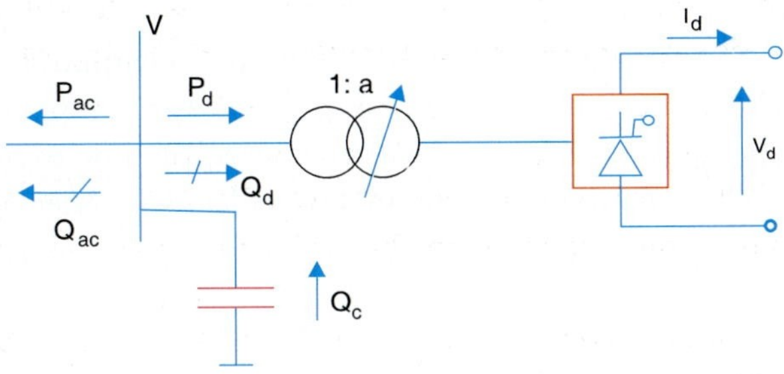

Fig. 10.1: A basic converter model

The DC current leaving a terminal is assumed to be positive. Hence is positive for a rectifier and negative for an inverter. The commutation resistance is negative for an inverter (with CEA control) and positive for a rectifier.

The power (P_d) and the reactive power (Q_d) supplied to a converter terminal are given by

$$P_d = p.V_dI_d, Q_d = |P_d|\tan\phi \tag{10.7}$$

where p = number of poles (1 or 2)

ϕ = power factor angle defined from $\cos\phi = \dfrac{V_d}{n_b V_{do}}$.

The expression for $\cos\phi$ follows from the approximate relation (based on assumption vii)

$$I_{AC} = atn_b \frac{\sqrt{6}}{\pi} I_d$$

where I_{AC} is the rms AC current phase.

10.3.3 A Flexible Per Unit System

It is convenient to express the DC system equations in per unit, as is common with AC systems. In general, for MTDC systems, we choose a common base power (not necessarily coinciding with the rating of a particular terminal). For connected DC network, it is essential that the base quantities for the DC voltage (and current) be identical for all the terminals (this will ensure that KCL and KVL apply also for the per unit quantities). However, it is not essential that base AC voltage be identical for all the terminals. We select the base AC voltage (line to line, rms) at each terminal as the rated AC voltage at the corresponding converter bus. The base quantities for other variables given by

$$I_{dB} = \frac{P_B}{V_{dB}}, Z_{dB} = \frac{V_{dB}}{I_{dB}} \tag{10.8}$$

$$I_{aBj} = \frac{P_B}{V_{aBj}} = \frac{V_{dB}}{V_{aBj}} I_{dB} \tag{10.9}$$

$$Z_{aBj} = \frac{V_{aBj}}{I_{aBj}} \tag{10.10}$$

Note that we have defined the base AC current (I_{aB}) at any terminal 'j', as $\sqrt{3}$ times the base line current. This simplifies the equations ($P_B = V_{aBj} \cdot I_{aBj}$) and retains the base impedance as identical to the base impedance per phase (if we considered the line to neutral voltage and the line or phase current). Thus, it is convenient to define I_{aj} ($\sqrt{3} \times$ line AC current per pole at terminal 'j') given by

$$I_{aj} = \sqrt{3} I_{ACj} = \sqrt{3} \cdot \frac{\sqrt{6}}{\pi} t_j \cdot n_{bj} \cdot a_j \cdot I_{dj}$$

$$= \frac{3}{\pi}\sqrt{2} t_j \cdot n_{bj} \cdot a_j \cdot I_{dj} \tag{10.11}$$

where subscript 'j' refers to the jth terminal.

From Eq. (10.9), we can express I_{aj} in per unit as

$$\bar{I}_{aj} = \frac{I_{aj}}{I_{aBj}} = k_j \cdot a_j \cdot \bar{I}_{dj} \tag{10.12}$$

where $\bar{I}_{dj}$ is the per unit DC current at terminal j.

We can also express Eq. (10.6) in per unit as

$$\bar{V}_{dj} = k_j \cdot a_j \cdot \bar{V}_j \cdot \cos\theta_j - \bar{R}_{cj}\bar{I}_{dj} \tag{10.13}$$

where

$$k_j = \frac{3\sqrt{2}}{\pi} t_j n_{bj} \frac{V_{aBj}}{V_{dB}}$$

$$\bar{R}_{cj} = \pm\frac{3}{\pi}\frac{X_{cj} n_{bj}}{Z_{dB}}$$

The positive sign applies to rectifiers and the negative sign to inverters.

The per unit system proposed here is more general than that used in references [2-6], since the base AC voltage (V_{aB}) can be chosen independent of the base DC voltage (V_{dB}). In other words the coefficient 'k' need not be equal to 1, in general. In addition, the turns ratio (t) and the number of bridges (n_b) can assume different values at different terminals just as base AC voltages. Base DC voltage is normally chosen equal to the rated voltage in a parallel connected MTDC system. From Eq. (10.13) the converter terminal can be represented by a Norton's equaivalent of a current source (I) in parallel with a commutating resistance (R_c) as shown in Fig. 10.2, where,

$$I = \frac{kaV\cos\theta}{R_c} \tag{10.14}$$

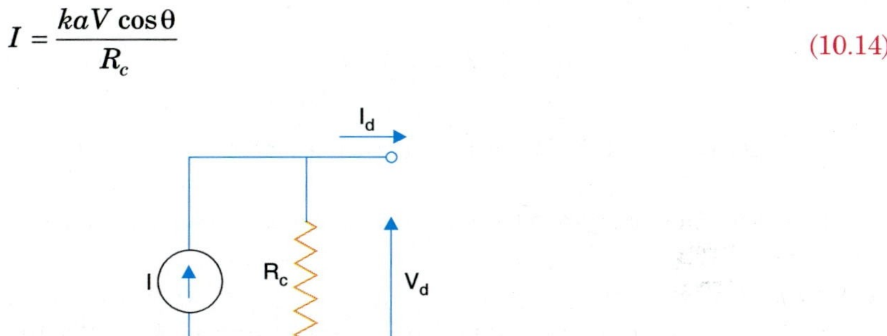

Fig. 10.2: Norton's equivalent circuit for a converter

the current source is proportional to the product $aV\cos\theta$. The subscript is dropped for convenience. It is to be noted that for rectifier $\theta = \alpha$ whereas for inverter θ is normally chosen as extinction angle γ in which case R_c is negative. However if θ is chosen as the angle of advance β, then R_c is positive. This follows from the relationship

$$V_d = kaV\cos\beta + R_c\,|I_d| \tag{10.15}$$

The equation (10.7) in per unit form is

$$\bar{P}_d = \bar{V}_d\bar{I}_d \cdot p, \bar{Q}_d = |\bar{P}_d|\tan\phi \tag{10.16}$$

The power factor at j^{th} terminal is given by

$$\cos\phi_j = \frac{\bar{V}_{dj}}{k_j a_j \bar{V}_j} \tag{10.17}$$

This follows from the equation $\bar{P}_{dj} = \bar{V}_{dj}\bar{I}_{dj} = \bar{V}_j\bar{I}_{aj}\cos\phi_j$, neglecting converter losses and substituting Eq. (10.12) for $\bar{I}_{aj}$.

Remarks

1. In a two terminal system, typically, there are two bridges per pole (to provide a 12 pulse converter) which are operated with identical delay/extinction angles to reduce the harmonic generation.
2. Similarly, in a bipolar HVDC link, the two poles are operated to circulate the DC current through the two conductors. Hence, under normal conditions, the ground current is zero. Thus, there are 4 identically operated bridges during bipolar and 2 bridges during monopolar operation.
3. If there are n_b identically operated bridges at a converter station, it is convenient to select the base DC voltage (V_{dB}) as ($n_b V_{dn}$) where V_{dn} is the rated (nominal voltage across a bridge. The base power (P_{dB}) is then chosen as ($n_b V_{dn} I_{dn}$) where I_{dn} is the rated DC current.
4. In chapter 3, it was shown that

$$\frac{V_{db}}{V_{don}} = a\bar{V}\cos\theta \pm \frac{\bar{X}_c}{2}\frac{I_d}{I_{dn}}$$

where V_{db} is the DC voltage across a bridge.

Defining $k = \frac{V_{don}}{V_{dn}}$, we can rewrite the above equation as

$$\bar{V}_d = \frac{V_{db}}{V_{dn}} = ka\bar{V}\cos\theta \pm \bar{R}_c\frac{I_d}{I_{dn}} = ka\bar{V}\cos\theta \pm \bar{R}_c\bar{I}_d$$

where $\bar{R}_c = \frac{k\bar{X}_c}{2}$

The above equation (in per unit) is unaltered if we considered n_b bridges connected in series. This is because we will define the DC voltage (V_d) as the voltage across the n_b bridges.

That is, $V_{dB} = n_b V_{db}$ and $V_{dB} = n_b V_{dn}$ and hence the expression $\bar{V}_d$ is identical to the expression obtained for a single bridge.

5. k is computed from

$$k = \frac{V_{don}}{V_{dn}} = \frac{1}{\cos\theta_n - \frac{\bar{X}_c}{2}}$$, θ_n is the nominal value of θ. A typical value of k is 1.15.

10.3.4 DC Network Equations

In references [2-7], the authors represent the DC network by the bus resistance matrix formed by treating the VST as the reference terminal. This can be cumbersome if VST can change. Modeling the converters by its Norton's equivalent enables the DC network equations to be expressed with ground as reference node. The DC lines and smoothing reactors are modeled as resistive networks.

Combining the equivalent circuits for all the terminals and the DC network, the following equations (expressed in per unit quantities) are obtained:

$$[G]\underline{V_d} = \underline{I} \tag{10.18}$$

where $[G]$ is a 'n' × 'n' conductance matrix for a 'n' terminal MTDC system. It is assumed that the non-converter buses are absent or are eliminated. From equation (10.18) we get,

$$\underline{V_d} = [R]\underline{I},\ [R] = [G]^{-1} \tag{10.19}$$

To ensure the existence of $[R]$ in all cases, it is convenient to treat R_c as positive at inverters by putting $\theta = \beta$. The $[R]$ bus matrix of Eq. (10.19) can also be directly formed by using the Z bus algorithm [9]. Equation (10.19) is always applicable independent of the choice of VST.

10.3.5 DC Control Equations

The DC network equations based on the bus resistance matrix are used to form the following controller equations:

$$W_{dc1}^{spec} = W_{dc1}(x_{dc}) \tag{10.20}$$

where the state vector x_{dc} is the vector of current sources I. At the VST (say i^{th} terminal), V_d is specified and is given by

$$V_{di} = \sum_{j=1}^{n} R_{ij} I_j = V_{di}^{spec} \tag{10.21}$$

where, R_{ij} is the element of the bus resistance matrix corresponding to i^{th} row and j^{th} column. At the remaining terminals, P_d or I_d is specified.

$$P_{dj} = V_{dj} \times I_{dj} = P_{dj}^{spec} \tag{10.22}$$

or

$$I_{dj} = I_{dj}^{spec} \tag{10.23}$$

In the above equations P_{dj} and I_{dj} can be expressed in terms of I; for example, I_{dj} can be written as

$$I_{dj} = I_j - \frac{V_{dj}}{R_{cj}} \tag{10.24}$$

therefore, there are 'n' control equations for a 'n' terminal MTDC system of the form given by Eqs. (10.21) to (10.23). They can be grouped together and written as

$$W_{dc1}(I) = W_{dc1}^{spec} \tag{10.25}$$

where W_{dc1}^{spec} is the vector of first set of specifications and 'I' is the vector of all current sources. Equation (10.25) is in general, a non-linear vector equation and can be solved by Newton's method. Once I is determined, V_d and I_d are solved using Eqs. (10.19) and (10.24).

The second set of control specification W_{dc2}^{spec} involve specifying either θ (α or γ) or Q_d or $\cos\phi$ or V. Specifying Q_d or $\cos\phi$ is equivalent to specifying θ because Q_d and $\cos\phi$ are functions of I and θ. This is shown by Eqs. (10.16) and (10.17). Q_{dj} is a function of I_j and $\cos\phi_j$ and $\cos\phi_j$ is a function of I_j and θ_j as

$$\cos\phi_j = \frac{V_{dj}}{k_j a_j V_j} = \frac{V_{dj}\cos\theta_j}{I_j R_{cj}} \tag{10.26}$$

Remarks

1. It is assumed here that a converter is connected between a pole (say positive) and the ground. Eliminating non-converter buses, the DC network has exactly 'n' buses where 'n' is the number of converters. This allows the numbering of the DC buses in accordance with the numbering of the converters (converter 1 connected to the DC bus 1, converter 'n' connected to the bus 'n'). However this assumption can be relaxed easily. In general, there can be M DC buses where M>n. This permits the consideration of series connected converter, (which are connected across two DC buses) in addition to the non-converter buses.
2. It is also possible to relax the assumption that the two poles are balanced. If unbalanced operation is to be considered, the converters in each pole has to be modeled separately. Note that a two terminal, bipolar HVDC link can be modeled using 4 converters connected at 4 different buses. The electrode line at any terminal provides a common path for currents in the two converters connected at that terminal (to the two poles).
3. A multi-infeed DC system consisting of two terminal (point to point) DC links is a special case where the resistance matrix [R], is block diagonal, each block corresponding to a particular (two-terminal) DC link. In this case, it is possible to define base DC voltage for each DC link independent of other links.

10.4 SOLUTION PROCEDURE

The choice of the state variables is an important aspect of solution procedure. In [2], [3] five variables per terminal V_d, I_d, 'a', $\cos\phi$ and θ are used to describe the operating state. This can be confusing as all these variables are not independent. In [4], Fudeh and Ong treat DC bus voltages as state variables and solve the DC network equations directly using Gauss-Seidel method. For AC systems, the state variables are V, δ and knowing these variables, the other quantities can be calculated. By analogy to AC system network, it is adequate to utilize a minimal set of variables to describe the operating state. Thus the state vector chosen here is the vector of Norton's equivalent current sources 'I'. The control equations corresponding to the first set of specifications viz., Eq. (10.25) are solved by Newton's method (or by Gauss-Seidel method) for the state vector 'I'. Once 'I' is solved, the other variables viz., I_d, V_d and P_d are obtained. At this stage, we also get the $aV\cos\theta$ product. Thus, the real power at the AC/DC interface is a function of the first set of specifications W_{dc1}^{spec}. Using the first and second set of specifications viz., W_{dc1}^{spec} and W_{dc2}^{spec}, Q_d is calculated. The value of 'aV' is also calculated since θ is known from W_{dc2}^{spec}. Treating P_d and Q_d as loads at the converter buses, the AC power flow is performed using Newton's method or its variants such as fast decoupled load flow (FDLF) method [1]. The AC power flow solution establishes the voltage magnitudes at all the AC buses. The off-nominal tap ratio of the converter transformer is then determined as the product 'aV' is already known.

The major steps in the power flow analysis of MTDC-AC system are summarized below:

1. Enter with system specifications, W_{dc1}^{spec}, W_{dc2}^{spec}, and the initial conditions are obtained for 'I' from Eq. (10.24) by assuming V_d at all terminals to be equal to V_d^{spec} at VST (This is equivalent to putting $I_d = 0$ initially. Note this is similar to 'Flat Start' in AC power flow analysis).
2. Solve the DC control equations Eq. (10.25) for I using only W_{dc1}^{spec}. This establishes the DC voltages, currents, powers and $aV\cos\theta$ product at all the terminals.
3. Using W_{dc2}^{spec} and P_d, calculate Q_d's at all the converter buses. Also the product 'aV' is found because θ is known from W_{dc2}^{spec}.
4. Knowing real and reactive power loadings at all buses, AC power flow can be performed. This establishes the voltage magnitudes at all the AC buses.
5. From the knowledge of 'aV' and 'V', the values of tap ratios at all the converter buses are found. If there are no limits on tap, then step 5 is the end of the solution. But if there are limits on tap, violation if any, are removed by using the procedure given below.

10.5 INCLUSION OF CONSTRAINTS

The published literature on MTDC power flow do not discuss procedures for systematically taking care of the various constraints. In [4], only constraint violation of tap ratio 'a' is considered. It is said that if the limits on 'a' are violated on any terminal, the DC voltage has to be rescheduled and the DC and AC power flow have to be performed iteratively until feasible solution is obtained.

Here, the constraints on each of the variables are systematically and successively handled. The importance is given to the tap limits (both upper and lower) and the minimum limit on θ (θ_{min}). The use of sensitivity in AC power flow analysis is well known [10]. There, the sensitivity information is used to enforce limits on different variables in the process of AC power flow analysis. Similar information can be used here also for handling different limits in AC-MTDC power flow analysis.

The sensitivity information is first obtained as follows. From the results of DC power flow, the value of the product $aV\cos\theta$ is known. Two arbitrary but practical values of θ are assumed for each terminal. These may preferably be some assumed limits of θ. Using these two values of θ, two values of Q_d are calculated from Eqs. (10.16) and (10.17). Two AC power flows are performed corresponding to these two values of Q_d at each terminal. As a result, we get two values of converter bus voltages corresponding to two values of 'aV'. Using these values of converter bus voltages a graph V versus $a'V = \frac{aV}{V_d}$ is plotted for each terminal which has a drooping characteristic as shown in Fig. 10.3. This graph is obtained for fixed values of the reactive injection (Q_c) at the converter buses. For constant power specifications at all the converter terminals, the graphs of V versus $a'V$ are invariant with respect to V_d. This is because, even if V_d changes, the AC voltage (V) depends only on the power factors at the converter bus (which is related to $\frac{aV}{V_d}$). Although the assumption of constant power at the VST is not strictly correct due to the losses in the DC network (losses vary with V_d), this can be neglected. The characteristic shown in Fig. 10.3 can be approximated as a straight line and can be expressed as

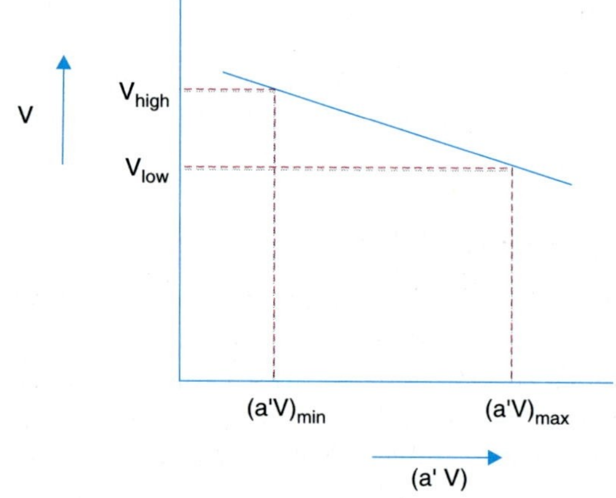

Fig. 10.3: Variation of V with (a' V)

$$V = -m(a'V) + c \tag{10.27}$$

where $a' = a/V_d$. From Eq. (10.27) we obtain,

$$a'V = \frac{a'c}{(1 + a'm)} \tag{10.28}$$

Note that θ is related to $a'V$ (for a given W_{dc1} specifications). Using Eq. (10.28) another curve is drawn between θ and a' as shown in Fig. 10.4. Referring to Fig. 10.4, let the values of a' corresponding to the minimum value of 'a' (a_{min}) and maximum value of 'a' (a_{max}) be a'_{min} and a'_{max} respectively. Let the values of $a'V$ corresponding to the values of a'_{min} and a'_{max} be $(a'V)_{min}$ and $(a'V)_{max}$ respectively. Then the values of θ corresponding to values of $(a'V)_{min}$ and $(a'V)_{max}$ are θ_{low} and θ_{high} respectively.

If the minimum values of θ (θ_{min}) is less than θ_{low}, then Fig. 10.4 shows the specification of θ in the range between θ_{low} and θ_{high} does not result in tap violation. Even if the specification is other than θ (V, Q_d or $\cos\phi$) it can be related to θ specification. If θ_{min} is greater than θ_{low} but less than θ_{high}, the feasible range for θ is θ_{min} to θ_{high}. The corresponding range for the AC voltage (V) is shown in Fig. 10.3.

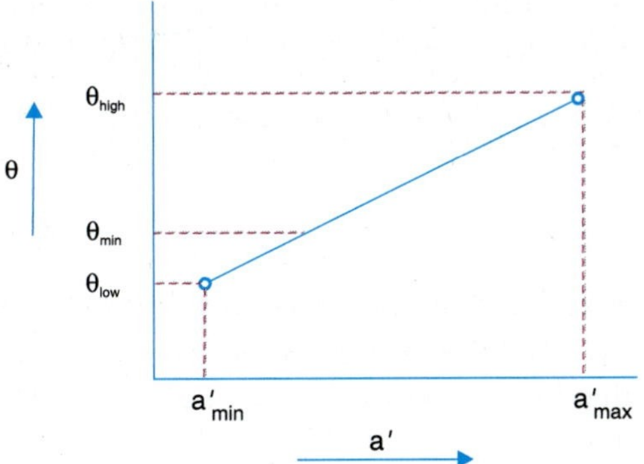

Fig. 10.4: Variation of θ with a'

What Figs. 10.3 and 10.4 illustrate are (*i*) by adjusting W_{dc2}^{spec} alone, it is possible to avoid tap violations and (*ii*) the W_{dc2}^{spec} can be selected within the feasible range to control the converter bus AC voltage in a range. If the desired AC voltage is beyond the range, the solution requires either (*a*) controlling the reactive injection (Q_c) at the converter bus or (*b*) altering W_{dc1}^{spec}.

It is to be noted that rescheduling active power at the converter bus is not desirable as the original specifications on power would be based on load or generation requirements influenced by economic and secure system operation. If W_{dc1}^{spec} needs to be modified, the first variable to be rescheduled would be the DC voltage at the VST. Rescheduling of V_d would be subjected to constraints on I_d (particularly when V_d has to be reduced). The maximum limits on I_d can be enforced, if necessary, by reducing the power specified at the non-VST where current violations are likely to occur.

Rescheduling of W_{dc1}^{spec} will be required if θ_{min} is greater than θ_{high}. In this case, rescheduling of W_{dc2}^{spec} alone will not be able to remove the tap violation. In such a case, V_d at the terminal having tap violation, needs to be reduced. This can be done as follows:

(*i*) Determine the new value of V_d required at the terminal from $V_{dnew} = (a_{max}/a^*).V_{dold}$ where a^* is the value of 'a' corresponding to θ_{min}.

(*ii*) Repeat step (*i*) for all the terminals where tap violation takes place.

(*iii*) Select V_d^{spec} at VST such that the following inequalities are satisfied. $V_{dj} \le V_{dnewj}$. The rescheduling of V_d^{spec} can be done with the knowledge of the sensitivity of V_{dj} with respect to V_d^{spec}.

(*iv*) If V_d^{spec} determined from the previous step is less than critical value (corresponding to maximum current limit at one or more terminals), reduce the power specifications at these terminals to satisfy the constraints on the DC current.

To summarise, the procedure for the elimination of the transformer tap violations is to reschedule W_{dc2}^{spec} first and if this is not adequate, reschedule V_{dc}^{spec}. Only in rare case, it may be necessary to reschedule P_{dc}^{spec} also.

It is to be noted that rescheduling of W_{dc2}^{spec} (or θ) results in change of AC voltage. In case the AC voltage has to be maintained at the previous value (corresponding to the original specifications W_{dc2}^{spec}) it may be necessary to reschedule W_{dc1}^{spec} as described above. (A simpler approach may be to adjust the reactive source at the converter AC bus, if this is feasible).

Remarks

1. The slope of the line (showing variation of the AC bus voltage with variation of $(a'V)$ is negative as $aV\cos\theta$ is fixed at each terminal from the W_{dc1} specifications. Hence, increase in aV implies reduction in $\cos\theta$ (increase in the value of θ). As the power factor reduces with the reduction of $\cos\theta$ (or increase in θ), the AC bus voltage tends to reduce with increase in θ, for a fixed power (P_d) injection.
2. If the losses in the DC network are ignored, then

$$\sum_{j=1}^{n} P_{dj=0}$$

(Note that P_{dj} is considered as positive when the converter is operating as a rectifier and P_{dj} is negative when the converter is operating as an inverter)

If $P_{dj} = P_{dj}^{spec}$, j = 1, 2,.... n – 1

then the power at the nth terminal is also specified or constant, even if we assume that DC voltage is specified at the nth terminal.

3. If, in addition to P_d, the power factor ($\cos\phi$) or reactive power Q_d is specified at all the terminals, then we can show that the violation of the upper limit on the tap at one terminal can be removed exactly by reducing the DC voltage at the terminal (where the tap violation takes place such that

$$V_{djnew} = \left(\frac{a_{j\max}}{a_{jold}}\right) V_{djold}$$

This will result in $a_{jnew} = a_{j\max}$ with no change in V_j (AC voltage).

(Note that in a lossless DC network, we can specify V_d at any terminal if $P_{dj} = P_{dj}^{spec}$,

j = 1,2,....n – 1)

The equation for V_{djnew} follows from noting that

$$V_{djold} = k_j a_{jold} V_j \cos\phi_j$$

and if $\cos\phi_j = \text{constant} = \cos\phi_j^{spec}$, we get

$$V_{djnew} = k_j a_{jnew} V_j \cos\phi_j,\ a_{jnew} = a_{jmax}$$

4. The reduction of DC voltage at any terminal (say n) also reduces voltage at other terminals. It can be shown that

$$\frac{\partial V_{dj}}{\partial V_{dn}} = \frac{R_{jn}}{R_{nn}} \tag{10.29}$$

where R_{jn} and R_{nn} are elements of the resistance matrix. If all the (resistance) element values of a connected DC network are positive, it is well known that all the elements of the [R] matrix are positive. Also,

$$R_{jj} \geq \sum_{k=1}^{n} R_{jk}, \quad k \neq j \tag{10.30}$$

Applying the property described by the above equation to Eq. (10.29) we see that

$$\frac{\partial V_{dj}}{\partial V_{dn}} > 0 \text{ and } \frac{\partial V_{dj}}{\partial V_{dn}} < 1 \tag{10.31}$$

Equation (10.29) is easily derived from noting that

$$\frac{\partial V_{dj}}{\partial I_n} = R_{jn}, \quad \frac{\partial V_{dn}}{\partial I_n} = R_{nn} \tag{10.32}$$

To summarize, the reduction of V_{dn} also reduces V_{dj}. However, this may not be adequate to limit the tap violation at the terminal j, the reduction of V_{dn} must cure tap violations not only at terminal 'n' but at all the terminals where tap violations take place. There is one or more critical terminals which determine the quantum of reduction in V_{dn}.

5. It is to be noted that constant ϕ does not imply constant θ or vice versa if V_d is rescheduled. In the case of θ specifications, computation of $V_{dj\text{new}}$ given earlier is only approximate. However, it can be assumed that the ratio of $\cos\phi$ to $\cos\theta$ is nearly constant for small changes in V_d (or I_d).

10.6 A CASE STUDY

The proposed method of DC-AC power flow was tested on a 30 bus, 5 terminal MTDC – AC system [8]. The 5 terminal MTDC system is shown in Fig. 10.5.

Description of the system

This test system was obtained by adding to the IEEE 30 bus test system, a 5 terminal DC network with two rectifiers (C1, C5) and three inverters (C2, C3 and C4). The AC system parameters for this system are the same as those of the IEEE 30 bus test system and given in [9].

The W_{dc1}^{spec} are P_d specifications at converters 1 to 4 and V_{dc} specification at converter 5.

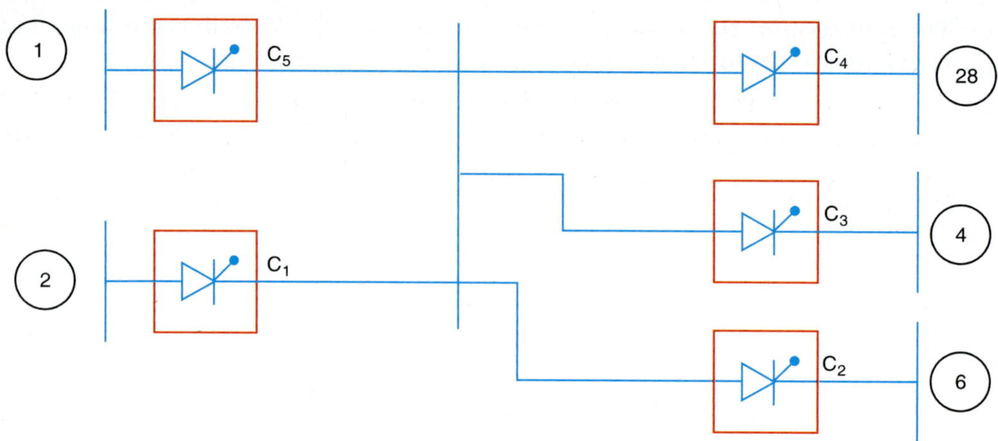

Fig. 10.5: 5 terminal MTDC system

Table 10.1: Base Case Results for the 30 Bus 5 Terminal MTDC–AC System

V_d	I_d	P_d	aVcosθ	Q_d	V	a	θ
1.0626	1.3176	1.4000*	1.09710	0.5380	1.0338	1.1011	15.47*
1.0310	–0.5820	–0.6000*	1.04626	0.2570	1.0155	1.1045	21.12*
1.0368	–0.6751	–0.7000*	1.05452	0.3050	1.0193	1.1096	21.19*
1.0369	–1.2538	–1.3000*	1.06971	0.6140	1.0002	1.1465	21.11*
1.0494*	1.1933	1.2523	1.08066	0.3460	1.0500	1.0369	7.0*

* indicates specified values

Terminal 5 was chosen as VST. W_{dc2}^{spec} are θ specifications as given in Table 10.1.

Using the values of P_d and Q_d at the respective converter buses as loads, fast decoupled power flow of Stott and Alsac [1] is preformed to get the AC voltage magnitudes. The voltages at the respective converter buses are shown in Table 10.1. Since 'aV' is already known, the value of tap is calculated. Table 10.1 also shows the specifications on P_d (at terminals 1 to 4 and V_d at terminal 5).

The minimum value of θ is taken as 15° for inverter and 7° for rectifier. The minimum and maximum tap limits are chosen as 0.9 and 1.1.

Removal of tap violation

From the values of taps obtained as shown in Table 10.1, we see that the taps at terminals 1,2,3 and 4 violate the maximum limit of 1.1. Therefore, it is required that the values of taps at these terminal be brought to the maximum limit of 1.1. Tap violation can be removed by

(*a*) adjusting of θ alone or

(*b*) by adjusting of V_d alone or

(*c*) by adjusting both

(a) By adjusting of θ alone:

The sensitivity information is obtained from the graph of V vs. $a'V$ where $a'V$ is equal to aV/V_d (V_d is at the respective terminal). A graph of V vs. $a'V$ is plotted as shown in Fig. 10.6. From Fig 10.6, the values of '*m*' and '*c*' are calculated for each terminal and are shown in Table 10.2. Table 10.2 also shows $(a'V)_{min}$, $(a'V)_{max}$, θ_{low}, θ_{high} and θ_{min} at each converter. These values are shown in Table 10.2.

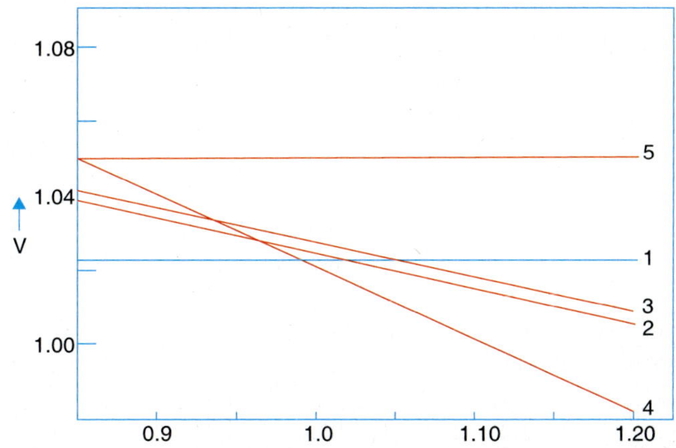

Fig. 10.6: Plot of V vs (a′V)

Table 10.2: Sensitivity Information

Terminal	m	c	$(a'V)_{min}$	$(a'V)_{max}$	θ_{low}	θ_{high}	θ_{min}
1	0.0000	1.0338	0.8756	1.0702	-	15.26	7
2	0.0930	1.1167	0.9016	1.0839	-	20.57	15
3	0.1041	1.1328	0.9018	1.0823	-	20.0	15
4	0.1958	1.2167	0.9027	1.0687	-	15.13	15
5	0	1.05	0.9005	1.1006	-	20.66	7

To remove tap violation at terminals 1,2,3 and 4, the values of θ at these terminals are adjusted to 15.26°, 20.57°, 20.0° and 15.13° respectively. With these values of θ , Q_d values are calculated and the AC power flow is performed. The results obtained at the end of this step are given in Table 10.3 which shows that the tap violations are removed.

(b) By adjusting V_d alone:

The values of V_{dnew} are calculated at the terminals at which the tap limits are violated. For each of these values of V_{dnew}, V_d at VST is calculated from sensitivity information. The lowest among them is chosen to be V_d^{spec} at VST.

CHAPTER 10

It turns out minimum $V_{d5} = 1.0086$ corresponding to tap violation at terminal 4. With this specification, the final DC-AC power flow results are given in Table 10.5 which shows the tap violations are removed.

Table 10.3: Final Results by Adjusting θ

Terminal	Q_d	θ	V	tap
1	0.534	15.26	1.0338	1.1000
2	0.251	20.57	1.0178	1.0979
3	0.290	20.0	1.0213	1.0988
4	0.490	15.13	1.0077	1.0996
5	0.346	7.0	1.05	1.0368

Table 10.4: Final Results by Adjusting V_d

V_d	I_d	P_d	aVcosθ	Q_d	θ	V	a
1.0217	1.3703	1.4000	1.05760	0.549	15.47	1.0338	1.0615
0.9888	–0.6068	–0.6000	1.004373	0.259	21.12	1.0152	1.0610
0.9949	–0.7036	–0.7000	1.01333	0.308	21.19	1.0191	1.0664
0.9949	–1.3066	–1.3000	1.02917	0.623	21.11	0.9996	1.1036
1.0086	1.2467	1.2567	1.04066	0.360	7.00	1.05	0.9986

Discussion

The power flow analysis of AC-DC systems presented here is based on a new formulation given in [11] and is distinct from the approaches reported in the previous literature. There are several shortcomings in the development of power flow analysis reported in the literature. These are stated below:

1. No clear cut distinction has been made between the scheduled variables and the control variables in the DC system. The control variables in the DC system are θ and a, and these are used to schedule some other variables – current (I_d), Power, (P_d), voltage (V_d), reactive power (Q_d) or reactive current (I_q).

   It is to be noted that, in the AC system, the control variables are generator power (P_g) and its terminal voltage (V_g). These are also the scheduled variables.

2. Scheduling the control variable $\theta = \theta^{spec}$ does not make full use of the flexibility in the operation of HVDC converters.

   With two control variables per terminal, it is possible to specify two variables – say P_d (or I_d) and Q_d (or I_q).

3. The number of variables which have to be solved during the iteration process are non-minimal. For example, in [2, 3], five variables per terminal are solved. It is to be noted that it is simpler to solve for the state variables (forming the minimal subset), the knowledge of which, readily gives other variables. For example in AC networks, the

knowledge of bus voltage magnitudes and angles is adequate to compute reactive power output (Q_g) of generators and line flows. While two variables per node are required for AC networks (some of which may be known in advance) one variable per node is adequate for the solution of DC networks. Although this idea is utilized in [4, 5], a systematic development is lacking. For example, the distinction between the network and controller equations is not explored.

4. A major shortcoming is the use of inflexible per unit system where base AC and DC voltages are related. As only one base DC voltage can be used in a connected MTDC system, this implies a fixed AC base voltage, which is inconvenient.
5. Finally, a systematic and general framework based on sensitivity analysis is lacking for rescheduling of specified variables to eliminate constraint violations on transformer taps and other variables. For example, it may be feasible to modify the specifications on control angles (rather than the DC voltage) to satisfy tap limits.

The formulation and analysis presented in this chapter are aimed at removing the lacunae described above and proposing a general formulation which is novel.

The major features of the formulation given here are (*a*) the clear demarcation between the control variables and the scheduled variables and (*b*) between the control equations and the network equations in the case of DC system and (*c*) the ability to handle the constraints on different variables such as converter transformer tap (a), firing angle (θ), AC voltage (V), DC voltage (V_d) and DC current (I_d). Use of sensitivity information for handling the constraints is another special feature. If there are no limits on tap, only one DC and one AC power flow solution is sufficient and there is no necessity to perform iterations.

10.7 ON-LINE POWER FLOW ANALYSIS FOR SECURITY CONTROL

During the operation of large power systems, several contingencies can occur – such as sudden tripping of a large generator or a AC transmission line due to a permanent fault. While spinning reserve can take care of the first type of contingency (loss of a generator or sudden switching of a load), the loss of a line can result in redistribution of power flow in the AC network resulting in overloading of one or more lines. To maintain static security (assuming the system is dynamically stable), it is necessary to alleviate the overloads in AC lines by control methods that act before the protection system trips the overloaded lines. Contingency analysis can also be used to anticipate potential problems and maintain secure operation by preventive control.

Generation rescheduling is the prevalent control method to alleviate AC line overloads. If phase shifting transformers are present in the system, they can also be used to redistribute power flow. However, generators are normally dispatched to minimize fuel costs and rescheduling will increase the costs. If HVDC links are present in the system, advantage can be taken of their fast controllability of power to alleviate AC line overloads. The control will be fast and the costs lower than the costs involved in rescheduling generator output power.

If reactive power constraints are not an issue, linear AC power flow model (called DC power flow – not to be confused with power flow in HVDC systems) can be used for contingency analysis and control. It is to be noted that linear AC power flow solution is equivalent to the solution available at the end of the first $P - \theta$ iteration of the Fast Decoupled Load Flow (FDLF) [1].

Although the linear AC power flow is approximate, it is reliable and computationally efficient when studying several contingencies. Also, superposition theorem can be applied for multiple contingency analysis, as the model is linear.

Problem formulation

Let us consider an AC/DC power system of 'n' buses and 'nl' lines in AC system and 'm' terminals in DC system. Let the nth bus be chosen as the reference bus (with bus angle equal to zero). Similarly let the mth terminal be the Voltage Setting Terminal (VST) in the DC system. All the buses are interconnected by transmission lines. The linear AC power flow is based on the following assumptions:

1. All the bus voltage magnitudes are assumed to be equal to 1 p.u.
2. The line resistances are neglected. Only the reactances are considered
3. The difference in the bus angles are assumed to be small.

   From these assumptions it follows that only real power flows need to be considered and are described by

$$\underline{P} = [B]\underline{\theta} \tag{10.33}$$

where $\theta = [\theta_1, \theta_2, ...\theta_{n-1}]^t$ is the vector of $(n - 1)$ bus angles (in radians)

$P = [P_1, P_2, ...P_{n-1}]^t$ is the vector of $(n - 1)$ power injections

$[B]$ is the susceptance matrix of dimension $(n - 1) \times (n - 1)$

$$B_{ii} = \sum_{j=1}^{n} \frac{1}{x_{ij}}$$

$$B_{ij} = -\frac{1}{x_{ij}}, \quad i \neq n, \quad j \neq n$$

x_{ij} is the reactance of the line connected between buses i and j.

The power injections P include the DC link powers at converter buses. At a rectifier bus, P is negative and at inverter bus, P is positive.

The power flow in any line 'ij' is given by

$$S_{ij} = \frac{1}{x_{ij}}(\theta_i - \theta_j) \tag{10.34}$$

where S stands for the power flow in line.

In matrix form Eq. (10.34) is written as

$$\underline{S_l} = [B_l]\underline{\theta} \tag{10.35}$$

where $[B_l]$ is a matrix of dimension $(nl \times (n - 1))$ with each of its row containing only two elements, one equal to $\frac{1}{x_{ij}}$ and other equal to $-\frac{1}{x_{ij}}$. From Eq. (10.33), θ can be written as

$$\underline{S}_l = [A]\underline{P} \tag{10.36}$$

where $[A] = [B_l][B]^{-1}$. The assumption of neglecting losses in the AC system is equally applicable to the DC system. Thus the power at the slack terminal of the DC system is negative of the

$$P_{dm} = -\sum_{i=1}^{(m-1)} P_{di} \tag{10.37}$$

As we desire that the line flows lie within tolerable limits of $S_l^{\min}$ and $S_l^{\max}$

$$S_l^{\min} \le S_l \le S_l^{\max} \tag{10.38}$$

$S_l^{\min}$ and $S_l^{\max}$ are usually of the same magnitude but of opposite sign.

The converters are designed to operate within limits which may also prevent the rectifier from becoming an inverter and vice versa.

$$P_{di}^{\min} \le P_{di} \le P_{di}^{\max}, \text{ for i=1, 2 …, (m − 1)} \tag{10.39}$$

The problem of finding the required changes in the DC network power transfers can now be formulated as that of finding the minimum of the objective function

$$f = \sum_{i=1}^{(m-1)} c_i |x_i| \tag{10.40}$$

Where $x_i = P_{di} - P_{di}^0$ for $i = 1, 2, \ldots, m - 1$. c_i are the weighting coefficients, P_{di} is the final DC power to be obtained at the end of the solution and P_{di}^0 is the initial DC power.

Minimizing the above objective function is equivalent to minimising the control effort required to change the DC powers, subject to the variables satisfying the AC network constraints, Eq. (10.38), and DC system constraints, Eq. (10.39).

The objective function can also be chosen as

$$f = \sum_{i=1}^{n} c_i x_i^2 \tag{10.41}$$

Linear or quadratic programming technique can be used to solve the problem depending on the objective function selected.

10.8 POWER FLOW ANALYSIS UNDER DYNAMIC CONDITIONS

The previous sections described power flow analysis of AC-DC systems in steady state(when the system is operating under equilibrium conditions which is represented by a point in the dynamic state space in which all the system trajectories lie). A disturbance followed by control actions results in the system leaving a stable equilibrium point and entering another stable equilibrium point. The transition may be accompanied by oscillations that are damped as the system settles to a new steady state. The transient or dynamic conditions may last for a few seconds which is too short for the OLTC to act. Hence, the control variable at a converter is only the firing or extinction angle (θ) and the off-nominal tap ratio 'a' remains constant at the initial value.

In solving the DC network and control equations, there are no major differences except the fact that we have now only one set of specifications. (W_{dc}^{spec}) as there is only one control variable per terminal. At non-VST terminals, we can select I_d or P_d as specified variables. However at VST, we can select one among V_d , θ , Q_d , I_q or V. The power balance equations at a converter bus (say k) are given below:

$$P_k(V,\delta)+P_{dk}(I,\theta_k)=0 \tag{10.42}$$

$$Q_k(V,\delta)+Q_{dk}(I,\theta_k)-Q_{ck}(V_k)=0 \tag{10.43}$$

The first terms in the above equations are obtained from the AC network, the second terms are computed from the converter equations. The third term in Eq. (10.44) represents the reactive power output of the shunt connected reactive source (typically filters, shunt capacitors and reactors).

Since θ_k is unknown and is linked to the AC voltage (V_k), we cannot decouple the DC system and AC system equations. One can use either sequential or simultaneous method to solve the AC-DC power flow equations.

Remarks

1. If DC voltage is specified at VST, P_{dk}, k = 1, 2, .., n will be a function of I alone and the DC system equations can be solved as discussed in the section 10.4.

   However θ_k is unknown and Q_{dk} cannot be computed in advance. Note that the knowledge of I_k gives us the product ($V_k \cos\theta_k$) as a_k is known (fixed).
2. The transient stability analysis involves the study of low frequency phenomenon (below 5 Hz). The models used, ignore the dynamics of AC and DC networks and also the controller dynamics sometimes. This implies that only algebraic equations are involved in the solution of AC and DC networks. This is equivalent to the use of power flow analysis under dynamic conditions with constant tap ratios.

An example

An inverter is operating with Constant Extinction Angle (CEA) and the rectifier is on Constant Current (CC) control. The tap 'a' is kept constant as I_d is varied. Compute V_d and P_d as functions of I_d if the inverter is connected to a voltage source $E\angle o$ through an impedance $(R+jX)$.

Data: E = 1.0, X = 0.5; $\dfrac{X}{R}=\tan 80^\circ$, k = 1.1288; X_c = 0.16; B_c = 0.5 (B_c is the susceptance at the fundamental frequency, of the shunt connected capacitors and AC filters).

a = (*i*) 0.9; (*ii*) 1.0; (*iii*) 1.1; (*iv*) 1.2

γ = (*i*) 15°; (*ii*) 18°; (*iii*) 21°; (*iv*) 30°

The variations of V_d and P_d as functions of I_d are shown in Fig. 10.7 for γ = 18° and different values of a. The variations of V_d and P_d for a = 1.2 and for different values of γ are shown in Fig. 10.8.

It is observed that power delivered by the inverter increases (*a*) as γ is reduced and (*b*) tap 'a' is increased. However, the peak value of power (P_d) occurs at reduced value of I_d as 'a' is increased whereas maximum power occurs at increased values of I_d as γ is reduced. The decrease in V_d with increase in I_d is slower for a = 0.9 compared to a = 1.2.

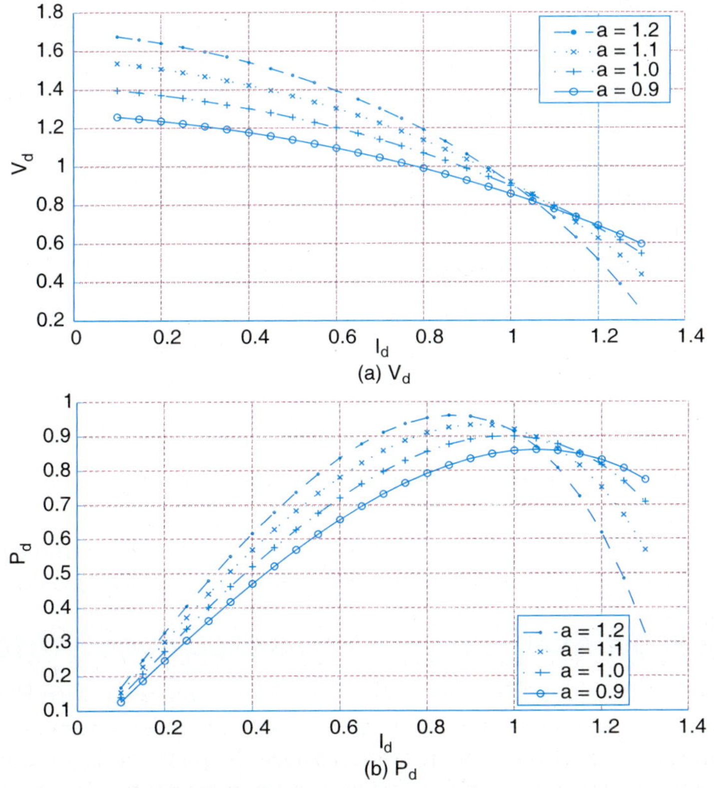

Fig. 10.7: Variation of V_d and P_d ($\gamma = 18°$)

If the ceiling on DC voltage at the rectifier is limited, it is not possible to operate the inverter at lower values of I_d if γ is kept constant. It becomes necessary to operate with constant V_d control (or constant AC voltage control). However, as I_d is increased for constant V_d, CEA control takes over as γ hits the lower limit. For example with V_d held constant at 1.0 p.u. (a = 1.2), $\gamma = 15°$ is reached at $I_d = 0.92$ p.u.

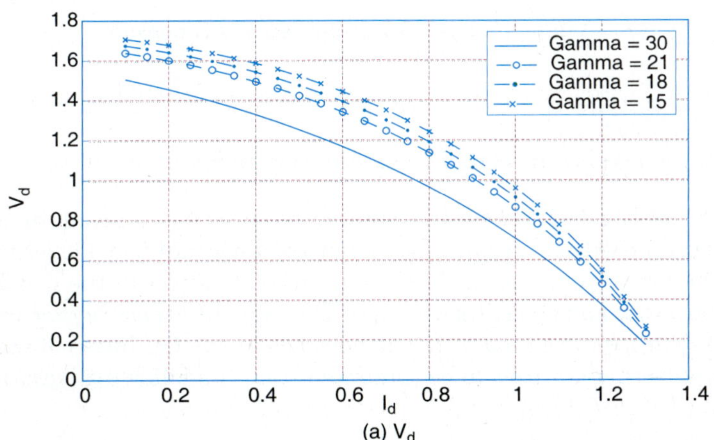

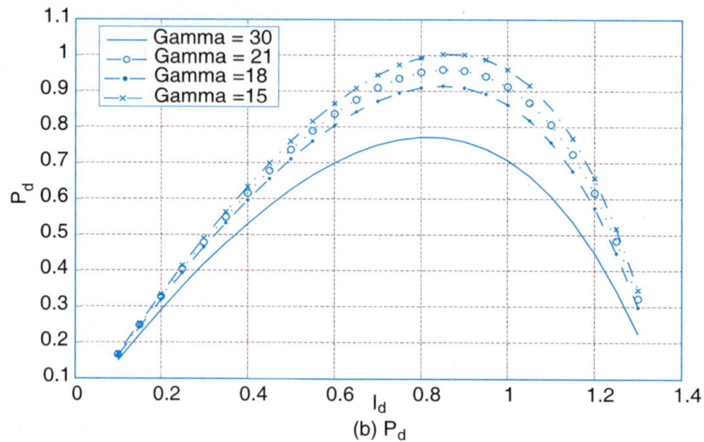

(b) P_d

Fig. 10.8: Variation of V_d and P_d (a = 1.2)

It is shown in Chapter 12, that if I_d is increased beyond the value corresponding to the Maximum Available Power (MAP), the system experiences voltage instability or collapse.

10.9 POWER FLOW ANALYSIS WITH VSC BASED HVDC SYSTEM

Figure 10.9 shows a VSC connected to the AC network. In comparing with Fig. 10.1, we note that the leakage reactance of the converter transformer (if any) is considered as part of the AC network. However, the interface bus with the AC network (where the power injections are considered) is the bus connected to the primary winding of the transformer (as in Fig. 10.1). The control variables for the VSC are the modulation index (m) and the control angle (α). Unlike in the

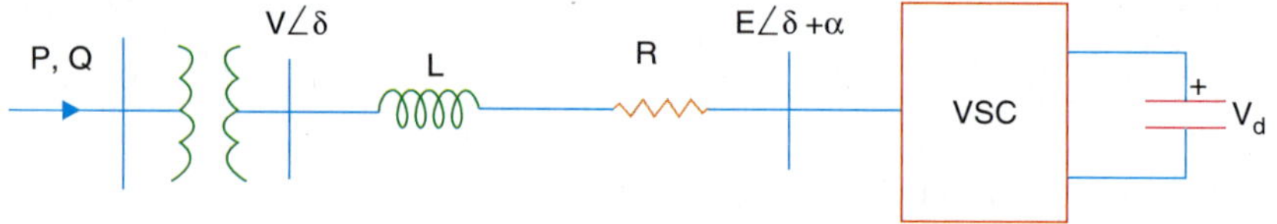

Fig. 10.9: A VSC connected to AC network

LCC, the injected voltage $\hat{E} = kmV_{dc}\angle(\delta + \alpha)$ where $k = \dfrac{\sqrt{6}}{\pi}$ for a two-level converter and $k = \dfrac{2\sqrt{6}}{\pi}$ for a three level converter. The modulation index m lies in the range $0 < m \leq 1$.

The solution of the DC network is similar as described earlier in section (10.4). In a multiterminal system with 'n' terminals, one of the terminal is selected as VST where V_d is specified. For the remaining terminals P_d is specified (subject to the constraints imposed by ratings of the converter and capacitor voltage). The only difference is that in a VSC, there is no commutation resistance to be modeled. The voltages at the DC buses of the VSC (where P_d is specified) are unknown quantities to be computed from the following equations:

$$[G]V_d = I_d \tag{10.44}$$

$$I_{dj} = \frac{P_{dj}^*}{V_{dj}}, j = 1,2,...(n-1) \tag{10.45}$$

$$V_{dn} = V_{dn}^* \tag{10.46}$$

there is no loss of generality in assuming VST as the nth terminal. $[G]$ is a $[n \times n]$ matrix which is singular. We can eliminate the row and column corresponding to VST and derive the following equation.

$$[G']V_{dr} = I_{dr} - g_n V_{dn} \tag{10.47}$$

where $[G']$ is a $(n-1) \times (n-1)$ matrix obtained from deleting a row and column corresponding to VST. V_{dr} and I_{dr} are column vectors of order $(n-1)$, obtained from V_d and I_d vectors, deleting elements V_{dn} and I_{dn} respectively. g_n is the reduced column vector of [G] (corresponding to VST).

The solution of the DC network gives the DC voltages at non-VST terminals and P_{dn}. With the specifications of Q_{dj}, $(j = 1, 2, ...n)$ AC power flow is carried out (without considering the internal bus shown in Fig. 10.7). The solution of the AC power flow gives V and δ at each converter bus. From this, it is possible to compute m and α. If the computed value of m exceeds 1, it is necessary to reschedule the specified variables. The first choice is Q_d which has to be increased (note that if Q_d is negative, (implying the injection of reactive power), its magnitude has to be reduced). If the constraint on m is not met by the rescheduling of Q_d alone, there is need to reschedule V_d^{spec} (at VST) and P_d^{spec} if necessary).

REFERENCES AND BIBLIOGRAPHY

1. Stott B. and Alsac O., '*Fast Decoupled Load Flow*', IEEE Trans., Vol. PAS-93, No.4, 1974, pp. 859–869.
2. Reeve J., Fahmy G. and Stott B., '*Versatile Load Flow for Multiterminal HVDC Systems*', IEEE Trans., Vol. PAS-96, No. 3, 1977, pp. 925–933.
3. Arrillaga J., Arnold C.P. and Harker B.J., '*Computer Modeling of Electrial Power Systems*', John Wiley, 1983.
4. Fudeh H. and Ong C.M., '*A Simple and Efficient AC-DC Load Flow Method for Multiterminal DC Systems*', IEEE Trans., Vol. PAS-100, No.11, 1981, pp. 4389–4396.
5. Ong C.M. and Hamzei-Nejad A., '*A General Purpose Multiterminal DC Load Flow*', IEEE Trans., Vol. PAS-100, No.7, 1981, pp. 3166–3174.
6. Mahseredjian J., Lefebvre S. and Mukhedkar D., '*A Multiterminal HVDC Load Flow with Flexible Control Specifications*', IEEE Trans., on Power Systems, Vol. 1, No. 2, 1986, pp. 272–282.
7. Ong C.M., Lu C.N. and Chen S.S., '*Incorporation of HVDC Equations in Optimal Power Flow Methods using Sequential Quadratic Programming Techniques*', IEEE Trans. on Power Systems, Vol. 3, No. 3, 1988, pp.1005–1011.
8. Opoku G. and Ong C.M., '*Coordination of Reactive Power Sources to Correct Bus Voltage Violations in an AC-DC System*', IEEE Trans., Vol. PAS-103, No. 6, 1984, pp.1128–1134.

9. Pai M.A., *'Computer Techniques in Power System Analysis'*, Tata McGraw-Hill, 1979.

10. Peschon J., Piercy D.S., Tinney W.F. and Tveit O.J., *'Sensitivity in Power Systems'*, IEEE Trans., Vol. Pas-87, No. 8, 1968, pp.1687–1696.

11. Padiyar K.R. and Kalyanararaman V., *'Power Flow Analysis in MTDC–AC Systems–A New Approach'*, Electric Machines and Power Systems, 23, pp. 37–54, 1995.

12. Xiao-Ping Zhang, *'Multiterminal Voltage Sourced Converter-based HVDC Models for Power Flow Analysis'*, IEEE Trans. on Power Systems, Vol. 19, No.4, pp. 1877–1884, November 2004.

13. Jiang X. et al., *'A Novel Approach for Modeling Voltage-Sourced Converter-Based FACTS Controllers'*, IEEE Trans. on Power Delivery, Vol. 23, No.4, pp. 2591–2598. October 2008.

14. Wood A.J. and Wollenberg B.F., *'Power Generation, Operation and Control'*, Second Edition (New York), John Wiley, 2003.

11 MODELING AND ANALYSIS OF AC-DC SYSTEM INTERACTIONS

11.1 INTRODUCTION

The development of converter technology in HVDC transmission was an offshoot of the developments in the industrial drives. The possibility of interactions between AC and DC systems was not anticipated in the early application of HVDC technology when the power ratings were modest. However, with the increase in the power ratings, several interactions surfaced which affect the design of controllers. The major types of AC-DC system interactions are:

1. Harmonic interactions or low order harmonic resonances at frequencies above fundamental.
2. Torsional interactions at subsynchronous frequencies (10-50 Hz).
3. Electromechanical interactions at frequencies below 5 Hz.
4. Control interactions caused by AC system resonances affecting converter control due to frequency transformation effect of converters.

The AC system strength at the converter bus has significant impact on these interactions. The Short Circuit Ratio (SCR) or Effective Short Circuit Ratio (ESCR) is an indicator of the AC system strength. SCR is defined as the ratio of the AC system three phase short circuit MVA at the converter AC bus to the capacity of the DC station. The ESCR takes into account the contribution of AC filters, shunt capacitors and reactors to the AC system impedance. Typically ESCR is (SCR-0.5). A minimum SCR level of 2.5 is often used as the lower limit for acceptable performance of the DC system.

Many system interactions were experienced after the commissioning of new HVDC projects which led to their study and the development of counter measures to overcome the problems.

The first occurrence of harmonic interactions was observed in some early HVDC schemes at New Zealand and Cross Channel. The phenomenon of severe harmonic distortion of the converter bus voltage, due to magnification of certain non-characteristic harmonics was observed. This problem prevented normal operation of the DC link due to commutation failures caused by severe harmonic distortion. This was termed on 'harmonic instability' and traced to the positive feedback loop, in which, the voltage distortion reaching the control system causes firing pulses which are not equidistant and exaggerate the original distortion. This has led to the development of Equidistant Pulse Control (EPC) firing scheme which superseded the Individual Phase Control (IPC) scheme. This has solved the problem of the 'harmonic instability' to a great extent. But subsequent studies point to the existence of harmonic interactions even with EPC firing control scheme. For example, the DC system resonance at or near the fundamental frequency resulted in 'Core Saturation Instability' [2] in Nelson River Scheme in Canada. A second harmonic instability [3] resulted from adverse harmonic interaction at Chateauguay BTB HVDC link in Canada which has a SVC at the converter bus for reactive power compensation.

The phenomenon of HVDC-turbine generator torsional interactions was discovered only in 1977 when field tests were carried out during the commissioning of Square Butte HVDC Scheme in USA [12]. The modification of the current controller solved the problem at Square Butte. However, in general, auxiliary controllers to damp subsynchronous oscillations may be required.

Several HVDC links built in late 1980's were found to operate at low SCR (below 2.0) during contingency conditions. The operation under low SCR resulted in voltage instability, power uncontrollability, dynamic overvoltages and voltage flicker (caused by switching shunt connected capacitor/filter banks). This led to the development of alternate control methods for inverters instead of Constant Extinction Angle (CEA) control. Constant reactive current, constant DC voltage or constant voltage control strategies were suggested and applied where necessary.

The analysis of AC-DC system interactions requires the development of appropriate system models. In general, the system models are nonlinear and time varying with power electronic controllers, the control action (application of firing pulses) is at discrete times and sampled data models have been suggested [21-23].

Small signal analysis is invariably used for preliminary design of controllers and checking the stability of the operating (equilibrium) point. Both time and frequency domain techniques are used for this purpose. The analytical techniques are not adequate in predicting performance under large disturbances. In such cases, only recourse is to simulate the system using mathematical or physical models. HVDC (physical) simulators have been used for studies in real time involving actual converter controls. Recent advances in digital signal processing technology and applications of parallel processing techniques have resulted in speeding up computations and carrying out simulation in real time. This has led to the development of Real Time Digital Simulators (RTDS) which are expected to replace physical simulators.

In this chapter, modeling, analysis and simulation of AC-DC systems for the study of major interactions are described. In particular, harmonic and torsional interactions are analyzed in detail. The electromechanical interactions (at low frequencies) are discussed in the next chapter.

11.2 SYSTEM MODELS

11.2.1 General

The system models are constructed from assembling the component models and identifying the interfaces between the various components. Here, we will primarily concentrate on the models for DC systems and the interfacing with the AC system models. The modeling of AC systems is relatively well established and hence will be discussed briefly. The DC system consists of converters, their controls and the DC network.

The modeling is dependent upon the nature of study. For study of electromechanical interactions at low frequencies, it is common to neglect the AC network dynamics. The voltages and currents in the AC network are assumed to be phasors that vary slowly in response to the low frequency oscillations (swings) of the generator rotors. The three phase AC network is also assumed to be balanced and only the positive sequences network (modeled as a single phase network) is considered for the analysis. The controller dynamics with time constants of 10 ms or higher are considered. The generator armature transients are neglected and the armature is considered as part of the AC network. It stands to reason that transients in the DC network can also be neglected although the literature survey shows there is no unanimity on this topic [24]. There are also some doubts about the requirements of the DC controller models.

For studies involving torsional interactions, it is well established that the AC network dynamics cannot be neglected. However, it is perfectly justified to ignore the presence of harmonics resulting from AC-DC converters. Hence the currents and voltages in the AC network are represented by 'dynamic' phasors that are governed by differential equations modeling the AC network dynamics in D-Q components (defined by transforming the three phase components with respect to a synchronously rotating reference frame). The transformation is defined by

$$\begin{bmatrix} f_a \\ f_b \\ f_c \end{bmatrix} = \sqrt{\frac{2}{3}} \begin{bmatrix} \cos\theta_0 & \sin\theta_0 & \frac{1}{\sqrt{2}} \\ \cos\left(\theta_0 - \frac{2\pi}{3}\right) & \sin\left(\theta_0 - \frac{2\pi}{3}\right) & \frac{1}{\sqrt{2}} \\ \cos\left(\theta_0 + \frac{2\pi}{3}\right) & \sin\left(\theta_0 + \frac{2\pi}{3}\right) & \frac{1}{\sqrt{2}} \end{bmatrix} \begin{bmatrix} f_D \\ f_Q \\ f_O \end{bmatrix} \tag{11.1}$$

where f represents either voltage or current, f_O is the zero sequence component which can be neglected under normal conditions. If the source currents or voltages are sinusoidal and balanced, and the network is linear, symmetric in three phases, then f_D and f_Q are constants in steady state.

If dynamic phasor representation is required for the AC network, obviously DC network and controller dynamics have to be modeled in detail. However, the harmonics generated by the AC-DC conversion can still be neglected. This also implies that firing angle controller does not require detailed modeling (involving gate pulse generation). The average converter DC voltage, expression (as given in Chapter 3) can be applied.

The detailed modeling based on dynamic phasors is also required in designing controllers taking into account instability introduced by controller interactions in the presence of AC and DC network resonances.

It is to be noted that the development of system models based on simplified or detailed component models also facilitates analytical study of system stability under small disturbances, in addition to simulation for large disturbances. However, these models do not permit accurate prediction of performance of AC-DC systems subjected to converter faults such as commutation failures and asymmetrical AC network faults. For accurate simulation of these events, it is necessary to model the switching action in the converters generated by firing angle controls. This involves detailed transient simulation of the system and representation of the AC network by three phase quantities. The Phase-Lock-Loop (PLL) required to synchronize the firing pulses to the AC converter bus voltages, also needs to be modelled. It is obvious that such detailed models consider harmonic generation by converters.

The detailed digital dynamic (or transient) simulation of AC-DC systems involves significant computation times. Hence physical HVDC simulators (similar to Transient Network Analyzers and hence also called as DC-TNA) were used extensively for design studies and factory system testing of DC controllers. With the advent of RTDS, it is expected that they may replace physical simulators.

Analytical studies can be carried out based on time-domain or frequency domain models and techniques. For example, the damping torque analysis based on frequency - domain is a fast screening tool for identifying torsional interactions. The applications of switching functions is a useful technique for the analysis of several interactions and in particular, harmonic resonances.

11.2.2 Converter Models

Here, we consider two types of converter models – (1) continuous time, average DC voltage (2) Discrete time model.

11.2.2.1 Simplified continuous time model

Consider a converter bridge fed from a transformer, which may be connected in star/star or star/delta (see Fig. 11.1). The equivalent circuit of the converter bridge is shown in Fig. 11.2 where

$$E_d = kaV \cos\theta \tag{11.2}$$

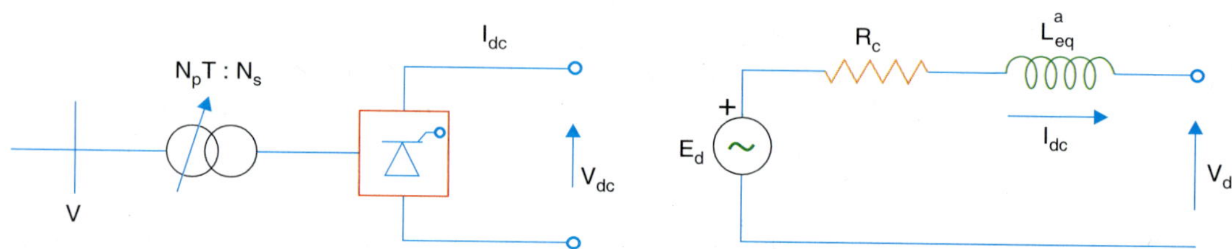

Fig. 11.1: A converter bridge

Fig. 11.2: Simplified continuous time equivalent circuit

where V is the converter bus AC voltage expressed in per unit on its own base (rated) voltage, θ is the delay angle, α for the rectifier and angle of advance, (β) for the inverter. $a = \frac{1}{T}$ is the off nominal tap ratio. k is defined as

$$k = \frac{3\sqrt{2}}{\pi} \frac{N_S}{N_P} \frac{V_{aB}}{V_{dB}}$$

where V_{aB} is the base AC voltage and V_{dB} is the base (rated) voltage.

By selecting rated DC current (I_{dn}) as the base DC current, we can express $\bar{R}_c$ (in per unit) as

$$\bar{R}_c = \frac{\bar{X}_c}{2} k \tag{11.3}$$

Note that R_c is the commutation resistance. The equivalent inductances, L^a_{eq} is the average inductance defined as

$$L_{eq}^{a} = \frac{X_c}{\omega_0}\left[2(1-d)+\frac{3}{2}d\right] \tag{11.4}$$

where $d = \frac{3u}{\pi}$, u = overlap angle in radians.

The transformer winding resistance and valve voltage drop can also be taken into account in a manner similar to the average inductance, L_{eq}^{a}.

The equivalent circuit of Fig. 11.2 is based on the following assumptions:

1. The harmonics in the DC voltage are neglected.
2. The AC voltages are assumed to be balanced and the transformer winding asymmetry is neglected.
3. The inclusion of inductance L_{eq}^{a} given by Eq. (11.4) is based on the state space averaging concepts [24–26].This is an average inductance calculated on the basis that when three valves conduct during the overlap period, the apparent inductance is $\frac{3}{2}\frac{X_c}{\omega_0}$ while during two valve conduction, the apparent inductance is $\frac{2X_c}{\omega_0}$.
4. The converter control is assumed to be continuous.

The last assumption can be relaxed and discrete time model can be derived. This is discussed next.

11.2.2.2 Discrete time converter model

The converter model is derived by considering the normal operation of a bridge converter with two and three valve conduction mode ($u < 60^\circ$). Figure 11.3 shows the circuit when valves 1, 2 and 3 are conducting. The converter transformer secondaries are represented as three voltage sources e_a, e_b and e_c in series with the leakage reactance. The expression for v_d is given by

$$v_{d_c} = \left(e_b - \frac{X_c}{\omega_0}\frac{di_b}{dt}\right) - \left(e_c - \frac{X_c}{\omega_0}\frac{di_c}{dt}\right) \tag{11.5}$$

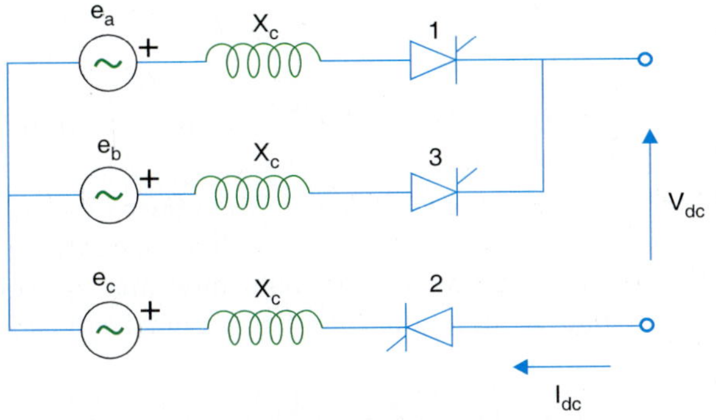

Fig. 11.3: A circuit showing 3 valve conduction

Consider that valve 3 begins conduction at time instant, t_k. The average DC voltage of the converter over the interval (t_k, t_{k+1}) is given by

$$V_d(k) = (1/\Delta t_k)\int_{t_k}^{t_{k+1}} v_{dc}(\tau)d\tau \quad (11.6)$$

where the instant t_{k+1} corresponds to the firing instant of the next valve. Assuming that the valves are fired at regular intervals of 60°,

$$\Delta t_k = t_{k+1} - t_k = \pi/3\omega_0 = h \quad (11.7)$$

Since the instant t_k coincides with the firing of valve 3, we have

$$e_b - e_c = \sqrt{2}E_s \sin[\omega_0(t - t_k) + \alpha_k + \pi/3] \quad (11.8)$$

where α_k is the delay angle at instant t_k. Substituting (11.8) in (11.5) and from Eq. (11.6), we get

$$V_d(k) = \left(\sqrt{2}E_s/h\omega_0\right)\cos\alpha_k - (2X_c/(\omega_0 h))I_{dc}(k+1) + ((X_c/(\omega_0 h))I_{dc}(k) \quad (11.9)$$

Note: The terminal conditions for the valve currents that are used in deriving Eq. (11.9) are

$$i_b(t = t_k) = 0,\ i_b(t = t_{k+1}) = I_{dc}(k+1)$$

$$i_c(t = t_k) = -I_{dc}(k),\ i_c(t = t_{k+1}) = -I_{dc}(k+1)$$

The equation (11.9) can be rewritten as

$$V_d(k) = (1/(h\omega_0))\left[\sqrt{2}E_s \cos\alpha_k - X_c I_{dc}(k)\right]$$

$$-(2X_c/(\omega_0 h))[I_{dc}(k+1) - I_{dc}(k)] \quad (11.10)$$

It is not difficult to see that if V_d is assumed to remain piecewise constant over the interval (t_k, t_{k+1}) the expression for V_d obtained from Fig. 11.2 would be identical to that in (11.10), by putting

$$L_{eq}^a = 2X_c/\omega_0 \quad (11.11)$$

and

$$\frac{dI_{dc}}{dt} = \frac{I_{dc}(k+1) - I_{dc}(k)}{h} \quad (11.12)$$

Remarks

1. It is to be noted that α_k is a variable defined at discrete times (corresponding to the generation of the firing pulses).

   The output of the converter control is α_k. The controller can be analog or digital. However, the AC and DC network are modeled as continuous time systems. The equations for these networks has to be transformed into discrete time (mathematical) models. Using state space description, the system model can be expressed in the general form

$$X(k+1) = f(X(k),\ U(k)) \quad (11.13)$$

$$Y(k) = g\big(X(k),\ U(k)\big) \tag{11.14}$$

2. The bandwidth of the converter controller (typically) does not exceed 30 Hz. Hence, continuous time models may be adequate for stability analysis.
3. The sample data model for the converter for a 12 pulse converter depends on the value of the overlap angle. There are two modes depending on (*i*) $u < 30°$ and (*ii*) $30° < u < 60°$

11.2.3 Model of Converter Controller

For sake of convenience, the overall control can be divided into two categories:

1. Power Control, Auxiliary Control and Voltage Dependent Current Order Limiter (VDCOL). The output of this block is the current order.
2. Constant Current (CC) and Constant Extinction Angle (CEA) controls. These are usually of feedback type. However, the extinction angle control can also be of predictive (open loop) type. The output of these controllers is a control voltage that determines the instant of gate pulse generation. The input is taken as the current order (generated locally or at the remote station) or the extinction angle reference (generated locally). The communication delay in transmitting the current order may have to be represented.
3. Gate pulse generator which has input from the CC or CEA controller and determines the instant of gate pulse generation for each valve. There are basically two types of firing control schemes. (*i*) Individual Phase Control (IPC) and (*ii*) Equidistant Pulse Control (EPC). The latter can be (*a*) Pulse Frequency Control (PFC) or (*b*) Pulse Phase Control (PPC).

The modeling of gate pulse generator in connection with detailed converter model will be taken up separately, in the next section on dynamic digital simulation. However, the effect of IPC or EPC can be included in the controller model in conjunction with the simplified converter model. The basic difference in the two schemes is that the change in the phase of the converter bus AC voltage will affect the delay angle in case of EPC, while it does not in the case of IPC.

The converter control is usually represented by block diagrams and specifying the transfer function of each block. A typical controller block diagram is shown in Figs. 11.4(*a*) and (*b*). (Note that Fig. 11.4 (*a*) is reproduced from Fig. 4.18) This does not give all the details but is adequate in system modeling. In Fig. 11.4(*b*), the switch S_1 is closed in the case of EPC and switch S_2 is closed in case the synchronizing circuit is provided. $\Delta\delta_v$ is the change in the phase of the converter bus voltage and the transfer function of the synchronizing circuit is typically

$$SYN(s) = 1/(1 + sT_{ss}),\ T_{ss} = 1/(2\pi B_w) \tag{11.15}$$

where B_w is the bandwidth of the synchronizing circuit in Hertz. As the bandwidth increases, the characteristics of EPC tend towards those of IPC.

The time delay T_d is introduced if continuous time model of the converter is used and is given by

$$T_d = 1/(2pf_o) \tag{11.16}$$

where p = pulse number, $f_o = \omega_0/2\pi$

This delay is to be ignored in case the discrete model of the converter is used, in which case, the delay angle is obtained by sampling the output of the controller at intervals corresponding to $2T_d$.

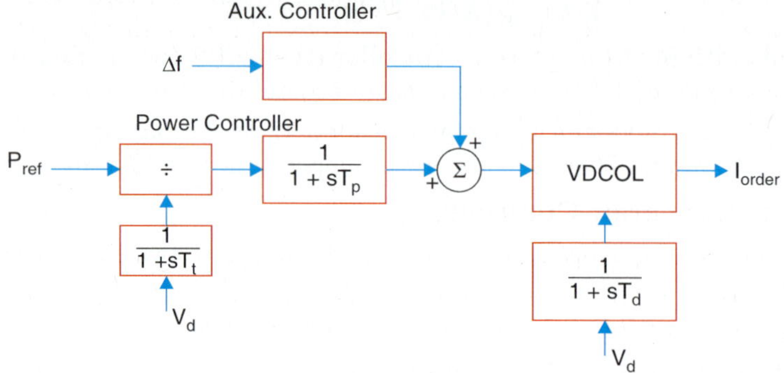

Fig. 11.4: (a) Power and auxiliary controller block diagram

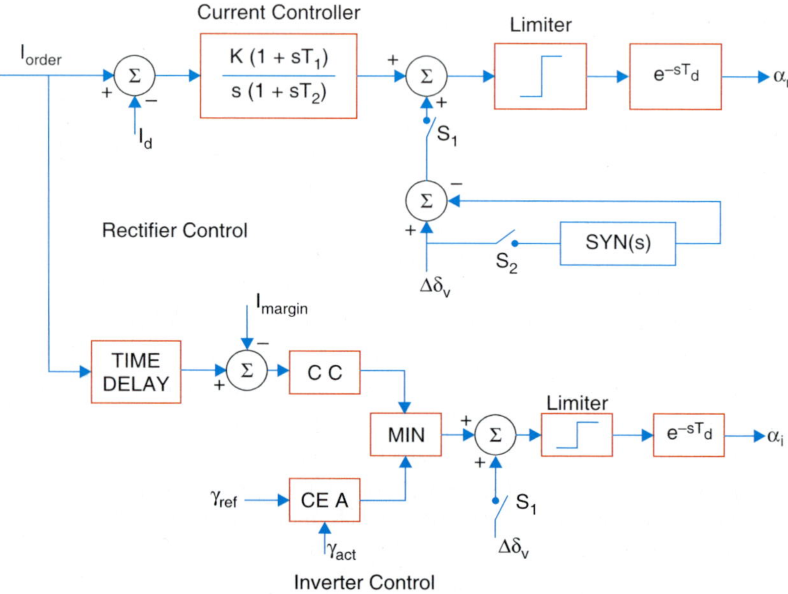

Fig. 11.4: (b) Rectifier and inverter controller block diagram

The representation of limiters in the control system requires care. The limiters can be of (*a*) wind up type, or (*b*) non-wind up type. In the former case, the limiter is treated as a separate block with the input variable unrestricted while the output variable is limited within a specified range. In the case of a non-wind up limiter, (which is normally applicable at the output of an integrator) the output variable of the block directly preceding the limiter is restricted. There is no need for a separate block to represent the limiter, in this case.

The major difference between the two types of limiters is that there is a time lag introduced in the case of the wind up limiter, when the output variable is at its limits. For example, if a wind up limiter is applied at the output of an integrator, the output of the limiter will change only after some time has elapsed following the reversal of the input to the integrator.

11.2.4 Modeling of DC Network

The DC network is assumed to consist of smoothing reactor, DC filters and the transmission line. The smoothing reactor and DC filters can be represented by lumped parameter linear elements. The DC line can also be modeled as of T or π equivalent if the higher frequency behaviour is not of interest.

As an example, consider a monopolar line with (ground return) which is terminated by converter stations with a single equivalent bridge in each station. Neglecting DC filters, the DC network is shown in Fig. 11.5 which also shows the simplified converter models (continuous time) as components. The smoothing reactors at the two ends are designated as L_{dr} and L_{di}.

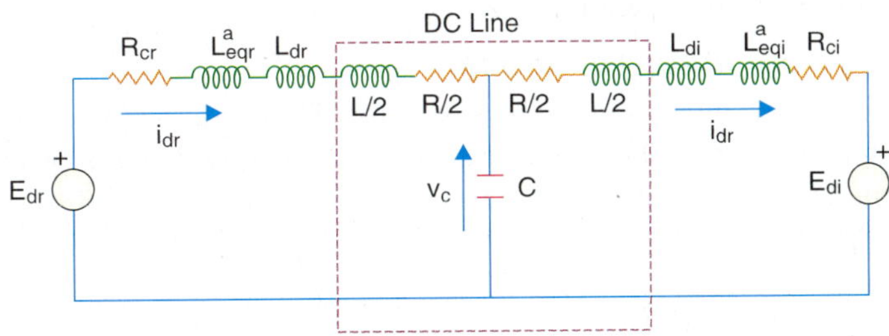

Fig. 11.5: Equivalent circuit of a two terminal DC system

The subscripts r and i refer to the rectifier and inverter terminal respectively. The series resistance of the smoothing reactor is not shown in the diagram but can be accounted quite easily. The state equations for the network shown in Fig. 11.5 are

$$di_{dr}/dt = -(R_{tr}/L_{tr})i_{dr} + (E_{dr} - v_c)/L_{tr} \tag{11.17}$$

$$di_{di}/dt = -(R_{tr}/L_{tr})i_{di} + (v_c - E_{di})L_{ti} \tag{11.18}$$

$$dv_c/dt = (1/C)(i_{dr} - i_{di}) \tag{11.19}$$

where
$$L_{tr} = L^a_{eqr} + L_{dr} + L/2,\ L_{ti} = L^a_{eqi} + L_{di} + L/2$$

$$R_{tr} = R_{cr} + R/2,\ R_{ti} = R_{ci} + R/2$$

In general, the state equations for any DC network can be written as

$$X_{DC} = A_{DC}X_{DC} + BU_{DC} \tag{11.20}$$

$$Y_{DC} = CX_{DC} \tag{11.21}$$

where

$$U^T_{DC} = [E_{dr}\ E_{di}]$$

$$Y^T_{DC} = [i_{dr}\ i_{di}]$$

11.2.5 Modeling of AC Network

11.2.5.1 General

For some types of analysis, the AC network can be assumed to in the steady state (say for power flow analysis or long term stability analysis). In this case, single phase representation is most often used. However, three phase representation is also possible for analysis on unsymmetrical networks.

In the case of transient analysis on symmetrical three phase networks it is adequate to consider only two phase representation using α, β components or *D-Q* components (defined on the synchronously rotating reference frame). The latter has the advantage of eliminating the time varying coupling between AC and DC system, and, hence, is convenient for analytical studies. It also has the advantage that, in steady state, *D-Q* components are constants and are related to the phasors in power flow analysis.

11.2.5.2 Formulation of state equations

Consider the α sequence network shown in Fig. 11.6. This network is excited by voltage and current sources at specific nodes. The contribution at the converter bus from the DC system is a current source dependent on the converter DC current. In general, the state equations for the network shown in Fig. 11.6 can be written as

$$\dot{x}_\alpha = [R]\,x_\alpha + [S_1]\;E_\alpha + [S_2]I_\alpha \tag{11.22}$$

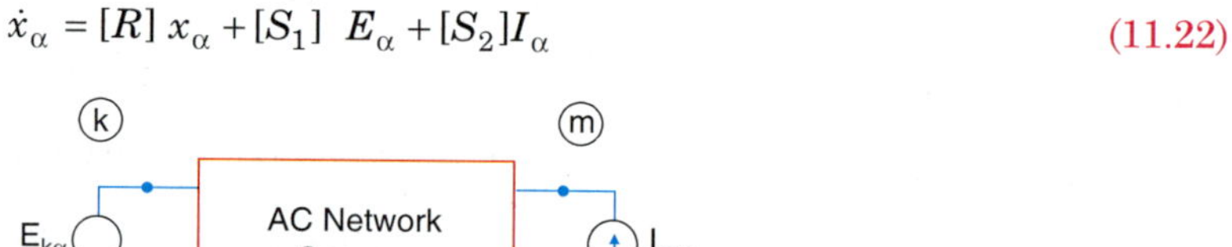

Fig. 11.6: 'α' Sequence AC network

where x_α is the vector of state variables corresponding to inductor currents in links and capacitor voltages in the tree branches. E_α and I_α are the injected voltage and current vectors $[R]$, $[S_1]$ and $[S_2]$ are constant matrices, functions of the network parameters. It is assumed that the network equations are linear.

The 'β' sequence network is also identical to the network shown in Fig. 11.6 except for the replacement of the sources E_α and I_α by E_β and I_β. Hence, the state equations for the 'β' sequence network are

$$\dot{x}_\beta = [R]\,x_\beta + [S_1]\;E_\beta + [S_2]I_\beta \tag{11.23}$$

Transformation to D-Q components

The transformation from α , β variables to *D-Q* variables is defined by the following equations:

$$\begin{bmatrix} x_\alpha \\ x_\beta \end{bmatrix} = \begin{bmatrix} U\cos\theta_0 & U\sin\theta_0 \\ -U\sin\theta_0 & U\cos\theta_0 \end{bmatrix} \begin{bmatrix} x_D \\ x_Q \end{bmatrix} \tag{11.24}$$

where U is the identity matrix of dimension equal to the order of x_α or x_β. θ_0 is the angle by which the D axis leads a stationary axis. It is to be noted that

$$d\theta_0/dt = \omega_0 \tag{11.25}$$

is assumed to be constant. Substituting Eq. (11.24) in Eqs. (11.22) and (11.23) leads to the following equations

$$\dot{x}_D = [R]\, x_D - \omega_0 x_Q + [S_1]\ E_D + [S_2] I_D \tag{11.26}$$

$$\dot{x}_Q = [R]\, x_Q + \omega_0 x_D + [S_1]\ E_Q + [S_2] I_Q \tag{11.27}$$

where E_D, E_Q and I_D, I_Q are the D-Q components of the voltage and current sources respectively.

An example

Consider the simple AC network connected to a converter bus, shown in Fig. 11.7. The source current components (through the inductor L) and the capacitor voltage components are the state variables. The AC network equations in the state space are

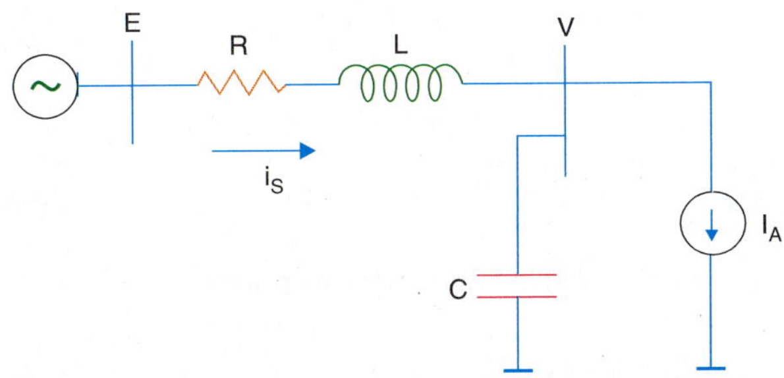

Fig. 11.7: A simple AC network

$$\frac{di_{SD}}{dt} = -\frac{R}{L} i_{SD} - \omega_0 i_{SQ} - \frac{v_D}{L} + \frac{E_D}{L}$$

$$\frac{di_{SQ}}{dt} = -\frac{R}{L} i_{SQ} + \omega_0 i_{SD} - \frac{v_Q}{L} + \frac{E_Q}{L}$$

$$\frac{dv_D}{dt} = -\omega_0 v_Q + \frac{i_{SD}}{C} - \frac{I_{AD}}{C}$$

$$\frac{dv_Q}{dt} = \omega_0 v_D + \frac{i_{SQ}}{C} - \frac{I_{AQ}}{C}$$

There is no loss of generality in assuming $E_Q = 0$. I_{AD} and I_{AQ} are the D and Q axis components of the AC current drawn by the converter.

11.2.5.3 Interface with DC system

The current injected into the AC network at the converter bus is proportional to the average DC current that flows in the link. If harmonics are neglected, then the fundamental component of the current I_A is given by

$$I_A = \pm kai_d \tag{11.28}$$

where i_d is the DC current flowing in the converter. The positive sign is associated with the rectifier where it is a load current, and the negative sign applies to the inverter where it is an injected current. As noted in Chapter 3, the equation (11.28) is not exact but accurate enough.

The current I_A in steady state, lags the converter bus voltage by an angle $\pm\phi$. The positive sign applies to the rectifier while negative sign applies to the inverter. The phasor diagram shown in Fig. 11.8 gives the relative position of the current and voltage phasors in D-Q reference frame for (a) inverter and (b) rectifier. From this, the following equations can be derived

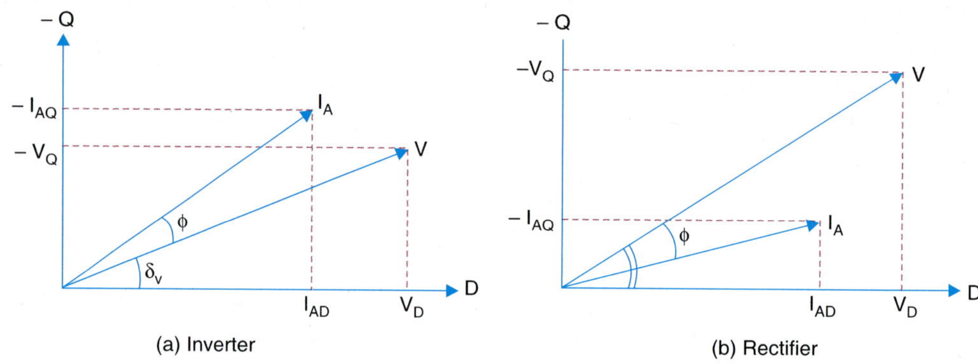

Fig. 11.8: Phasor diagrams

$$I_{AD} = I_A \cos(\delta_v \pm \phi) \tag{11.29}$$

$$I_{AQ} = -I_A \sin(\delta_v \pm \phi) \tag{11.30}$$

where positive sign applies to the inverter while negative sign applies to the rectifier. The angle δ_v by which the converter bus voltage phasor V leads the Q axis, is given by

$$\tan\delta_v = -\frac{V_Q}{V_D} \tag{11.31}$$

The angle ϕ is defined by

$$\phi = \cos^{-1}(V_d/kaV) \tag{11.32}$$

11.3 APPLICATION OF SWITCHING FUNCTIONS

The interaction between the AC and DC networks can be investigated by applying switching functions that relate the DC current to the three phase AC currents (and the DC voltage to AC phase voltages). In general, the phase currents drawn by a converter can be expressed as

$$i_a(t) = S_a(t)i_d(t) \tag{11.33}$$

$$i_b(t) = S_b(t)i_d(t) \tag{11.34}$$

$$i_c(t) = S_c(t) i_d(t) \tag{11.35}$$

The switching functions S_a, S_b and S_c are periodic in steady state and their waveforms (for a six pulse converter with a star/star connected converter transformer of ratio 1:1) neglecting overlap are shown in Fig. 11.9.

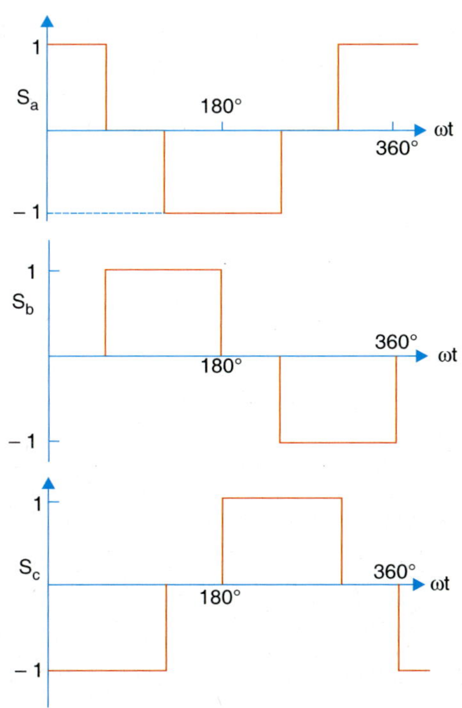

Fig. 11.9: Switching functions with star/star connected converter transformers

If the converter is assumed to be lossless, the AC and DC powers balance. Thus,

$$\begin{aligned} v_d i_d &= v_a i_a + v_b i_b + v_c i_c \\ &= (v_a S_a + v_b S_b + v_c S_c)\, i_d \end{aligned} \tag{11.36}$$

From the above, we get,

$$v_d(t) = S_a(t) v_a(t) + S_b(t) v_b(t) + S_c(t) v_c(t) \tag{11.37}$$

It is to be noted that

$$S_b(t) = S_a\left(t - \frac{2\pi}{3\omega_0}\right) = S_c\left(t + \frac{2\pi}{3\omega_0}\right) \tag{11.38}$$

The switching function, $S_a^{12}(t)$ for a 12 pulse converter (using per unit quantities) is shown in Fig. 11.10.

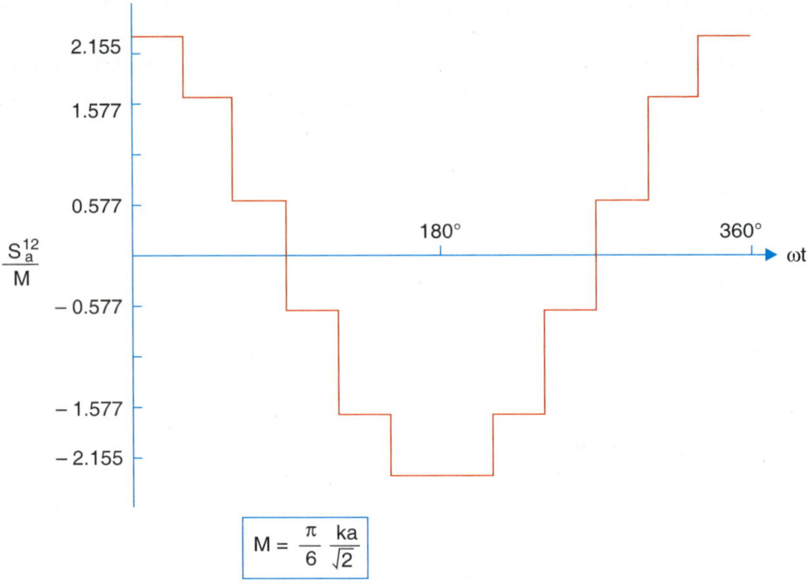

Fig. 11.10: Switching function for a 12 pulse converter

Fourier analysis of $S_a^{12}(t)$ for the twelve pulse converter gives

$$S_a^{12}(t) = \sqrt{\frac{2}{3}}ka\left[\cos\omega t - \frac{1}{11}\cos 11\omega t + \frac{1}{13}\cos 13\omega t - ....\right] \quad (11.39)$$

Since $S_b^{12}(t)$ also lags $S_a^{12}(t)$ by 120° and leads $S_c^{12}(t)$ by 120°, we get

$$S_b^{12}(t) = \sqrt{\frac{2}{3}}ka\left[\cos\left(\omega t - \frac{2\pi}{3}\right) - \frac{1}{11}\cos\left(11\omega t + \frac{2\pi}{3}\right) + \frac{1}{13}\cos\left(13\omega t - \frac{2\pi}{3}\right) - ...\right] \quad (11.40)$$

$$S_c^{12}(t) = \sqrt{\frac{2}{3}}ka\left[\cos\left(\omega t + \frac{2\pi}{3}\right) - \frac{1}{11}\cos\left(11\omega - \frac{2\pi}{3}\right) + \frac{1}{13}\cos\left(13\omega t + \frac{2\pi}{3}\right) - ...\right] \quad (11.41)$$

If the converter bus voltages are sinusoidal and balanced (positive sequence at fundamental frequency), it can be shown that the DC voltage contains a DC component of magnitude $(kaV\cos\theta)$ where θ is the angle by which the voltage leads the fundamental component of the AC current drawn by the converter. The DC voltage also consists of harmonics of the order (12 m, m = 1, 2...). For high impedance of the DC circuit at harmonic frequencies, $i_d(t)$ consist of only the DC component, say I_d . This results in a positive sequence, fundamental frequency current drawn by the converter. In addition, there are harmonic currents of order $h = 12\text{m} \pm 1$ where m is an integer. The harmonic is of negative sequence when $h = 12\text{m} - 1$ and of positive sequence when $h = 12\text{m} + 1$. These harmonics may be viewed as side bands which may be neglected for convenience.

The concept of switching functions is very useful in the analysis of harmonic and other interactions. For example, consider a case of injection of the fundamental frequency voltage in the DC link caused by magnetic coupling of a parallel AC line on the same tower carrying DC conductors. The induced AC voltage results in a fundamental frequency current in the DC link, whose magnitude is determined by the impedance of the DC circuit. This current results in a DC component and a second harmonic (positive sequence) current on the AC side. The DC component is restricted to flow only in the valve side (converter) transformer windings and results in the magnetic saturation of the transformer core. This causes large harmonics in the magnetizing current of even orders. One could say that the effect of magnetic saturation amplifies the second harmonic current injected by the modulation process of the converter. The impedance of the AC network, viewed from the converter terminals, determines the harmonic voltage at the converter bus. Even without quantitative analysis, it can be said that the parallel resonance of the AC network at second harmonic (which magnifies the AC impedance), coupled with the series resonance on the DC side (at fundamental frequency) can aggravate the problem of 'second harmonic instability'.

Another example is the presence of negative sequence, fundamental frequency voltages at the converter bus, caused by unsymmetrical faults. It can be shown that this results in second harmonic voltage on the DC side. This, in turn, results in the flow of second harmonic current in the DC link and third harmonic (positive sequence) currents in the AC network. This is also accompanied by negative sequence fundamental frequency current. If the AC network has parallel resonance at third harmonic and DC circuit has series resonance at second harmonic, this could be problematic.

11.4 SYSTEM SIMULATION

11.4.1 General

The planning, design and operation of power system requires the detailed study of various options and prediction of the system performance. The latter involves simulation of the system both in steady-state and transient conditions. For AC systems, the AC network analyzer has been replaced by digital programs to carry out load flow and stability studies. Although the use of Transient Network Analyzer (TNA), for the study of harmonic voltages and electromagnetic transients is still prevalent, digital programs such as EMTP are increasingly used.

The integration of HVDC systems into existing large AC systems requires understanding of the various AC-DC system interactions. The appreciation of such interactions has increased over the recent years. This is due to the discovery of new phenomena, the need to optimize the controller performance and reduction in the Short Circuit Ratio (SCR). In the early phase of the introduction of HVDC systems, there was a tendency to treat the HVDC systems in isolation and the studies were usually performed using a HVDC simulator. This is a miniature, physical model of the system with only a limited representation of the AC system. Analog and hybrid computers have also been used for system simulation, but the degree of detail in these are also necessarily limited.

At present, both digital computer programs which can handle large scale AC system simulation and physical model DC simulators or DC-TNA are used for a variety of studies and over a wide range of frequencies (in response) that are of interest. Generally, digital programs do not consider detailed DC system simulations while the DC simulator does not have detailed

AC system simulation. However, the trend appears to be towards detailed dynamic simulators using digital computers with increasing sophistication in DC system (particularly controller) models. With increasing confidence in the results of digital simulators validated by test results, it is to be expected that physical models may be obsolete except in manufacturing organizations where hardware testing is required.

This section reviews the general philosophy of system simulation and discusses the relative merits of variety of tools used for this purpose.

11.4.2 System Simulation: Philosophy, Tools and Applications

Large and complex systems are usually described by nonlinear models and thus are not tractable for analytical approaches (except in special cases, when the system can be simplified or described by linear models). Simulation is the only tool for predicting accurately the system performance under various operating conditions or when subjected to certain disturbances. For large systems, a great number of various case studies have to be carried out and there is a strong incentive to keep the costs and time required to carry out these studies minimum. The methodology that is normally employed is to keep the models simple enough to fit the type of studies without sacrificing accuracy. Thus, network reduction and simplified equipment or system models are invariably used wherever possible. For example, the control performance of a DC link under fault conditions may be adequately determined with the AC system represented by Thevenin's equivalent at the converter bus. Similarly, distant parts of a power system can be either reduced or represented by simple models in a stability study.

There are a number of tools that can be employed for the simulation of a dynamic system. These are as follows:

1. Physical simulator
2. Analog computer
3. Hybrid computer
4. Parity simulator
5. Digital computer.

Except for the first tool, the use of other tools requires knowledge of mathematical models of the various components of the system. The use of physical simulators is justified when there are doubts about the validity of mathematical models. Otherwise, they tend to be more costly and less flexible. Also, the scaling down can introduce modelling errors unless special measures are used to compensate for these errors. For example, the voltage drop in the thyristor valve of a simulator is much larger and requires electronic compensation to maintain the accuracy.

In HVDC simulation, physical simulators have been widely used. For control and protection studies, they are still considered to be invaluable tools. However, the advent of faster and cheaper digital computers has made the digital simulation an attractive tool, which could replace the physical simulators in future. Analog and hybrid computers have been used in the past on a limited scale but they are not popular. Parity simulation was proposed for simulating power electronic systems and has been applied to develop a compact DC simulator.

The requirements of good simulation tool are as follows:

1. Ease of maintenance
2. Accuracy of solution
3. Flexibility of use
4. Reduced cost
5. Ease of setting up
6. Real time simulation
7. Easy monitoring and control over the simulation process.

The last two factors are responsible for the continued use of physical and analog simulation over digital simulation. The 'physical feel' that is usually mentioned as an advantage of a real time simulator can be offset if digital simulation can be effected in real time and can be run in an interactive mode with control over stopping and restarting of the simulation (with or without modification of the parameters).

In recent times, the Real-Time Digital Simulator (RTDS) has been developed and is based on a parallel processing hardware architecture and utilizes many high speed Digital Signal Processors (DSP) [66].

11.4.3 System Studies

The design of a HVDC system requires a number of system studies, some of which are aimed at determining the following:

1. DC power transfer under various normal and contingency conditions
2. Reactive power requirements
3. Fundamental frequency over voltages
4. Post-fault DC power recovery
5. Stability improvements through power modulation or emergency controller
6. Subsynchronous torsional interactions
7. Insulation coordination
8. DC terminal arresters
9. AC and DC filter design
10. Controller requirements under normal and faulted system conditions

Some of these studies can be performed using:

1. AC/DC power flow program
2. AC/DC transient stability program.

Both these programs are conventional power flow and transient stability programs used for AC systems and have been modified to incorporate adequate representation of DC links. The major assumption in the transient stability program is that the AC network (and DC line) is assumed to be in steady state. This permits the consideration of fundamental frequency phenomenon, where the voltage and current magnitudes vary slowly at frequencies determined by electromechanical oscillations of the generator rotors (0.5-2 Hz). Another major assumption is that the AC system is balanced which permits the representation of AC networks on single phase basis. Where network transients are to be considered, for example in the study of sub-synchronous resonance (SSR), this can be done by developing a special purpose program. Analysis using linear models is usually performed for studying dynamic or small signal stability, damping of subsynchronous oscillations, controller design etc. This can also be done using special purpose programs.

The main point to be noted here is, for stability studies, the converter is usually modelled as a controllable voltage source in series with commutation resistance. The valve voltage drop can also be included, if necessary. This model ignores the switching or commutation phenomena in the converter, and it is not possible to simulate effects like commutation failure which can be caused by a sudden dip in the inverter side AC voltage. The detailed dynamic simulation of HVDC systems must consider the switching action of the individual valves. However, this introduces considerable complexity in the digital simulation, thus slowing it down considerably. Because of the complexity involved, simulation used in stability studies usually avoids detailed simulation of converters. This could introduce errors in predicting the DC system response to faults in the AC systems, particularly when the faults are close to the converter bus. An approach of combining detailed dynamic simulation has been suggested. However, the limited experimentation that has been carried out shows that transient stability models give reasonable results if they are modified suitably. In what follows, the detailed dynamic simulation of HVDC systems using various simulation tools is reviewed. In particular, physical model (HVDC simulator), digital simulator and parity simulator are considered.

11.4.4 Physical Model (HVDC Simulator)

HVDC simulator is similar to TNA (Transient Network Analyzer) which is used to determine overvoltages in AC systems due to switching surges and load rejection. Actually, the distinction between AC TNA and the HVDC simulator is increasingly disappearing as the VA ratings and components of the two become similar. The differences will further decrease with more advanced AC network representations in HVDC simulators and with simple converter representations being introduced in TNA's. A HVDC simulator has adequate numbers of scaled models of 6 and 12 pulse converters for two (or more) terminals of a bipolar HVDC scheme. The valve is modelled by a single thyristor. The DC simulator ratings are in the range of 20 to 100 V DC, 0.2 to 1A DC and it operates at power frequency, 50 or 60 Hz. The DC line/cable is represented by adequate number of pi (π) sections. The simulators include representation of sophisticated converter controls. The controls are usually capable of realistic performance during transients such as AC faults and commutation failures.

Some DC simulators include electronic models of synchronous machines with features representing detailed models of torsional oscillations. Scaled models of static compensators (Thyristor Controlled Reactor, (TCR) and Thyristor Switched Capacitor (TSC) models) are also provided. Surge arrestor and DC breaker models are also included. The models of transformer in a good DC simulator must consider accurate representation of saturation, hysteresis, remanent flux, and leakage reactances. There must also be provision for compensating losses by electronic circuits (which introduce negative resistances).

The schematic diagram of a typical model of a converter transformer for one phase is shown in Fig. 11.11. One section of an overhead HVDC bipolar line model is shown in Fig. 11.12.

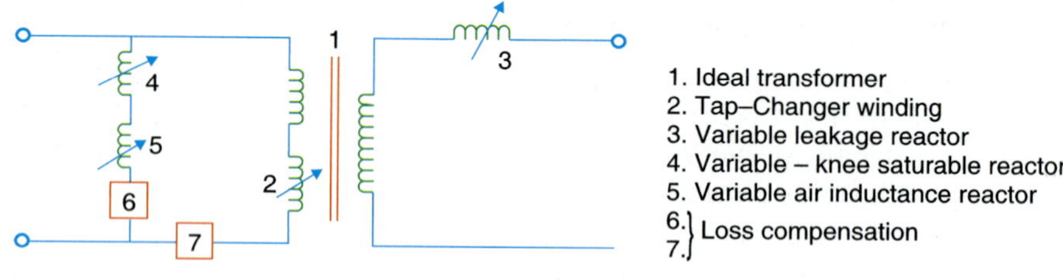

Fig. 11.11: Model of a converter transformer

Stray capacitances and inductances are not modelled in a HVDC simulator and it is primarily used to assess control system behaviour and temporary overvoltages of frequencies below 1000 Hz.

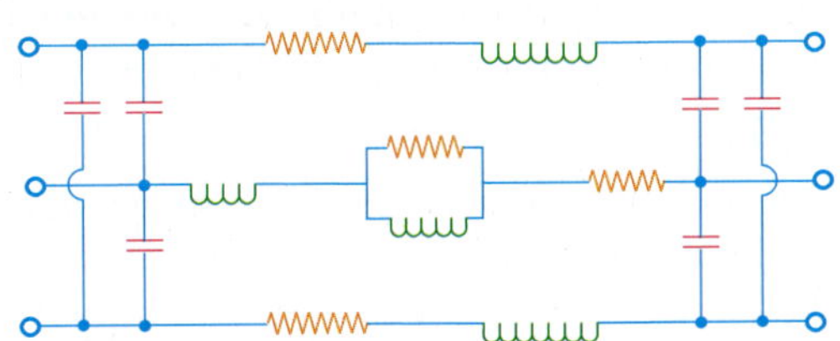

Fig. 11.12: Bipolar line section model

Some of the problems that can be studied using a DC simulator are as follows:

1. The development of concepts and equipment for control and protection of HVDC systems. This includes the following:
   (*i*) Control of power, current and extinction angle in two terminal and multiterminal DC systems.
   (*ii*) Evaluation of the control performance under AC and DC faults.
   (*iii*) Evaluation of overcurrent and overvoltage stresses in various components of the system, particularly valves, DC reactor, DC lines or cables.
2. Analysis of various AC/DC system interactions which include
   (*i*) HVDC operation with weak AC systems
   (*ii*) Subsynchronous frequency torsional oscillations
   (*iii*) Damping introduced through power modulation
   (*iv*) Analysis of AC and DC harmonics. (Harmonic interactions).

In addition, a HVDC simulator is also useful for (3) confirming computer models used in stability programs and (4) operator training and as an educational tool.

To summarize, the applications of a DC simulator are:

1. Insulation coordination
2. Testing of controllers and their optimization
3. Evaluation of surge arrestor ratings
4. Harmonic analysis
5. Development of new concepts and devices such as control and protection in MTDC systems, DC breakers, forced commutation, etc.

A modern DC simulator is equipped with data acquisition and processing system based on a micro computer. It also has an events sequencer to coordinate breakers, etc. for transient studies. Automated setting up of the simulation can speed up the studies that have to be performed.

The simulator is a flexible tool in that its component parameters can be easily adjusted, so that it is used to survey a host of transient conditions in order to identify the most severe situation. However, the physical hardware nature of the simulator forces the user to stay within the bounds of reality. The limited availability of adequate models of some of the system components restricts the scope of the studies that can be completed entirely by means of a simulator. Further investigations are necessary, particularly where accurate representation of damping effects of losses in filter coils and transformers are required. Higher frequency transients, lightning overvoltages, current and energy stresses in a surge arrestor are usually evaluated with a digital computer program such as EMTP.

11.4.5 Parity Simulator [35]

The principle of parity simulation has been used in electronic models of synchronous machines that are used with some DC simulators [37]. However, this principle was first formulated by J.G. Kassakian at M.I.T. in USA. According to him, 'a parity simulator is essentially a synthetic breadboard which electronically simulates the physical terminal characteristics of each network element rather than its mathematical input/output relationship'. For example, consider an inductor whose terminal behaviour is described by the relationship.

$$i = (1/L)\int v dt \tag{11.42}$$

An analog computer representation of this relationship is shown in Fig. 11.13(a) where the terminals 1-1′ and 2-2′ are input and output pairs, respectively. A parity simulator representation of the inductor is shown in Fig. 11.13 (b). A voltage to current converter is employed together with a local feedback loop to establish input current. The resulting model has only one terminal pair which exhibits one to one correspondence with the terminals of the inductor being represented. Such terminal equivalents can be established for other circuit elements such as transformers, thyristors, etc. Each equivalent is electrically floating with respect to others. A parity simulator, thus, consists of an interconnection of such terminal equivalents to represent the system being studied. This interconnection has the same topology as the network under study, that is, it exhibits topological parity with the system being simulated.

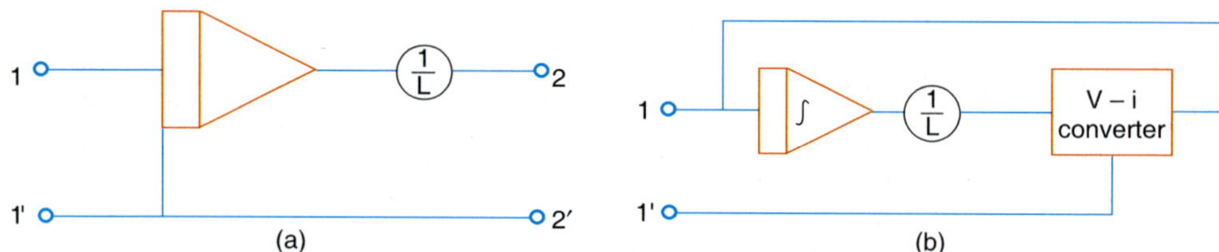

Fig. 11.13: (a) Analog computer representation of an inductor, (b) Parity simulation representation

The major advantages of a parity simulator are as follows:

1. It avoids the drawback of an analog computer simulation in which the identity of individual components is lost. The analog computer models the mathematical equations for the entire system. Change in the topology of the system requires reformulation of the system equations which can be quite tedious in the simulation of power electronic circuits.
2. The advantages of an analog computer in terms of change in time scaling are retained while eliminating the drawbacks mentioned above.
3. The principle of parity simulation permits hybrid structures, that is the combination of passive elements along with the parity simulated devices. For example, resistances and capacitors can be represented by scaled physical models in a parity simulator. This flexibility can simplify the arrangement of the circuits and reduce costs.

4. Parity simulator is more flexible than a physical simulator in terms of the range of component parameters, and yet it is very compact.

In 1981, the FGH, a research and development institute in the field of electric power transmission in Mannheim, Federal Republic of Germany commissioned their DC simulator with the bridge ratings of 25 mW, 2.5V and 10 mA. This simulator has been used for studies in HVDC and static var systems [36]. The frequency bandwidth of this simulator is around 5 kHz. The simulation of a thyristor controlled non-linear reactor is reported by Padiyar and Varma [38].

The major problems with parity simulator could be the same as that with an analog computer – the problem of offset voltages and currents of operational amplifiers used. In addition, since each component of a parity simulator is electrically isolated, there could be problems of interconnection and EMC (electromagnetic compatibility).

11.4.6 Digital Dynamic Simulation

Hingorani and his co-workers were the first to develop the dynamic simulation of HVDC systems [42, 53]. Since then, the work has been carried out in the direction of modelling and simulation of the converter circuit efficiently and to incorporate sophisticated controls.

In the initial stage of development, digital simulation was used mainly for specific studies such as fault analysis (to determine stresses on thyristor valves), harmonic analysis and design of controllers. The present trend is to develop digital simulation as a general tool for a variety of studies. Presently, digital simulation is considered as a supplement to the studies performed on a physical simulator.

The main advantages of a digital simulation are as follows:

1. ***Easy transportability and maintenance*:** Every utility can have a computer program, while only a few establishments can have physical simulator. However, it must be noted that the use of digital programs also requires some skill and experience in analyzing and interpreting the results and preparing the data.
2. ***Reduced cost of simulation*:** With the availability of inexpensive micro or mini computers, it is possible to have dedicated computers for DC simulation and the costs are drastically reduced.
3. ***Flexibility in terms of representing any component of the system***: It is possible to develop a digital simulator with most general model of a component, that is feasible. The parameters of this model can be varied over a wide range.

Some of the disadvantages are as follows:

1. ***Increased simulation time*:** Presently available software of simulation requires large amounts of computer time. The ratio of computer time to real time varies from 100 to 1000 depending on the extent of simulation. Improvements in hardware, computer architecture and numerical methods can reduce this ratio in future.

   Attempts to use sophisticated models of the system components can also increase the computer time and there is a need for determining the required level of sophistication in modelling.

   Even in RTDS, the minimum step size is above 50μ*s*. Simulation of VSC based HVDC links may require step sizes below 20 micro second. The computation time required to simulate 640 ms using EMTP (ATP20 Version 6) was stated to be 765 seconds [58].

2. ***Lack of adequate mathematical models***: This is particularly true of many nonlinear components such as transformers, control equipment, surge arrestors, etc. The validity of mathematical models can be verified by detailed comparison with results obtained from field tests or physical simulator.
3. ***Numerical problems*** can crop up in the simulation which may not be predictable in advance. Thus, there could be doubts on the final results. Use of numerically stable methods of integration, use of double precision arithmetic should go a long way in minimizing such problems.
4. ***Lack of interactive capability***: To get a feel of what is happening and have control over the process of simulation, it is necessary to develop interactive capabilities where the user can see the results during the progress of simulation, interrupt and restart. Some attempts are being made in this direction [57]. There is also need for simplifying the preparation of data and checking the errors.

The recent trends in digital simulation are to (*i*) develop a modular approach in modelling which allows flexibility in modelling individual components, (*ii*) use EMTP type of program, which has been already tested for simulating AC transients and modify it to include the representation of DC system, (*iii*) use of sophisticated control models, and (*iv*) development of interactive capability.

Some attempts have been made [44, 48, 49] to validate the results of digital simulation from field test results and this experience is expected to encourage the extensive use of digital simulation. It is claimed that digital simulation has resulted in the modification of inverter pole controls at Nelson River DC link to overcome the problem of core saturation instability. [44]

11.4.7 Modeling of DC Systems for Digital Dynamic Simulation

As the bandwidth of the transients considered is around 1000 Hz, it is quite in order to make the following assumptions:

1. The valve is represented by an equivalent thyristor device whose impedance in the off-state is very high (infinity) and whose impedance in the on-state is very low (zero). However, it is possible to include a constant on-state voltage drop across the valve.
2. The DC line is represented by a lumped parameter equivalent circuit. However, this assumption can be relaxed if required.

VALVE MODEL

A valve in a converter bridge can be represented either as (*i*) time-varying impedance, or (*ii*) ideal switch which is controllable or (*iii*) ideal switch in series with a constant voltage source (representing the on-state drop) when the switch is closed.

In EMTP type of transient simulation program, the first model is generally used. While this simplifies the formulation of the equations, numerical problems are experienced when the valve (in series with inductors) is opened. To overcome the numerical problems, it becomes necessary to model the snubber circuit in parallel with the valve. The combined representation of the valve in parallel with the snubber circuit is shown in Fig. 11.14. The resistance R_v is given by

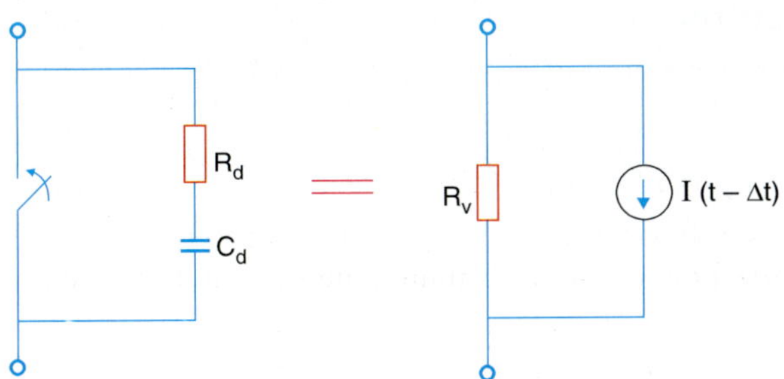

Fig. 11.14: Valve model

$R_v = 1\ \Omega$ during on-state

$= R_d + (\Delta t/2C_d)$ during off-state

It is to be noted that the representation of snubber circuit is usually not required in studies for predicting controller performance. It would be advantageous in such cases to model the valve as an ideal switch and employ graph theoretic approach to handle varying circuit topology.

It is assumed that the valve is turned on as soon as the controller generates a gate pulse, provided that the voltage across the valve is positive and exceeds a minimum limit. The valve turns off when the current falls below the holding current (around zero).

The valve is turned on again if the voltage across the valve goes positive within the turn-off time of the device (around 200 to 500 μs) after the cessation of the current.

GATE PULSE GENERATION

The firing controller generates gate pulses for individual valves in response to a control signal generated by (*i*) current controller or (*ii*) extinction angle controller or any other converter controller. These controllers can be analog or digital. However, as the generation of gate pulses is at discrete intervals, it is advantageous to develop a discrete time model for the overall converter control.

There are two principles of generation of firing pulses as given below:

1. Individual Phase Control (IPC)
2. Equidistant Pulse Control (EPC)
   (*a*) Pulse Frequency Control (PFC)
   (*b*) Pulse Phase Control (PPC)

These are distinguished by the reference point with respect to which the firing of a valve is delayed. In the first case, (of IPC) the reference point is the zero crossing (while going positive) of the valve (commutation) voltage. In the second case, the reference point for valve V_i is the firing instant of the previous valve V_{i-1}.

Individual phase control

In this case, the time delay t_{di} of gate pulse of valve V_i with respect to its reference point (zero crossing of valve voltage) is given by

$$t_{di} = f(e_c)$$

where e_c is the control voltage generated by the converter controller (CC or CEA) and which is sampled at regular intervals. The functions f may be linear or nonlinear.

Equidistant pulse control

In this case, the instant t_i of firing of valve V_i is given by

$$t_i = t_{i-1} + t_{di}$$

$$t_{di} = T/p + \phi_i \tag{11.43}$$

where t_{i-1} is the instant of firing valve v_{i-1}. T is the period of the AC voltage, p is the pulse number. For Pulse Frequency Control (PFC) scheme

$$\phi_i = f(e_{ci}) \tag{11.44}$$

In the case of Pulse Phase Control (PPC) scheme,

$$\phi_i = f(\Delta e_{ci}) \tag{11.45}$$

where $\Delta e_{ci} = e_{ci} - e_{c(i-1)}$

It is to be noted that the PFC has integral characteristic and in steady state, e_c is zero if the frequency is constant.

For current control, the firing control according to Eq. (11.43) is satisfactory. However, for constant extinction angle at the inverter, predictive type of control combined with EPC is desirable. Here, the reference is the zero crossing of the commutation voltage. The firing angle is determined from

$$\alpha_i = \pi - \gamma_o - u_i - \Delta\gamma_o \tag{11.46}$$

where γ_o is the reference value of the extinction angle, $\Delta\gamma_o$ is the change in the reference due to the modification of the inverter characteristics when the DC current varies from $(I_{ref} - I_m)$ to I_{ref} (I_m is the margin current). The overlap angle μ is to be predicted from the knowledge of the magnitudes of AC voltage and DC current.

In implementing this type of predictive control, there is need to use symmetrizing control to ensure equidistant operation. This involves the measurement of the extinction angle and modifying γ_o with the error signal. If the error $(\gamma_{ref} - \gamma_{mean})$ is positive, the correction is immediate, by altering the instant of next firing. However if the error is negative, the correction is done once in a cycle.

In feedback control of extinction angle using PFC, there is need for equalizing γ of all the valves by using the signals derived from the measurement of all the extinction angles in a cycle.

Reference [62] describes a DQO type phase locked firing control system for HVDC converters. Figure 11.15 shows the system. Here, two phase (v_α and v_β) voltages are generated from the commutating bus voltages v_a, v_b and v_c according to the following equations:

$$v_\alpha = \sqrt{\frac{2}{3}}\left(v_a - \frac{1}{2}v_b - \frac{1}{2}v_c\right)$$

$$v_\beta = \frac{1}{\sqrt{2}}(v_c - v_b)$$

An error v_Q signal is generated according to the following equation.

$$\text{Error} = v_Q = v_\alpha \sin\theta + v_\beta \cos\theta$$

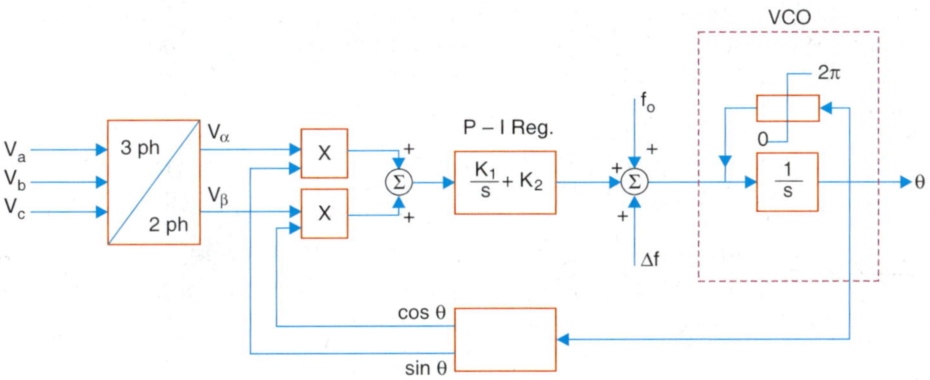

Fig. 11.15: D-Q type firing angle controller

The error is input to a PI controller which is followed by a Voltage Controlled Oscillator (VCO) to obtain a control signal (θ) for a sine-cosine oscillator. The output of the VCO, which limited between 0 and 2π radians, generates the timing sawtooth waveform which is used to generate the firing pulses for the converter valves according to the relation,

$$\theta = \left[\left(\frac{K_1}{s} + K_2\right)v_Q + (f_0 + \Delta f)\right] / s$$

The nominal frequency of VCO is f_0 and is modified by the variation Δf.

Generation of control voltage

The control voltage e_c which is the input to the firing controller is obtained as the output from current or extinction angle controllers. These controllers could be of analog or digital type. In either case, they can be described by discrete state equations. In general, the control voltage e_c is described by

$$z_{i+1} = Fz_i + Gu_i \tag{11.47}$$

$$e_{c(i+1)} = Hz_{(i+1)} + Ku_{(i+1)} \tag{11.48}$$

where u_i is the input (say current error) in the ith interval. z_i is the controller state vector corresponding to the ith interval. F, G, H and K are constant matrices. The magnitude of e_c may be limited.

The current reference is also limited by VDCOL which is described in Chapter 4.

CONVERTER MODEL

In modeling a converter, each three phase bridge converter (shown in Fig. 11.16) is represented individually. This enables the modelling of series and/or parallel connection of bridges constituting the converter unit.

In Fig. 11.16, the AC voltage sources e_a, e_b and e_c are determined as dependent sources derived from the AC system quantities. Similarly, the DC current i_d is obtained from the variables defined for the DC network. Generally, i_d is chosen as state variable. It will be shown later that the converter bridge can be represented by an equivalent circuit shown in Fig. 11.17 under the following assumptions:

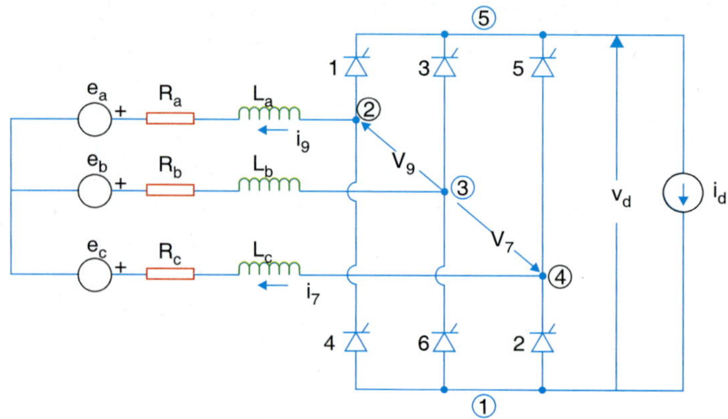

Fig. 11.16: A bridge converter

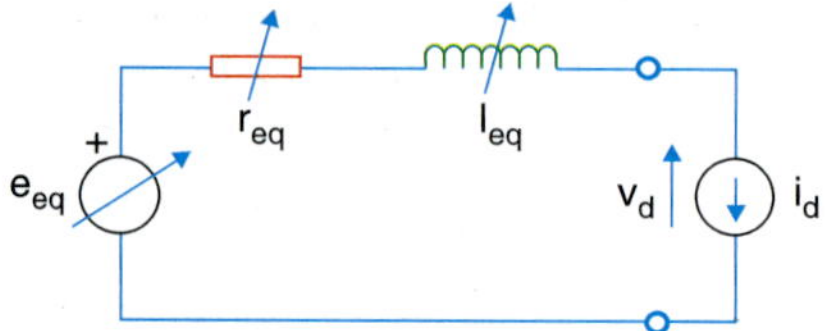

Fig. 11.17: Equivalent circuit of a bridge

1. All the valves in a bridge have identical characteristics
2. A valve offers infinite impedance in the reverse direction
3. The grading and damping circuits across the valves are ignored
4. The current i_d is assumed to be continuous and non-zero
5. L/R of each phase of the convener transformer is the same, i.e.

$$L_a/R_a = L_b/R_b = L_c/R_c = T$$

The quantities, e_{eq}, r_{eq} and l_{eq} in Fig. 11.17 are functions of the conduction pattern and change as a valve is fired (starts conducting) or turns off (ceases conduction). The firing instant of a valve is determined by the control system. However, the extinction of a valve is determined by solving for the valve currents and determining the instant when it passes through zero. This implies auxiliary equations for the independent valve currents in the conducting valves.

The existence of the equivalent circuit for the bridge is critically dependent on the last assumption–namely the time constants of all the phases are same. This assumption is not essential and can be relaxed. However, the simplicity of the equivalent circuit is lost. Also the existence of the equivalent circuit enables it to be combined readily with the DC network in deriving the DC system equations.

The parameters of the equivalent circuit shown in Fig. 11.17 can be derived from graph-theoretic considerations. This is because there can be up to (2^6) 64 patterns of conduction (depending on which valves are conducting) and it would be desirable to have general expressions applicable to any pattern of conduction. The assumption about continuous and non-zero i_d, reduces the number of conduction patterns to 49 as there is a restriction that at least one valve from each group (lower or upper) has to conduct. Even with this restriction, the total number of patterns is much higher than those corresponding to the normal conditions (which are 12 in number).

Graph theoretic analysis

The converter bridge shown in Fig. 11.16 can be represented by a directed graph of nine elements as shown in Fig. 11.18. The first six elements represent the six valves. The elements 7 and 9 correspond to the equivalent circuit representation of the secondary windings of the converter transformer. The element 8 represents the (dependent) current source i_d. The arrows on the graph indicate the directions of the current flow. There are 4 tree branches and 5 links in the graph. Although the choice of the tree need not be unique, for simplifying the analysis, the tree is chosen to include elements 7 and 9 and two conducting valves. Since the DC current is continuous and non-zero, at least one valve will conduct in each of the two groups of the bridge (from the upper group 1, 3 or 5, from the lower group 2, 4 or 6). If more than one valve is conducting in a group, any one can be chosen as a tree branch except for the case when two valves in an arm (say 1 and 4) are conducting. In this case, the conducting valves in the arm are chosen as tree branches. The choice of tree simplifies the formulation of the converter equations. The remaining four valves are included in the cotree.

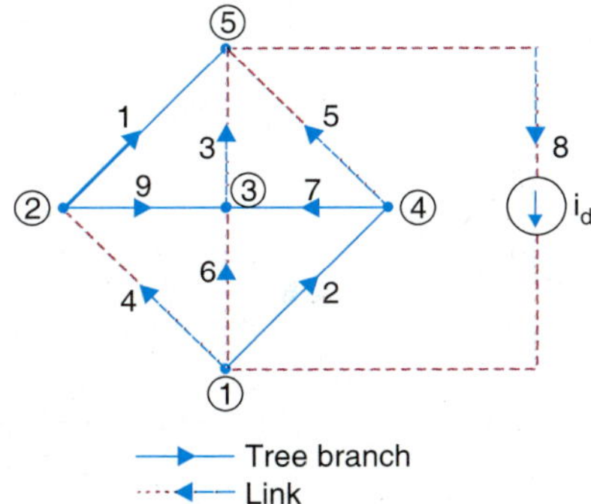

Fig. 11.18: Directed graph of converter bridge with valves 1 and 2 in tree.

Constituent equations

For elements 1 to 6, the following equations are applicable

$$V_g = V_D,\ g \in G \ : \textit{set of conducting valves} \tag{11.49}$$

where V_D is the forward voltage drop across a conducting valve

From assumption (2) given earlier,

$$i_l = 0,\ l \in L : \textit{set of non-conducting valves} \tag{11.50}$$

The constituent equations for the elements included in the tree can be written as

$$V_T = [Z_T]i_T + e \tag{11.51}$$

where

$$Z_T = [R_T] + p[L_T] \tag{11.52}$$

$$[R_T] = \begin{bmatrix} R_b + R_c & R_b & 0 & 0 \\ R_b & R_a + R_b & 0 & 0 \\ 0 & 0 & 0 & 0 \\ 0 & 0 & 0 & 0 \end{bmatrix} \tag{11.53}$$

$$[L_T] = \begin{bmatrix} L_b + L_c & L_b & 0 & 0 \\ L_b & L_a + L_b & 0 & 0 \\ 0 & 0 & 0 & 0 \\ 0 & 0 & 0 & 0 \end{bmatrix} \tag{11.54}$$

$$[e] = \begin{bmatrix} e_c - e_b \\ e_a - e_b \\ V_D \\ V_D \end{bmatrix} \tag{11.55}$$

$$p = d/dt,\ V_T^t = [V_7 \quad V_9 \quad V_x \quad V_y]$$

$$i_T^t = [i_7 \quad i_9 \quad i_x \quad i_y]$$

the superscript t indicates transpose, x and y are the conducting valves in the tree branches.

Circuit equations

For any network, the branch and link variables are related by

$$i_T = [B_L] i_L \tag{11.56}$$

$$V_L = -[B_L]^t V_T \tag{11.57}$$

where subscripts T and L refer to the tree and cotree (link) elements respectively. The matrix - $[B_L]$ is the component of the fundamental cutset matrix, corresponding to the link elements. For the graph shown in Fig. 11.18, B_L is of the order 4×5. Expanding Eq. (11.56)

$$i_T = B_d i_d + B_\alpha i_\alpha + B_\beta i_\beta + B_\gamma i_\gamma + B_\delta i_\delta \tag{11.58}$$

where B_d, B_α, B_β, B_γ and B_δ are the column vectors of $[B_L]$ corresponding to the element 8 and the valves in the cotree. i_α, i_β, i_γ and i_δ are the currents in the valves which form the cotree. If a valve is not conducting, the current in that is zero and the term involving that valve current in Eq. (11.58) can be dropped.

Although the total number of conduction patterns can be 49 as stated before, for the valve model considered here, only 39 patterns of possible modes of converter operation, are of interest. These are defined in Table 11.1. These exclude 7 possible modes with five and six valve conduction and 3 modes with 4 valve conduction. This is because, in these modes, there are 4 conducting valves that form two parallel paths and the currents in these paths are indeterminate as only the forward voltage drops in the valves are considered.

Table 11.1: Possible Modes of Converter Operation Assuming Continuous Direct Current

Two valve conduction		Three valve conduction		Four valve conduction	
Normal Modes	Abnormal modes	Normal modes	Abnormal modes	Normal Modes	Abnormal modes
1,2	1,4	1,2,3	1,2,4	1,2,3,4	1,2,3,5
2,3	3,6	2,3,4	1,2,5	2,3,4,5	1,2,4,6
3,4	2,5	3,4,5	1,3,4	3,4,5,6	1,2,5,6
4,5		4,5,6	1,3,6	4,5,6,1	1,3,4,5
5,6		5,6,1	1,4,5	5,6,1,2	2,3,4,6
6,1		6,1,2	1,4,6	6,1,2,3	2,4,5,6
			2,3,5		
			2,3,6		
			2,4,5		
			2,4,6		
			3,4,6		
			3,5,6		

Actually it can be said that the 10 excluded modes can be reduced to those belonging to the group of 39 modes by deleting the parallel path. For example, the mode in which valves 1, 4, 3 and 6 are conducting is equivalent to the mode in which only valves 1 and 4 conduct or the mode in which only valves 3 and 6 conduct. The mode in which 1, 2, 3, 4 and 5 conduct is equivalent to the mode in which only 1, 3 and 4 conduct or the mode in which 2, 3 and 5 conduct.

From the above discussion, the total number of conducting valves in a bridge can vary from 2 to 4. Hence the number of terms which are non-zero in Eq. (11.58) can vary from one to three. For the sake of convenience, it can be assumed that $i_\gamma = i_\delta = 0$ as there are at least two non-conducting valves in a cotree.

From Eqs. (11.49) to (11.58) it is possible to derive the converter equations for a specified conducting pattern or mode of operation. The possible number of modes of operation are divided into three categories as given below:

1. Two valve conduction
2. Three valve conduction
3. Four valve conduction

The derivation of the converter equations is given in Appendix B.

11.4.8 Transient Simulation of DC and AC Networks

The AC system consists of generators and their controllers, transmission lines, shunt reactors and capacitors, series capacitors, loads, static var systems, etc. Except for the generator and controllers, all the stationary elements are represented by their equivalent circuits forming the AC network. Three phase representation of the network (instead of sequence networks) is most desirable. For transient analysis of electrical networks, two approaches are available.

1. The use of trapezoidal rule of integration to transform the energy storage elements to resistive elements with current sources across them which represent the past history. This approach allows the transient analysis to be performed by solving a resistive network. This is used in EMTP (Electro Magnetic Transients Program) originally developed by Prof. H.W. Dommel and is widely used. The main advantages are: (*i*) ease of obtaining solutions particularly for piecewise linear components and

(*ii*) inclusion of models for distributed lossless elements such as long transmission lines.

2. Formulation of state equations for networks using standard techniques which first identify the state variables to be used and obtain equations from circuit theory. The equations are finally solved using anyone of the standard integration routines (explicit or implicit).

The DC network is augmented by the circuit models of the converter bridges and solved in a similar manner as the AC network. The interface between the DC and AC network is provided through dependent sources. The treatment of these sources in the solution of individual networks requires careful consideration to eliminate the errors and speed up solution.

EMTP representation of DC network

The electromagnetic transients program for the simulation of transients in AC networks is well established and is gaining ground in the simulation of transients in AC/DC systems. By EMTP, not a particular computer program (such as BPA-EMTP) is meant here but rather a generic type of program with the methodology of component modelling described by Dommel [50].

Lumped inductors and capacitors are modelled in EMTP as resistors in parallel with current sources dependent on the past history. This is based on applying the trapezoidal rule in the integration of the first order linear differential equation describing an inductor or a capacitor. For example, for an inductor L,

$$i(t) = (1/L)\int_{t-h}^{t} v(\tau)d\tau + i(t-h) \qquad (11.59)$$

By applying trapezoidal rule of integration

$$i(t) = (h/2L)[v(t) + v(t-h)] + i(t-h) \qquad (11.60)$$

The above equation represents an equivalent circuit shown in Fig. 11.19. Similarly, Fig. 11.20 represents a capacitor. It is to be noted that the equivalent circuits of both Figs. 11.19 and 11.20 are approximate as they are based on the approximate integration of the equations using trapezoidal rule. However, for the case of a lossless single phase line, it is possible to derive exact equivalent circuits of the similar type. Consider a single phase line connected between nodes m and k as shown in Fig. 11.21. The equivalent circuit of this line is shown in Fig. 11.22. Here Z_s is the surge impedance and $\tau = l/c$ is the travel time (l = length of the line, c = propagation velocity). For a multiphase line, similar equivalent circuits can be derived using modal analysis. For a bipolar DC line with mutual coupling, the analysis is similar and is given in Appendix C.

The modelling of synchronous generator with three phase representation is described in Appendix D.

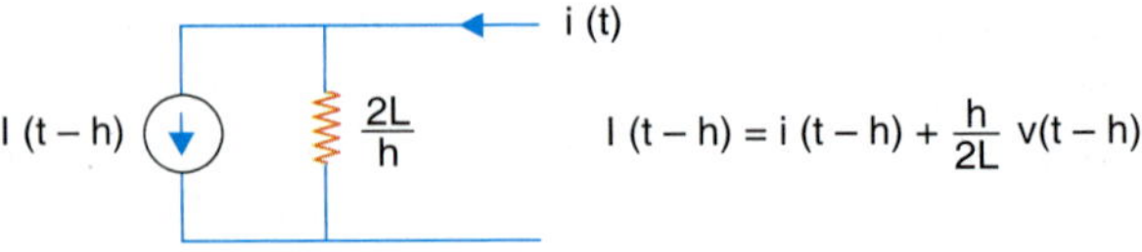

Fig. 11.19: Equivalent circuit for an inductor

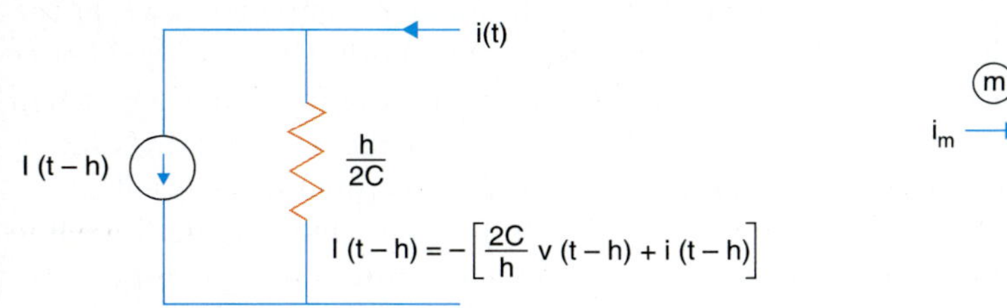

Fig. 11.20: Equivalent circuit for a capacitor

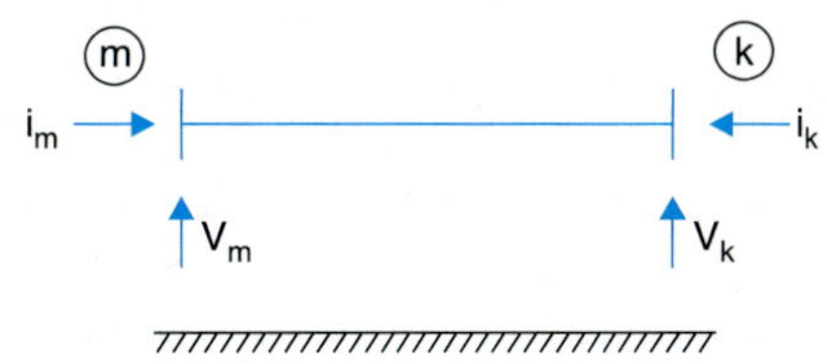

Fig. 11.21: A single phase transmission line

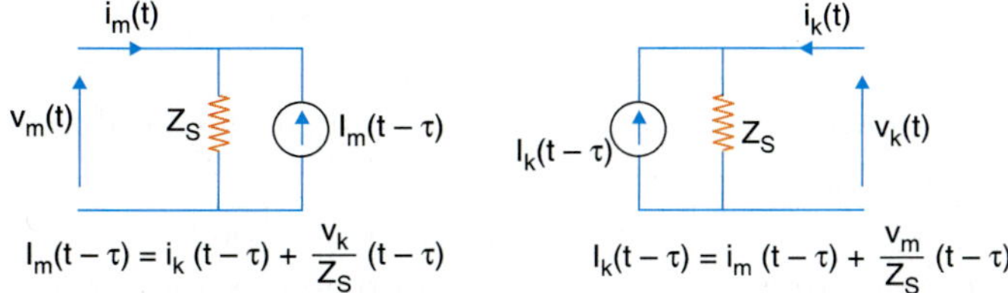

Fig. 11.22: Terminal equivalent circuits of a transmission line

11.5 TORSIONAL INTERACTIONS WITH HVDC SYSTEM

11.5.1 General

It has been known since quite some time that the rotor circuits of a synchronous generator contribute negative resistance at subsynchronous frequencies (frequencies below the fundamental frequency) to the generator impedance, when viewed from the stator terminals. This is due to the induction generator effect (at subsynchronous frequencies) and the phenomenon is termed self-excitation when sustained oscillations occur due to the system interactions, when a transmission line connected to the generator is compensated using series capacitors [20, 70]. The series capacitors introduce a series resonant frequency in the electrical circuit, comprising a generator stator, transformer and the transmission line. This frequency is given by

$$f_{er} = f_o\sqrt{\frac{X_C}{X'' + X_T + X_L}} \tag{11.61}$$

where X'' is the sub-transient reactance of the generator, X_T and X_L are the reactances of the transformer and the line respectively. X_C is the reactance of the series capacitor which is less than X_L. Hence the electrical resonant frequency (f_{er}) is less than f_o.

The self-excitation phenomenon can exist even when the generator inertia is considered to be infinite, ruling out rotor oscillations. However, in a realistic situation, there can be oscillations at the natural frequencies of the mechanical system comprising of generator and turbine rotors and elastic shafts. The rotor oscillation at a frequency of f_m, causes armature voltage components at the frequency of

$$f_e = f_o \pm f_m \tag{11.62}$$

The positive sign is associated with the supersynchronous frequency component, while the negative sign is to be used for the subsynchronous frequency component. (It is assumed that $f_m < f_o$). The subsynchronous resonance (SSR) due to the torsional interactions can occur when the frequency f_e is close to f_{er}. This phenomenon was first observed at Mohave generating station in December 1970 [69]. The SSR caused by the series compensation resulted in severe damage of the shaft. Incidentally, shaft damage can also occur due to fatigue when the peak torque exceeds a limit. In this case, the total duration of such high transient torques is to be minimized. This is affected by the net mechanical damping which is influenced by the torsional interactions. At the subsynchronous frequency of f_m, the negative damping (torque coefficient) introduced by the electrical circuit is given by [70]

$$-T_{de} = (f_e\ R)/(2\ f_m Z^2) \tag{11.63}$$

where R is the net resistance, and Z is the impedance of the electrical circuit at the subsynchronous frequency $(f_e = f_o - f_m)$.

Actually, this is an approximate relation. An exact relation is derived in Appendix E.

The negative damping due to the torsional interactions is more severe compared to the induction generator effect. Even if the resistances of the rotor circuits are neglected (assumed to be zero), the problem exists. There are many counter-measures suggested to overcome the problem of SSR [71].

The torsional interaction with HVDC converter controllers was discovered in 1977 at Square Butte, USA. This phenomenon, as will be explained later is similar to that caused by series capacitors. However, the problem of torsional interactions with HVDC system is much less severe and can be tackled by adding an auxiliary Subsynchronous Damping Controller (SSDC) to damp Subsynchronous (frequency) Oscillations (SSO).

The definitions of SSR and other terms are given in Appendix E. The damping torque analysis is also described there.

11.5.2 Factors Affecting the Phenomenon

The field tests carried out at Square Butte HVDC Project [12] and their subsequent analysis have pointed out the factors that affect the torsional interactions. The interaction at Square Butte occurred through two different control paths, namely (*i*) the auxiliary damping control, and (*ii*) the current control. In case of the auxiliary damping control (designed for damping low frequency oscillations), the interaction was due to the high gain and phase lag at the lower end of the subsynchronous torsional frequency range (at 11.5 Hz, which is the first torsional mode at Square Butte). In the second case, the current controller action was found to negatively damp the torsional oscillations. It was also observed that the Equidistant Pulse Control (EPC) firing scheme resulted in greater torsional interaction than the Individual Phase Control (IPC). In general, the possibility of adverse interaction due to the current controller exists in the frequency range of 10 to 20 Hz.

The recent torsional tests at the Intermountain Power Project (IPP) in USA [13] have confirmed the earlier conclusions regarding the major factors that affect the subsynchronous frequency interactions with the current controller. The overall conclusions are given below:

1. The torsional interaction is significant only at turbine-generators in the close vicinity of HVDC converters relative to other units.
2. Converter control has a major effect. The problem is more acute with EPC, than IPC, although this can be compensated by adding synchronizing circuit with sufficient bandwidth. The use of an auxiliary control in the current control loop to damp subsynchronous oscillations is beneficial.
3. The negative damping is of main concern for the torsional modes that lie in the frequency range of 10 to 20 Hz.
4. The negative damping increases with the power level.
5. For a given power level, the damping introduced by the DC system is sensitive to the level of the DC voltage. This effect can be traced to the changes in the operating values of the delay angle (α) caused by the voltage changes. The negative damping is less for smaller values of the delay angle.
6. The strength of the AC system adjacent to the converter is a major factor. The adverse torsional interactions are most significant for the radial operation of the DC link.
7. During tests at Intermountain Power Project, it was found that the damping is reduced for monopolar operation compared to the bipolar operation. However, it is unlikely that this conclusion can be generalized.

11.5.3 Analysis of HVDC–Turbine Generator Torsional Interactions

The analysis has been performed using (*i*) damping torque concept [14, 15], (*ii*) HVDC simulators [16], (*iii*) frequency scan [17], and (iv) eigenvalue analysis and digital simulation [18, 19]. In all the methods, except simulation, the analysis is performed on a linearized system which facilitates the prediction of the torsional interactions under various conditions.

The system modelling for the analysis of torsional interactions must consider (*i*) detailed generator (including the mechanical system) models, (*ii*) transients in DC and AC lines, and (*iii*) adequate controller models. The average converter models in continuous time or discrete time are adequate. For the stability analysis of AC-DC systems, the AC network equations are transformed using D-Q variables, which facilitate the elimination of time-varying coupling between the AC and the DC system.

For the analysis of torsional interactions, the mechanical system consisting of generator and turbine rotors, has to be adequately modelled. Two types of models in use are, (*i*) multiresonant model, and (*ii*) modal model. The first one is the basic and detailed model from which the second (modal) model is derived to simplify the analysis. These are described in Appendix D.

Although different types of analysis give similar results, the eigenvalue analysis using the detailed multiresonant model of the mechanical system yields better insights into the problem than the damping torque analysis. However, the latter is simpler and can be used as a screening tool.

The phenomenon of TI with HVDC system can be explained with reference to a turbine generator connected radially to a HVDC converter (rectifier) station. The analysis is based on the concept of damping torque.

The block diagram outlining the transfer functions from the generator rotor speed to the electrical torque is shown in Fig. 11.23. The following transfer function can be defined from Fig. 11.23.

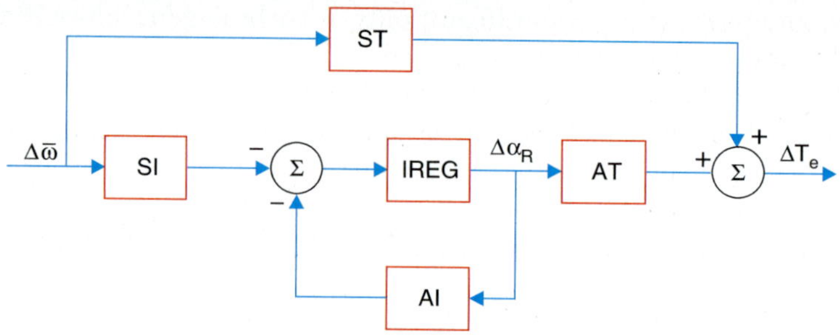

Fig. 11.23: Block diagram showing connection from rotor speed to electrical torque

$$ST(s) = \frac{\Delta T_e}{\Delta\bar{\omega}}\ |\Delta\alpha_R = 0,\ AT(s) = \frac{\Delta T_e}{\Delta\alpha_R}\ |\ \Delta\bar{\omega} = 0$$

$$SI(s) = \frac{\Delta I_d}{\Delta\bar{\omega}}\ |\Delta\alpha_R = 0,\ \ AI(s) = \frac{\Delta I_d}{\Delta\alpha_R}|\ \Delta\bar{\omega} = 0$$

IREG(s) is the transfer function of the current controller. The transfer function *ST(s)* is due to the AC transmission system and the HVDC system with no control.

The overall transfer function from the rotor speed to the electrical torque is given by

$$\frac{\Delta T_e(s)}{\Delta\bar{\omega}(s)} = \frac{\Delta T_e(s)}{\Delta S_m(s)} = ST(s) - \frac{SI(s)IREG(s)\cdot AT(s)}{1 + IREG(s)AI(s)} = STT(s) \tag{11.64}$$

The real part of this transfer function evaluated at the torsional mode frequency (ω_k) gives the damping torque coefficient (T_{Dk}) at that frequency. The decrement factor (σ_{ek}) is calculated from

$$\sigma_{ek} = \frac{T_{Dk}}{4H_k} \tag{11.65}$$

where subscript k refers to the kth torsional mode. T_{Dk} is positive if the phase of $STT(j\omega_k)$ lies in the range of the –90° to 90°. In deriving the transfer functions based on simplified models, the following assumptions are made:

1. The turbine-generator is supplying only the HVDC system (as its only load).
2. The commutating voltage (at the rectifier bus) magnitude and phase angle closely follow the generator internal voltage and angle.
3. In the generator,
   (*a*) field flux decay is neglected
   (*b*) damper windings are neglected

Based on the assumptions, we can express changes in the magnitude and phase of the voltage at the converter bus as

$$\Delta V_r = \Delta S_m V_{ro} \tag{11.66}$$

$$\Delta\delta_r = \Delta\delta = \frac{\omega_B}{s}\Delta S_m \tag{11.67}$$

The change in E_{dr} is obtained from linearizing Eq. (11.2) as

$$\Delta E_{dr} = k_r a_r (\Delta V_r \cos\alpha_0 - V_{r0}\sin\alpha_0 \Delta\alpha) \tag{11.68}$$

The subscript 'r' refers to the rectifier where $\theta = \alpha$. The change in the delay angle can be expressed as

$$\Delta\alpha = \Delta\alpha_R + k_p(s)\Delta\delta_r \tag{11.69}$$

where α_R is the firing angle command of current regulator and $k_p(s)$ is given by

$k_p(s) = 0$ or IPC scheme of firing

$= 1$ for EPC scheme of firing

$= 1 - \text{SYNC}(s)$ for EPC scheme with a synchronizing circuit whose transfer function is SYNC(s). Substituting Eqs. (11.66), (11.67) and (11.69) in (11.68), we get

$$\Delta E_{dr} = k_r a_r V_{r0}[\cos\alpha_0 - k_p(s)\frac{\omega_B}{s}\sin\alpha_0]\Delta S_m - k_r a_r V_{r0}\sin\alpha_0 \Delta\alpha_R \tag{11.70}$$

Assuming that $\Delta E_{di} = 0$, it is possible to write that

$$\Delta I_d = Y_d(s)\Delta E_{dr} \tag{11.71}$$

where $Y_d(s)$ is the admittance function of the DC network viewed from the rectifier internal bus.

The DC power is given by

$$\Delta P_d = I_{d0}\Delta E_{dr} + E_{dr0}\Delta I_d \tag{11.72}$$

The generator electrical torque (neglecting losses) is given by

$$\Delta T_e = \Delta P_d - P_{d0}\Delta S_m = [I_{d0} + E_{dr0}Y_d(s)]\Delta E_{dr} - P_{d0}\Delta S_m \tag{11.73}$$

Based on the above equations, the transfer functions defined earlier are given as

$$AT(s) = -k_r a_r V_{r0}\sin\alpha_0[I_{do} + E_{dr0}Y_d(s)] \tag{11.74}$$

$$AI(s) = -k_r a_r V_{r0}\sin\alpha_0 Y_d(s) \tag{11.75}$$

$$SI(s) = k_r a_r V_{r0}[\cos\alpha_0 - k_p(s)\frac{\omega_B}{s}\sin\alpha_0]Y_d(s) \tag{11.76}$$

Effect of the current control on damping torque

The influence of the current controller on the damping torque can be analyzed by considering only the second term on the RHS of Eq. (11.64). This can be expressed as

$$SIT(s) = -\frac{SI(s)AT(s)}{AI(s)}\frac{IREG(s)AI(s)}{1 + IREG(s)AI(s)} = -\frac{SI(s)AT(s)}{AI(s)}G_c(s) \tag{11.77}$$

where $G_c(s)$ is the transfer function of the closed loop current controller. For a suitably designed current controller, $G_c(s)$ approaches unity with zero phase shift at low frequencies. Hence the damping contribution of *SIT(s)* is determined by the transfer function *IT(s)* defined by

$$IT(s) = -\frac{SI(s)AT(s)}{AI(s)} \tag{11.78}$$

By substituting Eqs. 11.74 to 11.76 in 11.78, it can be seen that *AT(s)* and *AI(s)* terms nearly cancel to a constant, particularly at light loads ($I_{do} \simeq 0$) leaving the characteristics of *SI*(*s*) determine the damping action. For low frequencies below 20 Hz, the transfer function *SI*(*s*) is dominated by the integral characteristics ($k_p = 1.0$) resulting in a phase lag of 90°. It is also interesting to note that the integral term is proportional to sin α_0 which implies that the magnitude of the negative damping increases with α_0.

As mentioned earlier for a well designed current regulator, $G_c(s)$ has a phase shift between 0 and –90°. Thus, a current regulator actually contributes negative damping particularly at higher frequencies.

11.5.4 A Case Study

Description

The configuration of the system considered is shown in Fig. 11.24. To simplify the analysis, the inverter DC voltage is assumed to be constant. Figure 11.25 shows the electrical damping for

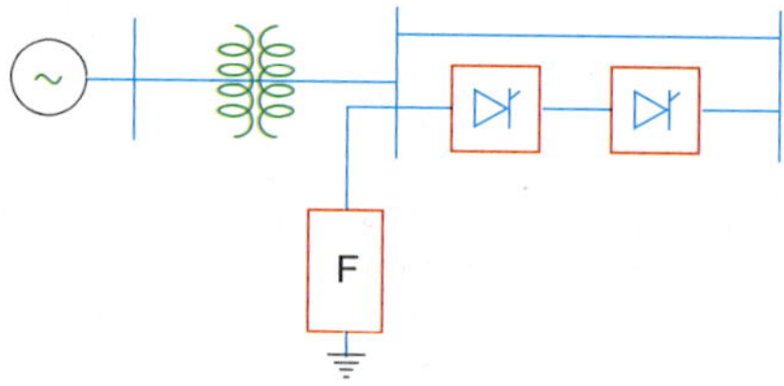

Fig. 11.24: System diagram

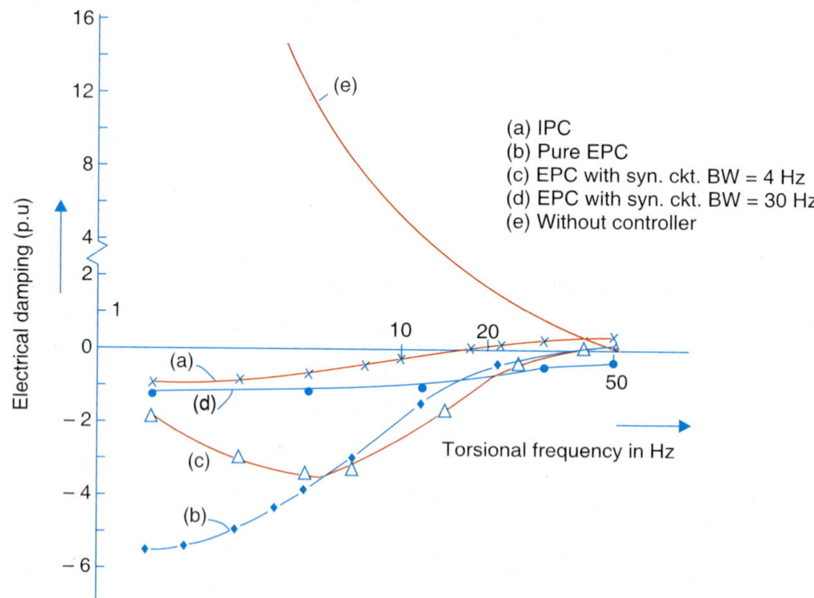

Fig. 11.25: Electrical damping for IPC/EPC with and without synchronising circuit

the case with and without controller [15]. Both *(a)* IPC and *(b)* EPC firing angle controls are considered. In the latter case, the effect of synchronizing circuit is examined. It is observed from Fig. 11.25 that the controller contributes negative damping which is higher for EPC. The synchronizing circuit with higher bandwidth improves damping.

An explanation of the phenomenon of HVDC torsional interaction

The torsional interactions with HVDC system have been previously explained in a simplistic manner, based on the fact that DC power is maintained constant and constant power type load introduces negative damping. However, this explanation does not explain why damping of higher frequency torsional modes increase with DC power. Also, the effects of controller and DC line parameters are not explained satisfactorily.

A better explanation is to consider the amplitude and phase modulation of the AC voltage at the rectifier bus, resulting from the torsional oscillations. This induces a voltage in the DC link, of the same frequency as the torsional mode, due to effective demodulation introduced by the converter. The effective impedance (also influenced by the converter controller) of the DC link determines the current oscillations in the DC link. These oscillations result in turn, the injection of sub and supersynchronous frequency currents in the ac system, that flow in the generator armature. In general, the subsynchronous frequency currents introduce negative damping torque while the supersynchronous frequency currents result in positive damping torque. Unlike in the case of series compensation, the magnitude of the two currents are equal. The net damping is determined from the phase of the two current components relative to the ac voltage components.

Analysis of a radial HVDC link

Since HVDC is the only load on the generator, the armature currents are determined by *(i)* the dc link impedance, and *(ii)* converter control. It will be assumed that the current controller is of pure integral type and the firing scheme is pure EPC.

The admittance function of the DC network (viewed from the rectifier internal bus) $Y_d(s)$ is assumed to be

$$Y_d(s) = \frac{1}{R + sL} \tag{11.79}$$

For pure EPC scheme, Eq. 11.70 can be expressed as

$$\Delta E_{dr} = k_r a_r V_{r0}[\cos\alpha_0 - \frac{\omega_B}{s}\sin\alpha_0]\Delta S_m - k_r a_r V_{r0}\sin\alpha_0 \Delta\alpha_R \tag{11.80}$$

The transfer function of the current regulator is given by

$$IREG(s) = -\frac{K_I}{s} \tag{11.81}$$

where K_I is the integral gain.

Equations (11.79) to (11.81) can be combined to represent the block diagram shown in Fig. 11.26. Here, A and B are defined as

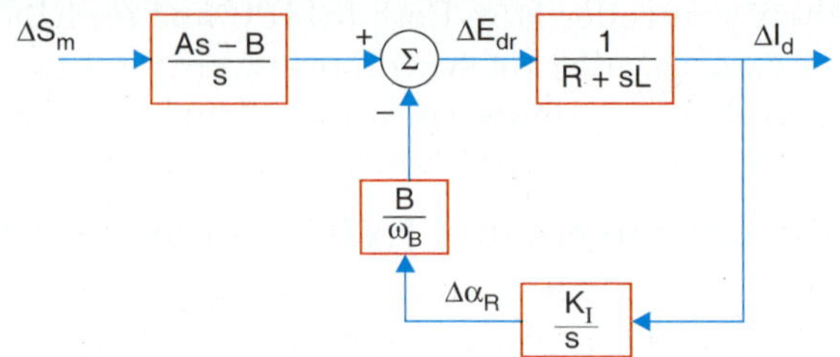

Fig. 11.26: Block diagram representing HVDC system

$$A = kaV_{r0}\cos\alpha_0 \tag{11.82}$$

$$B = \omega_B kaV_{r0}\sin\alpha_0 \tag{11.83}$$

The transfer function $SAI(s)$ between ΔS_m and ΔI_d is obtained from Fig. 11.26 as

$$SAI(s) = \frac{\Delta I_d(s)}{\Delta S_m(s)} = \frac{As - B}{L\left(s^2 + \dfrac{R}{L}s + \omega_n^2\right)} \tag{11.84}$$

where

$$\omega_n = \frac{1}{\sqrt{LC}}, \quad C = \frac{\omega_B}{K_I B} = \frac{1}{K_I kaV_{ro}\sin\alpha_0} \tag{11.85}$$

The transfer function $SA(s)$ relating ΔS_m and $\Delta\alpha_R$ is given by

$$SA(s) = \frac{\Delta\alpha_R(s)}{\Delta S_m(s)} = \frac{SAI(s)K_I}{s} \tag{11.86}$$

It is to be noted that the HVDC system has a predominant mode whose frequency is given by

$$f_{dc} = \frac{1}{2\pi}\sqrt{\omega_n^2 - \left(\frac{R}{2L}\right)^2} \tag{11.87}$$

Calculation of AC current

If it is assumed that ΔS_m or $(\Delta\bar{\omega})$ varies sinusoidally as given by the expression

$$\Delta S_m = \Delta\bar{\omega} = A_m \sin\omega_m t \tag{11.88}$$

the DC current in the link, ΔI_d can be expressed as (utilizing Eq. (11.84))

$$\Delta I_d = SAI(\omega_m)A_m \sin(\omega_m t + \phi_d) \tag{11.89}$$

where

$$SAI(j\omega_m) = |SAI(\omega_m)| \angle\phi_d(\omega_m) \tag{11.90}$$

Neglecting overlap angle and harmonics, the magnitude of the AC current is directly proportional to the DC current. The phase of the ac current lags the phase of the converter bus voltage by α.

Using $\alpha - \beta$ sequence networks, given the converter bus voltage as

$$v_{r\alpha}(t) = \overline{\omega} V_r \sin(\omega_0 t + \delta) \tag{11.91}$$

the AC current $i_{r\alpha}$ is given by

$$i_{r\alpha}(t) = kaI_d \sin(\omega_0 t + \delta - \alpha) \tag{11.92}$$

the change in $i_{r\alpha}$ is given by the expression

$$\Delta i_{r\alpha}(t) = ka\Delta I_d \sin(\omega_0 t + \delta - \alpha_0) + kaI_{d0}\cos(\omega_0 t + \delta - \alpha_0)[\Delta\delta - \Delta\alpha] \tag{11.93}$$

Since for the pure EPC scheme,

$$\Delta\delta - \Delta\alpha = -\Delta\alpha_R \tag{11.94}$$

we can rewrite Eq. (11.93) as

$$\Delta i_{r\alpha}(t) = ka\Delta I_d \sin(\omega_0 t + \delta_0 - \alpha_0) - kaI_{d0}\cos(\omega_0 t + \delta_0 - \alpha_0)\Delta\alpha_R \tag{11.95}$$

From Eqs. (11.86) and (11.88), $\Delta\alpha_R$ can be expressed as

$$\Delta\alpha_R = -\frac{K_I SAI(\omega_m)A_m}{\omega_m}\cos(\omega_m t + \phi_d) \tag{11.96}$$

Utilizing Eqs. (11.89) and (11.96), we can express $\Delta i_{r\alpha}$ in terms of sub- and supersynchronous frequency components as

$$\Delta i_{r\alpha} = \frac{kaSAI(\omega_m)A_m}{2}\left[1 + I_{d0}\frac{K_I}{\omega_m}\right]\cos[(\omega_0 - \omega_m)t + \delta_0 - \alpha_0 - \phi_d] \tag{11.97}$$

$$-\frac{kaSAI(\omega_m)A_m}{2}\left[1 - I_{d0}\frac{K_I}{\omega_m}\right]\cos[(\omega_0 + \omega_m)t + \delta_0 - \alpha_0 + \phi_d]$$

The damping torque coefficient is

$$T_{De} = T_{De}^{\text{sub}} + T_{De}^{\text{sup}} \tag{11.98}$$

where

$$T_{De}^{\text{sub}} = -\frac{kaE'SAI(\omega_m)}{2}[1 + \frac{I_{d0}K_I}{\omega_m}]\cos\phi_{\text{sub}} \tag{11.99}$$

$$T_{De}^{\text{sup}} = \frac{kaE'SAI(\omega_m)}{2}[1 - \frac{I_{d0}K_I}{\omega_m}]\cos\phi_{\text{sup}} \tag{11.100}$$

$$\phi_{\text{sub}} = \pi + \alpha_0 + \phi_d(\omega_m) \tag{11.101}$$

$$\phi_{\text{sup}} = \alpha_0 - \phi_d(\omega_m) \tag{11.102}$$

T_{De}^{sub}, T_{De}^{sup} and T_{De} are computed using the above expressions for the HVDC system, having the data,

$$L = 2.7H,\ R = 151.4\Omega,\ V_{r0} = E' = 1.0\text{pu}(500kV), \text{ka} = 1.18$$

$$\omega_B = 377,\ \alpha_0 = 18^\circ \text{ and } I_{d0} = 1.0\text{pu}(1\text{kA})$$

The damping torque coefficients are calculated for two values of the integral gain (K_I); (*i*) 85 rad/pu and (*ii*) 170 rad/pu. The variations of and T_{De}^{sub} and T_{De}^{sup} the total T_{De} as functions of f_m (the oscillation frequency in Hz) are plotted in Figures 11.27, 11.28 and 11.29 respectively.

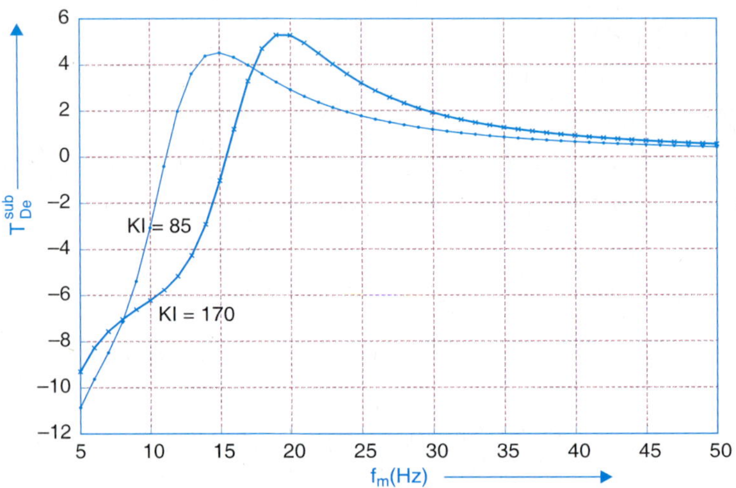

Fig. 11.27: Variation of T_{De}^{Sup} with f_m

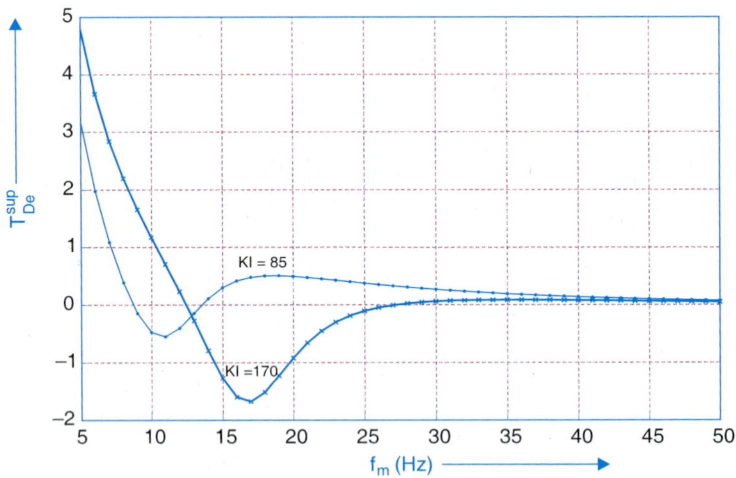

Fig. 11.28: Variation of T_{De}^{Sup} with f_m

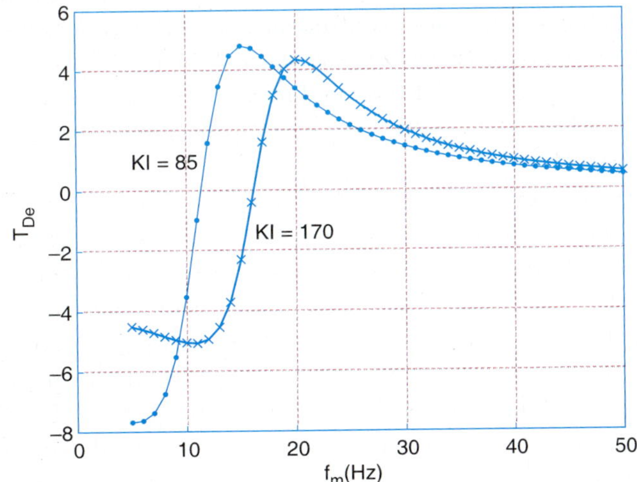

Fig. 11.29: Variation of T_{De} with f_m

It is interesting to observe that while T_{De}^{sub} becomes positive as the frequency f is increased, T_{De}^{sup} is negative for frequencies in the lower range. T_{De}^{sup} reaches a minimum value whenever

$$f_m \approx f_{dc}$$

where f_{dc} is defined by Eq. (11.87). The values of f_{dc} are (*i*) 11.2 Hz, and (*ii*) 16.4 Hz for values of $K_I = 85$ and 170 respectively.

To summarize, the results of the simplified damping torque analysis indicate that

1. The negative damping introduced by the HVDC system is mainly due to the converter current controller. Even here, the damping is negative only for $f_m < f_{dc}$
2. f_{dc} is reduced by
   (*a*) reducing the controller gain K_I.
   (*b*) increasing inductance L (increase in dc line length implies increase in L).
   (*c*) reducing operating value of firing angle α.
3. The peak value of negative damping is an order of magnitude less compared to the case with fixed series compensation.

11.5.5 Control of Torsional Interaction

It has been brought out from the case study and the simplified analysis, that the adverse torsional interaction is much less compared to that caused by fixed series compensation. Even here, the potential of adverse interaction exists only if radial operation of HVDC link connected to a turbine-generator is envisaged. There are two ways by which the adverse TI can be overcome.

1. Modification of the converter control.
2. Providing a Subsynchronous Damping Controller (SSDC).

The first option is feasible only if the modification does not affect the usual functions of the controller during normal and abnormal (contingency) conditions. A major requirement of the controller is to ensure satisfactory transient response during (*i*) DC line faults, and (*ii*) recovery from AC line faults. Whenever the first torisonal mode of the turbine generator has frequency higher than 15 Hz, the TI problem is minimal and can be solved with minor modifications of the existing converter controls.

The second approach of providing a SSDC to modulate the firing angle of the converter to damp subsynchronous oscillation is more flexible. The input signal can be taken from the rotor speed or bus frequency [16]. Alternatively, the signal representing the frequency of the generator internal voltage can be synthesized from voltage and current measurements [72].

The SSDC can be designed based on: (A) narrow bandwidth approach, or (B) wide bandwidth approach. In the former, sharp band pass filters are used to isolate the torsional mode of concern and has the advantage of minimizing the interference with existing HVDC converter controls. However, narrow band pass filters require precise tuning in the field, which may pose a problem. It can also introduce negative damping at frequencies on either side of the frequency for which the filter is tuned. The output of the SSDC goes directly to the firing-angle controller to modulate the delay angle α. In designing a SSDC, care must be taken to ensure that it does not amplify characteristic or non-characteristic harmonics.

11.5.6 Torsional Interactions with MTDC Systems

Padiyar and Geetha [68] report a case study of the analysis of torsional interactions in a multiterminal DC system; based on linearized system model. The major conclusions of this study are–

1. A rectifier on current control contributes to maximum negative damping of torsional modes, when the generator is connected to the rectifier terminal. The torsional mode with the lowest frequency (mode 1) is most affected. The voltage controller at the inverter contributes to the negative damping to some extent.
2. The torsional modes are better damped when the rectifier is chosen as Voltage Setting Terminal (VST) as opposed to the case when the inverter is chosen as VST.

11.5.7 Torsional Interactions with VSC-HVDC

A recent study of SSR with VSC based HVDC transmission systems [73] considered detailed system models and various operating modes of converters. The system diagram is shown in Fig. 11.30. The operating modes considered are shown in Table 11.2. The investigations were carried out based on linear (damping torque and eigenvalue) analysis and nonlinear transient simulation (modelling the switching action in converters).

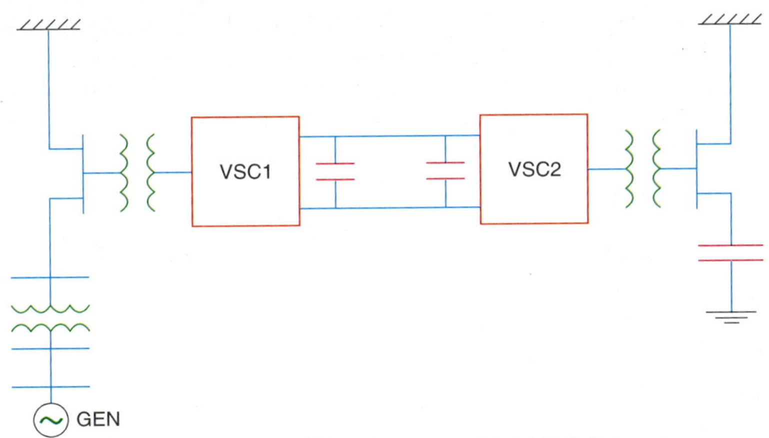

Fig. 11.30: System diagram of a VSC-HVDC link

The damping torque analysis considered classical model of the generator as Induction Generator Effect (IGE) is neglected. The assumption helps to speed up the computation of the damping torque [20]. However, the results of damping torque show good correlation with results from eigenvalue analysis based on detailed generator and HVDC system models. The eigenvalues corresponding to the torsional modes for the operating modes 3, 6 and 7 (see Table 11.2) are shown in Table 11.3. The decrement factor (σ) which is defined as the negative of the real past

Table 11.2: Operating Combinations of VSC Based HVDC

Case	VSC1 (Rectifier)		VSC2 (Inverter)	
1	Power	Reactive Current	DC Voltage	Reactive Current
2	Power	AC Voltage	DC Voltage	AC Voltage
3	DC Voltage	Reactive Current	Power	Reactive Current
4	DC Voltage	AC Voltage	Power	AC Voltage
	VSC1 (Inverter)		VSC2 (Rectifier)	
5	Power	Reactive Current	DC Voltage	Reactive Current
6	Power	AC Voltage	DC Voltage	AC Voltage
7	DC Voltage	Reactive Current	Power	Reactive Current
8	DC Voltage	AC Voltage	Power	AC Voltage

Table 11.3: Eigenvalues of the Detailed System with a VSC Based HVDC for ESCR1=2.5

Torsional Mode	Case 3	Case 6	Case 7
0	– 0.4571 ± j6.0803	– 0.8203 ± j6.1312	– 0.8595 ± j5.5268
1	– 0.2354 ± j98.9350	– 0.1450 ± j98.9030	– 0.2712 ± j98.928
2	– 0.0739 ± j127.010	– 0.0749 ± j127.000	– 0.0751 ± j127.01
3	– 0.6450 ± j160.570	– 0.6692 ± j160.55	– 0.6468 ± j160.56
4	– 0.3646 ± j202.910	– 0.3963 ± j202.91	– 0.3723 ± j202.91
5	– 1.8504 ± j298.170	– 1.8504 ± j298.1700	– 1.8504 ± j298.17

of the eigenvalue can also be computed form the damping torque analysis as $\sigma = \sigma_m + \sigma_e$, when σ_e is the contribution from the electrical system. σ_e is computed from Eq. (11.65). σ_m can be calculated from the eigenvalues of the mechanical system (poles of G(s)). See Fig. 11.31 that shows the interaction between the mechanical and electrical system.

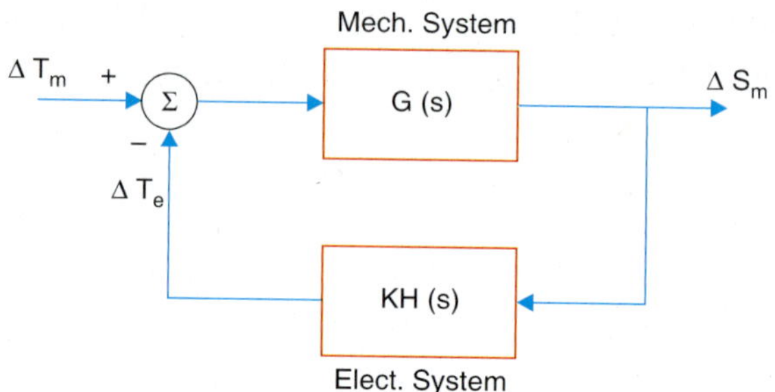

Fig. 11.31: Introduction between electrical and mechanical systems

Tables 11.4 and 11.5 show the decrement factors σ_e and σ_m for the two cases (operating mode 3 and 6 respectively). It is to be noted that σ_e for mode 5 is negligible as the modal inertia is very high.

Table 11.4: Correlation of Damping Torque and Eigenvalue Results (for Case 3)

Sl. No.	Torsional Mode	σ_e	σ_m	$\sigma = \sigma_e + \sigma_m$
1	Mode – 1	0.00765	0.21803	0.22568
2	Mode – 2	0.00033	0.07243	0.07276
3	Mode – 3	0.00061	0.64345	0.64406
4	Mode – 4	0.00279	0.36101	0.36380
5	Mode – 5	0.00001	1.85040	1.85041

Table 11.5: Correlation of Damping Torque and Eigenvalue Results (for Case 6)

Sl. No.	Torsional Mode	σ_e	σ_m	$\sigma = \sigma_e + \sigma_m$
1	Mode – 1	–0.06649	0.21803	0.15154
2	Mode – 2	0.00229	0.07243	0.07472
3	Mode – 3	0.02414	0.64345	0.66759
4	Mode – 4	0.03400	0.36101	0.39501
5	Mode – 5	0.00001	1.85040	1.85041

The conclusions based on the study reported in [73] are as follows:

1. DC Voltage Control of the VSC (rectifier or inverter) near the generator results in better damping of the torsional modes.
2. The constant power control of the rectifier station situated close to the generator contributes small negative damping.

3. The constant power control of the inverter station situated close to the generator can destabilize the torsional modes in a narrow range of lower frequencies.
4. Unlike the LCC-based HVDC systems, the problem of TI with VSC-HVDC is minor except when the converter is operated as an inverter with constant power control. However, DC voltage control of the VSC (rectifier or inverter) actually provides small positive damping. Thus, there may be no need for a SSDC.

It is to be noted that since the converters in VSC-HVDC system, also operate as STATCOM, the provision of a SSDC can help damp torsional oscillations caused by a series compensated AC line connected in parallel. The design of a SSDC with STATCOM to provide desired damping torques in a range of torsional frequencies is described in [76]. The controller uses a signal based on Thevenin Voltage (synthesized from the bus voltage and reactive current measurements) to modulate the reactive current reference of STATCOM.

11.6 HARMONIC INTERACTIONS

11.6.1 General

The converters in HVDC stations generate harmonics – both AC and DC. Some of these harmonics are called characteristic harmonics that will always be present even under ideal conditions, others are called non-characteristic (also called abnormal) harmonics which are caused by (*i*) firing angle errors, (*ii*) negative sequence components in the converter bus AC voltage, and (*iii*) unequal converter transformer leakage impedances.

While AC (and DC) harmonic filters are employed, invariably, to filter out the characteristic harmonics, the cost considerations do not allow the provision of filters at other harmonic frequencies except under special circumstances. If AC harmonic currents (at, say, third harmonic) are injected into the AC system, they can cause harmonic distortion of the converter bus voltage due to resonances. If the distortion is severe, this can lead to operational difficulties such as higher incidence of commutation failure, etc. Some of the major factors that affect the low order harmonic resonances are as follows:

1. Control system–generation of firing pulses.
2. Saturation in converter transformers.
3. The characteristics of system impedance (variation with frequency).
4. DC system characteristics–the impedance seen by the converter terminals.
5. Induction effects.

The possible solutions to the problem are as follows:

1. Modification of the control system.
2. Use of additional filters, say, at third harmonic. Also, provision of a C-type damped filter [84].
3. Use of synchronous condensers, SVC or STATCOM at converter station.

It is to be noted that harmonic distortion during transients is always present and is caused by the changes in the firing angles due to the controller action. The problem of harmonic instability (magnification) is a steady state phenomenon and is to be overcome by suitable control and/or design of the system.

Before presenting the basic aspects of the harmonic resonance problem and its analysis, it is useful to review the literature and describe some forms of the 'harmonic instability' that have occurred in actual HVDC installations.

11.6.2 Harmonic Instability with IPC Scheme of Firing Pulse Generation

This was first studied by Ainsworth [4] using a physical model simulator. He also derived expressions for harmonic magnification based on certain assumptions and approximations. The assumptions are as follows:

1. The DC current is constant–finite nature of the impedance of the DC system is neglected.
2. The overlap due to the leakage reactances is neglected.
3. The constant firing angle type of individual phase control scheme is considered.

For frequencies below the fifth harmonic, the system impedance can be viewed as inductive, while the filter impedance is capacitive (assuming the lowest order of the tuned AC filter is fifth). This shows that there is parallel resonance at a frequency in the range from second to fifth harmonic as the SCR varies from 2.0 to a high value. The net impedance seen by the converter can be high at the (parallel) resonant frequency component which may be present due to firing angle errors. The harmonic distortion is also a function of the impedance angle, the smaller angle leads to better damping and can avoid instability.

Ainsworth showed that the saturation of the converter transformer can aggravate the problem. This is due to the presence of DC currents in the transformer valve windings.

The major conclusion of the study are as follows:

1. The harmonic instability can be expected with systems having low SCR at the converter bus. The problem may be present even at moderate values of SCR if there is a resonance.
2. The firing control scheme has a major effect. The problem is worse with inverse cosine control scheme compared to the constant α control scheme (both schemes are based on IPC).
3. The performance with IPC schemes can be improved substantially using filters in the control system such that the commutation voltages derived from the bus voltage are free from the harmonics. However, there are certain problems with this method. The filters can cause errors and also slow down the response. Furthermore, the filtering may be ineffective due to the variation in the system frequency.
4. The problem of harmonic instability is substantially solved using the EPC scheme of firing pulse generation as this eliminates the firing angle errors that are caused by the shifting of the zero crossings of the commutation voltages (due to the harmonics).

In the cross-channel HVDC scheme, a third order, AC harmonic filter was provided to overcome the problem. In New zealand scheme, a ninth harmonic filter was added.

11.6.3 Core Saturation Instability [2]

This was observed on the Kingsnorth and Nelson River HVDC schemes. In the former, a 12th harmonic component in the magnetizing current of the converter transformer was magnified due to the instability and this also resulted in a circulating current between the 11th and 13th harmonic filters. The total filter current rose to values exceeding the filter rating and caused the protection to shut down the DC link. The build up of instability was slow and took upto 5 minutes.

The major causes for this type of harmonic instability are due to (*i*) the DC system having series resonance at or near the fundamental frequency, (*ii*) low short circuit ratio at the converter bus. If there is a second harmonic voltage at the converter bus, this causes fundamental frequency voltage on the DC side. Due to series resonance in the DC line, there will be relatively large fundamental frequency components in the current through the converters. This, in turn, results in DC components in the currents flowing in valve windings of the converter transformers causing magnetic saturation of the cores. This will reinforce the original second harmonic component of the voltage present at the converter bus.

The main feature of this instability is the presence of DC components in the magnetizing current of the converter transformers causing saturation. The harmonics generated due to the core saturation contain a large spectrum of even harmonics and if the system impedance has a large value at one of these frequencies (due to parallel resonance), the harmonic distortion in the bus voltage is aggravated. In the case of Kingsnorth scheme, the parallel resonance occurred at the 12th harmonic due to the presence of 11th and 13th harmonic (tuned) filters. In Nelson River scheme, the resonance occurred at the fourth harmonic.

It is to be noted that the core saturation instability can occur even with EPC scheme. This type of instability cannot be predicted unless the finite nature of the DC system impedance and its variation with frequency is taken into consideration. In the earlier analysis [4], the DC current was assumed to be constant.

The possible solutions to this instability are as follows:

1. Selecting smoothing reactor values to avoid the resonance in the DC system at or near the fundamental frequency.
2. Modification of the controller by adding an additional dc flux control loop [5]. The control signal is derived from the measured DC magnetizing current or the second harmonic component. This is used to modulate the control signal that is normally generated from the converter controllers. The modulating signal is limited to a low amplitude so that its effect on the normal operation is negligible.

The DC component in the converter transformers may also be caused by the induction of fundamental frequency currents in the DC line caused by adjacent AC circuit on the same right of way [6].

11.6.4 A Generalized Analysis of Harmonic Instability

It is shown by Yacamini et al. [7] that there can be problems of harmonic instability independent of the firing control scheme and transformer saturation. The problem of magnification of a harmonic of particular frequency (f_{ac}) can arise if the AC system (including the filters) has a parallel resonance at that frequency and the DC system (including the smoothing reactors) has a series resonance at the frequency (f_{dc}) such that the following relationship is satisfied.

$$f_{ac} = \pm f_{dc} + (np \pm 1) f_1 \qquad (11.103)$$

where f_1 is the fundamental frequency, p is the pulse number, n is any positive integer including zero. For example, if $f_{dc} = f_1$ and $f_{ac}/f_1 = 2$, 10, 12 or 14, there can be a problem of harmonic resonances, for a twelve pulse converter.

To illustrate the analysis, consider the DC system incorporating a single converter. The equivalent circuit for one phase of the AC network is shown in Fig. 11.32 (*a*) where the current $i_{ph}(t)$ is related to the DC current $i_{dc}(t)$. The equivalent circuit of the DC network is shown in Fig. 11.32 (*b*) where $Z_{dc}(s)$ is the total impedance (in the Laplace domain) consisting of the transformer, smoothing reactor and Thevenin impedance of the DC network viewed from the converter terminals (on the DC side). $v_{dc}(t)$ is related to $v_{ph}(t)$ from the equation

$$v_{dc}(t) = S_a(t)v_a(t) + S_b(t)v_b(t) + S_c(t)v_c(t) \qquad (11.104)$$

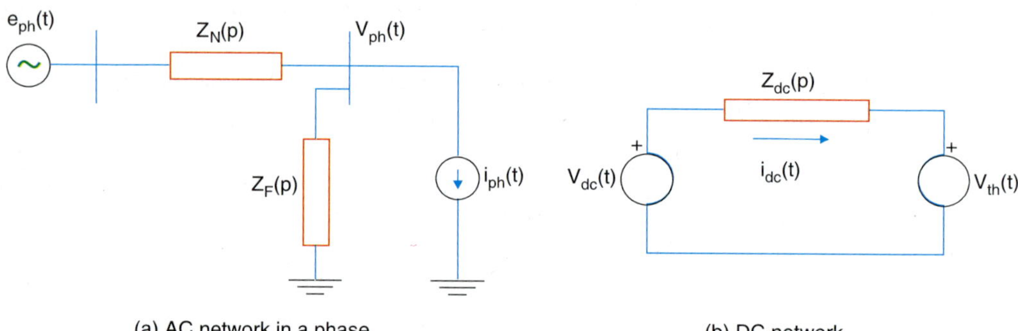

Fig. 11.32: Equivalent circuits

where $S_a(t)$, $S_b(t)$ and $S_c(t)$ are the switching functions defined in section 11.3. Approximating the switching functions by the fundamental frequency components, we have,

$$S_a(t) = \sqrt{\frac{2}{3}}\, ka \cos(\omega t + \delta - \alpha)$$

$$S_b(t) = \sqrt{\frac{2}{3}}\, ka \cos(\omega t + \delta - \alpha - \frac{2\pi}{3})$$

$$S_c(t) = \sqrt{\frac{2}{3}}\, ka \cos(\omega t + \delta - \alpha + \frac{2\pi}{3})$$

Assuming the converter bus voltages as

$$v_a(t) = \sqrt{\frac{2}{3}}\, V \cos(\omega t + \delta)$$

$$v_b(t) = \sqrt{\frac{2}{3}}\, V \cos(\omega t + \delta - \frac{2\pi}{3})$$

$$v_c(t) = \sqrt{\frac{2}{3}}\ V\cos\left(\omega t + \delta + \frac{2\pi}{3}\right)$$

We get the following expression for $v_{dc}(t)$

$$v_{dc}(t) = kaV\cos\alpha \tag{11.105}$$

The expressions for $i_{ph}(t)$ are

$$i_a(t) = \sqrt{\frac{2}{3}}ka\ i_{dc}(t)\cos(\omega t + \delta - \alpha)$$

$$i_b(t) = \sqrt{\frac{2}{3}}\ ka\ i_{dc}(t)\cos\left(\omega t + \delta - \alpha - \frac{2\pi}{3}\right)$$

$$i_c(t) = \sqrt{\frac{2}{3}}\ ka\ i_{dc}(t)\cos\left(\omega t + \delta - \alpha + \frac{2\pi}{3}\right)$$

It is to be noted that $v_{ph}(t)$ and $i_{ph}(t)$ are vectors of phase voltages and currents.

The voltages $v_{ph}(t)$ can be expressed as

$$v_{ph}(t) = F_{ph}(p)e_{ph}(t) - Z_{ph}(p)i_{ph}(t) \tag{11.106}$$

where $p = \frac{d}{dt}$ and $e_{ph}(t)$ is the vector of three phase source voltages. $F_{ph}(p)$ and $Z_{ph}(p)$ are scalars defined below:

$$F_{ph}(p) = \frac{Z_F(p)}{Z_F(p) + Z_N(p)},\ Z_{ph}(p) = F_{ph}(p)Z_N(p) \tag{11.107}$$

Using *D-Q* transformation (defined in Eq. (11.1)) we obtain *D*- and *Q*-axis components of voltages and currents. The *D*- and *Q*-axis components of $v_{ph}(t)$ are

$$V_D = V\cos\delta, \qquad V_Q = -V\sin\delta \tag{11.108}$$

The *D*- and *Q*- axis components of the current $i_{ph}(t)$ are

$$I_D = kai_{dc}\cos(\delta - \alpha),\ I_Q = -kai_{dc}\sin(\delta - \alpha) \tag{11.109}$$

From *D-Q* transformation, Eq. (11.106) can be expressed as

$$\begin{bmatrix} V_D \\ V_Q \end{bmatrix} = \begin{bmatrix} F_{DD} & F_{DQ} \\ F_{QD} & F_{QQ} \end{bmatrix}\begin{bmatrix} E_D \\ E_Q \end{bmatrix} - \begin{bmatrix} Z_{DD} & Z_{DQ} \\ Z_{QD} & Z_{QQ} \end{bmatrix}\begin{bmatrix} I_D \\ I_Q \end{bmatrix} \tag{11.110}$$

where the voltages and currents are in the Laplace domain. The elements of the $[F]$ and $[Z]$ matrices are transfer functions that are related to $F_{ph}(s)$ and $Z_{ph}(s)$ respectively.

An example

Consider the simple AC network shown in Fig. 11.7 where $R + sL$ represents Thevenin impedance and Cs represents the net shunt admittance at the converter bus. (AC filters, if any, are represented by equivalent capacitors). For this network, it can be shown that

$$F_{ph}(s) = \frac{\omega_n^2}{(s+\sigma)^2 + \omega_r^2}, \quad Z_{ph}(s) = \frac{(s+2\sigma)}{C[(s+\sigma)^2 + \omega_r^2]}$$

where $\sigma = \dfrac{R}{2L}$, $\omega_r^2 = \omega_n^2 - \sigma^2$, $\omega_n^2 = \dfrac{1}{LC}$

The elements of the [Z] matrix in D-Q variables can be expressed in terms of $Z_{ph}(s)$ as follows [20]

$$Z_{DD}(s) = Z_{QQ}(s) = \text{Re}[Z_{ph}(s + j\omega_0)] \tag{11.111}$$

$$Z_{DQ}(s) = -Z_{QD}(s) = \text{Im}\left[Z_{ph}(s + j\omega_0)\right] \tag{11.112}$$

Similar expressions apply for the elements of [F] matrix, in terms of $F_{ph}(s)$.

Although the D-Q transformation results in the time-invariant mathematical model of the system, it is still nonlinear as α is the output of the converter controller and δ is dependent on i_{dc}. To make the analysis tractable, it is necessary to linearize the system equations.

Analysis of AC induction in DC line

The harmonic interactions can result from either (*a*) induced voltage in the DC link either from a parallel AC transmission line (on the same Right-of-Way) or transferred from the converter at the other end of the DC line or (*b*) generation of the harmonic in the DC link by voltage unbalance or harmonics in the AC voltage source. In this case, even a fundamental frequency, negative sequence voltage on the AC side results in a second harmonic voltage on the DC side.

Here, the first case is considered where the AC harmonics result from the voltage induced in the DC line. In this case, we neglect harmonics in $e_{ph}(t)$ to simplify the analysis. Thus, E_D and E_Q are constants. Hence, linearizing Eq. (11.110), we get

$$\begin{bmatrix} \Delta V_D \\ \Delta V_Q \end{bmatrix} = -\begin{bmatrix} Z_{DD} & Z_{DQ} \\ Z_{QD} & Z_{QQ} \end{bmatrix}\begin{bmatrix} \Delta I_D \\ \Delta I_Q \end{bmatrix} = -[Z]\begin{bmatrix} \Delta I_D \\ \Delta I_Q \end{bmatrix} \tag{11.113}$$

We can also express V_{dc} in terms of V_D and V_Q on

$$V_{dc} = ka\left[V_D \cos(\delta - \alpha) - V_Q \sin((\delta - \alpha)\right] \tag{11.114}$$

If we assume, pure Equidistant Pulse Control (EPC) scheme of firing, we can express α as

$$\alpha = \delta + \alpha_c \tag{11.115}$$

where α_c is the output of the converter (current) controller. Linearizing Eqs. (11.109) and (11.114), we can derive

$$\Delta V_{dc} = g_1^t \Delta V_{DQ} + (g_2^t V_{DQ})\Delta\alpha_c \tag{11.116}$$

$$\Delta I_{DQ} = g_1 \Delta I_{dc} + g_2 I_{dc} \Delta\alpha_c \tag{11.117}$$

where

$$\Delta V_{DQ}^t = [\Delta V_D \ \ \Delta V_Q], \quad V_{DQ}^t = [V_D \ \ V_Q]$$

g_1 and g_2 are column vectors defined by

$$g_1^t = \left[ka\cos\alpha_c \quad ka\sin\alpha_c\right], \quad g_2^t = \left[-ka\sin\alpha_c \quad ka\cos\alpha_c\right]$$

From Eqs. (11.113) and (11.117), we can obtain ΔV_{dc} as

$$\Delta V_{dc} = -[g_1^t Z g_1]\,\Delta I_{dc} - [g_1^t Z g_2] I_{dc}\Delta\alpha_c + [g_2^t V_{DQ}]\Delta\alpha_c \tag{11.118}$$

It can be shown that

$$[g_1^t Z g_1] = k^2 a^2 Z_{DD}, \quad [g_1^t Z g_2] = k^2 a^2 Z_{DQ} \tag{11.119}$$

The equation for the DC network shown in Fig. 11.32(*b*) is given by

$$V_{dc} - V_{th} = Z_{dc}(s) I_{dc} \tag{11.120}$$

We derive the following linearized equation after substituting Eq. (11.118)

$$\begin{aligned}\Delta E_{dc} &= [g_2^t V_{DQ}]\Delta\alpha_c - k^2 a^2 Z_{DQ}(s) I_{dc}\Delta\alpha_c \\ &= \left[Z_{dc}(s) + Z_{dc}^{ref}(s)\right]\Delta I_{dc} + \Delta V_{th}\end{aligned} \tag{11.121}$$

Where $Z_{dc}^{ref}(s) = k^2 a^2 Z_{DD}(s)$ is the reflected impedance (on the DC side) from the AC network.

An example

For the AC network shown in Fig. 11.7, the elements of the [*Z*] matrix can be derived from Eqs. (11.111) and (11.112). For the case when $R = 0 = \sigma$ we get,

$$Z_{DD}(s) = Z_{QQ}(s) = \frac{Ls\omega_n^2\left(s^2 + \omega_n^2 + \omega_0^2\right)}{\left[s^2 + \left(\omega_n - \omega_0\right)^2\right]\left[s^2 + \left(\omega_n + \omega_0\right)^2\right]}$$

$$Z_{DQ}(s) = -Z_{DQ}(s) = \frac{(\omega_0 L)\omega_n^2\left(\omega_n^2 - \omega_0^2 - s^2\right)}{\left[s^2 + \left(\omega_n - \omega_0\right)^2\right]\left[s^2 + \left(\omega_n + \omega_0\right)^2\right]}$$

It is to be noted that the poles and zeros of $Z_{DD}(s)$ are simple and lie on the imaginary axis and have the interlacing property – a zero is followed by a pole and vice versa. In this case, the zeros are $s = 0$ and $s = \pm j\left(\omega_n^2 + \omega_0^2\right)^{\frac{1}{2}}$ and the poles are $s = \pm j\left(\omega_n - \omega_0\right)$ and $\pm j\left(\omega_n + \omega_0\right)$. $Z_{DD}(s)$ is purely imaginary for $s = j\omega$ as we assumed there are no losses in the AC network. The reactance of the reflected impedance in the DC circuit is inductive at frequencies below $\left(\omega_n - \omega_0\right)$ and capacitive above (until $\omega = \left(\omega_n^2 + \omega_0^2\right)^{\frac{1}{2}}$). Although the reflected impedance is zero at $\left(\omega_n - \omega_0\right)$, the net impedance in the DC circuit can be purely resistive at the resonant frequency determined by

$$\text{Im}\left[Z_{dc}(j\omega) + Z_{dc}^{ref}(j\omega)\right] = 0$$

Thus, even if $Z_{dc}(s) = R_d + sL_d$, the resonance can occur at a frequency slightly above $\omega = \omega_n - \omega_0$. This shows that resonance in AC network corresponding to second harmonic can bring about resonance in a simple DC network at a frequency close to the fundamental.

Both $Z_{DD}(s)$ and $Z_{DQ}(s)$ affect the controller design. The transfer function $Z_{DQ}(s)$ has a positive real zero (implying that it is non-minimum phase).

Remarks

1. The analysis given here defines explicitly the reflected impedance from the AC network, at all the harmonic frequencies. In reference [9], although the authors have presented for a sample system, second harmonic equivalent impedance, the general analysis is missing.

The analysis assumes that the switching function can be represented by its fundamental frequency component alone (ignoring the harmonic components). This assumption is used widely. However, it can be relaxed as given in [81-83]. Here, the analysis is performed in the Laplace domain (using complex frequencies). The Laplace transforms of the AC currents injected by the converter is obtained from the convolution of the transforms of the switching functions and DC current. Similarly, the transform of the voltage injected in the converter (on the DC side) by the AC network, is obtained from the convolution of the transforms of the switching function and the converter (AC) bus voltages. This is based on the result that the Laplace transform of product of two functions, $f_1(t)$ and $f_2(t)$ and is given by [95].

$$L\left[f_1(t)f_2(t)\right] = \sum_{k=1}^{q} \frac{A_1(s_k)}{B_1'(s_k)} F_2(s - s_k)$$

where $F_1(s) = \dfrac{A_1(s)}{B_1(s)}$ is the Laplace transform of $f_1(t)$ and $F_2(s)$ is the Laplace transform of $f_2(t)$. $B_1'(s_k)$ is the first derivative of $B_1(s)$ at the pole $s = s_k$. It is assumed that all the poles of $F_1(s)$ are simple.

If $f_1(t)$ is the switching function $S_a(t)$ shown in Fig. 11.9, $F_1(s)$ is given by

$$F_1(s) = \frac{1}{1 - e^{-sT}}\left[\frac{e^{-st_{1a}} - e^{-st_{2a}} - e^{-st_{3a}} + e^{-st_{4a}}}{s}\right]$$

t_{1a}, t_{2a} are the instants when the switching function transits from zero to positive maximum and returns to zero respectively. Similarly, t_{3a}, t_{4a} are the instants when the switching function transits from zero to negative value and returns to zero.

The poles of $F_1(s)$ are simple at $s = \pm jn\omega_1$, $\omega_1 = \dfrac{2\pi}{T}$ and n is any integer. It can be shown that

$$\frac{A_1(s_k)}{B_1'(s_k)}\Big|_{s_k=0} = \frac{(t_{1a} - t_{2a}) - (t_{3a} - t_{4a})}{T}$$

$$\frac{A_1(s_k)}{B_1'(s_k)}\Big|_{s_k \neq 0} = \frac{e^{-s_k t_{1a}} - e^{-s_k t_{2a}} - e^{-s_k t_{3a}} + e^{-s_k t_{4a}}}{s_k T}$$

2. The induction of fundamental frequency voltage in a DC line due to currents flowing in a parallel AC line is the most stringent case for adverse effects of harmonic interactions. If we assume ΔV_{th} is a sinusoidal voltage of fundamental frequency, this causes a fundamental frequency component in the DC currents (ΔI_{dc}) which is limited by the net impedance $(Z_{dc} + Z_{dc}^{ref})$; if we assume $\Delta\alpha_c = 0$. This current would then appear in the secondary winding of the converter transformer as a direct component and also a second harmonic (positive sequence) component superposed on the fundamental component [6, 78]. The direct current component does not flow in the primary windings and appears as a part of the magnetizing current that offsets the knee of the flux-current characteristic, leading to unsymmetrical saturation of the converter transformer. This results in increase in the audible noise and possible loss of life expectancy of the transformer [6]. Other reported problems are inaccurate measurements due to saturation of current transformers on the AC side. A typical value of 0.1% of the rated AC current for the DC offset is often cited as the limit.

   While DC side blocking filters have been suggested for reducing the fundamental frequency currents in the DC line, they can result in increased overvoltages. In reference [78] modulation of firing angle is suggested to eliminate transformer DC currents.

3. One approach for eliminating the DC currents is to eliminate the fundamental frequency current in the DC line. From Eq. (11.121), it is obvious that if we arrange $\Delta E_{dc} = \Delta V_{th}$, $\Delta I_{dc} = 0$. In reference [78], a case study is presented that demonstrates this result. However, it is mentioned that, the modulation that eliminates ΔI_{dc}, does not remove DC components. An alternative modulation scheme is presented in [78] that is not based on analysis.

   From Eq. (11.117) given above, it is obvious that even if $\Delta I_{dc} = 0$, $\Delta\alpha_c$ can result in DC components in the transformer. The control strategy to eliminate DC current components is presented next.

Elimination of transformer DC currents

Let $\Delta I_{dc} = I_m \cos(\omega t + \phi)$. The current in phase '$a$' ($\Delta i_a$) is given by

$$\Delta i_a(t) = \sqrt{\frac{2}{3}}\left[\cos(\omega_0 t)\Delta I_D + \sin(\omega_0 t)\Delta I_Q\right] \tag{11.122}$$

Substituting from Eq. (11.117), we get

$$\Delta i_a(t) = \sqrt{\frac{2}{3}}ka\left[\cos\alpha_c \cos\omega_0 t + \sin\omega_0 t \sin\alpha_c\right]\Delta I_{dc}(t)$$

$$+\sqrt{\frac{2}{3}}kaI_{dc}\left[-\sin\alpha_c \cos\omega_0 t + \cos\alpha_c \sin\omega_0 t\right]\Delta\alpha_c(t)$$

$$\Delta i_a(t) = \sqrt{\frac{2}{3}}kaI_m\left[\cos(\omega_0 t - \alpha_c)\cos(\omega_0 t + \phi)\right]$$

$$+\sqrt{\frac{2}{3}}kaI_{dc}\left[\sin(\omega_0 t-\alpha_c)\right]\Delta\alpha_c(t) \tag{11.123}$$

If we assume,

$$\Delta\alpha_c(t)=I_m^c\sin(\omega_0 t+\eta)$$

The DC component of the current in phase 'a' (i_{ao}) is given by

$$i_{ao}=\sqrt{\frac{2}{3}}\frac{ka}{2}\left[I_m\cos(\phi+\alpha_c)+I_m^c I_{dc}\cos(\eta+\alpha_c)\right] \tag{11.124}$$

To force $i_{ao}=0$, we have to select

$$I_m^c=\frac{I_m}{I_{dc}},\qquad \eta=180+\phi \tag{11.125}$$

It can be shown that i_{bo} and i_{co} will also be zero with the choice of $\Delta\alpha_c(t)$. (Note that I_{dc} and α_c are operating values which are constant).

Harmonic transfer in BTB links

A BTB link, generally, has low impedance as there is no DC line and the smoothing reactors have smaller inductances. If V_{th} in Fig. 11.32(*b*) has non-characteristic harmonic components, then it results in the transfer of the voltage harmonics on one side of the link to the other. For example, a second harmonic positive sequence component in the bus voltage at the inverter results in second harmonic positive sequence component in the bus voltage at the rectifier. The magnitude depends on the characteristics of the DC and AC impedances as well as control characteristics. Here also, the effective solution is to modulate the firing angle to eliminate the effect of the fundamental frequency component in the DC link current.

In an asynchronous HVDC link the operating frequencies at the two end may be different even though the nominal frequencies are same. In this case, there could be lower frequency components in the DC current. For example if the operating frequency at inverter is 49 Hz, this results in 588 Hz components in the DC current. This results in a positive sequence AC current component of frequency 38 Hz at the rectifier leading to the injection of a 12 Hz voltage and consequently, the 12 Hz current component in the DC link. (It is assumed that the operating frequency at the rectifier in 50 Hz). The current controller can magnify the current oscillations at this low frequency.

11.6.5 Harmonic Interactions in VSC-HVDC Systems

Harmonic interactions are also present in VSC based HVDC systems. However, the mechanism is different. The harmonics in the AC current are transferred to the DC side current with frequency translation. If switching functions for the VSC are approximated by neglecting the harmonic components, the presence of AC harmonic current order 'n' is manifested as DC harmonic of order (n – 1) if the AC harmonic current has positive sequence. If the AC harmonic has negative sequence, then DC harmonic is of order (n + 1).

If the DC voltage (on the capacitor) has a harmonic component of order 'n', it results in the AC voltage harmonics of order (n – 1) and (n + 1). The former has negative sequence while the latter has positive sequence. These rules can be mathematically stated as follows. The frequency f_{dc} of the harmonic component is the DC current is given by

$$f_{dc} = f_{ac} - 1 \quad (11.126)$$

Here, f_{ac} is treated as positive if the harmonic has positive sequence and negative, if it has negative sequence. Note that the sign of f_{dc} is not of consequence.

The frequency F_{ac} of the harmonic component in the AC voltage resulting from the harmonic component in the DC voltage, (having frequency F_{dc}) is given by

$$F_{ac}^{+} = F_{dc} + 1, \; F_{ac}^{-} = F_{dc} - 1 \quad (11.127)$$

Here F_{dc} is assumed to be positive. The superscript '+' and '–' indicate positive and negative sequence respectively.

If $f_{ac} = -1$, implying a negative sequence component of fundamental frequency, it results in a second harmonic component on the DC side which in turn contributes a third harmonic (positive sequence) voltage component in addition to the negative sequence, fundamental frequency component.

The feedback from the converter will be zero, if the harmonic impedance on the DC side is zero (the DC capacitor is infinitely high). It is to be noted that in a LCC, the harmonic feedback from the DC side is zero, if its harmonic impedance is infinite.

The choice of DC capacitor is affected by considerations of harmonic interactions. Also, the current oscillations in the capacitor can be overcome by modifying the converter control to inject a negative sequence voltage during phase unbalance.

11.7 CONTROL INTERACTIONS

Some aspects of control interactions are covered under harmonic interactions where it is shown that converter control can help to overcome problems of harmonic instability or oscillations.

HVDC control is quite complex with several levels of control and various modes of operation that affect system performance. For example, the low frequency damping (modulation) controller can adversely affect the damping of torsional oscillations. The bang-ramp controller that swiftly reduces the DC current for severe voltage reduction (bang) and then gradually ramp it to its initial value after time delay, can interact with VDCOL action [61]. In general, it can be said that there is need to reduce unnecessary complexity in control and optimize its performance. Although, the common practice is to use a simulator for fine tuning the controllers, there is need to investigate analytical issues to simplify the controller design and build robustness. Essentially, the controller must be adaptive and intelligent to provide the best response for various operating conditions.

It was shown in [74] that a shunt connected FACTS Controller for voltage control can interact with the AC network resonance. The conventional P-I controller can induce instability due to non-minimum phase of the plant transfer function given by

$$G(s) = \frac{\Delta V(s)}{\Delta I_R(s)} = Z_{QD}(s) \quad (11.128)$$

where $Z_{QD}(s)$ is the impedance function in the D-Q domain defined earlier. The solution proposed is to introduce a compensator in cascade with the voltage regulator. The transfer function of the compensator is similar to that of a notch filter which rejects the frequency component $(\omega_r - \omega_0)$ where ω_r is the AC network resonance frequency.

In a VSC-HVDC converter, the reactive current can be controlled to regulate the bus voltage, just as done in a STATCOM (a FACTS controller). Hence, this controller must take into account the network resonances.

A recent study [94] has shown that the design of DC and AC voltage controls in a BTB VSC-HVDC link involves handling non-minimum phase transfer functions even in the absence of network resonance. The problem arises when the DC voltage control is applied at the rectifier station. The analytical design of the DC and AC voltage controllers for a BTB, VSC-HVDC link is described in Appendix G.

REFERENCES AND BIBLIOGRAPHY

1. Padiyar K.R., '*HVDC-AC System Interactions*', IE (I) Journal –EL, Vol.70, pp. 88–96, August 1989.
2. Ainsworth J.D., '*Core Saturation Instability in the Kingsnorth HVDC Link*', CIGRE Study Committee 14, Winnipeg, Canada, June 1977.
3. Hammad A.E., '*Analysis of Second Harmonic Instability for the Chateauguay HVDC) SVC Scheme*', IEEE Trans., Power Delivery, Vol.7, No.1, pp.410-415, January 1992.
4. Ainsworth J.D., '*Harmonic Instability between Controlled Static Converters and AC Networks*', Proc. IEE(London), Vol.114, No.7, July 1967, pp. 949–958.
5. Ainsworth J.D., '*Development in the Phase-locked Oscillator Control System for HVDC and Other Large Converters*', Fourth Int. Conf. on 'AC and DC Power Transmission', IEE(London), 1985.
6. Larsen E.V., Walling R.A., and Bridenbaugh C.J., '*Parallel AC/DC Transmission Lines, Steady-state Induction Issues*', IEEE Trans., Power Delivery, Vol. 4, No.1, pp. 667–673, January 1989.
7. Yacamini R. and De Oliveria J.C., '*Instability in HVDC Schemes at Low-order Integer Harmonics*', Proc. IEE (London), Vol.127, Pt. C, No.3, May 1980, pp. 179–188.
8. Bahrman M.P., et al., '*DC System Resonance Analysis*', IEEE Trans., Vol. PWRD-2, No.1, January 1987, pp. 156–164.
9. Larsen E.V., Baker D.H. and McIver J.C., 'Low-order Harmonic Interactions on AC/DC Systems', IEEE Trans., Vol. PWRD-4, No.1 January 1989, pp 493-501.
10. Hatziadoniu G. and Galanos G.D., '*Interactions between the AC Voltages and DC Current in Weak ACI DC Interconnections*', Paper 88 WM 074-7, IEEE/PES Winter Meeting, New York, 1988.
11. Eggleston J.F., Arrilaga J. and Semlyen A., '*Analysis of the Harmonic Distortion Resulting from the Interaction between Synchronous Machines and HVDC Converters*', IEEE Trans., Vol. PWRD-3, No.1, January 1988, 325–332.
12. Bahrman M., et al., '*Experience with HVDC–Turbine Generator Torsional Interaction at Square Resonance*', IEEE Trans., Vol.PAS-99, May/June 1980, pp. 966–975.
13. Wu C.T., et al., '*The Intermountain Power Project Commissioning–Subsynchronous Torsional Interaction Tests*'. IEEE Trans., Vol. PWRD-3, No.4, October 1988.

14. Mortensen K., Larsen E.V, and Piwko R.J., *Field Tests and Analysis of Torsional Interaction between the Coal-Creek Turbine-Generators and the CU HVDC System*', IEEE Trans., Vol. PAS-100, January 1981, pp. 336–345.
15. Padiyar K.R. and Kothari A.G., '*Analysis of HVDC Turbine Generator Torsional Interactions*', Electric Machines and Power Systems, Vol.16, pp. 303–317, 1989.
16. Svensson S. and Mortensen K., '*Damping of Subsynchronous Oscillations by an HVDC Link: An HVDC Simulator Study*', IEEE Trans., Vol. PAS-100, March 1981, pp. 1431–1437.
17. Hedin R. and Stump K. '*Effect of Series Capacitors on HVDC Control System Torsional Interactions*', Int. Conf. '*DC Power Transmission*', Montreal, Canada, June 1984.
18. Kothari A.G., '*Analysis and Simulation of HVDC-Turbine Generator Torsional Interactions*', Ph.D. Thesis, I.I.T. Kanpur, October, 1985.
19. Padiyar K.R. and Kothari A.G., '*Study of HVDC Torsional Interactions through Digital Dynamic Simulation*', Electric Machines and Power Systems, Vol. 14, No.5, 1988.
20. Padiyar K.R., '*Analysis of Subsynchronous Resonance in Power Systems*', (Boston), Kluwer Academic Publications 1999.
21. Sucena–Paiva J.P., Hernandez R. and Freris L.L., '*Stability Study of Controlled Rectifers using a New Discrete Model*', Proc. IEE, Vol.119, September 1972, pp. 1285-1293.
22. Sucena-Paiva J.P. and L.L. Freris, '*Stability of a DC Transmission Link between Weak AC Systems*', Proc. IEE, Vol. 121, June 1974, pp. 508–515.
23. Padiyar K.R. and Sachchidanand, '*Stability of Converter Control for Multiterminal HVDC Systems*', IEEE Trans., Vol. PAS-104, No.3, March 1985, pp. 690–695.
24. Arabi S., P. Kundur and J.H. Sawada, '*Appropriate HVDC Transmission Simulation Models for Various Power System Stability Studies*', IEEE Trans., Power Systems, Vol. 13, No.4, pp. 1292–1297, November 1998.
25. Middlebrook R.D. and C. Cuk, '*A general Unified Approach to Modeling Switching Converter Power Stages*', Proc. IEEE Power Electronics Specialists Conf., pp. 18–34, 1976.
26. Maksimovic D., Stankovic A.M., Thottuvelid V.J. and Verghese G.C., '*Modeling and Simulation of Power Electronic Converters*', Proc. IEEE, Vol. 89, No.6, pp. 898–912, June 2001.
27. Padiyar K.R. and Ramshaw R.S., '*State Space Formulation for Transient and Dynamic Analysis for Power Systems*', IEEE Communication and EHV Conference, Montreal, November 1972.
28. Padiyar K.R. and Sachchidanand, '*Digital Simulation of Multiterminal HVDC System Using a Novel Converter Model*', IEEE Trans., Vol. PAS-102, No.6, June 1983, pp. 1624–32.
29. Padiyar K.R. and Kothari A.G., '*Modeling and Simulation of HVDC Bridge Converters*', Electric *Machines and Power Systems*, Vol. 13, 1987, pp. 285–298.
30. Hingorani N.G. and Burbery M.F., '*Simulation of System Impedance in HVDC System Studies*', IEEE Trans., Vol. PAS-89, No.5/6, May/June 1970, pp. 820–826.
31. Hudson J.E., Hunter E. M and Wilson D.D. '*EHV-DC Simulator*', Vol. PAS-85, No.11, pp. 1101–1107, November 1966.
32. MacCallum D.A. and Gavrilovic M. M., '*IREQ's HVDC Simulator: Advances in Automation of Study Procedure and Analysis*, Proceedings of Int. Conf on DC Power Transmission, Montreal, Canada, June 1984.
33. Bonner J.J., Grassel W.E. and McGranagan M.F., '*Development of a Real-Time HVDC Terminal Model*', Proceedings of Int. Conf on DC Power Transmission, Montreal, Canada, June 1984.
34. Breuer G.D. and Degneff R.C., '*Required System Studies for HVDC Terminal Equipment*', Proceedings of Int. Conf. on DC Power Transmission, Montreal, Canada, June 1984.

35. Kassakian J.G., '*Simulating Power Electronic Systems – A New Approach*', Proceedings of IEEE, Vol. 67, No.11, pp. 1428–1439, October 1979.

36. Joetten R., et al., '*A New Real Time Simulator for Power System Simulation*', IEEE Trans., Vol. PAS-104, No. 9, pp. 2604–11, September 1985.

37. Bowles J.P. and Jasmin. G., '*An Electronic Machine for Power System Simulation*', Proc. Canadian Comm. and EHV Conference, Montreal, 1972.

38. Padiyar K.R. and Varma R.K., '*Developement of Parity Simulation for Thyristor Controller Reactor*', Journal of Inst. of Engrs. (India), Vol. 67, Part EL4, Feb. 1987.

39. Heffernan M.D., Turner K.S., Arrillaga J. and Arnold, C.P. '*Computation of AC-DC System Disturbances Part I: Interactive Coordination of Generator and Converter Transient Models*', IEEE Trans., Vol. PAS-100, No.11, pp. 4341–4348, November 1981.

40. Clayton R.E., De Mello F.P., Johnson B.K. and Hedman D.E., '*Digital Modelling of AC/DC Systems*', Paper presented at IEEE Conference on '*Overvoltages and Compensation of Integrated AC-DC Systems*, Winnipeg, Canada, 1980.

41. CIGRE Study Committee No.33, '*Application Guide for Insulation Coordination and Arrester Protection of HVDC Converter Stations*', 33.83 (SC 03-21 WD).

42. Hingorani N.G., Kitchen R.H. and Hay J. L., '*Dynamic Simulation of HVDC Power Transmission Systems on Digital Computers*', IEEE Trans., Vol. PAS-87, No.4, pp. 989–996, April 1968.

43. Milias J. – Argitis, Giannakopoulos G. and Galanos G., '*Dynamic Simulation of Multiterminal HVDC Systems*', IEEE Trans., Vol. PAS-87, pp. 587–593, March/April 1978.

44. Woodford D.A., '*Validation of Digital Simulation of DC Links*', IEEE Trans. Vol. PAS-104, No. 9, Sept. 1985, pp. 2588–2594.

45. Woodford D.A., Gole A.M. and Menzies R.W., '*Digital Simulation of DC Links and AC Machines*', IEEE Trans., Vol. PAS-102, No.6, pp. 1616–1623, June 1983.

46. Reeve J. and Chen S.P., '*Versatile Interactive Digital Simulator Based on EMTP for AC/DC Power System Transient Studies*', Paper 84-200-2 Presented at IEEE PES Winter Meeting 1984.

47. Padiyar K.R., et al., '*Study of HVDC Controls through Efficient Dynamic Digital Simulation of Converters*', Paper 89, WM 113-2, PWRD Presented at IEEE/PES 1989 Winter Meeting, New York.

48. Ino T., Mathias R.M., Iravani M.R. and Sasaki S., '*Validation of Digital Simulation of DC Links*'-Part II: IEEE Trans., Vol. PAS-104, No.9, Sept. 1985, pp. 2596–2603.

49. Vithayathil J.J., '*Digital Dynamic Simulation*', Proc. of National Workshop on HVDC Transmission, I.I.T., Kanpur, January 1987.

50. Dommel H.W., '*Digital Computer Solutions of Electromagnetic Transients in Single-phase and Multiphase Networks*', IEEE Trans., Vol. PAS-88, April 1969, pp. 388–399.

51. Dommel H.W. and Mayer, W.S. '*Computation of Electromagnetic Transients*', Proc. IEEE Vol. 62, July 1974, pp. 983–993.

52. El-Serafi A.M., Shahata A., '*Digital Simulation of AC/DC Systems in Direct Phase Quantities*', IEEE Trans., Vol. P and S-95, March/April 1976, pp. 731–736.

53. Hay J.L. and Hingorani N.G., '*Dynamic Solution of Multi Converter HVDC Systems by Digital Computer*'-Part I: Mathematical model, Part II: Computer Program', IEEE Trans., Vol. PAS-89, Feb. 1970, pp. 218–228.

54. Hingorani N.G., Hay J.L. an Crosbie R.E., '*Dynamic Simulation of HVDC Systems on a Digital Computer*', Proc. IEE, Vol. 113, March 1966, pp. 793, 802.

55. Ramshaw R.S. and Padiyar K. R., '*Digital Simulation of a Fullwave Single-phase Converter System*', Proc. IEE, Vol. 117, No.11, Nov. 1970, pp. 2151–2158.
56. S. Bhattacharya, '*Simulation of Transient Phenomena in High Voltage Direct Current Converter Systems*', Ph.D. Thesis, University of British Columbia, Canada, March 1987.
57. Reeve J. and S.P. Chen, '*Versatile Interactive Digital Simulator Based on EMTP for AC/DC Power System Transient Studies*', Paper 84-200-2 presented at IEEE Winter Meeting 1984.
58. Lehn P., Rittiger J. and Kulicke B., '*Comparison of the ATP Version of the EMTP and the NETOMAC Program for Simulation of HVDC Systems*', IEEE Trans., Power Delivery, Vol. 10, No.4, pp. 2048-2053, October 1995.
59. Working Group on Dynamic Performance and Modeling of DC Systems and Power Electronics for Transmission Systems, '*Report on Test Systems for AC/DC Interaction Studies*', IEEE Trans. Power Delivery, Vol.10, No.4, pp. 2027–2034, October 1995.
60. Thallam R.S., '*Review of the Design and Performance Features, of HVDC Systems Connected to Low Short Circuit Ratio AC Systems*', IEEE Trans. Power Delivery, Vol.7, No.4, pp. 2065–2073, October 1992.
61. Hammad A.E., Gagnon J. and D. McCallum, '*Improving the Dynamic Performance of a Complex AC/DC System by HVDC Control Modifications*', IEEE Trans., Power Delivery, Vol. 5, No.4, pp. 1934–1943, November 1990.
62. Gole A. and Sood V.K., '*Performance Analysis of a DQO Type Phase Locked Grid Control System for HVDC Converters*', NPSC-VI, I.I.T. Mumbai, 1990.
63. Larsen E.V., Clark K. and London D.J., '*Stability Analysis of a Digitally based HVDC Firing – Pulse Synchronization Control*', IEEE Trans. Power Systems, Vol. 7, No.3, pp. 1415–1424, July 1992.
64. Durie R.C. and C. Pottle, '*An Extensible Real-Time Digital Transient Network Analyzer*', IEEE Trans. Power Systems, Vol. 8, No.1, pp. 4–8, Feb. 1993.
65. Wang X., Woodford D.A., R. Kuffel and R. Wierckx, '*A Real-Time Transmission Line Model for a Digital TNA*', IEEE Trans. Power Delivery, Vol. 11, No.2 pp. 10-92-1097, April 1996.
66. Duchen H., Lagerkvist M., Kuffel R. and Wierckx R., '*HVDC Simulation and Control Testing using a Real-Time Digital Simulator (RTDS)*' Conf. Proc., ICDS 1995, College Station, Texas, USA, April 1995, pp. 213–218.
67. Padiyar K.R., Sachchidanand and J. Senthil, '*Digital Computer Study of the Control of Torsional Interactions in HVDC Turbine-Generators*', Electric Machines and Power Systems, Vol. 22, pp. 87–103, 1994.
68. Padiyar K.R. and Geetha M K., '*Analysis of Torsional Interactions in MTDC Systems*', Elect. Power & Energy Syst., Vol. 17, No.4, pp. 257–266, 1995.
69. Hall M.C. and Hodges D.A., '*Experience with 500 kV Subsynchronous Resonance and Resulting Turbine Generator Shaft Damage at Mohave Generating Station*' , IEEE publication 76 CH 1066-O- PWR on 'Analysis and Control of Subsynchronous Resonance', 1976.
70. Kilgore L.A., Ramey D.G. and Hall M.C., '*Simplified Transmission and Generation System Analysis Procedures for Subsynchronous Resonance Problems*', IEEE Trans., Vol. PAS-96, No.6, Nov./Dec. 1977, pp. 1840–1846.
71. IEEE Working Group, '*Countermeasures to Subsynchronous Resonance Problems*', IEEE Trans., Vol. PAS-99, No.5, Sept/Oct. 1980, pp. 1810–1818.
72. Piwko R.J. and Larsen E.V., '*HVDC System Control for Damping of Subsynchronous Oscillations*', IEEE Trans. Power Apparatus and Systems, Vol.101, No.7, pp. 2203–2211, July 1982.

73. Nagesh Prabhu and Padiyar K. R., '*Investigation of Subsynchronous Resonance with VSC-Based HVDC Transmission Systems*', IEEE Trans. Power Delivery, Vol. 24, No.1, pp. 433–440, January 2009.

74. Padiyar K.R., '*FACTS Controllers in Power Transmission and Distribution*', (New Delhi) New Age Publishers, 2007.

75. Padiyar K.R. and Nagesh Prabhu, '*Modeling, Control Design and Analysis of VSC based HVDC Transmission Systems*', IEEE POWERCON 2004, Singapore, November 2004.

76. Padiyar K.R. and Nagesh Prabhu, '*Design and Performance Evaluation of Subsynchronous Damping Controller with STATCOM*', IEEE Trans. Power Delivery, Vol. 21, No.3, pp. 1398-1405, July 2006.

77. Yacamini R. and J.C. De Oliverira, '*Comprehensive Calculation of Converter Harmonics with Systems Impedances and Control Representation*', Proc IEEE, Vol. 133, Pt.C, pp. 95–102, March 1986.

78. Gole A.M., Verdolin R. and Kuffel E., '*Firing Angle Modulation for Eliminating Transformer DC Currents in Coupled AC-DC Systems*', IEEE Trans. Power Delivery Vol. 10, No.4, pp. 2040–2047, October 1995.

79. Farret F.A. and Freris L.L., '*Minimization of Uncharacteristic Harmonics in HVDC Convertors through Firing Angles Modulation*', IEE Proc., Vol. 137, Pt.C, No.1, pp. 45–52, January 1990.

80. Hu L. and Yacamini R., '*Calculation of Harmonics and Interharmonics in HVDC Schemes with Low DC Side Impedance*', IEE Proc.-C, Vol. 140, No. 6, pp. 469–476, November 1993.

81. Padiyar K.R. and Bhowmick A., '*Analysis of Harmonic Interactions in Back to Back DC Links*', 3rd Workshop & Conference on EHV Technology, IISc Bangalore, August 1995.

82. Padiyar K.R. and Bhowmick A., '*Modeling of Current Control in the Analysis of Harmonic Interactions in HVDC Systems*', 9th National Power Systems Conference (NPSC), IIT Kanpur, December 1996.

83. Padiyar K.R. and A. Bhowmick, '*Analysis of Harmonic Interactions in HVDC Systems*', Preprint.

84. Bodger P.S., Erwin G.D., and Woodford D.A., '*Controlling Harmonic Instability of HVDC Links Connected to Weak AC Systems*', IEEE Transactions on Power Delivery, 5(4):2039–2046, November 1990.

85. Xiao Jiang and Gole A.M., '*A Frequency Scanning Method for the Identification of Harmonic Instabilities in HVDC Systems*', IEEE Transactions on Power Delivery 10(4):1875-1881, October 1995.

86. Hu L. and Yacamini R., '*Harmonic Transfer through Converters and HVDC Links*', IEEE Transactions on Power Electronics, 7(3):514–525, July 1992.

87. Dickmander D.L., Lee S.M., Desilets G.L., and Granger M., '*AC/DC Harmonic Interaction in the Presence of GIC for the Quebec-New England phase II HVDC Transmission*', IEEE Transactions on Power Delivery, 9(1):68-78, January 1994.

88. Wood A.R. and Arrillaga J., '*The Frequency Dependent Impedance of an HVDC Converte*r', IEEE Transactions on Power Delivery, 10(3):1635–1641, July 1995.

89. Wood A.R. and Arrillaga J., '*Composite Resonance; A Circuit Approach to the Waveform Distortion Dynamics of an HVDC Converter*', IEEE Transactions on Power Delivery", 10(4):1882-1888, October 1995.

90. Rittiger J. and Kulicke B., '*Calculation of HVDC-Converter Harmonics in Frequency Domain with Regard to Asymmetries and Comparison with Time Domain Simulation*', IEEE Transactions on Power Delivery, 10(4):1944–1949, October 1995.

91. Padiyar K.R. and Senthil J., *'Development of a Versatile Package for the Digital Simulation of Transients in MTDC Systems'*, In Proc. of Sixth National Power Systems Conference, pp. 58–63, 1990.

92. Padiyar K.R., *'Power Systems Dynamics–Stability and Control'*, Second Edition (Hyderabad), B.S. Publications, 2002.

93. Jiang Y. and Ekstrom A., *'General Analysis of Harmonic Transfer through Converter'*, IEEE Trans. Power Electronics, Vol. 12, No.2, pp. 287–293, March 1997.

94. Abhishek Tyagi and Padiyar K.R., *'Dynamic Analysis and Simulation of a VSC based Back-to-back HVDC Link'*, IICPE 2006, Chennai, December 2006.

95. Lepage W.R., *'Complex Variables and the Laplace Transform for Engineers'*, (New Delhi), Tata McGraw-Hill Publishing Co., 1961.

12 STABILITY ANALYSIS AND POWER MODULATION

12.1 INTRODUCTION

In the previous chapter, the analysis of torsional and harmonic interactions was presented. In this chapter, the analysis of electromechanical interactions at low frequencies is presented. One aspect of these interactions are the problems of voltage instability that have been experienced when HVDC links are operating under low SCR (Short Circuit Ratio) when connected to weak AC systems. According to reference [1], there were 10 HVDC systems (commissioned before 1988) connected to weak AC systems. The minimum value of SCR under contingency conditions could be around 2.0 and ESCR around 1.5. Generally, BTB links tend to operate with low SCR or ESCR. In Miles City Converter station (a 200 MW, back-to-back DC link which is one of the four DC links connecting the Western and Eastern AC systems in U.S.A), the DC system recovery time after AC fault clearance, can be as much as 450 ms and the recovery power not exceeding 170 MW for east to west transfer. The problems of voltage instability and reduction of maximum power flow in the link under contingency conditions can be tackled using improved converter control and AC voltage regulation.

Although HVDC converter controls (in particular, constant extinction angle control) can aggravate voltage stability problems, they can also help in damping low frequency oscillations by power modulation.

Large interconnected power systems exhibit low frequency oscillations which are sustained at certain (usually high) loading conditions. The major factors that are responsible for this are generally ascribed to high gain automatic voltage regulators (AVR) and fast exciters. Power Systems Stabilizers (PSS) which are essentially auxiliary controllers that modulate the generator terminal voltage magnitudes are used for damping these oscillations. If there are DC links in the system, they can also be provided with auxiliary controllers to modulate the power flow in the link, which in turn, can help in stabilizing the system.

The stability of a power system when subjected to small perturbations (which are always present) is termed dynamic stability or small disturbance stability. (This is an extension of the concept of steady-state stability to include the effect of controllers in the system). It is obvious that a power system should always operate within the dynamic stability limits although transient stability limits may be occasionally violated. Thus, the analysis and improvement of dynamic stability using well designed auxiliary controllers assume importance.

The analysis of dynamic stability is invariably performed using linearized system models. This is valid as the disturbances are assumed to be small. The analysis is performed using block diagrams (for the study of simple systems) or the system matrix and the computation of eigenvalues. The choice of control signals, the design procedures for the auxiliary controllers are some of the issues which are of importance. Both classical and modern control theoretic concepts can be used in the design of controllers.

It was first shown by Uhlmann [2], that an AC tie can be stabilized by power modulation in the parallel DC tie. However, this was not considered practical with mercury arc valves which suffer from arc backs. The power modulation was first tried at the Eel River DC link in Canada, where thyristor valves were used for the first time. Although this is an asynchronous link, the power modulation was shown to provide damping of oscillations in remote AC ties.

The power or current modulation has now become a standard feature in all modern DC links. With weak AC systems at the inverters, it may become necessary to modulate gamma (extinction angle) also to maintain the effectiveness of power modulation and help to improve voltage stability.

In this chapter, some aspects of the power (including gamma) modulation will be discussed. The analysis of voltage stability with weak AC systems will be presented. The modelling and analysis of transient stability is also discussed.

12.2 POWER SYSTEM STABILITY—BASIC CONCEPTS

Power system stability is an important issue for secure system operation. Many major blackouts are characterized by system instability. Historically, transient instability has been a major problem. However, with the introduction of new technologies (such as static exciters with high gain electronic AVRs) and increased operation under highly stressed conditions have given rise to new forms of instability. The problems of voltage instability and collapse, undamped interarea oscillations have become more significant than transient instability. The issue of frequency control is also of concern when synchronism is lost and islands are formed with the loss of major tie lines.

A formal definition of power system stability is given in reference [3] by the IEEE/CIGRE Joint Task Force on Stability Terms and Definitions. They define 'Power System Stability is the ability of an electric power system, for a given operating condition, to regain a state of operating equilibrium after being subjected to a physical disturbance, with most system variables bounded so that practically the entire system remains intact'.

It is to be noted that power systems are subjected to a wide range of disturbances, small and large. Small disturbances in the form of random load variations occur continually. Large disturbances such as short circuit on a transmission line followed by tripping of the line (by the action of the protection system consisting of relays and circuit breakers) are major ones. The loss of a large generator or connection of a large load are also large disturbances. It is to be noted that malfunction of the equipment, overloading or failure of the protection system can result in disconnection of generators and/ or transmission lines. Power Systems are designed to handle credible contingencies without loss of system integrity. The operating margins in generation and transmission should take care of the load demand even under (credible) contingencies or disturbances. Thus, a power system operating in a 'normal' state (where load demands are met and equipment constraints are not violated) continues to operate in the normal state although the operating equilibrium may change; if the system is stable. Note that, in the new equilibrium state, the network topology may be different than the initial topology, due to the tripping of some generators and/ or transmission lines.

Mathematically, the steady-state equilibrium in a power system can be represented as a point in a suitably defined state space (of large dimensions). A disturbance (small or large) can result in the equilibrium shifting to another point in the state space. The trajectory of the system from the initial to the final equilibrium point must be bounded in the state space to prevent break up of the system, if it is to remain stable.

Classification of power system stability

There are three broad categories of power system stability.

A. **Rotor Angle Stability** refers to the ability of synchronous machines in an interconnected power system to remain in synchronism after being subjected to a disturbance. It depends on the ability to maintain/restore equilibrium between the electromagnetic torque and mechanical torque of each synchronous machine in the system. Instability that may result, occurs in the form of loss of synchronism with other machines.

The change in electromagnetic torque of a synchronous machine following a perturbation can be resolved into two components:

1. Synchronizing torque component; in phase with rotor angle deviation
2. Damping torque component, in phase with the speed deviation.

(Note there is an implicit assumption here, that the speed deviation is (initially) small and sinusoidal. System stability requires existence of both components of torque. Inadequate synchronizing torque results in aperiodic or non-oscillatory instability, whereas lack of damping torque results in oscillatory instability. It is to be noted that these definitions apply strictly for small disturbances.

For convenience of analysis, the rotor angle stability is divided into two subcategories:

(*a*) Small-disturbance (or small-signal) stability

(*b*) Large-disturbance (or transient) stability.

Mathematically, the small-signal stability (in the past, also called as dynamic stability) is the stability of equilibrium point. Linearized models are used to study this form of stability. Eigenvalue analysis is the basic approach to investigate the stability of various swing modes and other important modes. Depending on the frequency, the swing modes are loosely divided into three groups (*i*) Intraplant (*ii*) local and (*iii*) interarea. The intraplant modes have the highest frequency and as the name suggests, only synchronous generators within a plant participate. On the other hand, a large group of generators participate in interarea modes which have the lowest frequency. While PSS on individual generators have been used extensively for damping swing modes, their tuning requirements which depend on the system operating conditions, are still viewed as a problem [4]. It is in this context, the modulation controls are invariably applied in modern power electronic controllers involving HVDC and FACTS.

Transient stability depends on both the initial operating state of the system and the disturbances. Mathematically, the system is transiently stable if the trajectory approaches a stable equilibrium point after the transients die down. All stable equilibrium points have a region (in state space) of attraction where all the trajectories starting from any point in the region gravitate towards the (stable) equilibrium point. Hence, it is adequate to check whether the system trajectory enters the region of attraction for the post-disturbance equilibrium point (assuming it is stable), to determine whether the system is transiently stable. The time frame of interest in transient stability studies is usually 3-5 seconds following the disturbance. It may extend to 10-20 seconds with dominant interarea swings.

B. **Voltage Stability** refers to the ability of the power system to maintain steady voltages at all buses in the system following a disturbance. It depends on the ability to maintain/ restore equilibrium between load demand and supply from the power system. Voltage instability results in voltage collapse at some load buses. The trajectory from the initial operating state can be either aperiodic or oscillatory. The factors that affect voltage stability are primarily (*a*) reactive power constraints imposed by generator field current limiters, increased reactive power consumption in stressed transmission networks and (*b*) dynamics of load restoration whether through inherent dynamics (in an induction motor) or controller dynamics (HVDC converters, tap-changing transformers, thermostatically or electronically controlled loads). Voltage stability is also divided into two categories:

(*a*) Small disturbance (or small signal) voltage stability

(*b*) Large disturbance voltage stability.

In addition, the classification is also based on (*i*) Short term and (*ii*) Long term based on the study period of interest. In the former case, the period is in the order of several seconds. In the latter case, it may extend to several minutes.

C. **Frequency Stability** refers to the ability of a power system to maintain steady frequency following a significant imbalance between generation and load. It depends on the ability to maintain/ restore equilibrium between generation and load. Instability occurs in the form of sustained frequency swings leading to tripping of generators and/ or loads. Modelling for frequency stability analysis involves processes or protection not modelled in conventional transient stability analysis – such as boiler dynamics, Volts/ Hertz protection tripping of generators. The study of frequency stability becomes essential following splitting of the system into islands. Modelling for frequency stability includes a wide spectrum of control actions from fast acting underfrequency load shedding and generator controls to slow response of prime mover controls and load voltage regulators. Thus frequency stability may be a short-term or a long-term phenomenon.

12.3 POWER MODULATION: BASIC PRINCIPLES

12.3.1 Synchronous Link

Consider a two area system shown in Fig. 12.1. Each area is represented by an equivalent generator. Assuming that the generators are represented by classical models of voltage sources behind transient reactances, the linearized system can be represented by the block diagram shown in Fig. 12.2. It is assumed that the output of the auxiliary controller is added to the reference value at the summing junction before the power controllers. The total electrical torque T_e is assumed to be same as the power output (P) which is the sum of P_{AC} and P_{DC}.

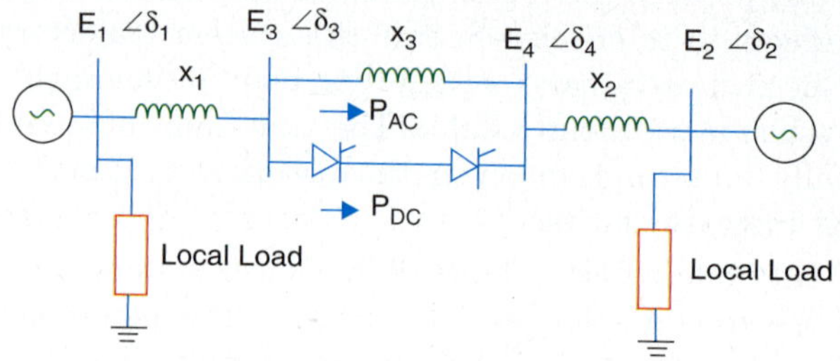

Fig. 12.1: A two area system

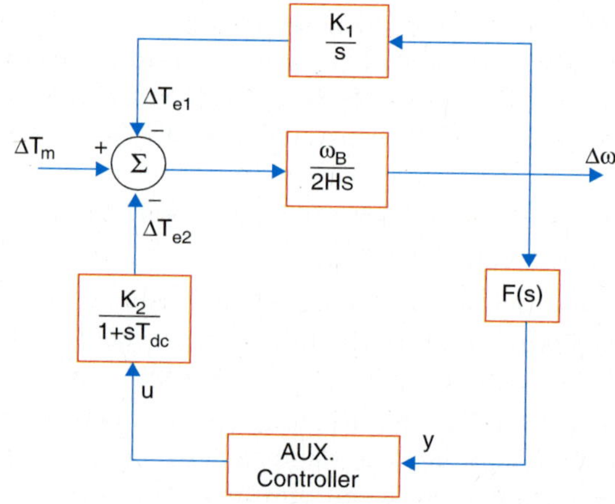

Fig. 12.2: System block diagram

The controller and system dynamics are represented by a single time constant (T_{dc}). It can be shown (with some approximations) that

$$\Delta P = \Delta T_e = \Delta P_{AC} + \Delta P_{DC} = K_1 \Delta\delta_{12} + K_2 \Delta P_{DC} \tag{12.1}$$

where

$$K_1 = (E_1 E_2 / X) \cos \delta_{12}^0, \; K_2 = X_3 / X \tag{12.1a}$$

$$X = X_1 + X_2 + X_3, \;\; H = H_1 H_2 / (H_1 + H_2), \Delta\omega = s\Delta\delta_{12} \tag{12.2}$$

The superscript 'o' indicates the value at the operating point, 's' is the Laplace variable.

The control signal y determines $F(s)$. If bus frequency deviation ($s\Delta\delta_{34}$) is used as the control signal, it can be shown that

$$F \approx K_2 \tag{12.3}$$

It is not feasible to use the signal $\Delta\omega$ directly. The difference in the frequency measured at the two converter buses is the most appropriate signal to be used. If the length of the DC line is appreciable, the control signal can be generated by differentiating the signal derived from the power in the AC line in parallel. This is because,

$$s \cdot \Delta\delta_{34} \approx s \cdot \Delta P_{AC} \tag{12.4}$$

Controller structure

If the time delay T_{dc} can be ignored, it is possible to improve the damping and synchronizing torques in the system by having controller structure as

$$G(s) = K_p + (K_I / s) \tag{12.5}$$

The first term in the above expression introduces damping torque, while the second term contributes to the synchronizing torque. The time delays in the feedback path due to the DC link controllers can be overcome by introducing phase lead compensators in series. If the objective is to only damp oscillations, K_I can be set equal to zero.

The controller gain K_p has to be chosen such that the damping of the low frequency rotor oscillations is maximum without destabilizing other modes. If an interarea mode (the dominant mode of oscillations in the tie lines between two areas) is to be damped, it may be required to filter the control signal such that other modes are eliminated before passing through the controller.

The analysis presented above is simplified to indicate the basic nature of the effects of DC power modulation on the system damping. In a large system, there are many generators and a two machine idealization may not be adequate in accurate prediction of the performance of the auxiliary controllers. However, even this simplified analysis shows that the effectiveness of the controller is reduced considerably if $X_3 << (X_1 + X_2 + X_3)$. The controller gain cannot be increased beyond a limit due to the DC link dynamics and reactive power constraints. The typical locus of the rotor mode with variation in the controller gain is shown in Fig. 12.3.

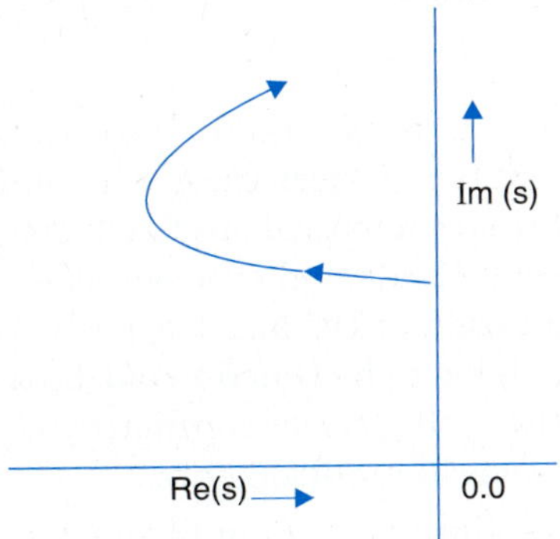

Fig. 12.3: Locus of rotor mode eigenvalue with variation in controller gain

12.3.2 Asynchronous Link

In the case of an asynchronous DC link, the control signal contains two components, derived from the frequency deviations at the two terminals of the link and can be combined in any

CHAPTER 12

proportion. The major objective here is to use the DC link for the frequency control of the two areas interconnected by the link. In addition, as the control is fast, it also helps to damp oscillations in AC ties which may be at some distance from the DC link.

Consider, for example, the system shown in Fig. 12.4. It is assumed that the areas 2 and 3 are very large compared to area 1. For the equivalent generator representing area 1, we have the following equation (considering classical model)

$$p^2(\Delta\delta_1) = -(\omega_b / 2H_1)(\Delta P_{AC} + \Delta P_{DC}) \tag{12.6}$$

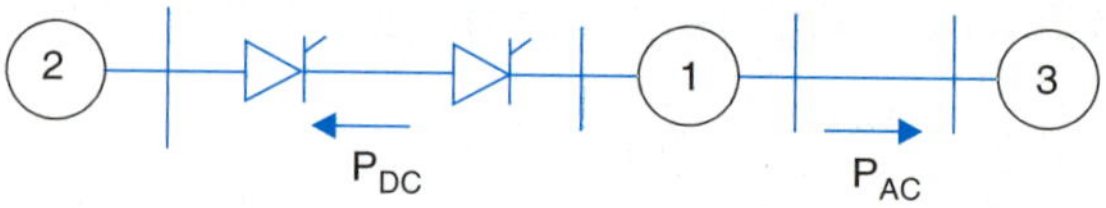

2 1 3 P_{DC} P_{AC}

Fig. 12.4: System with asynchronous DC link

where

$$p = d/dt$$

It can be assumed that

$$\Delta P_{AC} = K_1 \Delta\delta_1 \tag{12.7}$$

By proper design of the auxiliary controller, we can arrange to have

$$\Delta P_{DC} = K(p\Delta\delta_1) \tag{12.8}$$

In this case, it can be seen that the oscillations in $\Delta\delta_1$ and consequently in the power flow in the AC tie are damped. If the size of the area 3 is not very large, it would be necessary to obtain the feedback from the frequency deviation of area 3 also, to obtain satisfactory damping [5].

12.4 PRACTICAL CONSIDERATIONS IN THE APPLICATION OF POWER MODULATION CONTROLLERS

12.4.1 General

In the planning and design of the (auxiliary) stabilizing control for DC links, it is necessary to conduct studies and evaluate the system performance. The configuration and the parameters of the controller can be determined from analytical studies (using linearized system models). The performance of the controller is usually evaluated from simulation studies using transient stability programs incorporating representation of DC links. Finally, the controller has to be tested in the field and the parameters may have to be tuned to optimize the performance under various operating conditions. Usually, there are many conflicting considerations and a satisfactory solution has to be found to meet several requirements.

The power modulation can be classified as small signal or large signal, depending on the level of power changes ordered by the modulation controllers. The objective of the small signal modulation is to provide damping of one or more rotor modes of (low frequency) oscillations. The amplitude of the modulating signal need not be large to achieve the desired objective of damping. Even ±3% modulation of power at Pacific Intertie has resulted in damping oscillations at 1/3 Hz in the parallel AC link, thereby increasing the power transfer capability by more than 400 MW.

Small signal modulation does not impose additional requirements of telecommunications in the control of DC links.

Large signal modulation, on the other hand, is used to augment transient stability limit, thereby improving system security. Large changes in the power flow in the DC link are required to compensate for tripping of loads, generators or AC ties. While overload capability in DC links is useful, the limits imposed by ratings of the link usually do not curtail the benefits of power modulation. Hence, significant improvements can be expected out of the use of DC links in emergency control. The rapid response of DC link controllers makes it possible to arrest large fluctuations in the frequency by matching generation to the load in the area to which the DC link is connected.

Some of the aspects in the application of power modulation in DC links are discussed below.

12.4.2 Selection of Control Signals

It is to be noted that in large multimachine power systems, there are many modes of oscillations of generator rotors. The stabilizing control is usually employed to damp one or more of the predominant modes of oscillation. However, there is always a danger of undamping of some of the remaining modes. The controller itself can introduce new modes which may be undamped unless care is taken in the design. Examples of this phenomenon are given in the case studies of damping controller design at Nelson River [6] and Pacific Intertie [7]. Some of these problems can be solved by increasing the bandwidth of the controller but this can lead to the problems of noise interference.

It is desirable (from the cost and reliability considerations) to obtain the control signal locally. Some of the control signals that can be used are as follows:

1. Rotor frequency of adjacent generator
2. Frequency at the converter bus
3. Power or current in the adjacent, parallel AC tie

The telecommunication delay in the transmission of the control signal can be a problem and should be limited. An ideal control signal should have only the components of modes (of oscillation) which are to be damped. The extraneous components such as those at supply frequency and harmonics, subsynchronous frequency oscillations (SSO), local mode oscillations when the damping is desired for interarea modes, can pose problems. These are suppressed by suitably designed bandpass or notch filters. However, it is to be noted that the frequencies of some of the extraneous components do not remain constant and change with the loading conditions and system configuration.

It has been shown by Padiyar et al. [8] that the effectiveness of the DC power modulation (using bus frequency signals) is dependent on its location in the system. This aspect may be taken into account in the planning stage itself.

12.4.3 Controller Design

The classical control theory using frequency domain methods, has been traditionally employed for the design of controllers in power systems. For damping controllers, the objective is to provide adequate damping torque on the rotor at the dominant mode of oscillation of the rotor, without sacrificing synchronizing torque. The range of frequencies for the rotor mode normally lies in a narrow band. The optimization of the controller performance is usually done by trial and error.

The modern techniques of control synthesis using optimal control or pole assignment are more attractive and need to be considered when the system complexities have to be adequately represented. Here, the controller can be viewed as shown in Fig. 12.5 where the system description contains all the mathematical details necessary in determining the response. Kalman filter is used to generate the state estimates, which are fed to the regulator to generate the modulation signal [9].

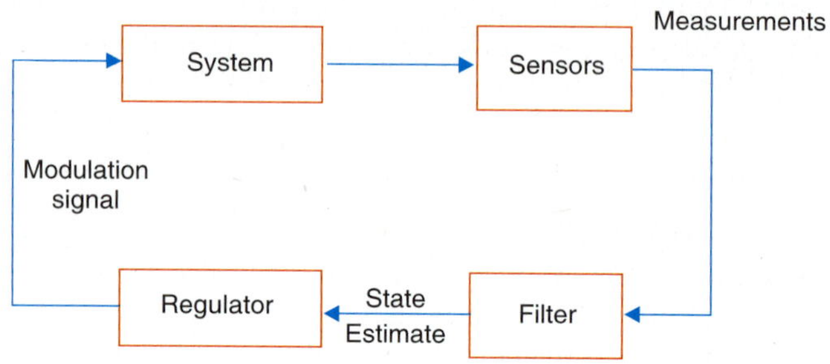

Fig. 12.5: Controller diagram

Adaptive or self tuning regulators can also be used for power modulation, such that the variations in the system configuration or loading conditions lead to automatic changes in the controller parameters. This is carried out by on-line identification of the system. The availability of microprocessors for converter control has made it feasible to apply these techniques.

12.4.4 Communication Requirements

The normal DC link control requires the power (or current) order from the master (usually the rectifier) station to be transmitted to the trailing station, for coordination of current controllers at the two ends. The current order at the rectifier station is higher than the current order at the inverter station by a fixed amount called the current margin.

If the power order changes slowly, the information can also be conveyed manually on telephone lines. While this is generally true, (as the power order is determined at the energy control centre and varies slowly with the change in the system loading), the presence of power or current modulation requires a faster, and reliable communication channel if the amplitude of modulation exceeds the current margin (around 0.1 p.u). However, because of low frequencies involved, the sampling rates required are such that a 1200 baud channel per pole is considered to be adequate. In the case of failure of the communication channel, large signal modulation has to be interrupted. However, small signal modulation of amplitude less than the current margin can be continued.

12.4.5 Examples of Power Modulation

Two examples are included here; one corresponds to an asynchronous DC link (Eel River, Canada), the second example is that of a DC tie in parallel with AC tie (synchronous link) at Pacific Intertie, USA.

Eel River

This is an asynchronous back to back DC tie of 320 MW rating, between Hydro Quebec (HQ) and New Brunswick Electric Power Commission (NBEPC) in Canada.

An external control system (ECS) is provided on each pole and provides modulation of the current reference. The block diagram of the ECS is shown in Fig. 12.6 [10]. The transfer functions of ECS were first developed using analog computer study and then verified based on the large-scale system simulation using a digital computer program. However, some of the parameters had to be readjusted at the time of installation of ECS.

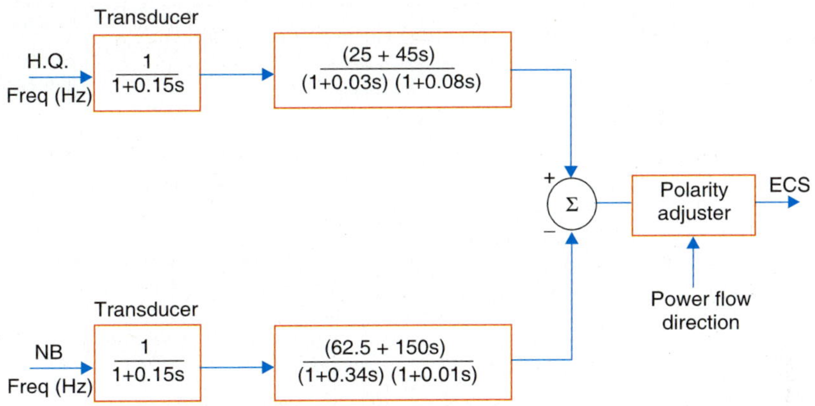

Fig. 12.6: Eel River controller (*Source*: Reference 10)

The tests on ECS have demonstrated the following:

1. The amplitude of power swings in major AC ties are reduced
2. The transient stability of associated AC system is improved
3. The frequency regulation is improved.

Pacific Intertie

The Pacific DC intertie was originally ± 400 kV (and now upgraded to ± 500kV) bipolar line, 850 miles long, connecting Pacfic Northwest to Los Angeles area. This DC tie is in parallel with two 500 kV AC ties of capacity 2000 MW. The presence of persistent power oscillations of 300 MW peak to peak, led to the development of small signal modulation of DC power [11].

Since the DC line is long, the time delay in transmitting the frequency signal from one end to the other end of the line is significant. Moreover, the frequency signal was found to be inherently a noisy signal containing strong local swing modes (1–2 Hz) between tightly coupled machines, while the mode of interest was of frequency 1/3 Hz. As a result, Bonneville Power Administration (BPA) has utilized the deviation of power in the AC tie as the control signal.

The block diagram of auxiliary control system is shown in Fig. 12.7. The power signal is filtered to remove 60 and 120 Hz components and the high frequency noise. The filtered output is differentiated and transmitted from John Day Substation located at the northern terminal of the AC intertie to Celilo converter station (operating as the rectifier) using wide band FM telemetry.

At Celilo, the signal is again filtered to remove the telemetry noise. Lead compensation is used to provide additional phase lead required (over and above that provided by the differentiator). The local mode frequencies (1 to 1.5 Hz) are removed with the notch filter. Finally a limiter is employed to limit the amplitude of the modulation signal applied to the current regulator. Initially the limits were put at ± 3% , but now the limits have been raised to ± 7% .

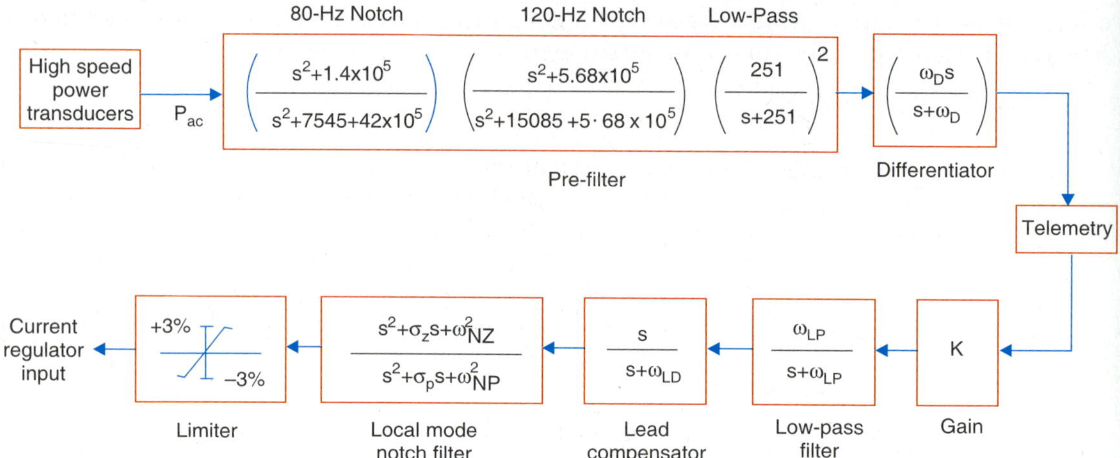

Fig. 12.7: DC Modulation at Pacific Intertie (*Source*: Reference 11)

The initial field tests of the control system showed that it produced considerable damping at 1/3 Hz. However, it was observed that the DC tie had power oscillations of 0.7 Hz of magnitude as high as 30 MW peak to peak. The presence of these oscillations was confirmed by a root locus study and was traced to a higher value of the gain. The modification of the controller parameters eliminated the problem.

12.4.6 Gamma or Reactive Power Modulation

With weak AC systems (to which DC links are connected), the effectiveness of power modulation reduces considerably. Consider, for example, modulation at the rectifier which results in increase of DC current. The voltage in the link is determined by the inverter control and the converter bus AC voltage. The increase in the DC current leads to increase in the reactive power requirements at the inverter station. For weak AC systems, this leads to the reduction of the converter bus voltage which then results in reducing the DC power below the level ordered by the modulation controller.

One solution to the problem is to introduce gamma modulation at the inverter such that the effectiveness of power modulation at the rectifier is maintained. The control signal is derived from the AC voltage (magnitude) at the converter bus. A typical block diagram of the gamma modulation controller is shown in Fig. 12.8. ltaipu scheme in Brazil uses gamma modulation with a control range of 13°.

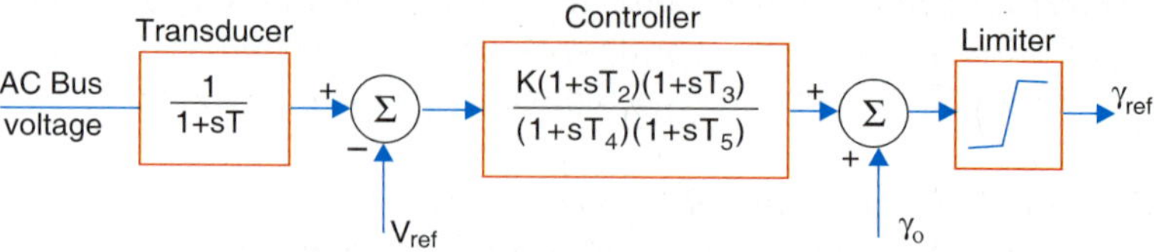

Fig. 12.8: Typical γ modulation controller

If a static var compensator (SVC) is provided at the converter bus, the modulation of reactive power output of SVC can also result in improved system response. A proper coordination is necessary in the operation of the two controllers for getting the best performance.

Gamma modulation may require the nominal value of the extinction angle to be increased to accommodate the excursions of gamma in both directions. This implies higher costs of converter valves, increased losses and reactive power requirements in steady state.

12.4.7 Power Modulation in MTDC Systems

The simulator and analytical studies carried out by many researchers have shown that power modulation, with or without reactive power modulation is beneficial in MTDC systems also (14-19). The controller design is more complicated because of the need for coordination of the controllers associated with each terminal. The need for simplicity and avoidance of excessive communication requirements may result in the use of decentralized control strategy and innovative controller design. There is a lot more work to be done in this area as the state of MTDC system operation is still in the stage of infancy.

12.5 VOLTAGE STABILITY IN AC/DC SYSTEMS

12.5.1 General

The incidence of heavily loaded long distance transmission lines coupled with the problems of reactive support can lead to problems of voltage instability – the system breakdown caused by slow voltage collapse. This problem has occurred in AC systems; the major power failure in France in 1978, being a well known incident.

The load characteristics also influence the voltage stability. The load characteristic, where the reactive power requirement reduces with the increase in the voltage, accentuates the stability problem.

The DC link may be viewed as dynamic loads at the converter buses. The converter operation at the constant value of the control angle (usually extinction angle at the inverter), combined with low short circuit ratio at the converter bus, can lead to the problem of voltage instability. Before investigating the problem of voltage stability in AC/DC systems, the explanation of the phenomenon is given below considering the example of a simple two bus system.

12.5.2 Explanation of Voltage Stability Problem

Consider a single machine system feeding power to a large load centre as shown in Fig. 12.9. As mentioned earlier, the voltage stability is related to the steady-state stability, which can be analyzed, as a first approximation, using the algebraic network equations.

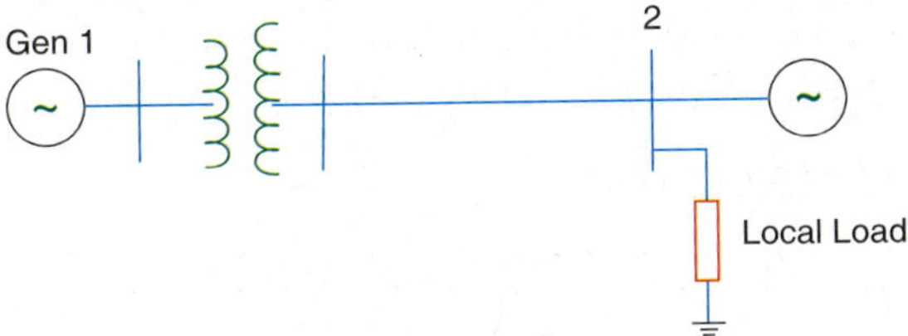

Fig. 12.9: A single machine system

It can be assumed that the generator is represented by a voltage source behind an equivalent reactance. The system can be represented by the simple equivalent circuit shown in Fig. 12.10, where bus 1 is the internal bus of the generator. If bus 2 is assumed to be an infinite bus, its voltage is constant both in magnitude and angle. In this case, the criterion for steady state stability is given by

$$K = (dP/d\delta) > 0 \tag{12.9}$$

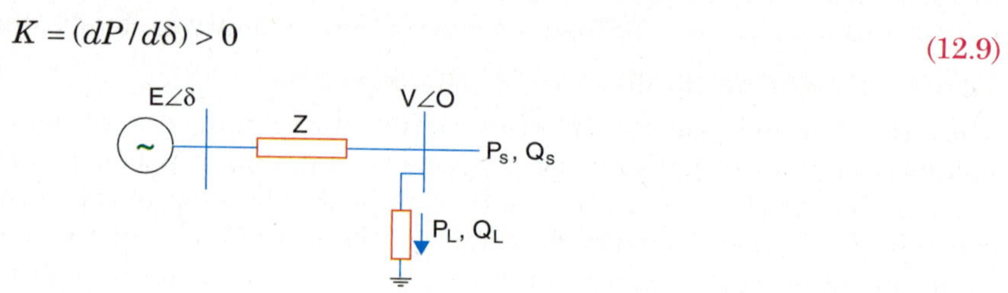

Fig. 12.10: A two bus system

The characteristics of the load connected to bus 2 has no effect on the stability.

In most of the cases, the assumption of an infinite bus is unrealistic. If the capacity of the generators connected to bus 2 is much larger than the capacity of generator 1, it is realistic to consider the frequency at bus 2 as constant. However, if the reactive power available at bus 2 is limited and has slow control (with manually switched capacitors/reactors) it is not realistic to assume the voltage magnitude as constant. In such cases, the bus 2 may be called as semi-infinite bus (SIB), where reactive power and angle δ are specified.

If the power output of the generator is varied slowly, allowing enough time for the control of voltage at bus 2 using manually switched capacitors/reactors, the maximum power that can be transmitted on the line is still given by Eq.(12.9). However, the system will lose stability prior to the power level at which $K = 0$. This can be shown as follows:

If the voltage at bus 2 is variable due to reactive power constraints, the criterion for stability now becomes

$$K_C = \frac{\Delta P}{\Delta \delta} = [(\partial P/\partial \delta) + (\partial P/\partial V)(\partial V/\partial \delta)] > 0 \tag{12.10}$$

The Jacobian of the power flow equations at bus 2, can be expressed as

$$\begin{bmatrix} J_1 & J_2 \\ J_3 & J_4 \end{bmatrix} \begin{bmatrix} \Delta\delta \\ \Delta V \end{bmatrix} = \begin{bmatrix} \Delta P \\ \Delta Q \end{bmatrix} \tag{12.11}$$

where

$$J_1 = (\partial P/\partial \delta), J_2 = (\partial P/\partial V), J_3 = (\partial Q/\partial \delta), J_4 = (\partial Q/\partial V)$$

For a lossless line, it can be shown that

$$P = \frac{EV \sin\delta}{X}, Q = \frac{EV\cos\delta - V^2}{X}$$

where X is the net reactance of the line.

J_1 and J_2 are normally positive while J_3 and J_4 are negative. Since

$$Q = Q_L - Q_s \tag{12.12}$$

where Q_s is the reactive power supplied by the local sources at bus 2, and if Q_s remains constant, we can obtain

$$\Delta P = (J_1 - J_2 J_4'^{-1} J_3)\Delta\delta \tag{12.13}$$

where

$$J_4' = (\partial Q/\partial V) - (\partial Q_L/\partial V)$$

From Eqs.(12.10) and (12.13), the steady-state stability criterion can be expressed as

$$K_C = \left(J_1 - J_2 J_4'^{-1} J_3\right) > 0 \tag{12.14}$$

As J_3 and J_4 are both negative,

$$K_C < K \tag{12.15}$$

as, $K = (\partial P/\partial\delta) = J_1$

If the load characteristics are such that $(\partial Q_L/\partial V) > 0$, the stability limit is increased as K_C will be increased. On the other hand, if $(\partial Q_L/\partial V) < 0$, the stability limit will decrease.

The voltage stability criterion is usually given as [22, 33]

$$VSF = (\Delta V/\Delta Q_s)\,|_{P=const} > 0 \tag{12.16}$$

when VSF is termed as voltage stability (or sensitivity) factor.

It can be shown that Eq.(12.16) is equivalent to Eq. (12.14). This follows from the fact that

$$VSF = (J_4' - J_3 J_1^{-1} J_2)^{-1} \tag{12.17}$$

The inequality (12.16) can be derived from Eq. (12.14) by multiplying the latter by $(J_4' J_1^{-1})$.

The equivalence between Eqs. (12.14) and (12.16) shows clearly, that the voltage instability occurrence coincides with the steady state instability under the reactive power constraints. The condition VSF < 0 implies that with the increase in the reactive power supply, the bus voltage reduces. This is contrary to what is normally expected. As all the reactive power controllers work on the principle of boosting the reactive power when the voltage drops, it is obvious that even with slow acting automatic voltage regulators, the bus voltage will collapse.

To overcome the voltage instability problem, fast acting reactive power controllers with adequate margins have to be provided such that the voltage support is available at all times (with varying load conditions of the system).

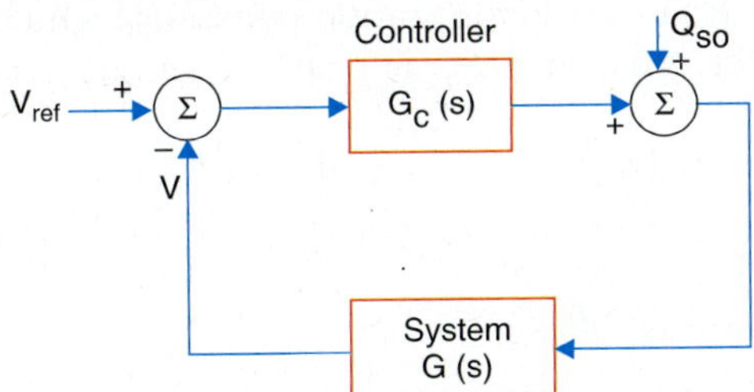

Fig. 12.11: Reactive power controller-block diagram

CHAPTER 12

In designing the reactive power controller, the system characteristics will influence the choice of the gain. This is illustrated from the block diagram shown in Fig. 12.11. As a first approximation, the system dynamics can be neglected and the transfer function becomes

$$G(s) = VSF = dV/dQ_s \tag{12.18}$$

The magnitude and the sign of VSF will affect the controller design. Thus, adaptive control in voltage regulation can overcome the problems of voltage instability.

12.5.3 Analysis of Voltage Stability in Asynchronous AC-DC System

Consider a two terminal HVDC link which interconnects two AC systems. The equivalent circuit at one of the terminals is shown in Fig. 12.12. The AC system is represented by Thevenin equivalent (at fundamental frequency) viewed from the converter bus. Although the converter shown appears as a rectifier, it can also represent an inverter if $I_d < 0$ and $P_d < 0$. Although, it is common to consider inverter terminal as a Voltage Setting Terminal (VST), it is also possible to have a rectifier station operate as the VST. Since On-Load Tap-Changers (OLTC) of converter transformers are very slow in action, we assume that the tap ratio 'a' is constant during a transient. For convenience, we assume VST is numbered 2 and non-VST is numbered 1.

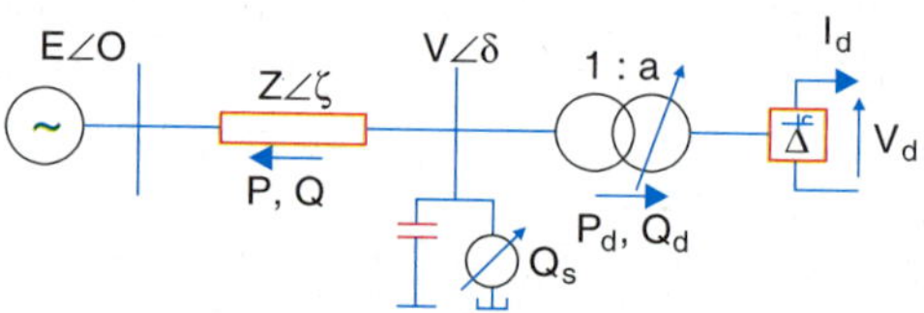

Fig. 12.12: Converter terminal connected to an isolated AC system

Converter equations

At converter terminal $j (j = 1 \text{ or } 2)$, the following equations are applicable

$$V_{dj} = K_j a_j V_j \cos\theta_j - R_{cj} I_{dj} \tag{12.19}$$

$$P_{dj} = V_{dj} I_{dj} \tag{12.20}$$

$$Q_{dj} = |P_{dj}| \tan\phi_j \tag{12.21}$$

$$\cos\phi_j = \frac{V_{dj}}{k_j a_j V_j} \tag{12.22}$$

All the variables are expressed in per unit.

The converter angle can be either the delay angle (α) for the rectifier or the extinction angle (γ) for the inverter terminal. R_{cj} is the commutation resistance which is positive for the rectifier and negative for the inverter. Similarly I_{dj} is positive for the rectifier and negative for the inverter.

DC network equations

For the two terminal system,

$$V_{d1} = V_{d2} + R_d I_{d1} \tag{12.23}$$

$$I_{d2} = -I_{d1} \tag{12.24}$$

R_d is the combined series resistance of the DC line and smoothing reactors.

Power flow equations at the converter bus

The power balance equations at terminal (j = 1, 2) are

$$P_j + P_{dj} = 0 \tag{12.25}$$

$$Q_j + Q_{dj} - B_{cj} V_j^2 = Q_{cj} \tag{12.26}$$

Note that Q_{cj} is the controllable reactive power source at terminal j

The expressions for P_j and Q_j are given below

$$P_j = \frac{V_j^2 \cos \zeta_j}{Z_j} - \frac{V_j E_j \cos(\delta_j + \zeta_j)}{Z_j} \tag{12.27}$$

$$Q_j = \frac{V_j^2 \sin \zeta_j}{Z_j} - \frac{V_j E_j \sin(\delta_j + \zeta_j)}{Z_j} \tag{12.28}$$

where $\tan \zeta_j = \dfrac{X_j}{R_j}$. $Z_j = \sqrt{R_j^2 + X_j^2}$

Controller equations

As the transformer tap 'a' is assumed to be fixed, the only control variable at each terminal is θ. The control strategy is to maintain one of the scheduled variables at a preset value.

At terminal 1 (Non-VST),

$$W_1 = P_{d1} \quad \text{or} \quad I_{d1} = W_1^{spec} \tag{12.29}$$

where W_1 is the specified variable at terminal 1. At terminal 2, the specified variable W_2 can be one of the following variables

(i) control angle (θ_2)
(ii) reactive power (Q_{d2})
(iii) reactive current (I_{q2})
(iv) DC voltage (V_{d2})
(v) converter bus (AC) voltage (V_2)

Note that the reactive current I_q is defined as

$$I_q = kaI_d \sin \phi$$

In general, at VST (terminal 2) we have

$$W_2 = W_2^{spec} \tag{12.30}$$

Linearized equations

Since, $P_{d1} = V_{d1} I_{d1}$ and $P_{d2} = V_{d2} I_{d2}$, we get

$$\Delta P_{d1} = I_{d1}\Delta V_{d1} + V_{d1}\Delta I_{d1}$$

$$\Delta P_{d2} = I_{d2}\Delta V_{d2} + V_{d2}\Delta I_{d2}$$

If $W_1 = I_{d1}$, then $\Delta I_{d1} = 0$. If $W_1 = P_{d1}$, then

$$\Delta I_{d1} = \frac{-I_{d1}}{V_{d1}}\Delta V_{d1}$$

Since, $I_{d2} = -I_{d1}$, $\Delta I_{d2} = -\Delta I_{d1}$

Linearizing Eq.(12.23), we get

$$\Delta V_{d1} = \Delta V_{d2} + R_d \Delta I_{d1} \tag{12.31}$$

Combining the above equations, we get

$$\Delta I_d = [y]\Delta V_{d2}, \ \Delta V_d = [z]\Delta V_{d2} \tag{12.32}$$

where ΔI_d, ΔV_d, $[y]$ and $[z]$ are column vectors of dimension two. $[y]$ is a null vector if $W_1 = I_{d1}$. If $W_1 = P_{d1}$, then $y_1 = \frac{-I_{d1}}{V_{d1}}/(1+\frac{R_d I_{d1}}{V_{d1}})$ and $y_2 = -y_1$, where y_1 and y_2 are elements of the vector $[y]$.

Equation (12.31) can be expressed as

$$\Delta V_{d1} = (1 + R_d y_1)\Delta V_{d2} = z_1 \Delta V_{d2}$$

where z_1, is the first element of vector $[z]$. The second element (z_2) is obviously equal to 1.

Linearizing Eqs. (12.25) to (12.28), we get

$$\begin{bmatrix} J1_j & J2_j \\ J3_j & J4_j \end{bmatrix}\begin{bmatrix} \Delta\delta_j \\ \Delta V_j \end{bmatrix} = \begin{bmatrix} -\Delta P_{dj} \\ -\Delta Q_{dj} \end{bmatrix} + \begin{bmatrix} 0 \\ 1 \end{bmatrix}\Delta Q_{cj} \tag{12.33}$$

where $J1_j = \frac{\partial P_j}{\partial \delta_j}, J2_j = \frac{\partial P_j}{\partial V_j}, J3_j = \frac{\partial Q_j}{\partial \delta_j}, J4_j = \frac{\partial Q_j}{\partial V_j}$, $j = 1, 2$

Solving for ΔV_j from the above equation, we have

$$\Delta V_j = d_{pj}\Delta P_{dj} + d_{qj}\Delta Q_{dj} + d_{cj}\Delta Q_{cj} \tag{12.34}$$

where, $d_{pj} = \frac{J3_j}{\det}$, $d_{qj} = \frac{-J1_j}{\det}$, $d_{cj} = \frac{-J1_j}{\det}$, $det = J1_j J4_j - J2_j J3_j$

Combining the equations for ΔV_1 and ΔV_2, we get

$$\Delta V = [D_p]\Delta P_d + [D_q]\Delta Q_d + [D_c]\Delta Q_c \tag{12.35}$$

From Eq. (12.32), we can get

$$\Delta P_d = [F_{pd}]\Delta V_{d2} \tag{12.36}$$

Linearizing Eqs. (12.21) and (12.22), we get

$$\Delta Q_d = [F_{qd}]\Delta V_{d2} + [F_{qv}]\Delta V \tag{12.37}$$

where,

$$[F_{pd}] = \text{Diag}[V_{d1}, V_{d2}]y + \text{Diag}[I_{d1}, I_{d2}]z$$

$$[F_{qd}] = [A_1]F_{pd} + [A_2]z$$

$$[A_1] = \text{Diag}[\pm\tan\phi_1, \pm\tan\phi_2]$$

$$[A_2] = \text{Diag}\left[-\frac{|P_{d1}|\sec^2\phi_1}{k_1 a_1 V_1 \sin\phi_1}, -\frac{|P_{d2}|\sec^2\phi_2}{k_2 a_2 V_2 \sin\phi_2}\right]$$

$$\left[F_{qv}\right] = \text{Diag}\left[-\frac{|P_{d1}|V_{d1}\sec^2\phi_1}{k_1 a_1 V_1^2 \sin\phi_1}, \frac{|P_{d2}|V_{d2}\sec^2\phi_2}{k_2 a_2 V_2^2 \sin\phi_2}\right]$$

The sign before $\tan\phi$ is positive for the rectifier terminal, negative for the inverter terminal.

Substituting Eqs. (12.36) and (12.37) in (12.35), we get,

$$\Delta V = [g]\Delta V_{d2} + [H]\Delta Q_c \tag{12.38}$$

where,

$$[g] = [I_2 - D_q F_{qv}]^{-1}\{[D_p]F_{pd} + [D_q]F_{qd}\}$$

$$[H] = [I_2 - D_q F_{qv}]^{-1}[D_c]$$

I_2 is the identity matrix of order 2.

Linearizing Eq.(12.30) we can obtain the following relationship

$$\Delta W_2 = c_1\Delta V_2 + c_2\Delta V_{d2} = 0 \tag{12.39}$$

(*i*) For θ control,

$$c_1 = 1, \qquad c_2 = \frac{-(1 + R_{c2}y_2)}{k_2 a_2 \cos\theta_2}$$

(*ii*) For Q_d control,

$$c_1 = 1, \qquad c_2 = \frac{V_2(k_2 a_2 V_2 F_{pd2}\sin\phi_2\tan\phi_2 - P_{d2}\sec^2\phi_2)}{V_{d2}P_{d2}\sec^2\phi_2}$$

(*iii*) For I_q control,

$$c_1 = 1, \qquad c_2 = \frac{k_2 a_2 V_2^2 \sin\phi_2\tan\phi_2 - V_2 I_{d2}}{V_{d2}I_{d2}}$$

(*iv*) For V_d control,

$$c_1 = 0, \qquad c_2 = 1$$

(v) For V control,

$$c_1 = 1, \; c_2 = 0$$

From Eq.(12.38) we have

$$\Delta V_2 = g_2 \Delta V_{d2} + H(2,2)\Delta Q_{c2} \tag{12.40}$$

(It is to be noted that [H] is a diagonal matrix)

Substituting Eq. (12.40) in (12.39) we can solve for ΔV_{d2} as

$$\Delta V_{d2} = \lambda_2 \Delta Q_{c2} = \lambda_2 e_2 \Delta Q_c \tag{12.41}$$

where $e_2 = [0 \;\; 1]$, $\Delta Q_c^t = [\Delta Q_{c1} \;\; \Delta Q_{c2}]$ and

$$\lambda_2 = \frac{-c_1 H(2,2)}{(c_2 + C_1 g_2)}$$

Substituting Eq. (12.41) in (12.38) we finally obtain

$$\Delta V = [\text{VSF}]\Delta Q_C \tag{12.42}$$

where $[\text{VSF}] = [H + \lambda_2 g e_2]$

Remarks

1. Note that [VSF] is a 2×2 matrix with 3 non-zero elements. The diagonal elements are self VSF at terminals 1 and 2, whereas the off diagonal element VSF (1, 2) is the mutual VSF $(\Delta V_1 / \Delta Q_{c2})$.
2. For DC voltage control at VST, $\lambda_2 = 0$ and hence the [VSF] matrix is diagonal, the elements of which are only functions of the SCR at the terminals. They are independent of the control strategy at the non-VST.
3. For control strategies other than constant DC voltage control, self VSF at VST is a function of the controls strategy at both terminals in addition to the SCR at VST.
4. The self VSF at non-VST (terminal 1) is a function of only SCR at terminal 1 and is independent of the control strategy at both terminals.
5. The mutual VSF ($\Delta V_1/\Delta Q_{c2}$) is a function of SCR at both terminals in addition to the control strategies at VST and non-VST.
6. The derivation given here can be generalized to a multiterminal asynchronous DC systems with 'n' terminals [35]. In this case, [VSF] is a $n \times n$ matrix with $2n$–1 non-zero elements. Assuming VST is at the last terminal 'n', the non-zero off diagonal elements of the [VSF] matrix are only in the last column $\left[\Delta V_j/\Delta Q_{cn}, j = 1,2...(n-1)\right]$. Here also, with DC voltage control at VST, the matrix is diagonal.

It can be shown that the self VSF at VST, for the case when the AC network is lossless (the impedance angle ζ is 90°) and I_d specified at the non-VST, is given by [35].

$$\frac{\Delta V}{\Delta Q_c} = \frac{Dx}{Ax^2 + Bx + C} \tag{12.43}$$

where $A = \frac{P^2}{V}, B = (K_2 - \frac{2Q_d}{V})P,$

$$C = -P\left(\frac{P}{V_n} + K_1\right), \qquad D = P^2$$

$$K_1 = \frac{-F_{pd}c_1}{c_2}, \quad c_2 \neq 0, \quad K_2 = F_{qv} - \frac{F_{qd}c_1}{c_2}, \quad c_2 \neq 0$$

$$K_1 = 0, \quad c_2 = 0, \quad K_2 = F_{qv}, \quad c_2 = 0$$

Note that subscript indicating the terminal number is dropped for convenience. The variable x is positive and is defined by

$$x = \pm(\tan\delta)^{-1} \tag{12.44}$$

The positive sign applies if VST is inverter and negatives sign applies if VST is rectifier. For a specified power and reactive power at the terminal, the angle δ is a function of the SCR as given below.

$$\text{SCR} = \frac{1}{X} = \frac{P}{VE\sin\delta}$$

Note that V is also a function of δ for a specified power and reactive power at the terminal.

The results of the case study presented in [35] indicate that

1. The voltage instability is more severe at the voltage setting terminal (VST) compared with other terminals irrespective of whether it is a rectifier or inverter. The voltage instability for a specified AC system strength is a function of the operating power level. On the other hand, for a specified power level, it is a function of the AC system strength. Generally, the AC system strength is higher at rectifier terminal compared to the inverter terminal. In such cases, it may be desirable to operate the rectifier as VST.
2. Constant control angle (θ) results in voltage instability at the highest value of the critical SCR (the value of SCR at which the VSF changes sign from positive to negative). The constant DC voltage control gives the best results (with lowest value of the critical SCR) followed by the constant reactive current control. Another advantage of constant DC voltage control is that mutual VSF terms are zero and hence the voltage instability at terminal other than VST is unaffected by the AC system strength at VST.
3. Although, constant AC voltage control at the VST improves voltage stability at that terminal, it does not help in overcoming voltage instability at other terminals.
4. The constant current control at non-VST, results in a lower value of critical SCR compared to the constant power control. It is to be noted that power control is made deliberately slow compared to the current control in practical HVDC installations.

Since voltage instability is strictly a dynamic phenomenon, it is necessary to apply dynamic analysis for accurate prediction of voltage stability problem. The analysis presented in [36] shows that the results predicted by algebraic methods such as computation of VSF are accurate only if the instability is due to a saddle-node bifurcation (when the critical eigenvalue that is responsible for voltage instability) is real. However, there are many cases with inverter as VST and the instability is caused by Hopf bifurcation (the critical eigenvalue is complex). This results in undamped voltage oscillations that are responsible for instability. The frequency of oscillations is typically less than 15 Hz. The application of VSF analysis gives inaccurate results in cases involving Hopf bifurcations.

Voltage oscillations induced by control interactions

Voltage oscillations at the converter bus may be induced by the interactions between the current controller and VDCOL [34]. However, this is not to be confused with voltage instability problem.

If the current is raised by giving a step order for the current controller (at the rectifier), the AC voltages at both converter stations fall as the DC current increases. The drop in the AC voltage results in a drop in the DC voltage. If this level is such that VDCOL break point is reached, the consequence is the decrease in the current order at both stations. This, in turn, results in a drastic drop in the DC power and a very fast rise in the AC voltage. With the rise in the voltage, VDCOL ramps up the DC current to its ordered value, causing again a voltage dip and restarting the same process. The inverter voltage dips are related to the fast advances of the extinction angle.

Actually, a limit cycle is established by the hunting between the current controller and the VDCOL adjustments. This problem can be cured by simple modifications in the set points and time constants of the VDCOL function.

12.6 SIMULATION OF ELECTROMECHANICAL TRANSIENTS

This is also termed as the transient stability analysis that is used to investigate the stability of a power system under sudden and large disturbances such as faults (single or three phase) followed by their clearing under the action of protective relays. There are several digital computer programs which are available for the study of large scale systems consisting of hundreds of generators, thousands of buses and transmission lines. The present day programs have many features which lend themselves for the simulation of low frequency transients in the systems. Some of these features are as follows:

1. Detailed generator models including accurate representation of amortisseur circuits, excitation and turbine-governor systems.
2. Voltage and frequency dependent nonlinear load models.
3. Representation of protection system to simulate the switching of lines under the action of the protective relays.

Dommel and Sato [37] describe a methodology for the solution of system equations (differential equations for the dynamic system including generator and controllers and algebraic equations describing the network). A major assumption in the transient stability analysis is to neglect line transients and consider only the fundamental frequency behaviour of the AC network. This assumption is valid for the simulation of low frequency (below 5 Hz) transients, although not applicable for the simulation of subsynchronous frequency transients.

It is well recognized that a properly controlled DC link can enhance the stability characteristic of the AC system in which it is embedded. Hence it is necessary to model the DC link in the stability analysis. The transient stability programs written for the AC systems can be modified to include DC link representation.

12.6.1 Converter Model

There are two basic types of converter models—

Simplified model

Here the valve switching is neglected and the converter is represented by average DC voltage equation

$$V_d = kaV\cos\theta - R_c I_d$$

where θ is either delay angle (α) for a rectifier or extinction angle (γ) for the inverter. V is the converter bus AC voltage. This model is similar to that used in power flow analysis. However, there are some differences. The transformer tap is assumed to be constant (at the value prior to the initiation of the disturbance) as the tap changer dynamics is very slow.

The power or reactive power is not specified as the power controller is not fast acting. Instead, the dynamics of power (and auxiliary) controller including VDCOL, are represented. The current order (reference value) is obtained as the output of the power controller.

During a transient, it is possible to reverse the power flow (in a two terminal DC link) under the action of an emergency controller. The converter may be blocked following fault and unblocked after a time delay. Mode shifts (shifting of current control from rectifier to inverter and back) can also occur and the converter representation must include these features.

The drawbacks of using a simplified converter model during dynamic simulation are as follows:

1. Inaccuracy caused by the inapplicability of the steady-state converter equation during severe disturbances such as an AC fault close to the converter terminals.
2. Inability to handle the phenomenon of commutation failures and the recovery of the inverter from these failures.
3. Inability to predict correctly the converter behaviour during unsymmetrical faults. Although the positive sequence voltages in the AC network can be correctly calculated by the use of equivalent impedances inserted at appropriate places, the effect of negative sequence components on the converter behaviour can be substantial. The simplified converter model does not consider these effects.

Detailed model

Here, the valve switching is represented and the model is free from the drawbacks associated with the simplified model. However, the transients simulation of converters now require integration step sizes as small as 50–100 μs; whereas the transient stability simulation usually employs step sizes not smaller than 10ms.

This implies that heavy computational effort is involved in using the detailed converter model. Furthermore, it becomes necessary to extract the fundamental frequency components of the current and voltage at the converter terminals, as the harmonic components cannot be handled by the standard AC system transient stability programs. (Also it is not meaningful to include the effect of harmonics in the study which is mainly concerned with low frequency transients). For this reason, Heffernan et al. [38] suggest the case of the detailed model only for a short duration (say 0.2 sec) immediately after the occurrence of the disturbance. As soon as the control action in the DC link settles down to a steady-state and the harmonic distortion in the converter bus voltage disappear, the detailed converter modeling can be replaced by the simplified model.

The use of detailed converter model requires the representation of AC network transients. To limit the computational burden, it is obvious that some sort of AC system equivalent is required in transient converter simulation. Heffernan et al. [38] recommend the use of time-varying Thevenin's equivalent.

In weak AC systems, the harmonic behaviour of AC systems needs to be accurately modeled to evaluate correctly the harmonic interaction between AC-DC systems. In such cases, it is advantageous to shift the interface buses further away from the converter buses [39, 40] although the transient converter simulation will involve more computational effort.

12.6.2 Converter Controller Models

Response type model

In this case, the dynamics of the current/extinction angle and firing controllers are neglected and only the steady-state controller ($V_d - I_d$) characteristics are represented. This type of controller model is advantageous in the planning studies where the design aspects of the controller can be neglected.

The main feature of this type of controller model is that the configuration and the parameters of the controller are not required in the study. It is the manufacturer who is charged with the responsibility of designing suitable controllers which achieve the desired objective of maintaining the current or extinction angle at the reference values determined by the higher level controllers. The assumption of neglecting the converter control dynamics is justified as the responses of these controllers are generally very fast.

Detailed converter controller representation

This is required with transient simulation of converters using the detailed model. However, the detailed controller models are optional when simplified converter models are used.

A major problem in the use of detailed converter controller models is the lack of standard functional controller blocks unlike in the case of excitation or turbine-governor systems. For an existing DC installation, the accurate representation of current controllers would involve the analysis of actual control circuitry and the establishment of a dynamic equivalent with a frequency response which matches closely the actual controls in the control band of 0 to 20 Hz.

The detailed controller models are required for (*i*) evaluating the response of the DC system under abnormal operating conditions (such as operation under low short circuit ratios) and (*ii*) study of converter response to DC system faults.

12.6.3 DC Network Models

Resistive network

The use of response type controller models imply that the DC current is not obtained as a state variable belonging to the DC network. Thus, it is possible to represent the DC network as a resistive network (a single resistance in the case of a two terminal DC link) ignoring energy storage elements. This approach is valid when the DC lines are short or for back to back HVDC links and when the smoothing inductors are moderate in size.

For longer lines, this approach may still be used. However, the inductive voltage drops in the line may have to be considered for correct evaluation of reactive power consumption of the converters.

Mode shift logic

The network solution assumes that one of the terminal operates at constant (usually minimum) delay (or extinction) angle and the rest of the terminals operate at constant currents. It is necessary to calculate the angles at current controlling terminals and if it turns out that these angles are outside the limits specified, a mode shift is indicated. The terminal having maximum violation in cosθ (where θ is delay or extinction angle) will then operate at the limit value of θ and the previously angle controlled terminal takes over current control. The reference value of this controller is determined by the current margin.

For a two terminal DC link, the mode shift logic is as follows:

Compute

$$I_d^M = \frac{V_{dor}\cos\alpha_{\min} - V_{doi}\cos\gamma_{\min}}{R_{cr} + R_d - R_{ci}} \tag{12.45}$$

If,

$$I_d^M > I_{d1}, \quad I_d = I_{d1} \tag{12.46}$$

$$I_d^M < I_{d2}, \quad I_d = I_{d2} \tag{12.47}$$

$$I_{d2} \le I_d^M \le I_{d1}, \quad I_d = I_d^M \tag{12.48}$$

where I_{d1} is the current reference at the rectifier and I_{d2} is the current reference at the inverter ($I_{d1} = I_{d2} + I_m$, where I_m is the current margin).

The condition (12.46) corresponds to the mode of current controller (CC) at the rectifier, condition (12.47) corresponds to CC at the inverter and the condition (12.48) corresponds to the rare mode of minimum angles at both rectifier and the inverter.

Transfer function representation

For a two terminal DC link with the response type controller model, an alternative representation of the DC network is to use a transfer function instead of a resistance (see Fig. 12.13). In this case, the time constant T_{dc} represents the delay in establishing the DC current after a step change in the order is given.

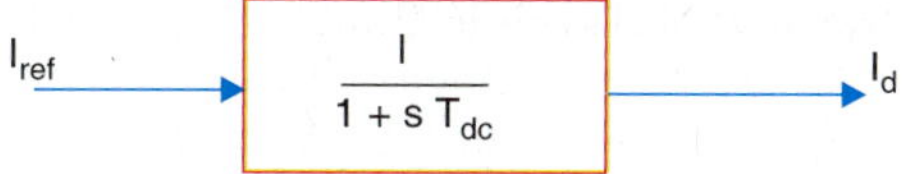

Fig. 12.13: Transfer function

It has been observed that this representation is more accurate than the earlier network model and is an approximation to the representation of network or line transients that will be considered next.

Dynamic representation using equivalent circuits

As the frequency bandwidth of response considered in the transient stability studies is modest, it is adequate to represent the DC network by simple equivalent circuits of the type shown in Fig. 12.14. Even here, the shunt branches may be neglected. This type of network representation is to be used when detailed controller representation is employed.

Fig. 12.14: Equivalent circuit

12.6.4 AC/DC System Interface: Simplified Converter Model

In all the cases, the AC network is represented by an admittance or impedance matrix defined at the fundamental frequency. The DC converter can be viewed as controllable current source whose magnitude is dependent on the DC current and the power factor dependent on the delay (or extinction) angle. Alternatively, the DC link can be viewed as dynamic loads at the converter buses. Figure 12.15 shows the AC network with current drawn at a converter bus. The current magnitude, I_j; and the phase η_j are defined by the following equations.

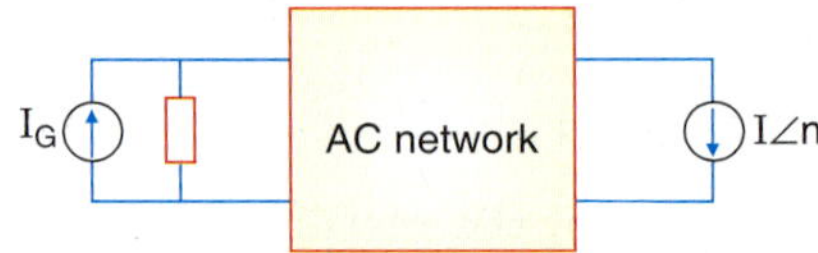

Fig. 12.15: Current at converter bus

$$I_j = k_j a_j I_{dj}, \quad \eta_j = \delta_j \pm \phi_j$$

$$\phi_j = \cos^{-1}\left(\frac{V_{dj}}{k_j a_j V_j}\right) = \cos\theta_j - \frac{R_{cj} I_{dj}}{k_j a_j V_j}$$

Note that I_{dj} is negative and the sign associated with ϕ_j is positive at an inverter.

At any given instant of time, I_{dj} and the delay (extinction) angles are known from the DC system dynamic equations. However, V is to be solved from the AC network equation and the above equation introduces non-linearity. But this does not pose any serious problem as, in general, most the loads in the AC system have non-linear characteristics. The network equations can be solved using an iterative approach described by Dommel and Sato [37]. Here, the network equations are formulated using the admittance matrix is as follows

$$\begin{bmatrix} Y_{GG} & Y_{GL} \\ Y_{LG} & Y_{LL} \end{bmatrix} \begin{bmatrix} V_G \\ V_L \end{bmatrix} = \begin{bmatrix} I_G \\ I_L \end{bmatrix} \tag{12.49}$$

where the subscripts G and L refer to the generator and load buses respectively. Both I_G and I_L are voltage dependent and this requires an iterative solution of the voltages. However, as the admittance matrix remains constant and sparse, it needs to be factored only once. Only repeat solutions are required as the right hand side vector changes. The number of iterations required are few (4 to 5).

The inclusion of DC link in the AC system calculations does not change the structure of Eq. 12.49 as the current sources due to the link are merged in I_L. The only difference is the need to update this current at each time step by solving the dynamic and algebraic equations of the DC system (including controllers, converters and the DC network).

12.6.5 Interface Using Detailed Converter Model

The method for the solution of the AC system does not change even when detailed converter model is considered. However, the current $I\angle\eta$ is not directly available from the transient converter simulation as this calculates only the instantaneous values. In reference [38], it is suggested to calculate the fundamental frequency components of voltage (and current) using discrete Fourier transforms on the samples taken from the waveforms. Although there are problems of accounting for discontinuity, frequency mismatch (slight variations in the frequency) and spectral leakage (due to non-periodic noise) these can be resolved satisfactorily. For example, spectral analysis can be avoided by calculating the rms value of the voltage (which is approximated to the value of the fundamental component if the harmonics are small in magnitude) and the rms power (approximated to the power at fundamental frequency) as

$$V_{rms} = \left[1/N)\sum_{n=1}^{N}((V_n)^2\right]^{1/2} \tag{12.50}$$

$$P_{rms} = (1/N)\sum_{n=1}^{N} V_n I_n \tag{12.51}$$

The rms current can be calculated from Eqs. (12.50) and (12.51) as follows:

The transient converter simulation is based on the reduced (Thevenin's equivalent) AC system representation as shown Fig. 12.16. From this, it can be shown that

$$I\angle\eta = (E\angle\xi - V\angle\delta)/(Z\angle\phi) \tag{12.52}$$

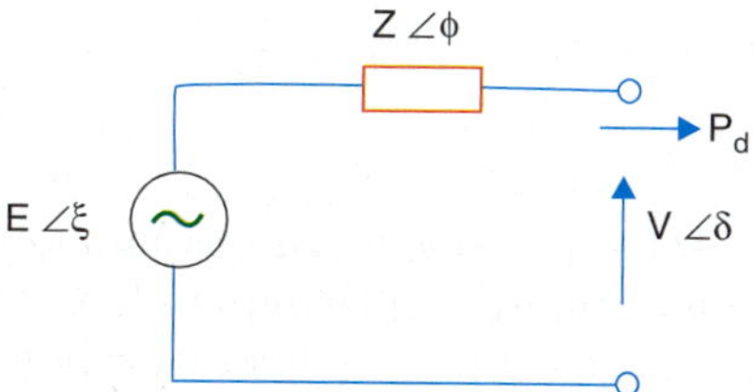

Fig. 12.16: Simplified AC system representation

The angle of V can be calculated from

$$\delta = \xi - \phi - \cos^{-1}\{(P_d Z)/(VE) + (V\cos\phi)/E\} \tag{12.53}$$

E, ξ, Z and ϕ are calculated at each step of the AC system solution and are known. V and P_d are calculated either from spectral analysis or from Eqs. (12.50) and (12.51).

12.7 DIRECT METHODS FOR STABILITY EVALUATION

The equal area criterion is the first direct method employed for determining the transient stability of two machine systems, without having to integrate the system equations beyond the clearing of the fault. This criterion is applicable for the analysis of first swing stability. The machines are represented by classical models with negligible damping.

The attempts to extend direct methods to multi-machine systems using Lyapunov or energy functions have been made and are fairly successful [51]. The current directions of research in this area involve the development of accurate methods with efficient numerical computations. The concept of structure preserving energy functions has led to the incorporation of detailed load and generator models [49, 50].

The basic procedure in applying energy functions for direct stability evaluation is outlined below:

The post-fault system, defined by

$$\dot{x} = f(x,y) \tag{12.54}$$

$$0 = g(x,y) \tag{12.55}$$

has an associated energy function $W(x,y,t)$ such that

$$W = (\partial W / \partial x)\dot{x} + (\partial W / \partial y)\dot{y} + (\partial W / \partial t) = 0 \tag{12.56}$$

It is possible that some components of W may be path dependent. This will be the case when the transmission line losses, non-constant power loads and excitation controllers are considered.

The system is said to be stable if

$$W(t_{c1}) < W_c \tag{12.57}$$

where t_{c1} is the clearing time (of the fault) and W_c is the critical energy. A major computational burden lies in the evaluation of W_c which is obtained from

$$W_c = W(x_u^*, y_u^*) \tag{12.58}$$

$$0 = g(x_u^*, y_u^*) \tag{12.59}$$

where x^* is termed as the controlling unstable equilibrium point (UEP) which is closest to the fault trajectory. In equation (12.58), the path dependent terms in W are ignored. The determination of UEP can be complex as there are several of them in large systems. An alternate approach is to employ Potential Energy Boundary Surface (PEBS) method where W_c, is determined from

$$W_c = \max W_p(x_f, y_f) \tag{12.60}$$

where W_p is the potential energy (the sum of the components of W which do not depend on the rotor velocities) and the subscript 'f' indicates the maximum of W to be found along the fault trajectory. Actually, this is an approximate method of finding W_c, but computationally attractive.

The applications of direct methods for AC-DC systems have been few [52-54]. The use of structure preserving energy functions (SPEF) is attractive as the DC link can be viewed as dynamic loads at converter buses. The contribution of the DC link to the energy function is the term given below

$$W_{dc} = \sum_{k=1}^{N} [P_{dk}(\delta_k - \delta_{ko}) + \int_{V_{ko}}^{V_k} (Q_{dk}/V_k) dV_k] \quad (12.61)$$

where P_{dk}, Q_{dk} are the power and reactive power at converter bus k, V_k, δ_k are the bus voltage magnitude and the angle respectively. It is assumed that there are N converter buses. The integral may have to be determined approximately.

12.8 TRANSIENT STABILITY IMPROVEMENT USING DC LINK CONTROL

The rapid controllability of power in a DC link can be used to advantage in improving the transient stability of the AC system in which the DC link is embedded. The power flow can even be reversed in a short time (less than 0.25 sec). Thus, DC link control can be viewed as an alternative to fast valving or braking resistor.

The stabilizing control is implemented through large signal modulation of power in response to a control signal derived from the AC system variables (such as frequency). The effectiveness of the control can be enhanced by increased overload rating of the converters which permit short-term overloads.

A VSC-HVDC transmission system performs better than LCC-HVDC system as there are independent controls over active and reactive power flow and the VSC can help in voltage regulation. In addition, the speed of control is much faster with VSC-HVDC. It can help in speedy restoration of the grid following a blackout. If equipped with a small diesel generator feeding auxiliary power (or power from another grid in asynchronous interconnection) it can provide black start capability. The VSC-HVDC link can be used to stabilize both frequency and voltage during the grid restoration.

REFERENCES AND BIBLIOGRAPHY

1. Thallam R.S., '*Review of the Design and Performance Features of HVDC Systems Connected to Low Short Circuit Ratio AC Systems*', IEEE Trans., Power Delivery, Vol. 7, No. 4, pp. 2065–2073, October 1992.
2. Uhlmann E., '*Stabilization of an AC Link by a Parallel DC Link*', Direct Current, August 1964, pp. 89–94.
3. IEEE/CIGRE Joint Task Force on Stability Terms and Definitions, '*Definition and Classification of Power System Stability*', IEEE Trans. Power Systems, Vol. 19, No. 2, pp. 1387–1401, May 2004.
4. Kamwa I., Grondin R. and Trudel G., '*IEEE PSS2B Versus PSS4B: The Limits of Performance of Modern Power System Stabilizers*', IEEE Trans. Power Systems, Vol. 20, No. 2, pp. 903–915, May 2005.
5. Uhlmann E., '*AC Network Stabilization by DC Links*', CIGRE Paper No.32-01, 1970.
6. Davies J.B. and Chapman D.G., '*Recent AC Control Enhancements of the Nelson River HVDC Links*', Proc. Int. Conf. on DC Power Transmission, pp. 86-90, June 1984.
7. Cresap R.L. and Mittelstadt W.A., '*Small Signal Modulation of the Pacific HVDC Intertie*', IEEE Trans., Vol. PAS-95, No. 2, March/April 1976, pp. 536–541.

8. Padiyar K.R., Pai M.A. and Radhakrishna C., '*Analysis of DC link Control for System Stabilization*', IEEE Conf. Publ. No. 205, on '*Thyristor and Variable Static Equipment for AC and DC Transmission*', (London), 1981, pp. 145–148.
9. Grund C.E., Pohl R.V. and Reeve J. '*Control Design of an Active and Reactive Power HVDC Modulation System with Kalman Filtering*', IEEE Trans., Vol. PAS-101, No. 10, October 1982, pp. 4100–4111.
10. Patterson W.A., '*External Control Systems on Eel River HVDC Scheme*', Canadian Conf. on Automatic Control, University of New Brunswick, Sept. 1973.
11. Cresap R.L., Mittelstadt W.A., Scott D.N. and Taylor C.W., '*Operating Experience with Modulation of the Pacific HVDC Intertie*', IEEE Trans., Vol. PAS-97, No. 4 July/Aug. 1978.
12. Peixoto C.A.O., Frontin S.O. and Jardini J.A., '*Engineering Studies for Itaipu Converter Station Design*', IEEE Proac. (London), Vol. 130, C. No.1, January 1983, pp. 41–50.
13. Tamm M. et al., '*Itaipu HVDC Transmission System, Principal Aspects of AC/DC Interaction*' CIGRE Conf., Paper 14-03, 1988.
14. Jotten R., et al., '*Control in HVDC Systems–The State of the Art, Part II-Multiterminal Systems*', CIGRE, Report 14-07, 1980.
15. Caroll D.P. and Ong C.M., '*Coordinated Power Modulation in Multiterminal HVDC Systems*', IEE Trans., Vol. PAS-100, No. 3, March 1981, pp. 1351–1359.
16. Lefebvre S., Caroll D.P. and DeCarlo, R.A., '*Decentralized Power Modulation of Multiterminal HVDC Systems*', IEEE Trans., Vol. PAS-100, No.7, July 1981, pp. 3331–3339.
17. Chaudhry M.A. and Caroll D.P., '*Coordinated Active and Reactive Power Modulation of Multiterminal HVDC Systems*', IEEE Trans., Vol. PAS-103, 1984, pp. 1480–1485.
18. Emarah A.S., Chaudhry M.A. and Galanos G.D., '*Design of Optimal Modulation Controllers for Multiarea AC/DC Systems using Eigenvalue Sensitivities*', IEEE Trans., Vol. PWRS-2, No.3, 1987, pp. 522–528.
19. Lee R.L., et al., '*Enhancement of AC/DC System Performance by Modulation of a Proposed Multiterminal DC System in the Southwestern U.S.*' IEEE Trans., Vol. PWRD-3, No.1 January 1988, pp. 307–316.
20. Grund C.E., et al., '*Dynamic Performance Characteristics of North American HVDC Systems for Transient and Dynamic Stability Evaluations*', IEEE Committee Report, IEEE Trans., Vol. PAS-100, No.7, July 1981, pp. 3356-3364.
21. IEEE Committee Report, '*A Description of Discrete Supplementary Controls for Stability*', IEEE Trans., Vol. PAS-97, No.1, Jan./Feb. 1978, pp. 146–165.
22. Hammad A.E. and Kuhn W., '*A Computation Algorithm for Assessing Voltage Stability at AC/DC Interconnections*', IEEE Trans., Vol. PWRS-1, No.1, February 1986, pp. 209–216.
23. Yoshida Y., '*Development of a Calculation Method of AC Voltage Stability in HVDC Transmission Systems*', Electrical Engg. in Japan, Vol. 94, No.2, 1974, pp. 77–85.
24. Garvrilovic A., '*Interaction between AC and DC Systems*', CIGRE Paper No. 14-09, 1986.
25. Krishnayya P.C.S. et al., '*DC Transmission Terminating at Low Short Circuit Ratio Allocations*', Report of a Panel Discussion, IEEE Trans., Vol. PWRD-1, No.3, July 1986, pp. 308–318.
26. Szechtman M., et al., '*Advanced Scheme for AC Voltage Control Technique for Stabilization of Weak Power Systems*', IEEE Trans., Vol. PAS-103, No. 8, August 1984, pp. 2244–48.

27. Hammad A., et al., '*Advanced Scheme for AC Voltage Control at HVDC Converter Terminals*', IEEE Trans., Vol. PAS-104, No.3, March 1985, pp. 697–703.
28. Padiyar K.R., Pai M.A. and Radhakrishna C., '*A Versatile System Model for Dynamic Stability Analysis of Power Systems Including HVDC Links*', IEEE Trans., Vol. PAS-100, April 1981, pp. 1871–1880.
29. Smed T. and Andersson G., '*Utilizing HVDC to Damp Power Oscillations*', Paper 92, WM 006-7 PWRD, IEEE.
30. Grund C.E. et al, '*Square Butte HVDC Modulation System Field Tests*', IEEE Trans., Power Delivery, Vol. 5, No.1, pp. 351–357, January 1990.
31. Martin D. et al., '*Modulation Controls for the New Zealand DC Hybrid Project*', IEEE Trans., Power Delivery, Vol. 6, No. 4 pp. 1825–1830, October 1991.
32. Klenk E., Povh D., Torgerson D.R. and Weaver T.L., '*Analysis and Operating Experiences of the HVDC Converter Station AC Bus Voltage Control Scheme*', CIGRE Paper 14-103, 1990.
33. Franken B. and Andersson G., '*Analysis of HVDC Converters Connected to Weak AC Systems*', IEEE Trans., Power Systems, Vol. 5, No.1, pp. 235–242, February 1990.
34. Pilotto L.A.S., Szechtman M. and Hammad A.E., "*Transient AC Voltage Related Phenomena for HVDC Schemes Connected to Weak AC Systems*", IEEE Trans., Power Delivery, Vol. 7, No.3, pp. 1396–1404, July 1992.
35. Padiyar K.R. and Kalyana Raman V., "*Study of Voltage Collapse at Converter Bus in Asynchronous MTDC-AC Systems*", Electrical Power & Energy Syst. Vol. 15, No.1, pp. 45–53, February 1993.
36. Padiyar K.R. and Rao S.S., "Dynamic Analysis of Voltage Instability in AC-DC Systems", Electrical Power & Energy Syst., Vol. 18, No.1, pp. 11–18, 1996.
37. Dommel H.W.and Sato N., '*Fast Transient Stability Solutions*', IEEE Trans., Vol. PAS-91, No. 8, 1972, pp. 1643–1650.
38. Heffernan M.D., Turner K. S., Arrillaga J. and Arnold C. P., '*Computation of AC-DC System Disturbance*'–Part I: '*Interactive Coordination of Generator and Convener Transient Models*', IEEE Trans., Vol. PAS-100, No.11, 1981, pp. 4311–4348.
39. Reeve J. and Rambabu Adapa, '*A New Approach to Dynamic Analysis of AC Networks Incorporating Detailed Modeling of DC Systems*' –Part I: '*Principles and Implementation*', IEEE Trans., Vol. PWRD-3, 1988.
40. Rambabu Adapa and Reeve J., '*A New Approach to Dynamic Analysis of AC Networks Incorporating Detailed Modeling of DC Systems*', Part II: Application to Interaction of DC and Weak AC Systems', IEEE Trans., Vol. PWRD-3, 1988.
41. EPRI Final Report EL 4345- '*Methodology for Integration of HVDC Links in large AC Systems-* Phase II: Advanced Concepts, 1987.
42. Breuer G.D., Luini J. R. and Young C.C., '*Studies of Large AC-DC Systems on Digital Computer*', IEEE Trans., Vol. PAS-85, Nov. 1966, pp. 1107–1115.
43. Clifford J.F. and Schmidt A.H., Jr., '*Digital representation of DC Transmission System and Its Controls*', IEEE Trans., Vol. PAS-89, Jan. 1970, pp. 97–105.
44. Marshall W.K., Padiyar K.R., Denton L.M., Smolinski W.J. and Hill E. F., '*A Simplified HVDC Link Representation for System Stability Studies*', C 74-434-7, Paper presented at IEEE-PES Summer Meeting 1974, Anaheim, California.

45. Carter G.K., Grund C.E., Happ H.H. and Pohl R.V., '*The Dynamics of AC/DC Systems with Controlled Multiterminal HVDC Transmission*', IEEE Trans., Vol. PAS-96, March 1977, pp. 402–413.
46. Sato N., Dravid N.V., Chan S.M., Burns A.L., and Vithayathil J.J., '*Multiterminal HVDC System Representation in a Transient Stability Program*', IEEE Trans., Vol. PAS-99, No.5, Sept/Oct. 1980, pp. 1927–1935.
47. IEEE Committee Report, '*Functional Model of Two-terminal HVDC Systems for Transient and Steady-state Stability*', IEEE Trans., Vol. PAS-103, No.6, June 1984, pp. 1249–1255.
48. Proulx R. et al., '*User Oriented Simulation of HVDC Control in a Transient Stability Program*', IEEE Trans., Vol. PAS-104, No.7, July 1985, pp. 1609–1616.
49. Padiyar K.R. and Sastry H.S.Y., '*A Topological Energy Function Analysis of Stability of Power Systems*', Int. J. Elec. Power and Energy Syst., Vol. 9, No.1, January 1987, pp. 9–16.
50. Padiyar K.R. and Ghosh K.K., '*Direct Stability Evaluation of Power Systems with Detailed Generator Models using Structure-preserving Energy Functions*', Int. J. Elec. Power and Energy Syst., Vol. 11, No.1, January 1989, pp. 47–56.
51. Pai M.A., '*Power System Stability-Application by Direct Method of Lyapunov*', North Holland, Systems and Control Series, Vol. 3, 1981.
52. Pai M.A., Padiyar K.R., and Radhakrishna C., '*Transient Stability Analysis of Multi-machine AC/DC Power System via Energy Function Method*', IEEE Trans., Vol. PAS-100, December 1981, pp. 5627–5635.
53. Padiyar K.R. and Sastry H.S.Y., '*A Structure–Preserving Energy Function for Stability Analysis of AC-DC Systems*', Sadhana, Vol. 18, Part 5, pp. 787–799, September 1993.
54. Yi-Xin Ni and Fouad A.A., '*A Simplified Two-terminal HVDC Model and its use in Direct Transient Stability Assessment*', IEEE Trans., Vol. PWRS-2, No.4, November 1987, pp. 1006–1012.
55. Lefebvre S. et al., '*Considerations for Modeling MTDC Systems Intransient Stability Programs*', Paper 90 SM 351-7 PWRD, IEEE/ PES Summer Meeting 1990.
56. Wong W.K., et al., '*Experience with Modeling MTDC Systems in Transient Stability Programs*', Paper 90SM 352-5, PWRD, IEEE/ PES Summer Meeting, 1990.
57. Jiang C., Vittal V., Ejebe G.C. and Irissari G.D., '*Incorporation of HVDC and SVC Models in the Northern State Power Company (NSP) Network for On-Line Implementations of Direct Transient Stability Assessment*', IEEE Trans., Power Systems, Vol. 10, No.2, pp. 898–906, May 1995.
58. L.A.L. Pilotto et al., '*Synchronizing and Damping Torque Modulation Controllers for Multi-Infeed HVDC Systems*', IEEE Trans., Power Delivery, Vol. 10, No.3, pp. 1505–1514, July 1995.
59. K.R. Padiyar, '*Power System Dynamics–Stability and Control*', Second Edition, (Hyderabad), B.S. Publications, 2002.
60. S.G. Johansson et al., '*Power System Stability Benefits with VSC-DC Transmission Systems*', CIGRE, Paper No. 134–204, 2004.

APPENDICES

APPENDIX-A

THYRISTOR AND IGBT VALVES

A.1 INTRODUCTION

HVDC converters are an assembly of valves which have the property of conducting in the forward direction and blocking in the reverse direction. The term 'valve', carried over from mercury arc valve days, is applied even now for thyristor valves which are made up of series and parallel connection of many thyristor cells or devices. While thyristor valves are used in Line Commutated Converters, IGBT (Insulated Gate Bipolar Transistor) valves are used in Voltage Source Converters.

The major problem with the mercury valves is the occurrence of arc backs (or back fire) which results in the destruction of the rectifying property of the valves. Arc backs are random phenomena which result in failure to block in reverse direction. Although the incidence of arc backs can be reduced by carefully controlling the factors that influence them, complete elimination is impossible and the valve cost is also increased. Furthermore, arc backs are non-self-clearing and result in line to line faults which stress transformer windings and anodes in the valve. The maintenance requirements for the valves go up and lead to poor reliability. Thyristor valves which were developed in late sixties have eliminated all these problems. They have now completely displaced mercury arc valves in HVDC transmission.

A.2 THYRISTOR DEVICE

A.2.1 Description

Thyristor is now defined as a generic term applicable to the whole range of four layer (PNPN) semiconductor switches. It is also known commercially as silicon controlled rectifier (SCR).

The structure of a thyristor with the three terminals and its electrical symbol are shown in the Fig. A.1. The device can carry current only in one direction from anode to cathode and the instant of initiation of conduction can be controlled by the gate.

At present, the blocking capability of thyristors has reached up to 12 kV and the power capacity (using a 6-inch wafer) is 32kVA (with blocking voltage of 8 kV and average current of 4 kA). While the current ratings available now are adequate, thus avoiding the necessity of parallel connections, the voltage rating is insufficient to make up a high voltage valve. Thus, a series connection of devices to make up a thyristor valve is necessary and introduces some problems that have to be considered in the design of valves and protection.

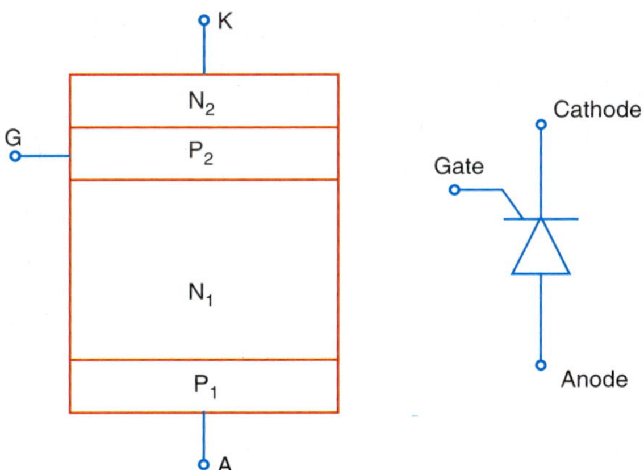

Fig. A.1: Symbol and structure of a thyristor

Increasing the voltage rating of a thyristor is feasible, but it is at the cost of increased losses, turn-off times and reduced peak allowable junction temperatures. The capability of a thyristor to withstand high voltages critically depends on the quality of silicon crystal from which the device is made—the more uniform the crystal, the better the voltage withstand capability. The irradiation of silicon crystal using low energy neutron or gamma rays results in light, precisely controlled doping.

A.2.2 Device Characteristics

The device can be in one of the three following states:

(*i*) Forward biased and blocking

(*ii*) Forward biased and conducting

(*iii*) Reverse biased and blocking.

The transition from the first to the second state is called turn-on, while the transition from the second to the third state is called turn-off. The characteristics of the device refer to the parameters of the device both in steady-state and transient conditions (during the transition of state).

Steady state characteristics

Off-State: The volt-ampere characteristics of the device are shown in Fig. A.2. During the off-state (both forward and reverse blocking), only a small magnitude of leakage current flows (of the order of 100 mA). The blocking capability with gate open is specified in terms of limiting repetitive peak forward (V_{DRM}) or reverse (V_{RRM}) voltages.

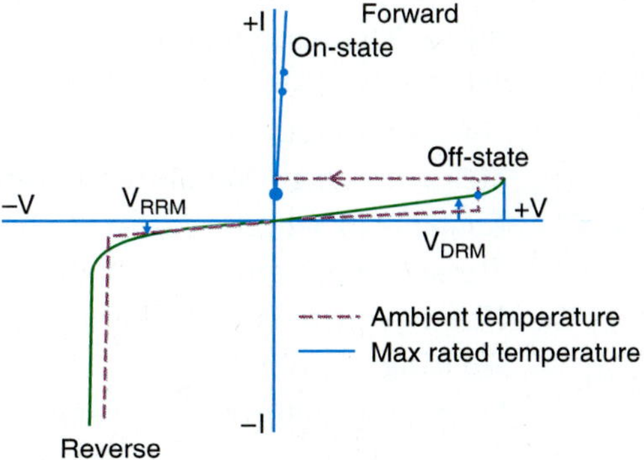

Fig. A.2: Characteristics of a thyristor

There is also a non-repetitive peak reverse voltage rating (V_{RSM}) which is specified. The voltage ratings are specified for power frequency (50 or 60 Hz) half-cycle sinusoidal voltages and rated junction temperature of the thyristor (typically 125C). The variation of the voltage ratings with junction temperature is shown in Fig. A.3 for defferent levels of rated reverse current. Here, M is the maximum rated junction temperatute. The effect of increasing the reverse (leakage) current is to increase the rating.

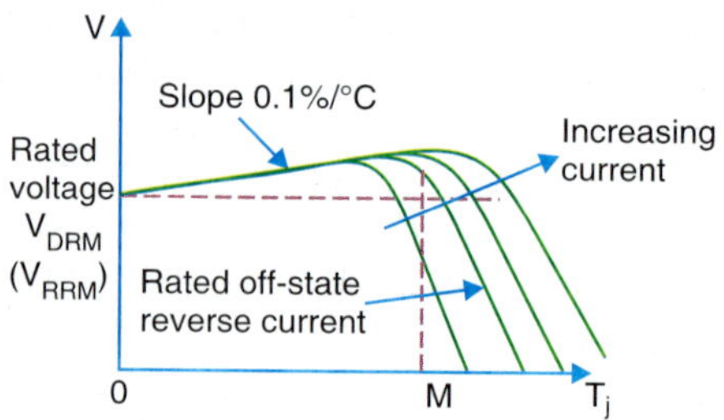

Fig. A.3: Typical thyristor off-state voltage characteristic as a function of junction temperature

The behaviour of thyristors under transient voltages is not well understood. However, according to one particular study [10, 11], the following conclusions can be drawn:

1. The transient breakover voltage of a thyristor is independent of its voltage rating.
2. The forward breakover voltage of a thyristor under a transient voltage may be lower than its voltage rating and decreases with increasing junction temperature.
3. The instant of forward breakover of a thyristor can frequently occur on the tail of the applied voltage wave. This may be due to the requirement of the minimum charge and inherent time delay of the thyristor.
4. The reverse breakover voltage levels of a thyristor under slow transients (30/600 μs) can be significantly lower compared to those obtained under fast transient voltages (1.2/50 μs). This behaviour has been attributed to the statistical and formative time lags of avalanche formulation.

The thyristor capability in the reverse direction is related to the permissible energy losses which in turn is dependent on the variation of the reverse avalanche current with the voltage magnitude of the reverse voltage. The transient voltage blocking capability of a thyristor is perhaps related to the critical power needed to damage the device. Cumulative effects due to transients with less than critical power may result in degradation the device.

On-State: There are a number of electrical and thermal parameters that characterize the on-state behaviour. Some of these are as follows:

1. On-state voltage (V_{TM})
2. Mean (average) on-state current (I_{TAV}).
3. Root mean square value of the on-state current (I_{TRMS}).
4. Surge (non-repetitive) on-state current (I_{TSM}).
5. Non-repetitive survival rating ($\int i^2 dt$)
6. Holding current (I_H)
7. Operating temperature range
8. Junction to case thermal resistance (R_{THJC})
9. Contact thermal resistance.

The on-state voltage is the anode to cathode voltage of a thyristor in the forward conducting state. It is also referred to as the forward voltage drop. This is an important characteristic affecting the power losses during on-state and the parallel operation of thyristors.

On-state voltage depends, upon a number of factors such as the width of various regions, life time and mobility of minority carriers, the physical mechanisms of recombination, etc. The power loss at current densities around 100A/cm^2 (conditions corresponding to maximum continuous rating) is due to recombinations, whereas at current densities around 1000A/cm^2 (surge conditions), it is mainly due to ohmic heating.

Silicon wafers of small thickness and long carrier lifetime give rise to low on-state voltages. However, increasing the carrier lifetime also increases turn-off time. Trying to optimize both may result in high reverse recovery current.

On-state current ratings are determined by the junction temperature which must be kept below the value necessary to ensure that it can block the recovery voltage after a worst case credible overcurrent. Figure A.4 shows the temperature build-up in a thyristor valve. (The failure of a thyristor following an overcurrent is essentially a high temperature voltage failure produced by intense local heating induced by excessive leakage current at high voltage).

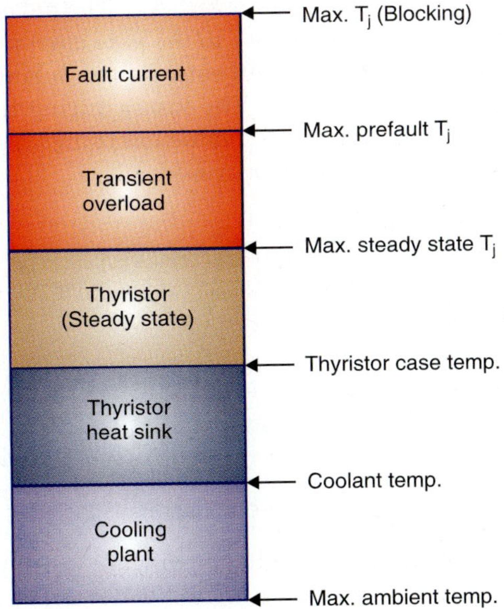

Fig. A.4: Temperature build-up in a thyristor

The surge current capability of a thyristor is based on its filamentation temperature at which mesoplasmas are formed. However, the requirement to have surge suppression capability (voltage blocking following the surge current) will result in the operation of the thyristor at reduced junction temperature. The maximum junction temperature attained due to the cumulative effects caused by the passage of repeated current surges should be below the thyristor filamentation temperature. The holding current, I_H is defined as the minimum current required to maintain the thyristor in the on-state. It is forward current below which it will cease to conduct. As the forward current is reduced, the turn-off occurs when recombination causes the minority carrier level to fall below the base-region doping level. The holding current reduces with increase in junction temperature.

Switching characteristics

Turn-on: When a gate drive is applied with the forward voltage above latching voltage (the minimum anode to cathode voltage that will successfully turn-on a thyristor with a given gate drive), turn-on occurs. Because of the finite sheet resistance of the *p*-base region, only those regions of the cathode nearest to the gate are influenced by the gate current. Regenerative switching action is initially restricted to these regions. The establishment of the equilibrium current flow over the cathode area follows by outward spreading from this conducting plasma, by diffusion. The plasma spreading is relatively slow and occurs with a typical velocity of 0.1mm/µs. When the area of conduction is small, the voltage across the device is considerable and excessive local heating will occur if the rate of rise of current is high. The device has an upper limit on the di/dt, which in modern amplifying gate thyristors is upto 500A/µs. Saturable reactors in series with the thyristors are used to limit di/dt, particularly arising from the discharge of current due to stray capacitances and snubber circuits.

There are three phases of turn-on shown in Fig. A.5. The delay time is associated with the establishment of regenerative action in response to the gate current. Its duration depends upon the level of the gate drive. Regeneration is well established during the rise time. The current continues to increase during the spread-time. The losses during the spread time can result in increase of 10-20% of that predicted for normal conduction using on-state voltage of the thyristor. The spreading phase may last over hundred microseconds.

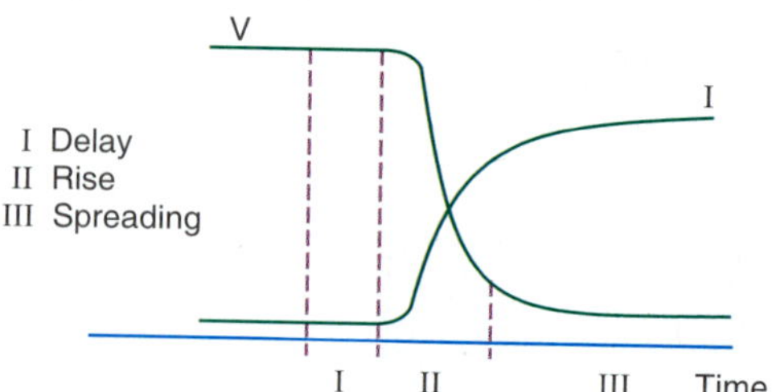

Fig. A.5: Turn-on characteristics

Ungated turn-on can occur due to over voltage, dv/dt or incomplete forward recovery. This is to be avoided because of the damage it can cause to the device. Protective turn-on by applying a gate pulse whenever the possibility of ungated turn-on is detected, is usually adopted.

Turn-off: All the three junctions are forward biased during on-state and the base regions contain excess minority and majority charge. This charge must either be swept out by an electric field or decay through regenerative processes within the silicon.

When the circuit voltage is reversed, the current falls to zero at a certain rate. Once the current reaches zero, the flow reverses, since the minority carrier concentration at the junctions can support this current by diffusion without build-up of depletion layer. The peak value of this reverse current is reached when the excess hole concentration at the anode junction has fallen to zero. At this time, the voltage across the thyristor reverses with the development of the depletion layer and the current decays in a near exponential manner as a result of charge recombination within the *n*-base region. The decay of current is dependent on the mean life-time of carriers in the *n* base region. The turn-off process is shown in Fig. A.6.

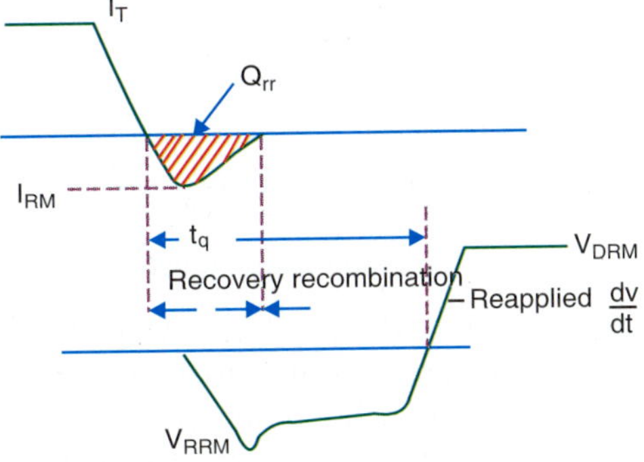

Fig. A.6: Turn-off characteristics

Immediately after current zero, a thyristor is unable to support forward voltage. Gradually, the thyristor acquires some forward blocking capability but its ability to withstand forward *dv/dt* is severely limited. Off-state *dv/dt* capability is attained only after a millisecond or so has elapsed from current zero. This characteristic is circuit and temperature dependent. Figure A.7 also shows the current and voltage waveforms for a recovery and spontaneous turn-on for a fixed *dv/dt* transient.

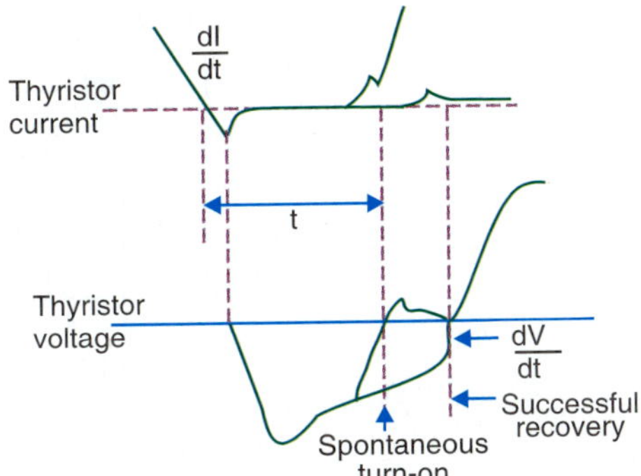

Fig. A.7: Current and voltage waveforms during positive recovery

A.2.3 Gate Drive

A hard drive with sharp rise time is necessary to turn-on the device quickly with large initial *di/dt*. A single short pulse is adequate to turn-on the device, provided the current in the device does not go below the holding current during the conduction period (about 120°) of the device. Sometimes, long pulses (or a pulse train lasting the required conduction period) are applied to avoid the blocking of the device due to discontinuous conduction.

The gate *v-i* characteristics are shown in Fig. A.8. The locus of possible gate trigger points is bounded by two lines *A* and *B*. There are also lower and upper limits on direct gate voltage and current. The upper limits are functions of the junction temperature.

APPENDIX-A

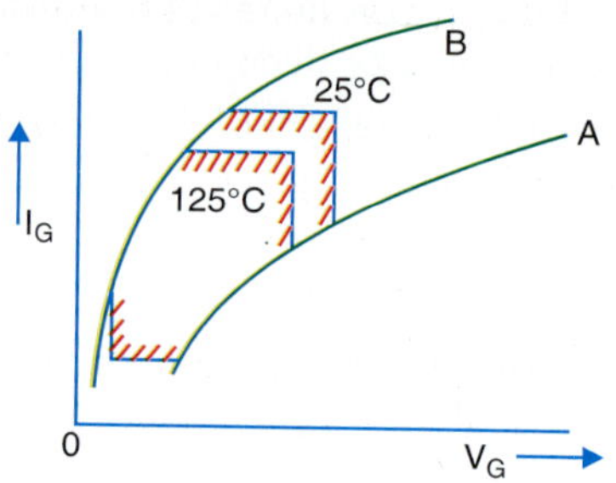

Fig. A.8: Gate characteristics

The thyristors in a modern HVDC valve are fired from the optical signals sent from the circuits at the ground potential. These signals in turn are generated from the converter controller. The source of firing energy is provided at each module by a power supply unit which is charged from the forward voltage across the device when it is not conducting. The diagram of typical module with one thyristor per module is shown in Fig. A.9.

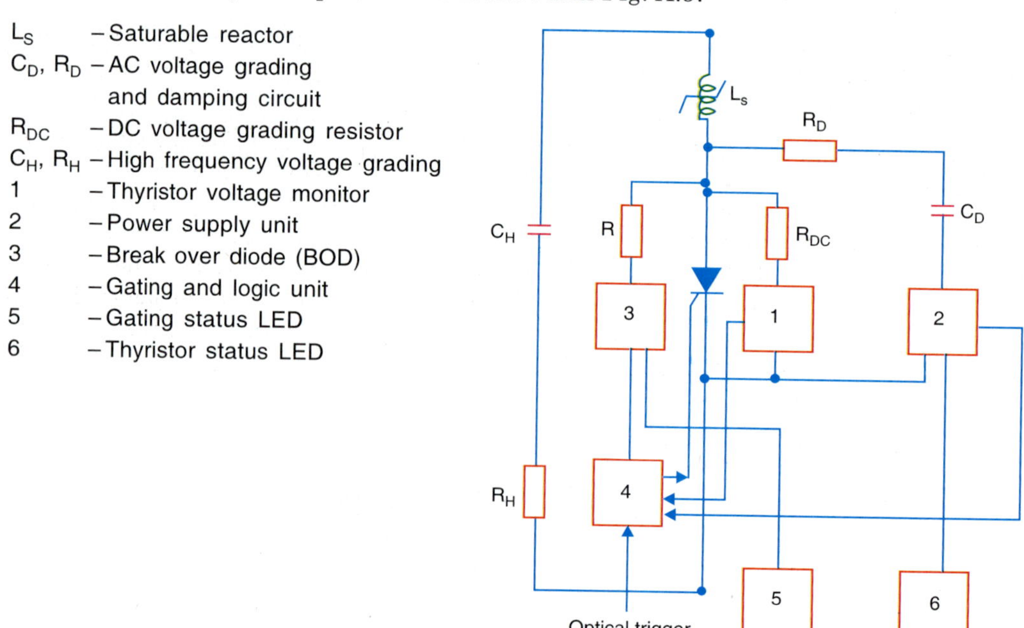

Fig. A.9: Typical valve module (Source: Reference 22)

The gate pulses are generated in the gating and logic units in response to optical signals, transmitted via fiber-optic light guides separately to each module. The gating and thyristor status are also monitored at the ground level from the signals sent via another set of light guides. The breakover diode (BOD) is used to sense overvoltage across the device and protect it by generating a gating pulse. BOD firing is also used as a back-up if the gating and logic unit fails. AC voltage grading is provided by resistor R_D and capacitor C_D in series. This circuit also damps the oscillations produced by step changes in the voltage, which occur during commutation of current from one valve to the next. It also limits the rate of rise of voltage on a late firing module.

The resistor R_{DC} provides for DC voltage grading and this circuit is also used for voltage measurement. The saturable reactor is used to protect the device against di/dt, C_H and R_H are used as grading circuits for high frequency voltages.

A.2.4 Light Triggered Thyristor

Both in USA and Japan, high power, high voltage Light Triggered Thyristors (LTT) have been developed and tested. LTT has the following advantages over electrically triggered thyristors (ETT).

1. Infinite gate isolation
2. Total noise immunity for the control circuit
3. Faster turn-on time
4. Elimination of high voltage pulse transformers and auxiliary power supplies.

The light sources used are either gallium arsenide light emitting diodes (LED) or laser diodes.

One of the problems of LTT is the reduced di/dt rating. This problem can be solved by (*i*) separate gate LTT to fire an ETT or (*ii*) controlled LTT turn-on. The electrical equivalent of the latter is shown in Fig. A.10.

Although the power supply circuits (at the thyristor level) in LTT are eliminated, the overvoltage protection circuits are still required. To improve the system reliability, it would be advantageous to eliminate these also. With this objective, a novel LTT with overvoltage self-protection structure has been reported by Hitachi, Japan [8]. Self- protecting thyristors are of two types: (*i*) using an avalanche breakdown phenomenon, (*ii*) using punch-through phenomenon. It is claimed in reference [8] that the variation in breakover voltage (V_{BO}) of the device is kept small and the temperature dependence is about 10% when the temperature is varied from 23°C to 100°C. The device turns on safely at 7300V.

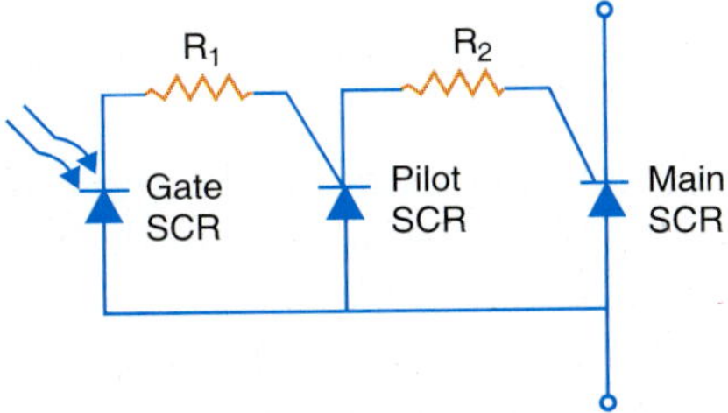

Fig. A.10: Light triggered thyristor

Recent developments [19] have aimed at overcoming the two main obstacles – 1) lack of suitable light source and 2) individual forward overvoltage protection. By using laser diodes, the light pulse transmission system can be greatly simplified as compared to ETT. Laser diodes provide sufficient light power to trigger up to 14 thyristors by one diode. Hence, one fiber-optic cable from ground potential to a valve module is sufficient. In practice, two redundant laser diodes and cables are used. For a typical 500 kV thyristor valve, the number of triggering cables is reduced by 85%. Also, as these cables are made from low attenuation monofibers, they can have lengths up to 100 meters. This makes it possible to move the cubicle housing the light sources (Valve Base Electronics) to the control room, away from the vicinity of the thyristor valves. In the module, the incoming pulse is spread out to the thyristors by Multimode Star Coupler (MSC). The outputs of the MSC are then connected to the individual thyristors by short fiber optic cables which are placed in a groove in the thyristor housing. Direct Light-triggered thyristors require gate pulses with a duration of 10μs and a minimum light power of 10 mW; it is normally operated with 40 mW as opposed to several watts needed for ETT. It has a forward breakover function integrated into the central gate structure. This also makes the thyristor immune to *dv/dt* damage because initiated breakover occurs in a controlled manner in the gate area. The main characteristics of a 125 mm LTT are shown in Table A.1.

Table A.1: Characteristics of a LTT

Repetitive peak reverse voltage (V_{RRM})	7500 V
Non repetitive peaks reverse voltage (V_{RSM})	8300V
On-state voltage (V_T)	2.3 V (max)
di/dt capability	300 A/μs
dv/dt capability	3500V/μs
Surge on-state current (10 ms) (I_{TSM})	63 kA
Turn-off time (t_q)	350 μs (max)
Recovery charge (Q_r)	6900 μAs (max)
Thermal resistance	4.3 K/ kW

A.3 THYRISTOR VALVE

A.3.1 General

A thyristor valve is made up of a number of devices connected in series to provide the required voltage rating and also of devices connected in parallel to provide the required current rating. With modern devices of high current ratings, the need for parallel connection does not arise. The number of series connected thyristors is determined by device ratings, transient overvoltages and protection philosophy.

The valves are usually placed indoor in a valve hall which is protected against contamination and dust. Valves are generally base mounted in a single, double or quadrivalve configuration. The latter results in a compact valve hall.

The valves are usually air insulated and cooled using air, water, oil or freon. The water cooling is normally used in modern converter stations as the power losses are reduced. In a valve, the heat sinks and damping resistors are cooled by the water flowing in ducts. The main objective of the cooling system is to reduce the total thermal resistance for the heat sink.

A.3.2 Valve Firing

The basic valve firing scheme is shown in Fig. A.11. The valve control generates firing signals. Each thyristor level receives the signal directly from a separate fiber-optic cable.Thus each thyristor level is independent, sharing only a duplicated light source at the ground potential.

The valve control unit also includes many monitoring and protective functions. The return pulse system coupled with short pulses firing scheme is used in present day valve control units.

A separate light guide is used to send a return pulse whenever the voltage across the thyristor is sufficient and the power supply unit is charged. If at that time, firing pulses are demanded from the valve control, the light signals are sent to all the thyristor control units simultaneously.

During normal operation, only one set of light pulses are generated in a cycle for each valve. However, during operation at low direct currents, many light pulses are generated due to discontinuous current.

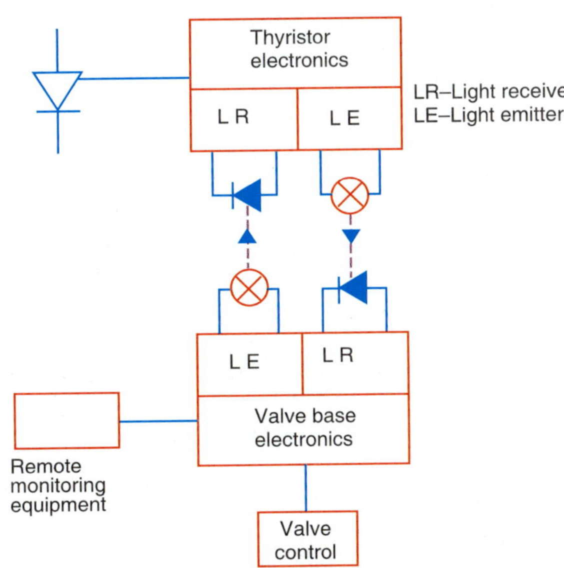

Fig. A.11: Valve firing scheme (*Source*: Reference 23)

A.3.3 Valve Design Considerations

The design of the valve must consider the voltage and current stresses that occur during normal and abnormal conditions. The overvoltages across a valve may be generated internally during switching action or the result of external causes such as short circuits on AC and DC systems. The fundamental frequency dynamic overvoltage is crucial in determining the voltage ratings and this arises when the load is thrown off due to blocking of converters. The magnitude of the dynamic overvoltage is dependent on the AC system strength.

The overcurrents in a valve arise from short circuits across a valve or a converter bridge. The surge current rating of a valve is limited both by transformer leakage and system impedances. The transformer leakage impedances have been reduced to 0.12 pu with higher rated devices.

The overload rating of a valve is a function of the size of the device as well as ambient temperature and the cooling system. For example, it is possible to increase the direct current by 20% for a decrease in ambient temperature by 8°C, with water cooling. The losses in a valve include (*i*) the losses during on-state and switching losses, (*ii*) damper and grading circuit losses and (*iii*) losses due to auxiliary power required for cooling. The resistances of the damper and grading circuits are also watercooled. Their losses are functions of converter delay and overlap angles.

The calculation of internally generated overvoltages and the design of valve components such as damper circuits is facilitated by digital computer programmes which can model the non-coherent turn-on and turn-off of the thyristor devices in a valve. For example, the non-coherent turn-on of a valve due to protective firing of individual levels results in a voltage waveform across the last level to be fired, as shown in Fig. A.12. The differences in the stored charges result in varying commutation margins in individual devices in an inverter valve. This is shown in Fig. A.13 which also shows that the device with the largest stored charge will experience a premature voltage zero crossing which can be even hundreds of microseconds in advance of the valve voltage which is usually used as the reference for inverter extinction angle control.

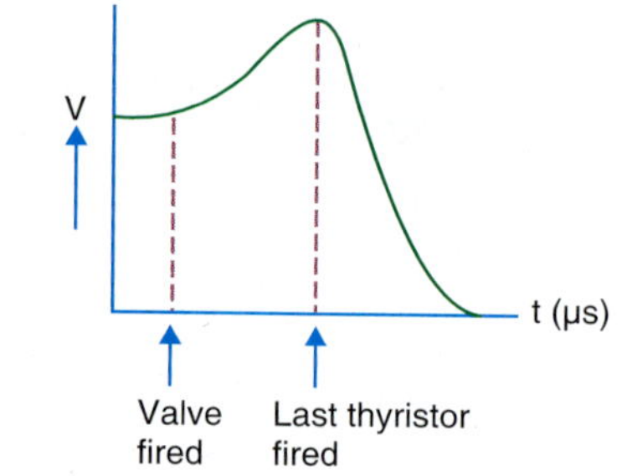

Fig. A.12: Voltage on last firing thyristor

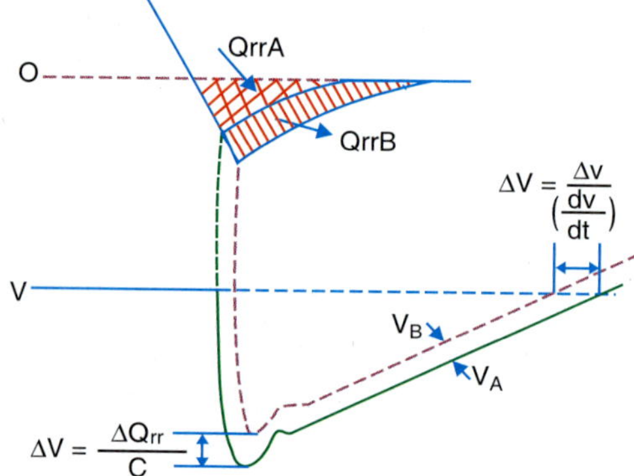

Fig. A.13: Effect of mismatch of reverse recovery charges during turn-off

The recovery voltage across the valve that has turned-off shows an overshoot that is due to the charging of the stray capacitance through the source inductance (see Fig. A.14). The overshoot is controlled by the proper choice of damper circuit resistance and capacitances. This overvoltage is maximum for the delay angle in the vicinity of 90°.

The reduction in Short Circuit Ratio (SCR) tends to reduce the maximum value of fault current in a valve. However, this does not necessarily reduce the stress on a valve as the device failures occur due to overvoltage (following the clearing of the fault). The low SCR can also result in non-sinusoidal voltage at the converter bus which can give rise to commutation failures.

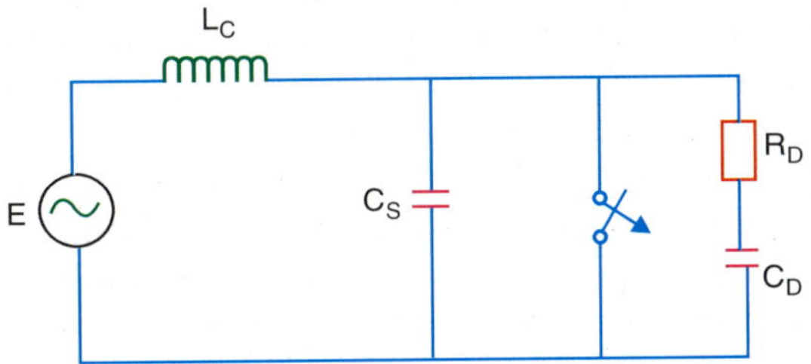

Fig. A.14: Simplified equivalent circuit during turn-off

The valve can be subjected to high stress commutation resulting from high *di/dt* when operating at low bridge currents. This is due to the premature current extinction of a valve due to discharging of stray capacitance. The discontinuous conduction can also result in high voltages across a valve due to the lightly damped oscillations caused by smoothing reactor exchanging energy with the line capacitance. Discontinuous operation is avoided by putting a minimum limit on the bridge direct current.

The control of electrostatic and electromagnetic fields surrounding a valve is essential to avoid corona discharges and interference with sensitive electronic circuits.

The damper circuit design is also affected by the losses in the damper as these are significant components of the overall losses. The damper losses are primarily due to the discontinuity in the (ideal) valve voltage waveform shown in Fig. A.15. There are 8 voltage jumps in a cycle. The voltage jumps increase in the magnitude with the increase in the delay angle.

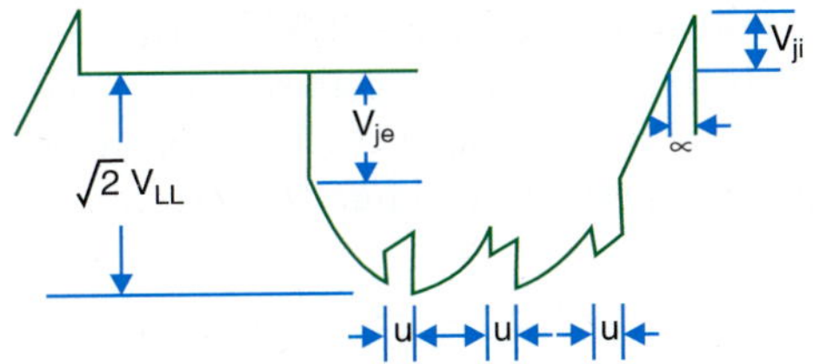

Fig. A.15: Valve voltage waveform

Suspended valve

The use of quadri-valve configuration results in a compact valve hall. However, quadri-valves are tall structures (upto 16 meters tall). Where high seismic forces have to be withstood (upto ± 0.5g acceleration), it is convenient to suspend the valve from the ceiling instead of base mounting. The suspension arrangement has spring and damper mechanism for mechanical isolation of the valve from the building vibrations. The valve structure is made flexible and the connections to the wall bushings and between the valves are also made by flexible bus.

Water cooling [20]

Water has replaced air for cooling of the valves about 30 years ago. The use of freon as a cooling medium was also given up due to environmental considerations.

Water has several excellent qualities as a good coolant: high specific heat, high thermal conductivity, high heat transfer capability, low viscosity and low density. However, in the presence of high voltages, it shows natural disassociation into ions and becomes an electrolytic conductor. It is also a solvent for various chemical substances.

In an HVDC valve, more than 95% of the heat losses are produced in the thyristors, snubber circuit resistors and the valve reactors. Water cooling is used to remove heat from the components and transfer it from high electrical potential to ground. The cooling circuit needs to be optimized with respect to 1) thermal/hydraulic and 2) electrical performance. The latter includes high voltage and electrochemical aspects.

Pure water has excellent dielectric properties that match transformer oil. But in contrast to the mineral oil, the conductivity of the water is rather high. To serve as an effective insulating medium, and to limit electrolysis currents and losses, the conductivity needs to be maintained at theoretical minimum (0.05 μS/cm at 20°C). Also, the cross-section of all piping should be kept as small as possible to provide high resistance. Due to conductivity of the water, the piping of the cooling circuit functions as a resistive network shunting thyristor, snubber etc. Advantage can be taken of this effect by suitable layout of the pipework to provide resistive voltage grading circuit that supplements the DC grading resistors of the valve. However, care must be taken to avoid partial discharges that result in long term degradation of the plastic material. Special grading electrodes are used in strategic locations to control potentials and eliminate partial discharges.

Reference [20] advocates that laws of electrochemistry and basic physics must be the governing criterion when finalizing the design of the cooling circuit for the HVDC valve. This results in a preference for a parallel cooling circuit configuration in the module, avoiding any flow of electrolytic current to metal parts (other than, those made from precious metals). The basic configuration of the cooling circuit for a HVDC thyristor valve group is shown in Fig. A.16. The heat exchanger is placed on the pressure side of the pump to minimize the stress on the plastic piping in the valves and increase the life expectancy. The quantities monitored are pressure (P), conductivity (λ), flow (Q), leakage (L) and water head (H).

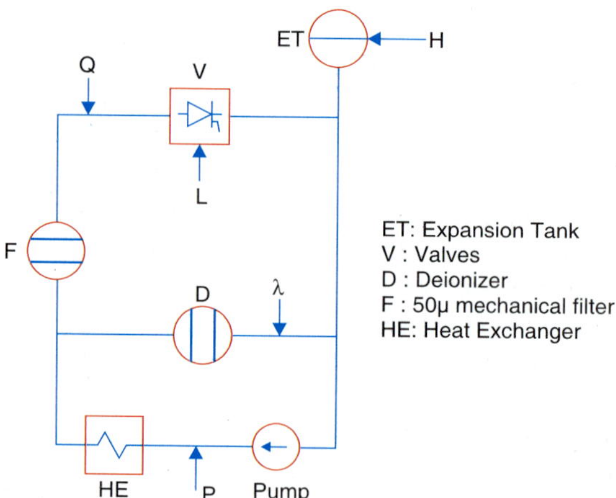

Fig. A.16: De-ionized water circuit for cooling

A.3.4 Valve Protection

Overvoltage protection

A HVDC valve must be designed to withstand (*a*) internal and (*b*) external overvoltages. The causes for the internal voltages have been mentioned in the previous section. The external sources of overvoltage across a valve can be due to (*i*) lightning and switching surges, (*ii*) dynamic overvoltages caused by load rejection, (*iii*) low order harmonic resonance in the AC system, and (*iv*) injection of AC voltage on the DC line due to converter faults.

Zinc oxide, gapless DC surge arresters across a valve can protect the valve against transient overvoltages. The overvoltage in the forward direction can be controlled by protective firing of the thyristors in a valve. This is achieved by utilizing the voltage sensitive switching action of a high voltage break-over diode across each thyristor. The protective firing thus is independent for each thyristor level. If there is some malfunction in the main firing system, the protective firing circuit is designed to work as a back-up, working continuously until the next maintenance period.

The overvoltages in the forward direction can also arise from partial recovery of a valve following turn-off process, (in inverters). In bypass pair mode operation, uneven distribution of voltage in a series connected string can arise due to dispersion in the holding current (I_H) of the thyristors. In such cases, forward *dv*/*dt* protection is also required by sensing *dv*/*dt* measured across a level and turning on the device if the threshold is exceeded. The threshold *dv*/*dt* level can be made a function of the junction temperature.

Overtemperature protection

The current rating of a valve is determined by the ability of thyristors to withstand the recovery voltages following a worst case, credible fault current. This in turn is determined by the junction temperature.

It is possible to have reliable thermal and electrical models of a thyristors from which an accurate prediction of junction temperature can be made, if pre-fault and post-fault current waveforms are specified.

The thermal model of a thyristor, usually characterized by its transient thermal impedance R_{TH}, is described by the following equation [6].

$$T(t) = \sum A_n \Phi_n(t) \tag{A.1}$$

$$d\Phi_n/dt + B_n\Phi_n = B_n W(t) \tag{A.2}$$

where

T is the temperature rise

$W = VI$

A_n, B_n = parameters determined from tests

V = on-state voltage drop

$$= V_0(1+\alpha T+\beta T^2)+R_0(1+\delta t+\varepsilon T^2)I$$

$$+E_0(273+T)\ln I + F_0(1+\gamma T+\theta T^2)\sqrt{I} \tag{A.3}$$

The last equation is a semi-empirical one, where $V_0, R_0, E_0, F_0, \alpha, \beta, \gamma, \varepsilon, \delta$ and θ are constants for a device based on its construction and physical parameters.

When $W(t)$ is a unit step function,

$$T(t) = R_{TH}(t)$$

$$= \sum A_n [1 - \exp(-B_n T)] \tag{A.4}$$

From the real-time modeling of the device, it is possible to predict on a continuous basis the instantaneous junction temperature of the worst-case thyristor. This enables to continuously monitor the overload capability of a valve. Also, when the junction temperature exceeds a critical level, (above which the recovery voltages may not be blocked), the optical firing pulses to all the thyristors in a valve are inhibited. If high temperature persists, then the converter pole is blocked and the AC breaker is tripped. If a heavy current fault occurs, it may be necessary to order continuous firing of thyristors in the affected valve until the breaker trips.

If two or more thyristors are connected in parallel, current sharing protection is provided to ensure that in case of a transient disturbance, one of the parallel thyristors does not extinguish. This is done by regating the thyristors whenever the valve current rises above a level chosen such that on-state voltage drop across the conducting thyristor is more than the pick-up voltage of the non-conducting thyristor.

A.4 IGBT VALVE

Bipolar Junction Transistors (BJT) and Metal-Oxide-Semiconductor Field Effect Transistors (MOSFET) have characteristics that complement each other. BJTs have lower conduction losses in the on-state, but have longer switching times (especially at turn-off). On the other hand, MOSFETs can be turned on or off much faster, but their on-state conduction losses are much larger. Hence, attempts to combine BJT and MOSFET monolithically in the same silicon wafer have led to the development of Insulated Gate Bipolar Transistor (IGBT). The circuit symbols for IGBT are shown in Fig. A.17.

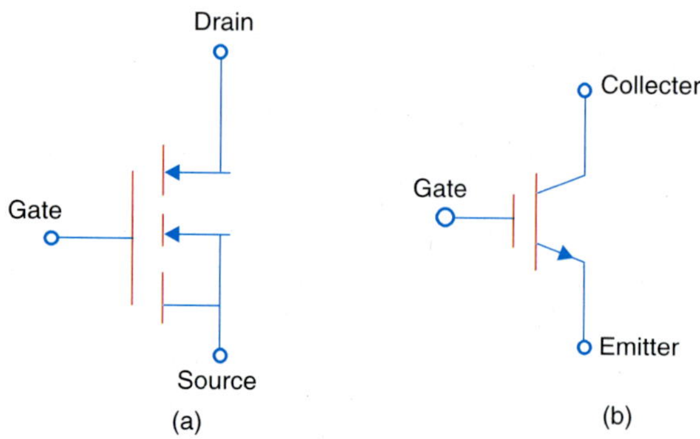

Fig. A.17: Circuit symbols of IGBT

IGBT combines a high impedance, low-power gate input with the power handling capacity of normal bipolar transistors. The control is achieved by using a pattern of MOS transistors distributed on the surface of the device. The bipolar effect of the IGBT gives a self-limiting current through the device in case of short circuit at the phase output. The device is protected by turning it off within a few microseconds. This feature is specific to IGBTs; by comparison GTOs do not provide any current limitation.

Another attractive feature of IGBT is the linear control through the gate. $\frac{di}{dt}$ and $\frac{dv}{dt}$ can be controlled during commutations. In particular, $\frac{di}{dt}$ limiting reactor is not required. Also, active voltage damping by the gate is often used to assist voltage sharing across series connected devices.

IGBT was invented in 1982. The development of high voltage IGBT was initiated a decade later. A 6.5 kV device was made available in 2002. IGBT valves are invariably used in VSC-HVDC converter stations. To obtain rated current capability, the IGBT is made of a number of chips connected in parallel in the same housing.

An anti-parallel Free Wheeling Diode (FWD) is often integrated in the same semiconductor package to ensure current capability in the reverse direction and to prevent the application of reverse voltage. The IGBT chip, if used separately, is designed to have only forward blocking capability, since the anti-parallel diode protects against voltage in the reverse direction. The IGBT and diode have identical voltage capabilities.

The IGBT is preferred choice for VSC based HVDC converters due to the following reasons:

1. It allows active voltage sharing.
2. Low power control which is advantageous at very high voltage levels.
3. Transistor action that enables the converter to be turned off even in short circuit conditions.
4. High switching speeds, thus making high switching frequency feasible.

Valve design considerations

To ensure reliability, an IGBT valve must have few redundant devices to enable confirmed operation of the valve in case of failures. A faulty device cannot be allowed to create an open circuit. Instead, a faulty device must enter into a short-circuit mode and carry current until it is replaced during the scheduled maintenance period. This capability of Short Circuit Failure Mode (SCFM) operation is very critical for series connected IGBT devices.

An important aspect of IGBT is their capability to turn off current. This capability is defined in the Switching Safe Operating Area (SSOA) shown in Fig. A.18.

Each single IGBT valve may have 100 or more IGBT devices connected to series to provide the required voltage rating. To ensure proper voltage sharing, traditional snubber circuits may be used, but this increases power loss in the valve in addition to the complexity. It is possible to design an IGBT valve that maintains voltage sharing during switching and blocked states by gate control.

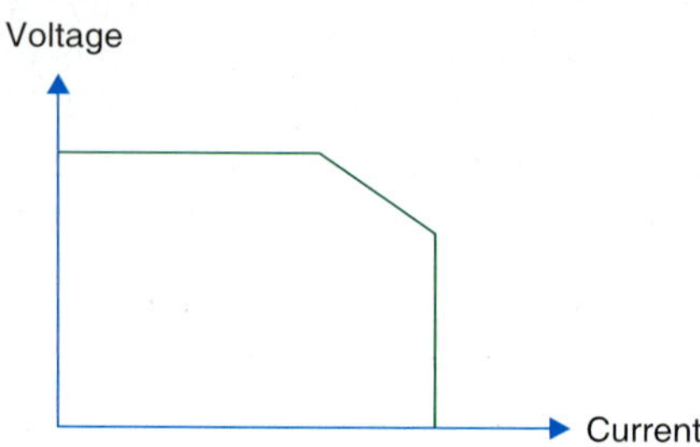

Fig. A.18: A typical SSOA for IGBT

To minimize switching losses; it is important to make the switching as fast as possible with high $\frac{di}{dt}$. However, the problem of stray inductance in the commutating circuit limits $\frac{di}{dt}$. To minimize stray inductance, care has to be taken in the layout of the converter circuit. To summarize, the valve design involves trade off between losses, $\frac{di}{dt}$ and voltages across the devices. Power losses can be reduced by the selection of 3-level converters that operate at lower switching speeds. For example, a 2 level converter operating at a switching frequency of 1950 Hz has a power loss of 3% for the VSC substation as opposed to a 3 level converter (operating at a switching frequency of 1260 Hz) has a power loss of 1.8%. The comparable loss figure for a LCC HVDC scheme is 0.8%. The power loss includes valve, filter and transformer losses in all cases considered.

The design of cooling system for IGBT and FWD is similar to that of thyristor valves. The only difference is the increased losses in the IGBT valves. The losses in IGBT consist of on-state conduction loss, turn-on and turn-off loss. The losses in the diode are similar except that it has negligible turn-on loss. The highest on-state loss for IGBT appears in inverter operation whereas for the diodes, it is during rectifier operation. Switching speed is critical in limiting the turn-on and turn-off losses. This leads to the requirement of high $\frac{di}{dt}$ and $\frac{dv}{dt}$ during switching to minimize the losses. The $\frac{di}{dt}$ capability is limited by the presence of stray inductances whereas the $\frac{dv}{dt}$ capability is limited by the necessity to operate series connection of IGBTs.

As with thyristor valves, IGBT valves must be installed in specially designed valve halls or separate valve enclosures. However, special attention is required for EMC and grounding to eliminate high frequency noise due to fast switching of IGBTs.

REFERENCES AND BIBLIOGRAPHY

1. Gandhi S.K., '*Semiconductor Power Devices*', New York, Wiley, 1967.
2. Jayant Baliga B., '*Modern Power Devices*', New York, Wiley, 1987.
3. Temple V.A.K., '*Advanced Light Triggered Thyristors for Electric Power Systems*', IEEE Conf. Publication 205, pp. 86–91, 1981.
4. Lowry L.R., Guygi L., Hingorani N.G. and Nylander P.L., '*Field Testing of Light-Triggered Thyristors in Electric Utility Applications*', *ibid* pp. 92–95, 1981.
5. Diesko N.L. et al., '*Application of a Digital Computer Program to Transient Analysis and Design of HVDC and AC Thyristor Valves*', *ibid*, pp. 167–170, 1981.
6. Gibson H. et al., '*Characterization, Evalution and Modelling of Thyristors for H.V.D.C. Convener Valves*', IEEE Conf. Publication 255 London, pp. 320–324, 1985.
7. Harrison R.E. , Shemie R.K., and Krishnayya P.C.S., '*A Proposed Test Specification for HVDC Thyristor Valve*', IEEE Trans., Vol. PAS-97, No.6, Nov/ Dec. 1978, pp. 2207–2218.
8. Shimzu Y. et al., '*A High Voltage Light, Activated Thyristors with Novel Overvoltages Self-Protection Structure*', IEEE Conference Proceedings (PESC) pp. 99–103, 1987.
9. Chen D.Y., '*Power Semiconductors: Fast, Tough and Compact*', IEEE Spectrum, September, 1987.
10. Krishnayya P.C.S., '*Important Characteristics of Thyristors of Valves for HVDC Transmission and Static Var Compensator*', CIGRE Paper 14–10, Paris, 1984.
11. Chowdhuri P., '*Behaviour of Thyristors under Transient Overvoltage*', Paper No. A80, 114-9, Presented at IEEE PES Winter Meeting, Feb. 1980.
12. Woodhouse M.L., '*Voltage and Current Stresses on HVDC Valves*', IEEE Trans. on Power Delivery, Vol. PWRD-2, No.1, pp. 199-206, 1987.
13. Karady G. and Gilsig T., '*The Calculation of Turn-off Overvoltages in a High Voltage DC Thyristor Valve*', IEEE Trans., Vol. PAS-90, pp. 2802–2811, 1971.
14. Karady G. and Gilsig T., '*The Calculation of Turn-on Overvoltages in a High Voltage DC Thyristor Valve*', IEEE Trans., Vol. PAS-91, No.2, pp. 565–573, 1972.
15. Beausejour Y. and Karady G., '*Valve Damping Circuit Design for HVDC Systems*', IEEE Trans., Vol. PAS 92, No.5, pp. 1615–1621, 1973.
16. Rambabu Adapa and Padiyar K.R., '*Alternative Evaluation of Damping Circuit Losses in HVDC Thyristor Valves*', IEEE Trans., Vol. PWRD-3, No.4, pp. 1823–1831, 1988.
17. Mehta H. and Temple V.K., '*Advanced Light Triggered Thyristors*', IEEE Conf. Publication 255, London, pp. 310–314, 1985.
18. Nishizawa I. and Tamamushi T., '*New Thyristor Application to DC Power Transmission*', *ibid*, pp. 304–309, 1985.
19. Lips H.P. et al., '*Simplifying HVDC Thyristor Valves with Direct-Light-Triggered Thyristors*', Paper No.14-110, CIGRE, Paris 2000.
20. Lips H.P., '*Water Cooling of HVDC Thyristor Valves*', IEEE Trans., Power Delivery, Vol. 9, No. 4, pp. 1830–1837, October 1994.
21. Gil Addis et al., '*Advanced HIVDC Valve*', IEEE Conf. Publication 255, London, pp. 340–345, 1985.

22. Goodrich F.G. and Tozer A.W., '*HVDC Converter Valves for the New Cross-Channel Link*', Electronics and Power, IEE (London), pp. 375–378, May 1982.

23. Lips P. and Thiele G., '*Design and Testing of Thyristor Valves for the HVDC Back-to-back tie Chateaguay*', Proceedings of Int. Conf. DC Power Transmission, Montreal, Canada, pp. 228–233, June 1984.

24. Jayant Baliga B., '*The Future of Power Semiconductor Device Technology*', Proc. IEEE, Vol. 89, No.6, pp. 822–832, June 2001.

25. Satoh K. and Yamamoto M., '*The Present State of the Art in High-Power Semiconductor Devices*", Proc. IEEE, Vol. 89, No. 6, pp. 813–821, June 2001.

26. Chokhawala R., Danidsson B. and Angquist L., '*Power Semiconductors in Transmission and Distribution Applications*', Proc. of the 2001 Int. Symp. on Power Semiconductor Devices & ICs (ISPSD), Osaka, Japan, pp. 3–10, 2001.

APPENDIX-B

DERIVATION OF CONVERTER EQUATIONS FOR TRANSIENT SIMULATION

Two Valve Conduction: In this case,

$$i_T = B_d\ i_d \tag{B.1}$$

From Eq. (11.57), we get

$$V_d = -B_d^t V_T \tag{B.2}$$

where V_d is the voltage across element 8, which is also the voltage across the bridge

Substituting Eq. (11.51) in (B.2), we get

$$V_d = -B_d^t(Z_T i_T + e) \tag{B.3}$$

From Eqs.(B.1), (B.3) and substituting for Z_T from (11.52)

$$V_d = -R_1(1+T_p)i_d - B_d^t e \tag{B.4}$$

where

$$R_1 = B_d^t[R_T]B_d \tag{B.5}$$

From Eq. (B.4), the parameters of the equivalent circuit can be derived as

$$e_{eq} = -B_d^t e,\ r_{eq} = R_1,\ l_{eq} = TR_1 \tag{B.6}$$

The voltage across the non-conducting valves are given by

$$V_i = B_i^t Z_T B_d i_d - B_i^t e \qquad i = \alpha, \beta \text{ or } \delta \tag{B.7}$$

Three Valve Conduction: In this case,

$$i_T = B_d i_d + B_\alpha i_\alpha \tag{B.8}$$

where i_α is the current in the conducting valve belonging to the link,

From the circuit equations,

$$V_d = -B_d^t Z_T i_T - B_d^t e \tag{B.9}$$

$$V_\alpha = -B_\alpha^t Z_T i_T - B_\alpha^t e = V_D \tag{B.10}$$

Substituting for Z_T and from (B.8), we get

$$V_d = -\left[R_1(1+T_p)i_d + R_2(1+T_p)i_\alpha\right] - B_d^t e \tag{B.11}$$

$$(1+T_p)i_\alpha = -(R_2/R_3)\,(1+T_p)i_d - (1/R_3)B^t e - V_d/R_3 \tag{B.12}$$

where

$$R_2 = B_\alpha^t\ R_T\ B_d,\ \ R_3 = B_\alpha^t\ R_T\ B_\alpha \tag{B.13}$$

From Eqs. (B.11) and (B.12), we get

$$V_d = -(R_1 - R_2^2/R_3)\,(1+T_p)i_d + \left((R_2/R_3)\,B_\alpha^t - B_d^t\right)e + (R_2/R_3)V_D \tag{B.14}$$

The equivalent circuit parameters can be identified as

$$r_{eq} = \left(R_1 - R_2^2/R_3\right),\ l_{eq} = Tr_{eq}$$

$$e_{eq} = \left((R_2/R_3)\,B_\alpha^t - B_d^t\right)e + (R_2/R_3)V_D \tag{B.15}$$

Equation (B.12) can be rewritten as

$$pi_\alpha = -i_\alpha/T - \left(R_2(1+T_p)i_d\right)/(R_3T) - 1/(TR_3)\,B_\alpha^t\ e - V_d/(TR_3) \tag{B.16}$$

The voltage across the non-conducting valves are given by

$$V_i = -B_i^t Z_T\ B_d i_d - B_i^t\ Z_T\ B_\alpha i_\alpha - B_i^t e,\qquad i = \beta, \gamma \text{ or } \delta \tag{B.17}$$

Four Valve Conduction: In this case,

$$i_T = B_d i_d + B_\alpha i_\alpha + B_\beta i_\beta \tag{B.18}$$

where α and β are the conducting valves in the cotree.

Also, the two conducting valves which are connected to the same phase are included in the tree. This implies that

$$B_d = [0\ 0\ 1\ 1]^t \tag{B.19}$$

$$V_d = -B_d^t V_T = -2V_D \tag{B.20}$$

The equivalent circuit parameters are identified as

$$r_{eq} = 0,\ l_{eq} = 0,\ e_{eq} = -2V_D \tag{B.21}$$

The expression for i_α and i_β are obtained by applying circuit equations. The final expressions are

$$\begin{bmatrix} pi_\alpha \\ pi_\beta \end{bmatrix} = -1/T\begin{bmatrix} i_\alpha \\ i_\beta \end{bmatrix} - (\frac{1}{T})[A_3]^{-1}\begin{bmatrix} B_\alpha^t \\ B_\beta^t \end{bmatrix} e - \frac{1}{T}[A_3]^{-1}\begin{bmatrix} 1 \\ 1 \end{bmatrix} V_D \tag{B.22}$$

where

$$R_4 = B_\alpha^t \ R_T \ B_\beta, \qquad R_5 = B_\beta^t \ R_T \ B_\beta$$

$$[A_3] = \begin{bmatrix} R_3 & R_4 \\ R_4 & R_5 \end{bmatrix} \tag{B.23}$$

The voltage across the non-conducting valves are

$$V_i = -B_i^t \ Z_T \ B_d \ i_d - B_i^t \ Z_T \ B_\alpha i_\alpha - B_i^t \ Z_T \ B_\beta i_\beta - B_i^t e \tag{B.24}$$

$$i = \gamma \text{ or } \delta$$

Example I

In this example, a 3 valve conduction mode is taken up. It is assumed that valves 1,2 and 3 are conducting. The elements 7, 9 and 1, 2 form a tree and the element 3 is in cotree. The matrix B_L is given by

$$[B_L] = \begin{array}{c} \\ 7 \\ 9 \\ 1 \\ 2 \end{array} \begin{array}{c} \begin{array}{ccccc} 8 & 3 & 4 & 5 & 6 \end{array} \\ \begin{bmatrix} 1 & 0 & -1 & -1 & -1 \\ -1 & 1 & 1 & 1 & 0 \\ 1 & -1 & 0 & -1 & 0 \\ 1 & 0 & -1 & 0 & -1 \end{bmatrix} \end{array}$$

$$R_1 = R_a + R_c, \ R_2 = -R_{a,} \ R_3 = R_{a,} + R_b$$

The parameters of equivalent circuit are

$$r_{eq} = R_c + R_a \ R_b /(R_{a,} + R_b), \ l_{eq} = T \ r_{eq}$$

$$e_{eq} = \frac{R_b e_a + R_a e_b}{R_a + R_b} - e_c - 2V_D$$

The valve current i_3 is given by

$$p \ i_3 = (-1/T)i_3 + (R_a /(R_{a,} + R_b))(i_d /T) + (R_a /(R_{a,} + R_b))p \ i_d$$

$$+(1/(R_{a,} + R_b))(e_b - e_a)/T$$

The tree branch currents are

$$i_7 = i_d, \ i_9 = -i_d + i_3, \ i_1 = i_d - i_3, \ i_2 = i_d$$

The voltage across non-conducting valves are

$$V_4 = (R_{a,} + R_c)(1 + T_p)i_d - R_a(1 + T_p)i_3 + e_c - e_a + V_D$$

$$V_5 = V_4$$

$$V_6 = R_c(1+T_p)i_d + R_b(1+T_p)i_3 + e_c - e_b + V_D$$

It can be shown that

$$V_6 = V_5 = V_4 = r_{eq}(1+T_p)i_d - e_{eq} - V_D$$

Example II

In this example, a four valve conduction mode is considered. The conducting valves are assumed to be 1, 2, 3 and 4.

The equivalent circuit parameters are (as already shown earlier)

$$r_{eq} = 0,\ l_{eq} = 0,\ e_{eq} = -2V_D$$

The conducting valve currents are defined by

$$i_1 = i_d - i_3,\ i_4 = i_d - i_2$$

$$pi_2 = (-1/T)i_2 + 1/(TR)\ [R_a(e_b - e_a) + (R_a + R_b)\ (e_a - e_c) + (R_b + 2R_a)V_D]$$

$$pi_3 = (-1/T)i_3 + 1/(TR)\ [(R_a + R_c)(e_b - e_a) + R_a(e_a - e_c) + (R_c + 2R_a)V_D]$$

where

$$R = R_a\ R_b + R_b\ R_c + R_c R_a$$

The voltages across non-conducting valves are

$$V_5 = (1+T_p)[(R_a + R_c)i_2 - R_a i_3] + e_c - e_a + V_D$$
$$V_6 = (1+T_p)[-R_a i_2 + (R_a + R_b)i_3] + e_a - e_b + V_D$$

Generation of Cutset Matrix

The formulation of converter equations described in the previous section is centered around the availability of the columns of the cut-set matrix component, – $[B_L]$. This is not a constant matrix, but varies with the choice of the tree, which has two elements chosen from the conducting valves (one belonging to each group of the bridge).

During the converter simulation using digital computers, the matrix B_L can be generated automatically as follows. The reduced incidence matrix A_I for the graph shown in Fig. 11.18 with node 1 as reference, is given below.

$$A_I = \begin{array}{c} \\ 2 \\ 3 \\ 4 \\ 5 \end{array} \begin{array}{ccccccccc} 7 & 8 & 9 & 1 & 2 & 3 & 4 & 5 & 6 \end{array} \atop \begin{bmatrix} 0 & 0 & 1 & 1 & 0 & 0 & -1 & 0 & 0 \\ -1 & 0 & -1 & 0 & 0 & 1 & 0 & 0 & -1 \\ 1 & 0 & 0 & 0 & -1 & 0 & 0 & 1 & 0 \\ 0 & 1 & 0 & -1 & 0 & -1 & 0 & -1 & 0 \end{bmatrix}$$

Node	7	8	9	1	2	3	4	5	6
2	0	0	1	1	0	0	–1	0	0
3	–1	0	–1	0	0	1	0	0	–1
4	1	0	0	0	–1	0	0	1	0
5	0	1	0	–1	0	–1	0	–1	0

(Note the rows correspond to the nodes, columns correspond to the elements).

The matrix $[A_I]$ is stored. The columns of this matrix are rearranged in two groups-branches and links, as given by

$$[A_I] = [A_T \mid A_L]$$

The matrix $[B_L]$ is obtained as

$$-[B_L] = [A_T]^{-1}[A_L]$$

The matrix $[B_L]$ changes each time a new valve begins to conduct or a valve ceases conduction. A valve ceases conduction when the current in that valve goes to zero. A valve is put into conduction when the following two conditions are met (*i*) when a gate pulse or signal is present (*ii*) when the valve voltage is positive.

Discussion

The converter model in the form of a time-varying equivalent circuit is simple and provides an elegant and computationally efficient approach for the dynamic simulation of converters. The varying topology is handled through automatically generated $[B_L]$ matrix and all possible modes of operation are considered. The forward conduction drop in a valve and dissymmetry in the leakage impedances of the converter transformer are considered in the analysis. The only assumption used is about the constant ratio of L/R in each phase. When this ratio is not constant, it is possible to realize the equivalent circuit for the converter with two and four valves conduction modes only. The equivalent circuit realization is not possible for normal modes of operation with three valve conduction. In such cases, the equations for i_d and i_α are coupled and have to be obtained from the knowledge of the network external to the converter. For example, consider a 12 pulse converter connected to the DC network as shown in Fig. B.1. If it is assumed that valves 1, 2 and 3 are conducting in the first bridge, valve 1′ and 2′ in the second bridge, the converter equations in i_d and i_3 are obtained as follows:

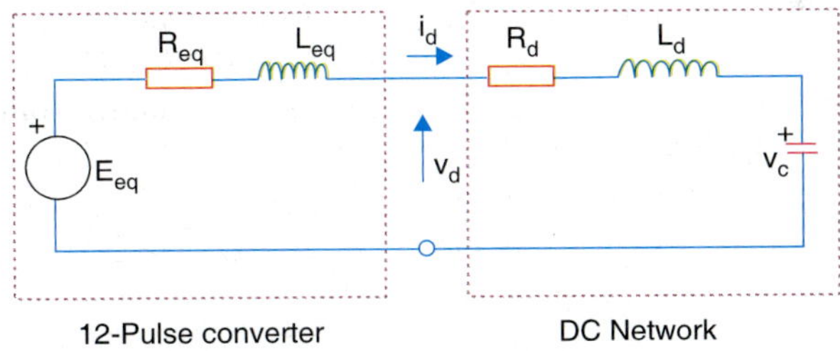

Fig. B.1: Representation of converter and DC network

$$v_{d1} = -(R_a + R_c)i_d + R_a i_3 - (L_a + L_c)p\, i_d + L_a p\, i_3 + e_a - e_c - 2V_D \tag{B.25}$$

$$0 = -R_a i_d - (R_a + R_b)i_3 + L_a p\, i_d - (L_a + L_b)p\, i_3 + e_b - e_a \tag{B.26}$$

$$v_{d2} = -(R_a' + R_c')i_d - p\,(L_a' + L_c')i_d + e_a' - e_c' - 2V_D \tag{B.27}$$

$$v_{d1} + v_{d2} = (R_d + L_d p)i_d + v_c = v_d \tag{B.28}$$

If valve 3′ is also assumed to be conducting in the second bridge, there will be an additional equation in i_3' and there will be three coupled differential equations in i_d; i_3 and i_3'. The major advantage of having the equivalent circuit for a converter is the ease in setting up equation for i_d. For example, for the network shown in Fig. B.1, the differential equation for i_d is directly obtained as

$$pi_d = \frac{-(R_{eq} + R_d)}{(L_{eq} + L_d)} i_d + \frac{(E_{eq} - V_c)}{(L_{eq} + L_d)} \tag{B.29}$$

where

$$R_{eq} = r_{eq1} + r_{eq2},\ L_{eq} = l_{eq1} + l_{eq2}$$
$$E_{eq} = e_{eq1} + e_{eq2}$$

APPENDIX-C

ANALYSIS OF A BIPOLAR DC LINE

The major assumption is that the losses in the HVDC line are neglected. The line is represented by a distributed parameter network where the voltages and currents are functions of the distance, x (measured from the sending end) and time, t.

The equations for a bipolar DC line are:

$$\partial/\partial x\begin{bmatrix} v_p \\ v_n \end{bmatrix} = -\begin{bmatrix} L_s & L_m \\ L_m & L_s \end{bmatrix}\partial/\partial t\begin{bmatrix} i_p \\ i_n \end{bmatrix} \tag{C.1}$$

$$\partial/\partial x\begin{bmatrix} i_p \\ i_n \end{bmatrix} = -\begin{bmatrix} C_s & C_m \\ C_m & C_s \end{bmatrix}\partial/\partial t\begin{bmatrix} v_p \\ v_n \end{bmatrix} \tag{C.2}$$

where subscripts p and n refer to positive and negative poles respectively. The voltages are with respect to the ground

Transforming the currents and voltages to the pole and ground components defined by

$$\begin{bmatrix} i_p \\ i_n \end{bmatrix} = [P]\begin{bmatrix} i_1 \\ i_0 \end{bmatrix}, \begin{bmatrix} v_p \\ v_n \end{bmatrix} = [P]\begin{bmatrix} v_1 \\ v_0 \end{bmatrix} \tag{C.3}$$

and

$$[P] = 1/\sqrt{2}\begin{bmatrix} 1 & 1 \\ -1 & 1 \end{bmatrix} \tag{C.4}$$

It can be shown that Eqs. (C.1) and (C.2) would be transformed to

$$\partial/\partial x\begin{bmatrix} v_1 \\ v_0 \end{bmatrix} = -\begin{bmatrix} L_s - L_m & 0 \\ 0 & L_s + L_m \end{bmatrix}\partial/\partial t\begin{bmatrix} i_1 \\ i_0 \end{bmatrix} \tag{C.5}$$

$$\partial/\partial x\begin{bmatrix} i_1 \\ i_0 \end{bmatrix} = -\begin{bmatrix} C_s - C_m & 0 \\ 0 & C_s + C_m \end{bmatrix}\partial/\partial t\begin{bmatrix} v_1 \\ v_0 \end{bmatrix} \tag{C.6}$$

v_1, i_1 are 'pole components' and v_0, i_0 are termed as ground components.

The pole and ground components are analogous to positive sequence (or α sequence) and zero sequence components used in three phase AC system analysis. These are defined by

$$v_1 = \frac{v_p - v_n}{\sqrt{2}} \tag{C.7}$$

$$v_0 = \frac{v_p + v_n}{\sqrt{2}} \tag{C.8}$$

The factor of $\sqrt{2}$ is used to obtain power invariance, which results from

$$[P]^{-1} = [P]^t \tag{C.9}$$

The solution of Eqs. (C.5) and (C.6) using Bergeron's method gives

$$v_1(t,x) + Z_1 i_1(t,x) = v_1(t-(x/c_1),0) + Z_1 i_1(t-(x/c_1),0) \tag{C.10}$$

$$v_1(t,0) - Z_1 i_1(t,0) = v_1(t-(x/c_1),x) + Z_1 i_1(t-(x/c_1),x) \tag{C.11}$$

$$v_0(t,x) + Z_0 i_0(t,x) = v_0(t-(x/c_0),0) + Z_0 i_0(t-(x/c_0),0) \tag{C.12}$$

$$v_0(t,0) + Z_0 i_0(t,0) = v_0(t-(x/c_0),x) + Z_0 i_0(t-(x/c_0),x) \tag{C.13}$$

where

$$c_1 = 1/\sqrt{L_1 C_1},\ c_0 = 1/\sqrt{L_0 C_0}$$
$$L_1 = L_s - L_m,\ L_0 = L_s + L_m$$
$$C_1 = C_s - C_m,\ C_0 = C_s + C_m$$
$$Z_1 = \sqrt{L_1 / C_1},\ Z_0 = \sqrt{L_0 / C_0}$$

Denoting the sending end of the line as m and the receiving end as k we note that

$$v_{m1}(t) = v_1(t,0), \quad i_{m1}(t) = i_1(t,0)$$
$$v_{k1}(t) = v_1(t,l), \quad i_{k1}(t) = -i_1(t,l)$$
$$v_{m0}(t) = v_0(t,0), \quad i_{m0}(t) = i_0(t,0)$$
$$v_{k0}(t) = v_0(t,l), \quad i_{k0}(t) = -i_0(t,l)$$

where l is the length of the line; i_k is defined as the current entering the line at node k and the subscripts 1 and 0 denote the pole and ground components.

From Eqs. (C.10) to (C.13), we can obtain

$$\begin{bmatrix} i_{m1}(t) \\ i_{m0}(t) \end{bmatrix} = \begin{bmatrix} Y_1 & 0 \\ 0 & Y_0 \end{bmatrix} \begin{bmatrix} v_{m1}(t) \\ v_{m0}(t) \end{bmatrix} - \begin{bmatrix} I_{m1}(t-\tau_1) \\ I_{m0}(t-\tau_1) \end{bmatrix} \tag{C.14}$$

$$\begin{bmatrix} i_{k1}(t) \\ i_{k0}(t) \end{bmatrix} = \begin{bmatrix} Y_1 & 0 \\ 0 & Y_0 \end{bmatrix} \begin{bmatrix} v_{k1}(t) \\ v_{k0}(t) \end{bmatrix} - \begin{bmatrix} I_{k1}(t-\tau_1) \\ I_{k0}(t-\tau_1) \end{bmatrix} \tag{C.15}$$

where

$$Y_1 = 1/Z_1,\ \ Y_0 = 1/Z_0,\ \tau_1 = l/c_1,\ \tau_0 = l/c_0$$

$$I_{m1}(t-\tau_1) = i_{k1}(t-\tau_1) + [v_{k1}(t-\tau_1)/Z_1]$$

$$I_{m0}(t-\tau_0) = i_{k1}(t-\tau_0) + [v_{k0}(t-\tau_0)/Z_0]$$

$$I_{k1}(t-\tau_1) = i_{m1}(t-\tau_1) + [v_{m1}(t-\tau_1)/Z_1]$$

$$I_{k0}(t-\tau_0) = i_{m0}(t-\tau_0) + [v_{m0}(t-\tau_0)/Z_0]$$

From Eqs. (C.14) and (C.15), we can obtain two sets of equivalent circuits in modal domain (pole and ground components) similar to that shown in Fig. 11.22

Transforming back to the actual variables, Eq. (C.14) is transformed to

$$\begin{bmatrix} i_{mp}(t) \\ i_{mn}(t) \end{bmatrix} = \frac{1}{2}\begin{bmatrix} Y_1 + Y_0 & Y_0 - Y_1 \\ Y_0 - Y & Y_1 + Y \end{bmatrix} - \begin{bmatrix} v_{mp}(t) \\ v_{mn}(t) \end{bmatrix} + \frac{1}{\sqrt{2}}\begin{bmatrix} I_{m1}(t-\tau_1) + I_{m0}(t-\tau_0) \\ -I_{m1}(t-\tau_1) + I_{m0}(t-\tau_0) \end{bmatrix} \quad \text{(C.16)}$$

Similar equation is obtained for the node k.

APPENDIX-D

MODELLING OF SYNCHRONOUS GENERATOR

In this Appendix, we consider the modelling of a synchronous machine (stator and rotor circuits) and the mechanical mass-spring system made up of the generator rotor, turbines and the rotating excitor masses and the connecting elastic shafts.

D.1 SYNCHRONOUS MACHINE MODEL

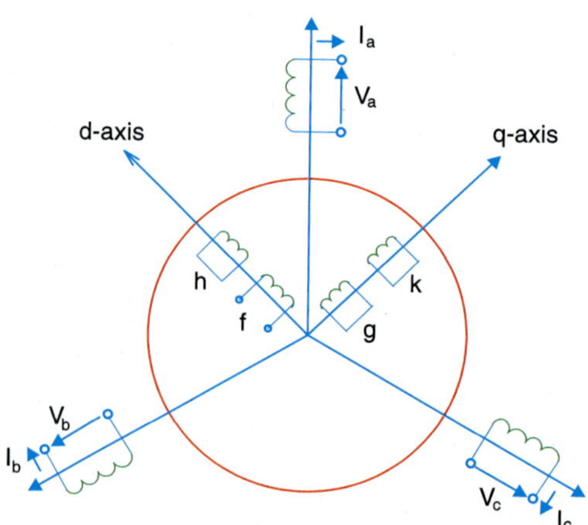

Fig. D.1: Synchronous machine

The synchronous machine considered is shown in Fig. D.1. This shows three phase armature windings (a, b and c) on the stator and four windings on the rotor including the field winding 'f'. The amortisseur (or damper) circuits in the salient pole machine or the eddy current effects in the solid rotor are represented by a set of coils. Three damper coils, 'h' in the d-axis and g, k in the q-axis are shown in Fig. D.1. Generally, three damper windings are the most detailed model of the rotor electrical circuits that represent both transient and subtransient effects in the machine. If subtransient effects are not important, one damper winding in the q-axis may be adequate.

Modelling of the Stator

Three representations of the stator circuits are considered here depending on the requirement

1. Transient simulation using three phase model
2. Transient simulation using *d-q* coordinates
3. Stability simulation neglecting stator transients

1. Three Phase Equivalent Circuit of the Stator

The synchronous machine equations are, in general, nonlinear and time varying. The solution of synchronous machine equations along with the stationary network equations requires a suitable interface technique.

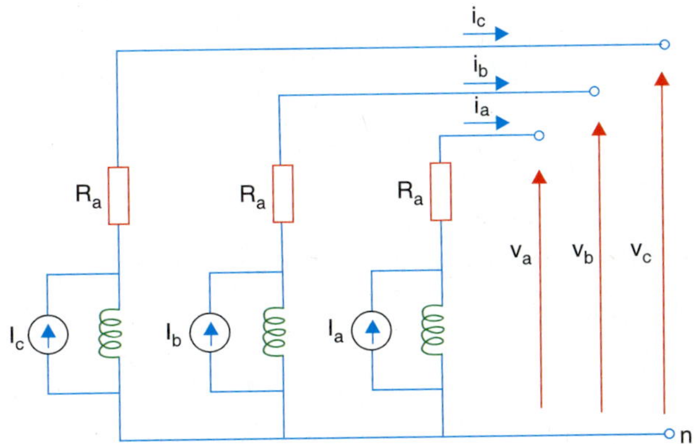

Fig. D.2: Equivalent circuit of stator

If the stator of a synchronous machine can be represented by an equivalent circuit, this will simplify considerably, the problem of interfacing the machine and the network. Ramshaw and Padiyar [3] presented such a model where the stator of a synchronous generator is represented by constant (substransient) inductances $[L_s^{"}]$ in parallel with current sources I_s which are dependent on the rotor flux linkages (see Fig. D.2). The inductances matrix $[L_s^{"}]$ is naturally a constant matrix if $L_d^{"} = L_q^{"}$ (there is no subtransient saliency). However, if $L_d^{"} \neq L_q^{"}$, the matrix can still be made constant by introducing a dummy coil on the rotor along the q-axis. The mutually coupled stator inductances can be included as part of the external network in the same manner as any other mutually coupled three phase branch.

The current source I_s is given by

$$I_s = [I_a I_b I_c]^t = I_d C + I_q S \tag{D.1}$$

where

$$C^t = \sqrt{\frac{2}{3}}\left[\cos\theta \; \cos\left(\theta - \frac{2\pi}{3}\right) \; \cos\left(\theta + \frac{2\pi}{3}\right)\right] \tag{D.2}$$

$$S^t = \sqrt{\frac{2}{3}}\left[\sin\theta \sin\left(\theta - \frac{2\pi}{3}\right) \sin\left(\theta + \frac{2\pi}{3}\right)\right] \quad \text{(D.3)}$$

$$I_d = -(C_1\psi_f + C_2\psi_h) \quad \text{(D.4)}$$

$$I_q = -(C_3\psi_g + C_4\psi_k + C_5\psi_c) \quad \text{(D.5)}$$

$$\theta = \omega_0 t + \delta \quad \text{(D.6)}$$

$$C_1 = \frac{x_{df}x_h - x_{dh}x_{fh}}{x''_d\left(x_f x_h - x_{fh}^2\right)}, \quad C_2 = \frac{x_{dh}x_f - x_{fh}x_{df}}{x''_d\left(x_f x_h - x_{fh}^2\right)}$$

$$C_3 = \frac{x_{qg}x_k - x_{qk}x_{gk}}{x''_d\left(x_g x_k - x^2_{gk}\right)}, \quad C_4 = \frac{x_{qk}x_g - x_{gk}x_{qg}}{x''_d\left(x_g x_k - x^2_{gk}\right)}$$

$$C_5 = \frac{1}{x''_d}\sqrt{\left(\frac{x''_q - x''_d}{x_c}\right)}$$

x_c is the self reactance of the dummy coil which can be chosen as $x_q \cdot \psi_c$ is the flux linkage of the dummy coil. $[L''_s]$ is defined by

$$\left[L''_s\right] = \frac{L_0}{3}\begin{bmatrix}1 & 1 & 1\\ 1 & 1 & 1\\ 1 & 1 & 1\end{bmatrix} + \frac{2L''_d}{3}\begin{bmatrix}1 & -\frac{1}{2} & -\frac{1}{2}\\ -\frac{1}{2} & 1 & -\frac{1}{2}\\ -\frac{1}{2} & -\frac{1}{2} & 1\end{bmatrix} \quad \text{(D.7)}$$

where L_0 is the zero sequence inductance and L''_d is the d-axis subtransient inductance.

ω_0 is the operating speed of the rotor (corresponding to a two pole machine). δ is the rotor angle measured with respect to a synchronously rotating reference frame.

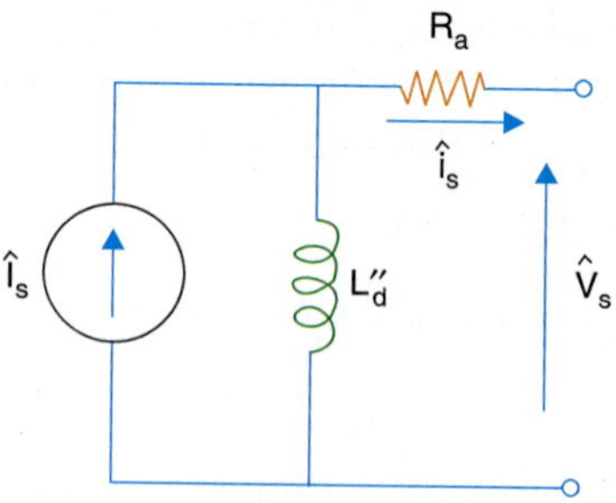

Fig. D.3: Equivalent circuit using *d-q* coordinates

2. Equivalent Circuit Using *d-q* Coordinates

If the network (connected to the stator) is represented using *d-q* coordinates it is adequate to model the stator also using *d-q* coordinates (with respect to a rotating reference frame fixed to the rotor). The single phase equivalent circuit of the stator is shown in Fig. D.3.

In this single phase circuit, $\hat{I}_s$, $\hat{i}_s$ are all complex variables defined by

$$\left.\begin{aligned} \hat{I}_s &= \left(I_d - jI_q\right)e^{j\theta}, \quad \hat{i}_s = \left(i_d - ji_q\right)e^{j\theta} \\ \hat{v}_s &= \left(v_d - jv_q\right)e^{j\theta} \end{aligned}\right\} \tag{D.8}$$

The complex equation for the equivalent circuit in Fig. A.3 is given by

$$L''_d \frac{d}{dt}\left(\hat{I}_s - \hat{i}_s\right) - R_a\hat{i}_s = \hat{v}_s \tag{D.9}$$

Substituting Eq. (D.8) in (D.9) and separating the real and imaginary parts, we get

$$-\frac{d\psi_d}{dt} - \dot{\theta}\psi_q - R_a i_d = v_d \tag{D.10}$$

$$-\frac{d\psi_q}{dt} + \dot{\theta}\psi_d - R_a i_q = v_q \tag{D.11}$$

where,

$$\psi_d = -L''_d\left(I_d - i_d\right), \quad \psi_q = -L''_d\left(I_q - i_q\right)$$

$$\dot{\theta} = \frac{d\theta}{dt} = \omega_0 + \frac{d\delta}{dt}$$

3. Equivalent Circuit in *d-q* Coordinates, Neglecting Stator Transients

For transient stability simulation, it is usual to neglect stator transients and the variations in $\dot{\theta}$. What this implies is that the derivative terms in Eqs. (D.10) and (D.11) are neglected and $\dot{\theta}$ is

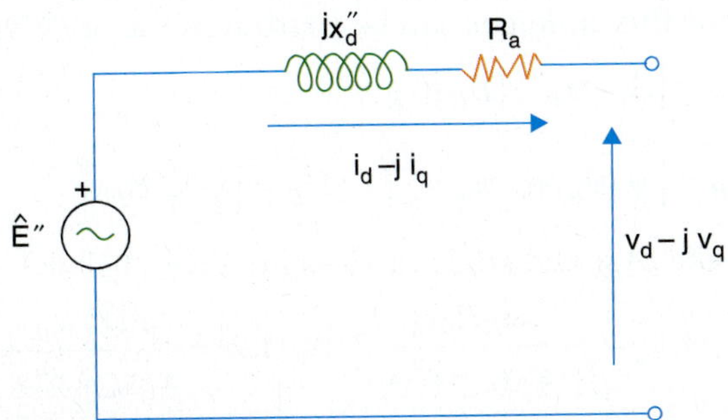

Fig. D.4: Equivalent circuit neglecting transients

assumed to be equal to ω_B (the rated speed). With these approximations, it is possible to derive an equivalent circuit shown in Fig. D.4 where the complex voltage source $\hat{E}''$ is defined as

$$\hat{E}'' = E_d'' - jE_q'' \tag{D.12}$$

where

$$E_d'' = x_d'' I_q \tag{D.13}$$

$$E_q'' = x_d'' I_d \tag{D.14}$$

If subtransient phenomenon is not considered, then only one damper winding in the q-axis is adequate. In this case, the equivalent circuit is shown in Fig. D.5. Here, E_d' and E_q' are defined by

$$E_d' = -\frac{x_{qg}}{x_g}\psi_g - x_d' C_5' \psi_c, \quad E_q' = \frac{x_{df}}{x_f}\psi_f \tag{D.15}$$

$$C_5' = \frac{1}{x_d'}\sqrt{\frac{(x_q' - x_d')}{x_c}} \tag{D.16}$$

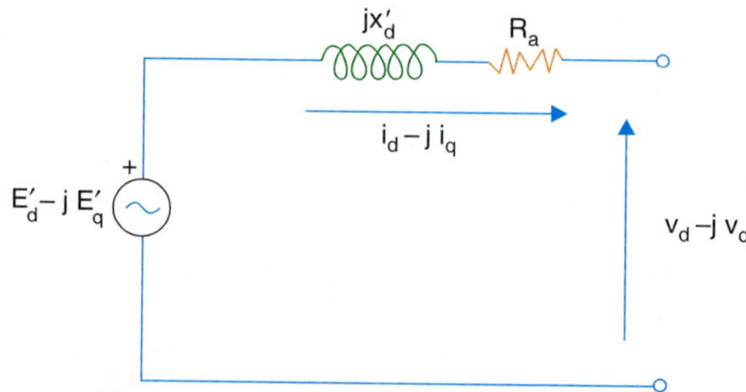

Fig. D.5: A simplified equivalent circuit (with one damper winding)

Rotor electrical equations

The equations for the rotor flux linkages can be written in the state space form as

$$\dot{\psi}_R = [A_R]\psi_R + [B_R]U_R \tag{D.17}$$

where $$\psi_R = \left[\psi_f \psi_h \ \psi_g \ \psi_k \ \psi_c\right]^t, \quad U_R = \left[i_d \ i_q \ E_{fd}\right]^t \tag{D.18}$$

The nonzero elements of the $[A_R]$ and $[B_R]$ matrices are given below,

$$A_R(1,1) = -\frac{\omega_B R_f x_h}{x_f x_h - x_{fh}^2}, \quad A_R(1,2) = \frac{\omega_B R_f x_{fh}}{x_f x_h - x_{fh}^2}$$

$$A_R(2,1) = -\frac{\omega_B R_h x_{fh}}{x_f x_h - x_{fh}^2}, \quad A_R(2,2) = -\frac{\omega_B R_h x_f}{x_f x_h - x_{fh}^2}$$

$$A_R(3,3) = -\frac{\omega_B R_g x_k}{x_g x_k - x_{gk}^2}, \quad A_R(3,4) = \frac{\omega_B R_g x_{gk}}{x_g x_k - x_{gk}^2}$$

$$A_R(4,3) = \frac{\omega_B R_k x_{gk}}{x_g x_k - x_{gk}^2}, \quad A_R(4,4) = -\frac{\omega_B R_k x_g}{x_g x_k - x_{gk}^2}$$

$$A_R(5,5) = -\frac{1}{T_c}$$

$$B_R(1,1) = \frac{\omega_B R_f (x_h x_{df} - x_{dh} x_{fh})}{x_f x_h - x^2{}_{fh}}, \quad B_R(1,3) = \frac{\omega_B R_f}{x_{df}}$$

$$B_R(2,1) = \frac{\omega_B R_h (x_f x_{dh} - x_{fh} x_{df})}{x_f x_h - x^2{}_{fh}}$$

$$B_R(3,2) = \frac{\omega_B R_g (x_k x_{qg} - x_{qk} x_{gk})}{x_g x_k - x_{gk}^2}$$

$$B_R(4,2) = \frac{\omega_B R_k (x_g x_{qk} - x_{gk} x_{qg})}{x_g x_k - x_{gk}^2}$$

$$B_R(5,2) = \sqrt{x_c \left(x''_q - x''_d\right)} / T_c$$

In these expressions, x_f, x_h, x_g and x_k represent the self reactances of the windings 'f', 'h', 'g' and 'k' respectively, The reactance x_{ij} represents the mutual reactance between the winding i and j. Without loss of generality, we can assume $x_{df} = x_{dh}$ and $x_{qg} = x_{qk} = x_{gk}$. However, $x_{fh} \neq x_{df}$ in general.

T_c is the time constant of the dummy coil in the q-axis given by $T_c = \dfrac{x_c}{\omega_B R_c}$. T_c can be arbitrarily chosen for the required accuracy. For stability studies, T_c can be chosen not smaller than 10 ms for adequate accuracy. For transient simulation, $T_c = 0.1$ ms gives accurate results. (Note that $T_c = 0.0$ for exact results).

D.2 MODELLING OF TURBINE GENERATOR MECHANICAL SYSTEMS

D.2.1 General

The rotor of a Turbine Generator (T-G) unit is a complex mechanical system made up of several rotors of different sizes, each with mechanical shaft sections and couplings. Turbine sections contain a number of discs which may be integral or attached to the rotor. The length of the rotor system may extend over 50 m and weigh several hundred tons. The system also has a number of smaller components including turbine blades, rotor coils, retaining rings, blowers and pumps.

While an exact analysis of the rotor system may require an advanced continuum model, for the study of torsional interactions, lumped multimass model is adequate [4]. Here, each major rotor element (generator, various turbine stages and rotating exciter) is considered to be a rigid mass connected to adjacent elements by shafts that are modelled as massless springs. This lumped multimass model has natural resonant frequencies below the system (electrical) frequency. The torsional mode oscillations, induced by a transient (such as a sudden change in the air gap torque caused by network switching) are lightly damped even when torsional interaction with the transmission network is neglected. The total damping forces are due to (*A*) steam pressure on turbine blades (*B*) bearing friction and windage on shaft elements (*C*) hysteritic damping in the rotor steel and (*D*) electrical damping due to generator, exciter and the transmission network. The first component is dependent on the generator loading while the last component can be negative. The hysteritic damping arises due to the energy dissipated in the material subjected to alternating stress.

D.2.2 State Equations

The state equations for the mechanical system can be expressed either based on (a) coupled multimass model or (b) decoupled modal model. Both approaches will be illustrated by taking an example of a four mass system made up of High Pressure (HP), Intermediate Pressure (IP), Low Pressure (LP) turbines and the generator rotor. (see Fig. D.6). There are four masses and three shafts sections. The number of state variable is eight and defined by,

$$X_m = [\delta\, S_m\, T_{LG}\, S_{LP}\, T_{IL}\, S_{IP}\, T_{HI}\, S_{HP}]^t$$

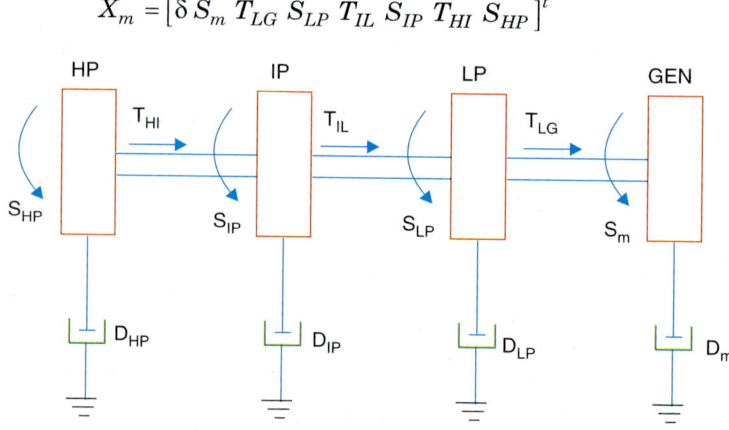

Fig. D.6: A four mass system

T_{ij} is the torque in the shaft connecting rotor masses i and j. S_i is the per unit slip of the rotor mass defined by

$$S_i = \frac{\omega_i - \omega_B}{\omega_B}$$

The state equations are given below:

$$\frac{d\delta}{dt} = \omega_B S_m \tag{D.19}$$

$$2H_m \frac{dS_m}{dt} = -(D_m + D_{LG})S_m + T_{LG} + D_{LG}S_{LP} - T_e \tag{D.20}$$

$$\frac{dT_{LG}}{dt} = \omega_B K_{LG}(S_{LP} - S_m) \tag{D.21}$$

$$2H_{LP}\frac{dS_{LP}}{dt} = -(D_{LG} + D_{LP} + D_{IL})S_{LP} + D_{LG}S_m + D_{IL}S_{IP} + T_{IL} - T_{LG} + F_{LP}T_m \tag{D.22}$$

$$\frac{dT_{IL}}{dt} = \omega_B K_{IL}(S_{IP} - S_{LP}) \tag{D.23}$$

$$2H_{IP}\frac{dS_{IP}}{dt} = -(D_{HI} + D_{IP} + D_{IL})S_{IP} + D_{HI}S_{HP} + D_{IL}S_{LP} + T_{HI} - T_{IL} + F_{IP}T_m \tag{D.24}$$

$$\frac{dT_{HI}}{dt} = \omega_B K_{HI}(S_{HP} - S_{IP}) \tag{D.25}$$

$$2H_{HP}\frac{dS_{HP}}{dt} = -(D_{HP} + D_{HI})S_{HP} + D_{HI}S_{IP} - T_{HI} + F_{HP}T_m \tag{D.26}$$

where D_{HI}, D_{IL} and D_{LG} are mutual damping terms associated with shaft sections connecting HP-IP, IP-LP and LP-GEN respectively. T_m is the net mechanical torque produced by the prime mover. F_{HP}, F_{IP} and F_{LP} are the fractions of the mechanical torque (or power) produced by HP, IP and LP turbines respectively. H_i is the inertia constant of rotor mass i. It is assumed that initial slip (S_{mo}) is zero.

Eqs. (D.19) to (D.26) can be expressed in the compact form as

$$\dot{X}_m = [A_m]X_m + [B_{m1}]T_m + [B_{m2}]T_e \tag{D.27}$$

where A_m is a matrix of order $2m \times 2m$ where m is the number of masses, B_{m1} and B_{m2} are column vectors of order $2m$.

State equations using modal quantities

As mentioned earlier, it is possible to transform the system equations into decoupled form using modal inertias, spring constants and damping parameters. These equations for the four mass system shown in Fig. D.6 are given below. To simplify the notation, the modal quantities are denoted by subscripts.

$$\frac{d\delta_0}{dt} = \omega_B S_0 \tag{D.28}$$

$$2H_0 \frac{dS_0}{dt} = -D_0 S_0 + (T_m - T_e) \tag{D.29}$$

$$\frac{d\delta_1}{dt} = \omega_B S_1 \tag{D.30}$$

$$2H_1 \frac{dS_1}{dt} = -D_1 S_1 - K_1 \delta_1 + (T_m^1 - T_e) \tag{D.31}$$

$$\frac{d\delta_2}{dt} = \omega_B S_2 \tag{D.32}$$

$$2H_2 \frac{dS_2}{dt} = -D_2 S_2 - K_2 \delta_2 + (T_m^2 - T_e) \tag{D.33}$$

$$\frac{d\delta_3}{dt} = \omega_B S_3 \tag{D.34}$$

$$2H_3 \frac{dS_3}{dt} = -D_3 S_3 - K_3 \delta_3 + (T_m^3 - T_e) \tag{D.35}$$

$$S_m = S_0 + S_1 + S_2 + S_3 \tag{D.36}$$

$$\delta = \delta_0 + \delta_1 + \delta_2 + \delta_3 \tag{D.37}$$

where H_0, H_1, H_2 and H_3 are modal inertias corresponding to modes zero, one, two and three respectively. The frequency of mode zero is zero and the mode three has the highest frequency. D_0, D_1, D_2 and D_3 are modal dampings. T_m^1, T_m^2 and T_m^3 are defined by

$$T_m^1 = q_1^t F T_m, T_m^2 = q_2^t F T_m, T_m^3 = q_3^t F T_m \tag{D.38}$$

$$F^t = [F_{HP} \; F_{IP} \; F_{LP} \; 0]$$

q_1, q_2 and q_3 are columns of $[Q]$ matrix (eigenvectors of the matrix $[M^{-1}][K]$ corresponding to the torsional modes 1,2 and 3). $[M]$ and $[K]$ are defined below:

$$[M] = \frac{1}{\omega_B}\begin{bmatrix} 2H_{HP} & 0 & 0 & 0 \\ 0 & 2H_{IP} & 0 & 0 \\ 0 & 0 & 2H_{LP} & 0 \\ 0 & 0 & 0 & 2H_m \end{bmatrix}$$

$$[K]=\begin{bmatrix} K_{HI} & -K_{HI} & 0 & 0 \\ -K_{HI} & (K_{HI}+K_{IL}) & -K_{IL} & 0 \\ 0 & -K_{IL} & (K_{IL}+K_{LG}) & -K_{LG} \\ 0 & 0 & -K_{LG} & K_{LG} \end{bmatrix}$$

The eigenvectors of $[M]^{-1}[K]$ are chosen such that their last elements (corresponding to generator rotor angle) are all equal to unity.

It is to be noted that

$$D_0 = D_T = D_{HP} + D_{IP} + D_{LP} + D_m$$

$$H_0 = H_T = H_{HP} + H_{IP} + H_{LP} + H_m$$

The modal dampings are normally obtained directly from test results or assumed.

D.3 RELATIONSHIP BETWEEN PARK AND KRON TRANSFORMATION

It is to be noted that the three phase, transient model of the synchronous machine described in section D.1, is derived from first principles without utilizing Park's transformation. However, the application of Park's transformation that involves a rotating reference frame fixed to the machine rotor, results in the equivalent circuit of the stator shown in Fig. D.3 or D.4 depending on whether stator transients are considered or neglected. The zero sequence current component in the armature windings has no effect on the electromagnetic torque produced.

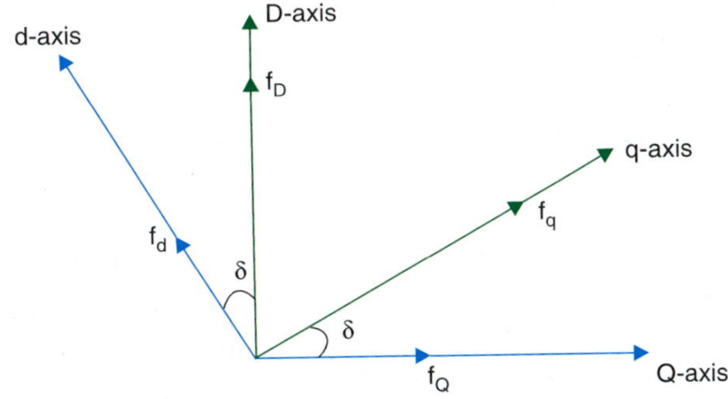

Fig. D.7: Phasor diagram showing relationship between *D-Q* and *d-q* components

For the stationary network, it is convenient to utilize a synchronously rotating reference frame with a constant speed. In power system literature, the associated transformation is termed as Kron's transformation. (In some papers by power electronics specialists, this is also (mistakenly) labelled as Park's transformation). The relationship between the *d-q* (Park's components) and *D-Q* (Kron's components) is given by

$$(f_Q + jf_D) = (f_q + jf_d)e^{j\delta} \qquad \text{(D.39)}$$

where f can be a voltage or current (dynamic) phasor. (Note that lower case letters refer to Park's components while the upper case letters refer to Kron's components). Typically, the reference phasor is considered to be along Q-axis. The machine q-axis leads Q-axis by δ (rotor angle). See Fig. D.7. Typically, the infinite bus voltage E_b is assumed to be along the Q-axis.

REFERENCES AND BIBLIOGRAPHY

1. Padiyar K.R., *'Analysis of Subsynchronous Resonance in Power Systems'*, Kluwer Academic Publishers, Boston, 1999.
2. Padiyar K.R., *'Power System Dynamics–Stability and Control'*, (Second Edition), B.S. Publications, Hyderabad, 2002.
3. Ramshaw R.S. and Padiyar K.R., *'Generalized System Model for Slipring Machines'*, Proc. IEE, (London), Vol. 120, No. 6, 1973, pp. 647–658.
4. Ramey D.G., Sismour A.C. and Kung G.C., *'Important Parameters in Considering Transient Torques on Turbine-generator Shaft Systems'*. IEEE Trans., Vol. PAS.99, No. 1, 1980, pp. 311–317.

APPENDIX-E

SUBSYNCHRONOUS RESONANCE

E.1 INTRODUCTION

The turbo generators have several rotors corresponding to steam turbines (High Pressure, Intermediate Pressure and Low Pressure) in addition to the generator and rotating exciter (if any), all connected by elastic shafts which can be modeled as springs. Thus, the mechanical system of rotors and shaft sections is a lightly damped mass-spring system analogous to the electrical network of capacitors (analogous to rotor inertias) and inductors (analogous to the springs). This system has as many torsional modes as the number of rotors (masses), of frequencies ranging from 0 to 50 Hz (in the subsynchronous range). These frequencies are computed considering the mechanical system in isolation. The torsional mode corresponding to the zero frequency is termed as 'mode zero' in which all the rotors participate equally. (The eigenvector has identical elements). The external electrical system consisting of the generator stator coils connected to the transmission network can affect the frequency and damping of the torsional modes. While the change in the frequency is significant for mode zero, the changes in the frequencies of the remaining modes are slight and can be neglected. (It is to be noted that the frequency of the 'mode zero' in the combined system varies in the range 0.2 to 2.0 Hz and the damping of the oscillations corresponding to this low frequency mode has been discussed extensively in the literature since last 40 years and still continues to be important for secure system operation). The damping of the remaining torsional modes is significantly affected by the characteristics of the electrical system. A series compensated transmission line connected to a turbogenerator has a (electrical) resonance frequency (f_{er}) given by

$$f_{er} = f_o \sqrt{\frac{x_C}{x'' + x_T + x_E}} \tag{E.1}$$

where x'' is the subtransient reactance of the generator, x_T is the leakage reactance of the transformer, x_E and x_C are external reactances of the series connected inductor and capacitor respectively. Since $x_C < x_E, f_{er} < f_o$ (f_o is the system frequency). When the torsional mode frequency (f_m) is a complement of the electrical resonance frequency (f_{er}), that is

$$f_m = f_o - f_{er}$$

there is a danger of Subsynchronous Resonance (SSR) which has the primary consequence of growing torsional oscillations due to the instability of the torsional mode caused by the large negative damping introduced by the electrical system. This problem has been termed as Torsional Interaction (TI) that results in self excitation (steady state SSR problem). Even if the net damping (due to the mechanical and the electrical system) is positive, there can be high amplitude transient subsynchronous frequency oscillations that can damage the shafts due to fatigue. Fatigue is a cumulative process and only after 100% fatigue life expenditure, an observable defect such a crack is initiated. The transient torque oscillations are induced by large disturbances such as faults followed by their clearing.

An IEEE Committee Report (1985) has defined SSR as follows: "Subsynchronous resonance is an electric power system condition where the electric network exchanges energy with a turbine generator at one or more of the natural frequencies of the combined system below the synchronous frequency of the system". The two aspects of the SSR problem are:

1. Self excitation (also called as steady state SSR)
2. Transient torques (also called as transient SSR)

The self excitation problem includes (*a*) Induction Generator Effect (IGE). Here the mechanical system is not modeled. The subsynchronous frequency (positive sequence) armature currents produce a rotating mmf which moves slower than the generator rotor. The resistance of the rotor circuits appear to be negative (viewed from the armature terminals) due to the negative slip of the machine (corresponding to the subsynchronous frequency currents). If the net resistance is zero or negative, self excitation occurs. (*b*) Torsional Interaction (TI). This is due to the interplay between the electrical and mechanical system and is a much more serious problem compared to IGE. The problem was discovered accidentally after the shaft damage experienced at Mohave generating station in 1970 and 1971.

The analysis of steady state SSR or self excitation problem can be carried out by linear system methods of checking for the stability of an operating point (steady state equilibrium). There are two major approaches

1. Damping torque analysis based on frequency domain technique. This is a heuristic, simpler approach for checking for TI and gives reasonably accurate results.
2. Eigenvalue analysis based on the state space model of the linearized system equations. This gives accurate information about the stability of all the system modes including torsional modes.

The damping torque analysis can be used as a screening tool for fast evaluation of a host of system conditions for torsional interactions. Since IGE is not a major problem, it can be neglected. This permits the use of simple 'classical model' of the generator which simplifies the computation of the damping torque.

The analysis of the transient torques requires detailed system simulation based on the nonlinear models. In the presence of HVDC and FACTS controllers, three phase models considering switching action of the thyristor or other power semiconductor devices are most accurate. However, for SSR analysis, the modeling using D-Q variables (based on synchronously rotating reference frame) is found to be adequate. The use of D-Q variables implies that the harmonics generated by the FACTS devices have little effect on the SSR performance. It must be noted that unlike in the case of study of low frequency phenomena involving swing modes, the analysis of SSR requires representation of the network dynamics (by differential equations rather than algebraic equations involving phasors). Thus, it is not unusual to employ EMTP

(Electromagnetic Transients Program) type software which was originally developed for the study of lightning and switching transients.

Following a disturbance, the turbine-generator rotors will oscillate relative to one another at one or more of the mechanical natural frequencies called 'torsional mode frequencies'. The relative amplitude and phase of the individual rotors are fixed and are also called as 'mode shapes' of torsional motion. The mode shapes are also eigenvectors corresponding to the individual torsional modes. These modes are numbered sequentially according to mode frequency (and the numbers of phase reversals in the mode shape). In general, mode 'n' has a frequency above those corresponding to (n –1) modes and its mode shape has 'n' phase reversals. Note that mode zero mentioned earlier is not a torsional mode. The total number of torsional mode is (N – 1) where N is the number of rotor masses.

E.2 A SIMPLIFIED SSR ANALYSIS

Consider the classical model of a generator where the field flux decay, damper circuits and transient saliency are ignored. Assuming that the generator rotor oscillates (about a constant speed of ω_0) sinusoidally, the per unit speed ($\bar{\omega}$) is given by

$$\bar{\omega} = \bar{\omega}_0 + A \sin \omega_m t \tag{E.2}$$

where ω_m is the oscillation frequency of the rotor (about a synchronously rotating axis) in radians per second. The induced voltages in the armature in the $\alpha - \beta$ sequence are given by

$$e_\alpha = \bar{\omega} E' \sin(\omega_0 t + \delta) \tag{E.3}$$

$$e_\beta = \bar{\omega} E' \cos(\omega_0 t + \delta) \tag{E.4}$$

Since,

$$\frac{d\delta}{dt} = (\bar{\omega} - \bar{\omega}_0)\omega_B \tag{E.5}$$

we get

$$\delta = \delta_0 - \left(\frac{A\omega_B}{\omega_m}\right) \cos \omega_m t \tag{E.6}$$

If the amplitude (A) of the rotor oscillations is very small, the induced voltages consist of three sinusoidal components, one of frequency f_0 (in Hz), the other two components of frequency, $f_0 \pm f_m$. We can get this by substituting Eqs. (E.2) and (E.6) in (E.3) and (E.4). Thus,

$$e_\alpha = \bar{\omega}_0 E' \sin(\omega_0 t + \delta_0) - \frac{AE'}{2\omega_m}(\omega_0 - \omega_m)\cos[(\omega_0 - \omega_m)t + \delta_0]$$
$$- \frac{AE'}{2\omega_m}(\omega_0 + \omega_m)\cos[(\omega_0 + \omega_m)t + \delta_0] \tag{E.7}$$

$$e_\beta = \bar{\omega}_0 E' \cos(\omega_0 t + \delta_0) + \frac{AE'}{2\omega_m}(\omega - \omega_m)\sin[(\omega_0 - \omega_m)t + \delta_0]$$
$$+ \frac{AE'}{2\omega_m}(\omega_0 + \omega_m)\sin[(\omega_0 + \omega_m)t + \delta_0] \tag{E.8}$$

If the network impedance (in the Laplace domain) is

$$Z(s) = R + Ls + \frac{1}{Cs} \tag{E.9}$$

we can obtain the currents flowing in the α and β sequence networks for the applied voltages given by Eqs. (E.7) and (E.8). These can be converted to the d-q (in Park's reference frame) components from the following equation

$$(i_q + ji_d) = e^{-j(\omega_0 t + \delta)}(i_\beta + ji_\alpha) \tag{E.10}$$

The electrical torque (T_e) is given by

$$T_e = E' i_q \tag{E.11}$$

i_q contains a constant component and two other components (both oscillating sinusoidally at frequency f_m) due to the (*i*) subsynchronous components and (*ii*) supersynchronous components of the armature currents. The oscillating component (of frequency, f_m) of the electrical torque consists of two components, T_e^{sub} and T_e^{sup} corresponding to the sub-and supersynchronous components of the armature currents. These can be derived as

$$T_e^{\text{sub}} = E' i_q^{\text{sub}} = \frac{-A(E')^2}{2\omega_m Z_{\text{sub}}}(\omega_0 - \omega_m)\sin(\omega_m t + \phi_{\text{sub}}) \tag{E.12}$$

$$T_e^{\text{sup}} = E' i_q^{\text{sup}} = \frac{A(E')^2}{2\omega_m Z_{\text{sup}}}(\omega_0 + \omega_m)\sin(\omega_m t - \phi_{\text{sup}}) \tag{E.13}$$

The damping torque coefficient (T_{De}) is given by

$$T_{De} = \text{Re}\left[\frac{\Delta T_e(j\omega_m)}{\Delta\bar{\omega}(j\omega_m)}\right] \tag{E.14}$$

where $\Delta T_e = T_e^{\text{sub}} + T_e^{\text{sup}}$ and $\Delta\omega = A\sin\omega_m t$. It can be shown from Eqs (E.12) to (E.14) that

$$T_{De}(\omega_m) = -\frac{(E')^2}{2\omega_m}\left[\frac{(\omega_0 - \omega_m)}{Z_{\text{sub}}}\cos\phi_{\text{sub}} - \frac{(\omega_0 + \omega_m)}{Z_{\text{sup}}}\cos\phi_{\text{sup}}\right] \tag{E.15}$$

It is to be noted that $Z_{\text{sub}}\angle\phi_{\text{sub}}$ and $Z_{\text{sup}}\angle\phi_{\text{sup}}$ are the impedances of the network calculated at the sub- and supersynchronous frequencies. That is,

$$Z_{\text{sub}}\angle\phi_{\text{sub}} = Z[s = j(\omega_0 - \omega_m)]$$

$$Z_{\text{sup}}\angle\phi_{\text{sup}} = Z[s = j(\omega_0 + \omega_m)]$$

Remarks

1. The damping due to the subsynchronous frequency currents is negative while that due to the supersynchronous frequency currents is positive. This results from the fact that both ϕ_{sub} and ϕ_{sup} can vary in the range –90° to 90° and cosϕ is positive in this range.
2. If the electrical network is resonant at a frequency (f_{er}) which is close to the frequency, $f_0 - f_m$, then $Z_{sub} \ll Z_{sup}$ for typical values of the network impedances. Actually if $f_{er} = f_0 - f_m$ then $Z_{sub} = R$ and $\phi_{sub} = 0$. Also, $\phi_{sup} \simeq 90°$. Thus T_e^{sub} has a much larger magnitude compared to T_e^{sup} and T_{De} tends to be negative. If we make the approximation, $T_e^{sup} = 0$, then,

$$T_{De}(\omega_m) = \frac{-(E')^2}{2\omega_m}\frac{(\omega_0 - \omega_m)}{Z_{sub}}\cos\phi_{sub}$$

$$\simeq \frac{-(\omega_0 - \omega_m)}{2\omega_m}G(\omega_0 - \omega_m) \quad \text{(E.16)}$$

where

$$G(\omega_0 - \omega_m) = \text{Re}[Z^{-1}(j(\omega_0 - \omega_m))] = R/Z^2 \quad \text{(E.17)}$$

is the conductance of the network viewed from the generator internal bus.

It can be observed that Eq. (E.16) is identical to Eq. (11.63) and is derived big ignoring the damping torque contribution of supersynchronous frequency armature currents.

The significance of the damping torque coefficient (T_{De}) is that with some approximations, it is related to the change in the decrement factor ($\Delta\sigma_e$)brought about by the electrical system. We can show that for negligible mechanical damping, the electrical system results in the decrement factor (for oscillation at frequency f_m)

$$\sigma_e \simeq \frac{T_{De}(\omega_m)}{4H_m} \quad \text{(E.18)}$$

where H_m is the 'modal' inertia constant which is associated with the swing equation in the decoupled modal variables given by

$$\frac{2H_m}{\omega_B}\frac{d^2\Delta\delta^m}{dt^2} + K_m\Delta\delta^m = -\Delta T_e \quad \text{(E.19)}$$

(Note that since the damping torque is related to the linearized systems, we are dealing with the linearized swing equation in the modal domain. In deriving (E.19), it is assumed that the mechanical torques are constants). The relationship between the generator rotor angle ($\Delta\delta$) and the modal variables ($\Delta\delta^m$, m = 0, 1, 2, ...(N – 1)) is given by

$$\Delta\delta = \sum_{m=0}^{m=N-1} \Delta\delta^m \quad \text{(E.20)}$$

where N is the number of rotor masses, $m = 0$ corresponds to mode zero.

The modal frequency (f_m) is given by

$$f_m = \frac{1}{2\pi}\sqrt{\frac{\omega_B K_m}{2H_m}} \tag{E.21}$$

For $m = 0$, $K_m = 0$ and $f_m = 0$.

Remarks

1. The concept of modal inertias is quite significant in judging the effect of the electrical system. For large modal inertias, the electrical system has negligible effect (see Eq. (E.18)).
2. For mode zero, the modal inertia is the sum of all the rotor inertias.

   Figure (E.1) shows the block diagram depicting the interaction between the electrical and the mechanical (in the modal domain) systems.

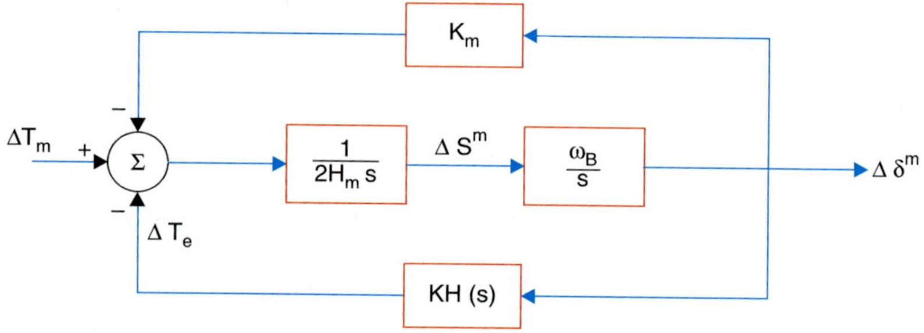

Fig. E.1: Block diagram

Here, $H(s)$ is defined by

$$H(s) = \frac{\Delta T_e(s)}{\Delta\delta^m(s)} \tag{E.22}$$

and K is parameter introduced for convenience. The value of K is obviously equal to 1. However, $K = 0$ indicates that the electrical system is not considered. In the absence of the electrical system, the eigenvalues of the system are imaginary given by

$$\lambda = \pm j\omega_m \tag{E.23}$$

It can be shown that T_{De} (ω_m) can be expressed as

$$T_{De}(\omega_m) = \text{Re}\left[\frac{\omega_B}{j\omega_m} H(j\omega_m)\right] \tag{E.24}$$

The sensitivity of the eigenvalues ($\lambda = -\sigma \pm j\omega$) with respect to the parameter K, evaluated at $K = 0$ is given by

$$\frac{\partial\lambda}{\partial K} \mid K = 0 = -RH(s)\mid_{s=\pm j\omega_m} \tag{E.25}$$

where R is the residue corresponding to the initial eigenvalues $\pm j\omega_m$, given by

$$R = \pm \frac{\omega_B}{4jH_m\omega_m} \qquad \text{(E.26)}$$

(Note that for $K = 0$, $\lambda = \pm j\omega_m$)

The sensitivity of the decrement factor σ, is given by

$$\frac{d\sigma}{dK}\Big|_{K=0} = \text{Re}[RH(j\omega_m)] = \frac{T_{De}(\omega_m)}{4H_m} \qquad \text{(E.27)}$$

If the sensitivity remains constant as K varies from 0 to 1, then Eq. (E.18) applies.

REFERENCES AND BIBLIOGRAPHY

1. IEEE Subsynchronous Resonance Working Group, '*Proposed Terms and Definitions for Subsynchronous Oscillations*', IEEE Trans., Power Apparatus and Systems, Vol. 99, No. 2, March/April 1980.
2. IEEE Subsynchronous Resonance Working Group, "*Terms, Definitions and Symbols for Subsynchronous Oscillations*", IEEE Trans., Power Apparatus and Systems, Vol. 104, No. 6, pp.1326–1334, June 1985.
3. Padiyar K.R., "*Analysis of Subsynchronous Resonance in Power Systems*", (Boston), Kluwer Academic Publishers, 1999.
4. Padiyar K.R., "*FACTS Controllers in Power Transmission and Distribution*", (New Delhi), New Age International Publishers, 2007.

APPENDIX-F

CIGRE BENCHMARK MODELS

F.1 INTRODUCTION

Ainsworth [1] proposed a benchmark model in 1985 which was later adopted by CIGRE SC-14 to carry out analysis and studies on the DC control aspects, aiming at the improvement of information exchange on system simulation. In the following years, a comparison of one simulator and four digital models was carried out and documented by a CIGRE Working Group [2]. The system considered in the Benchmark Model is a two terminal HVDC cable transmission (of about 100 km length) system. The main circuit values proposed are arbitrary and do not represent any particular HVDC scheme.

Subsequently, in 1994, a new Benchmark Model (here labeled as Number 2) was proposed, that considers a long distance overhead line [3].

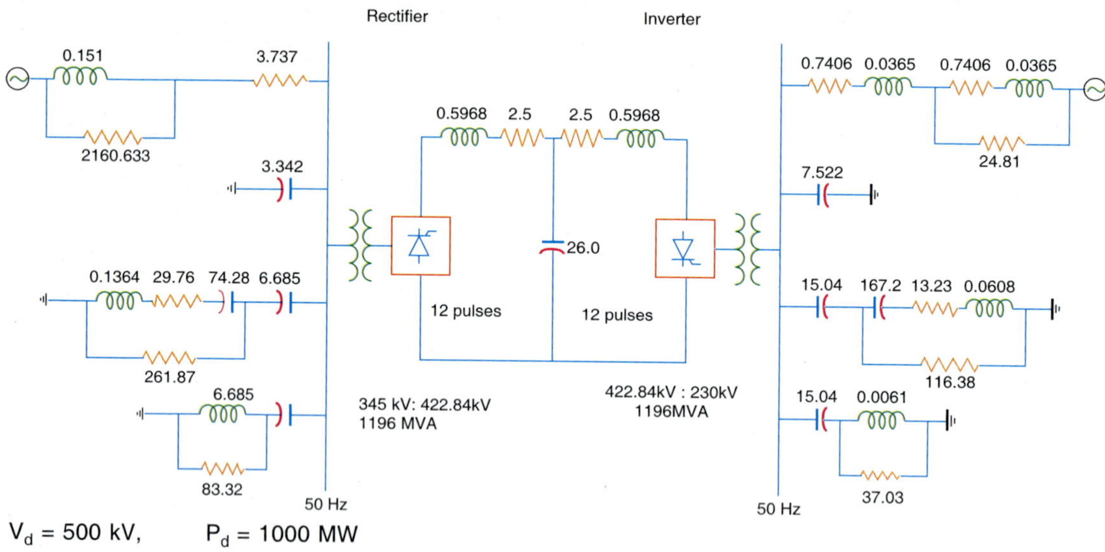

Note : 6.685 applies to the adjacent capacitor. The inductor value is 0.0136

Fig. F.1: Benchmark ac/dc system configuration

F.2 BENCHMARK MODEL 1

The system configuration is shown in Fig. F.1. (Note: The resistances are in ohms, inductances in mH and capacitances in μF.) A short circuit ratio (SCR) of 2.5 has been chosen for both ends, giving an effective ratio (ESCR) of 1.9. The selection of the parameters was to introduce some degree of difficulty – by tuning the DC network to have a series resonance around the fundamental frequency and the AC network having parallel resonance around second harmonic (base frequency = 50 Hz). The converters at both ends are 12 pulse and the rated DC voltage and power are 500 kV and 1000 MW respectively. There are AC filters at both ends, consisting of a double tuned and a high pass filter. There are no DC filters.

F.3 BENCHMARK MODEL 2

This is also a two terminal system and the configuration (of a monopolar equivalent of a bipolar HVDC line of 850 miles long) is shown in Fig. F.2. The main circuit parameters are given in Table F.1 (note the resistances are in ohms, inductances in mH and capacitances in μF). The DC line exhibits a series resonance close to 50 Hz. The inverter – end AC system has a parallel resonance at second harmonic. The rated DC voltage is 500 kV and the rated power is 1000 MW. The length of the line is 850 km and its parameters are given below:

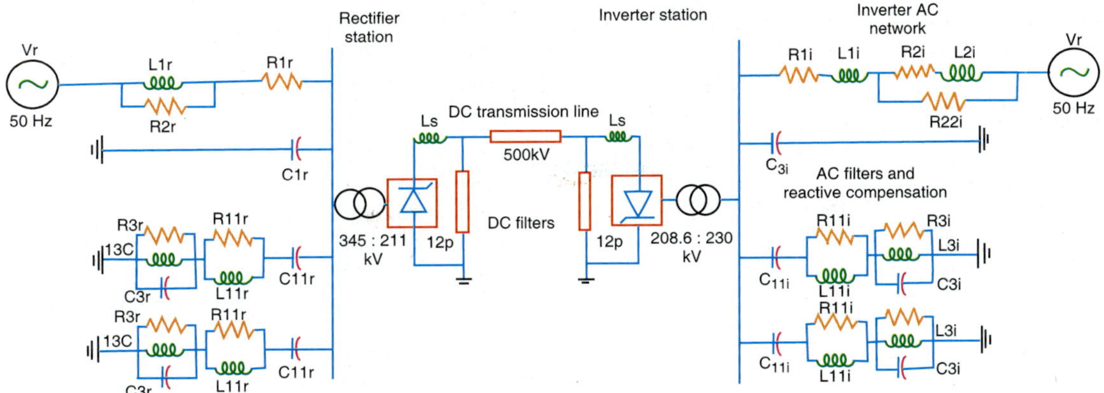

Fig. F.2: New benchmark configuration

$$\text{Resistance} = 0.0449 \text{ ohm/mile}$$

$$\text{Inductance} = 3.5989 \text{ mH/mile}$$

$$\text{Capacitance} = 0.0161 \mu\text{F/mile}.$$

The DC filter configuration is shown in Fig. F.3. The parameter data are given in Table F.2. The DC filters are identical at both ends and consist of a single tuned filter (for 12th harmonic) and a high pass filter.

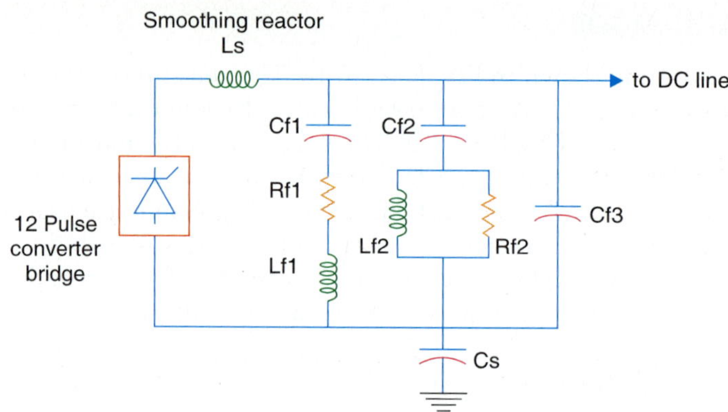

Fig. F.3: DC filters configuration

Table F.1: New benchmark parameters

Rectifier (345kV)	Inverter (230kV)
Vr = 1.0824/58.4° *pu*	Vi = 0.9373/-6.0° *pu*
R1r = 3.737 Ω	R1i = 0.7406 Ω
R11r = 83.0 Ω	R2i = 0.7406 Ω
R2r = 2160.0 Ω	R22i = 24.81 Ω
R3r = 50.0 Ω	R11i = 36.9 Ω
L1r = 150.8 mH	R3i = 22.2 Ω
L11r = 15.1 mH	L1i = 36.5 mH
L3r = 9.69 mH	L2i = 36.5 mH
C1r = 3.342 μf	L11i = 6.71 mH
C11r = 6.57 μf	L3i = 4.31 mH
C3r = 78.9 μf	C1i = 7.52 μf
Ls = 597 mH	C11i = 14.78 μf
	C3i = 177.5 μf
	Ls = 597 mH

Table F.2: DC filter (Rectifier and Inverter) data

Rf_1 = 6.3 Ω
Rf_2 = 1000.0 Ω
Lf_1 = 336.0 mH
Lf_2 = 0.84 mH
Cf_1 = 0.209 μF
Cf_2 = 3.0μF
Cf_3 = 0.84 μF
Cf_4 = 36.0 μF

REFERENCES AND BIBLIOGRAPHY

1. Ainsworth J.D., '*Proposed Benchmark Model for Study of HVDC Controls by Simulator or Digital Computer*', presented at CIGRE SC-14, Colloquium, on HVDC with Weak AC Systems, UK, September 1985.
2. Szechtman M., et al., '*First Benchmark Model for HVDC Control Studies*', Electra, No. 135, pp. 54–67, April 1991
3. Working Group 14.02, '*The CIGRE Benchmark Model–A New Proposal with Revised Parameters*', Electra, No. 157, pp. 61–66, December 1994.
4. Lehn P., Rittiger J. and Kulicke B., '*Comparison of the ATP Version of the EMTP and the NETOMAC Program for Simulation of HVDC System*', IEEE Trans., Power Delivery, Vol. 10, No. 4, pp. 2048–2053, October 1995
5. Babu Narayanan M.M. and Parameswaran S., '*Study of Control Strategies for New CIGRE HVDC Benchmark Model*', Paper 14–113, CIGRE, Paris, 1998.

APPENDIX-G

DESIGN OF DC AND VOLTAGE CONTROLLERS FOR VSC-HVDC LINKS

In Chapter 4, the decoupled control of active and reactive currents drawn by a VSC is described. The active current reference is obtained as the output of either (*i*) Power Controller or (*ii*) DC Voltage Controller. The reactive current reference can be obtained from the output of AC Voltage Controller if it is desired to regulate the AC Voltage at the converter bus. While the reactive currents at the two ends of a two terminal VSC - HVDC link can be controlled independently, the active currents have to be coordinated. This is easily achieved by regulating DC voltage at one end (rectifier or inverter) and controlling power at the other end.

Design of DC Voltage Controller

The design of DC voltage controller in a VSC - HVDC link is illustrated by considering a BTB link shown in Fig. G.1 The terminal 1 is connected to a voltage source while terminal 2 is connected to a power system represented by Thevenin equivalent. The AC filters are neglected in the analysis as they are not required to supply the fundamental frequency reactive power. It is assumed that the terminal 2 controller regulates the DC voltage while terminal 1 controls the DC power flow in the link. While there is no need for the control of AC voltage at terminal 1, it can be provided at terminal 2 if required. Although terminal 2 functions as an inverter under normal conditions, it is also possible to operate it as a rectifier terminal.

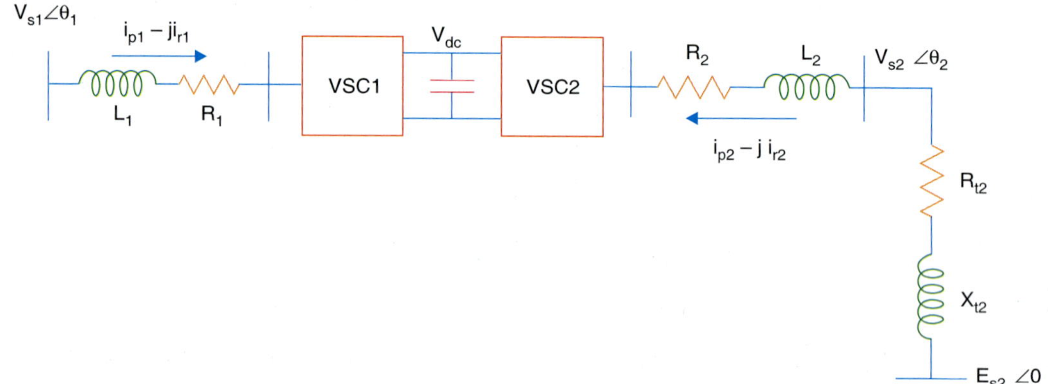

Fig. G.1: A BTB VSC–HVDC system diagram

In reference [1], it is suggested that both terminals can contribute to the DC voltage control in a BTB HVDC link. Thus, active current references at the two terminals are expressed as

$$i_{p1}^{*} = \frac{P_1^{*}}{V_{s1}} + K_c(V_{dc}^{*} - V_{dc}) \tag{G.1}$$

$$i_{p2}^{*} = K_v(V_{dc}^{*} - V_{dc}) + i_{p20} \tag{G.2}$$

where P_1^{*} is the reference value of the power at terminal 1. i_{p20} is the value of i_{p2} at the operating point.

The equations for i_p and i_r (in *p-r* variables) are given by

$$V_{sk} - e_{pk} = (R_k + sL_k)i_{pk} + X_k i_{rk} \tag{G.3}$$

$$-e_{rk} = (R_k + sL_k)i_{rk} - X_k i_{pk} \tag{G.4}$$

From decoupled control of i_p and i_r, we get (see Eq. (4.28))

$$i_{pk} = \frac{1}{(1+sT_i)} i_{pk}^{*},\quad i_{rk} = \frac{1}{(1+sT_i)} i_{rk}^{*} \tag{G.5}$$

The subscript k can be 1 or 2 according to whether terminal 1 or 2 are considered. The superscript* (star) represents the reference value.

Assuming the converters and DC link to be lossless, from the power balance, we get

$$V_{dc} i_{dck} = e_{pk} i_{pk} + e_{rk} i_{rk},\quad k = 1,2 \tag{G.6}$$

The equation for the DC link voltage (across the capacitor C) is given by

$$CsV_{dc} = i_{dc1} + i_{dc2} \tag{G.7}$$

Linearizing Eqs. (G.3) to (G.6) we get

$$\Delta i_{dck}(s) = T_{pk}(s)\Delta i_{pk}^{*}(s) + T_{rk}(s)\Delta i_{rk}^{*}(s) + G_k \Delta V_{sk}(s) \tag{G.8}$$

where

$$T_{pk}(s) = \frac{a_k s + b_k}{1 + T_i s},\ T_{rk}(s) = \frac{c_k s + d_k}{1 + T_i s},\ G_k = \frac{i_{pk}}{V_{dc}}$$

$$a_k = \frac{-i_{pk} L_k}{V_{dc}},\quad b_k = \frac{L_k}{T_i}\cos\alpha_k - \frac{i_{pk} R_k}{V_{dc}} + \frac{i_{rk} X_k}{V_{dc}}$$

$$c_k = \frac{-i_{rk} L_k}{V_{dc}},\quad d_k = \frac{-L_k}{T_i}\sin\alpha_k - \frac{i_{pk} X_k}{V_{dc}} - \frac{i_{rk} R_k}{V_{dc}}$$

α_k is the angle by which the voltage injected by the converter leads the converter bus voltage.

The block diagram of the DC voltage control system is shown in Fig. G.2. Δi^*_{rk} If is assumed to be zero, for $k = 1$ and 2 and in addition if $\Delta V_{sk} = 0$ (k = 1, 2) then we obtain the following closed loop transfer function

$$\frac{\Delta V_{dc}(s)}{\Delta V^*_{dc}(s)} = \frac{K\left[s(a_1 + a_2) + (b_1 + b_2)\right]}{s^2 CT_i + s\left[C + K(a_1 + a_2)\right] + K(b_1 + b_2)} \tag{G.9}$$

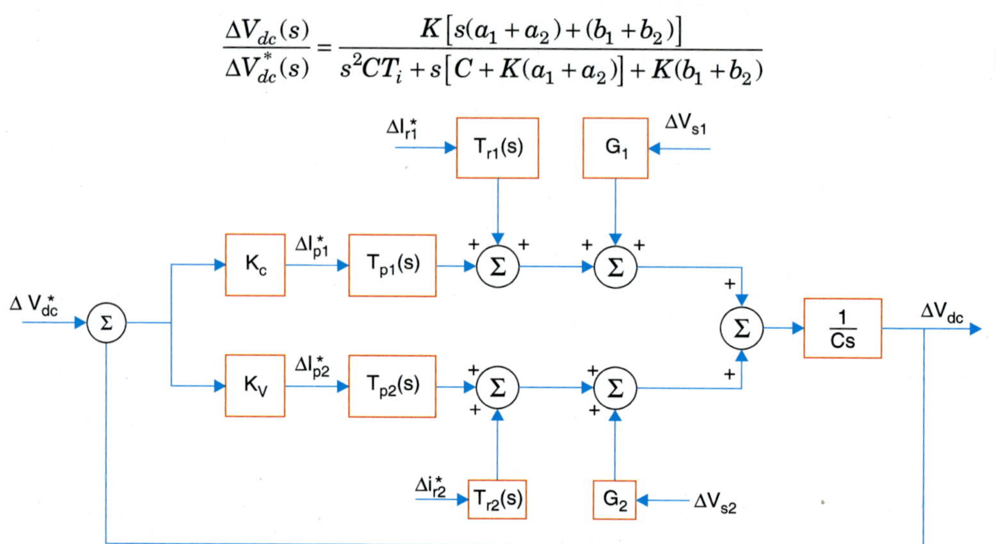

Fig. G.2: DC voltage control

If $L_1 = L_2$, then $a_1 + a_2 \simeq 0$ as $i_{p1} \simeq -i_{p2}$. Hence we simplify the closed loop transfer function to

$$\frac{\Delta V_{dc}(s)}{\Delta V^*_{dc}(s)} = \frac{1}{(CT_i)} \quad \frac{K(b_1 + b_2)}{\left[s^2 + s(\frac{1}{T_i}) + K\frac{(b_1 + b_2)}{CT_i}\right]} \tag{G.10}$$

Designing for a damping ratio $(\xi) = 0.707$, the controller (proportional) gain is obtained as,

$$K = -\frac{C}{2T_i(b_1 + b_2)}$$

When we compare with expression (G.1) and (G.2) we note that

$$K_c = K_v = K$$

Remarks

1. The choice of $K_c = K_v = K$ ensures that both rectifier and inverter contribute equally to the DC Voltage regulation. This is feasible in a BTB Link
2. The value of G_k for an inverter is negative as i_{pk} is negative . This implies that reduction in V_{sk} at the inverter results in increase in V_{dc} (unlike in the case of LCC - HVDC Link). It can be shown that $\Delta V_{dc} = -\frac{\Delta V_{sk}}{2K}$.

3. The DC Voltage control also depends on the AC voltage control as the reactive current influences the control design.

Design of AC Voltage Controller

Since the terminal 2 (Fig. G.1) is connected to a power system represented by a Thevenin equivalent, it is possible to regulate the terminal voltage V_{s2} by AC voltage controller using integral control. This is shown in Fig. G.3. Here, ΔV_{s2} is expressed as

$$\Delta V_{s2} = (a_3 s + b_3)\Delta i_{p2} + (c_3 s + d_3)\Delta i_{r2} + G_3 \Delta E_{s2} \tag{G.11}$$

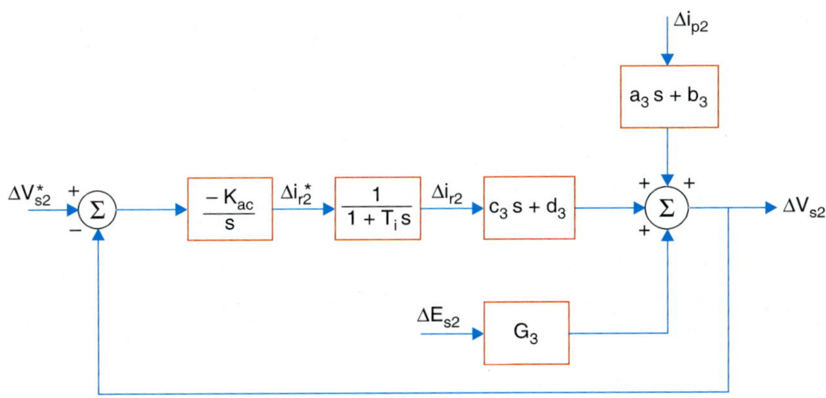

Fig. G.3: Block diagram of AC voltage controller

where

$$a_3 = \frac{-X_{t2}}{\omega_B}, b_3 = X_{t2}\tan\theta_2 - R_{t2}, c_3 = -\frac{X_{t2}}{\omega_B}\tan\theta_2$$

$$d_3 = -(X_{t2} + R_{t2}\tan\theta_2),\ G_3 = \frac{1}{\cos\theta_2}$$

The above expression can be derived from the following Eq. (in *p-r* frame of reference) for the network connected to terminal 2.

$$L_{t2}\frac{d}{dt}(i_{p2} - ji_{r2}) + j\omega_0 L_{t2}(i_{p2} - ji_{r2}) + R_{t2}(i_{p2} - ji_{r2}) = (E_{p2} - jE_{r2}) - V_{s2} \tag{G.12}$$

where $E_{p2} = E_{s2}\cos\theta_2,\qquad E_{r2} = E_{s2}\sin\theta_2$

Linearizing (G.12) and eliminating $\Delta\theta_2$, we can derive

$$\Delta V_{s2} = \frac{1}{(\cos\theta_2)}\Delta E_{s2} - \left(sL_{t2} - R_{t2} + X_{t2}\tan\theta_2\right)\Delta i_{p2}$$

$$-\left(sL_{t2}\tan\theta_2 - X_{t2} - R_{t2}\tan\theta_2\right)\Delta i_{r2} \tag{G.13}$$

Note that $X_{t2} = \omega_B L_{t2}$, where ω_B is the base frequency assumed to be equal to ω_0 (the operating frequency). Since the block diagrams shown in Fig. G.2 and G.3 are coupled, the closed loop transfer function of the AC voltage controller can be obtained as

$$\frac{\Delta V_{s2}(s)}{\Delta i_{r2}(s)} = \left(\frac{N_1}{D_1}\right) \frac{s^3 + N_2 s^2 + N_3 s + N_4}{s^3 + D_2 s^2 + D_3 s + D_4} \tag{G.14}$$

It can be shown that the transfer function given above is non-minimum phase when the terminal 2 operates as a rectifier. The closed loop transfer function of the DC voltage controller also changes when AC voltage controller is considered. The modified transfer function is given by

$$\frac{\Delta V_{dc}(s)}{\Delta V_{dc}^*(s)} = \frac{K}{s} \frac{\left(A_1 s^3 + A_2 s^2 + A_3 s + A_4\right)}{\left(B_1 s^3 + B_2 s^2 + B_3 s + B_4\right)} \tag{G.15}$$

This transfer function is also a non-minimum phase type when terminal 2 is operating as a rectifier. It is necessary to select the gains of the DC and AC voltage controller to ensure that system remains stable.

REFERENCES AND BIBLIOGRAPHY

1. Hagiwara M., Fujita M. and Akagi H., '*Performance of a Self-Commutated BTB HVDC Link Under a Single-Line-to-Ground Fault Condition*', IEEE Trans., Power Electronics, Vol. 18, No. 1, pp. 278–285, January 2003.
2. Yazdani A. and Iravani R., '*An Accurate Model for the DC – Side Voltage Control of the Neutral Point Diode Clamped Converter*', IEEE Trans., Power Delivery, Vol. 21, No.1, pp. 185–193, January 2006.
3. Tyagi A. and Padiyar K.R., "*Dynamic Analysis and Simulation of a VSC based Back-to-Back HVDC Link*", Proc. of India Int. Conf. on Power Electronics, pp. 232–238, Chennai, December 2006.

APPENDIX-H

MODULAR MULTILEVEL CONVERTER (MMC) BASED VSC-HVDC

H.1 INTRODUCTION

The VSC-HVDC has several advantages over LCC-HVDC. Among them, the following are significant which enable novel applications for HVDC transmission.

1. There is no need for reactive power compensation as the VSC does not require reactive power. It can even supply reactive power.
2. VSC-HVDC can be used to supply even passive loads as there is no requirement of AC voltage source at the inverter
3. The power reversal in the DC link does not require voltage reversal.

The implication of the first advantage is that the converter station can be compact with a small foot print. The benefit of the second advantage is that VSC-HVDC can be used to provide black start capability. In other words, no auxiliary power source is required to start generation that has been shut down. And finally, the significance of the third advantage is that cross linked polyethylene (XLPE) cables which are less costly, lighter and smaller in diameter can be used rather than traditional mass impregnated cables which can handle polarity reversal.

These advantages are useful in the application of VSC-HVDC for connecting offshore platforms. Since 2005, VSC-HVDC has been successfully deployed to power offshore oil and gas platforms from shore. Since 2010, VSC-HVDC is being applied to transmit power from offshore wind generators to shore. Already 5 GW of transmission of offshore wind generators located in North Sea has been awarded to be connected to the German shore up to 2015 [7]. There are plans to tap 40 GW of wind power in Europe by 2020.

Since the introduction of the experimental line project (3 MW at ±10 kV) in Hallsjon in 1997, only two or three level voltage source converters (VSC) have been applied in all VSC-HVDC projects until 2010. In a two level VSC, the output AC voltage has only two levels $\left(+V_{dc} \text{ and } -V_{dc}\right)$. In a three level converter, the output AC voltage has a third level of zero voltage. To reduce low frequency harmonics, it is necessary to have pulse-width modulation (PWM). The modulation frequencies up to 2 kHz have been used to ensure elimination of low frequency harmonies. However, the filtering of high frequency harmonics is still required. The use of PWM implies higher switching losses (2 to 3%) compared to 0.7% in thyristor based LCC. The IGBT valves used in two or three level converters are made up of a large number (over hundred) IGBT devices which have to switch on or off simultaneously at the same instants. This introduces problems of voltage sharing among individual devices and generation of high frequency noise causing electromagnetic interference (EMI) when large voltage steps are generated (when a valve is switched). To overcome these problems, modular multilevel converter (MMC) topology has been proposed and applied for the first time in 2010 by Siemens in the Transbay cable HVDC PLUS project, (rated at 400 MW, ±200kV) providing dedicated connection between the East Bay and San Francisco in USA. The objectives of this HVDC link are improved network security and reliability due to grid enhancement, voltage support and reduction in the system losses. The city of San Francisco relies mainly on AC grids which run along the lower part of the bay. With the new HVDC link, the power flows directly into the centre of San Francisco and closes the loop of already existing 'Greater Bay Area' transmission. This will increase the system security and avoid or postpone the building of additional power plant in the city. The HVDC link is also expected to reduce grid congestion in the East Bay.

H.2 MODULAR MULTILEVEL CONVERTER TOPOLOGY

Instead of two or three levels, several levels of injected AC voltage by the VSC, results in reducing the size of voltage steps and the related voltage gradients. With a high number of levels, the switching frequency of individual IGBT devices can be reduced.

The basic idea is to generate an approximate sine wave using several steps as shown in the Fig H.1 which shows the waveform of the injected AC voltage by the MMC. An MMC consists of six converter arms with two each in one phase unit as shown in Fig H.2. Each arm comprises a high number of power modules (PM), also called as sub modules (SM) and one converter reactor connected in series. Each power module contains (*a*) an IGBT half bridge as a switching element and (*b*) a DC capacitor unit for energy storage. The sub-module (SM) is shown in Fig H.3. The half bridge consists of two switches (T_1 and T_2) each consisting of an IGBT and a freewheeling diode connected in antiparallel. In normal operation, only one switch (T_1 or T_2) is on and the other is off. Depending in the current direction, the capacitor can charge or discharge.

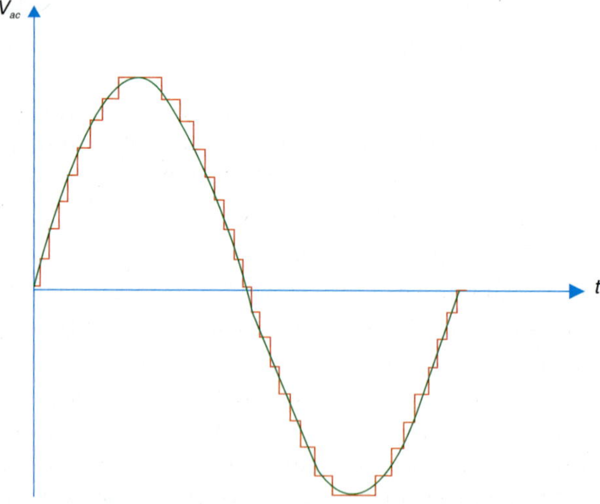

Fig. H.1: Waveform of injected AC voltage by MMC

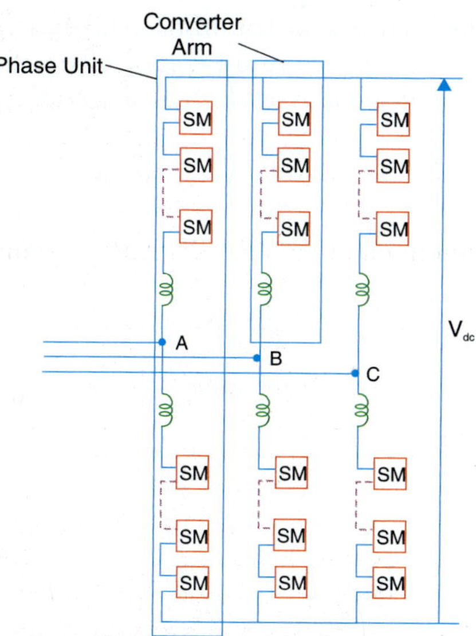

Fig. H.2: MMC topology

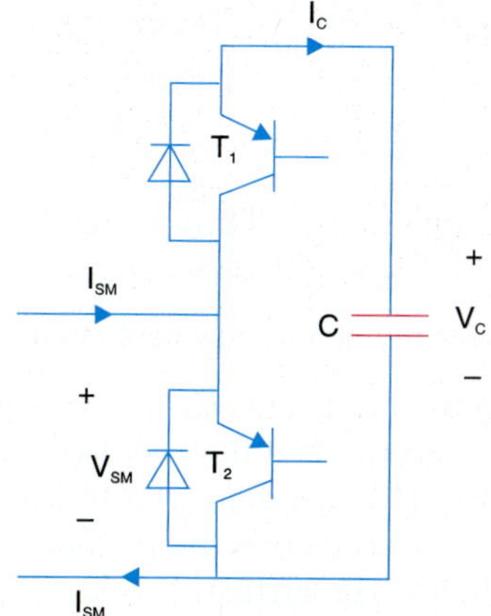

Fig.H.3: Submodule circuit

There are 3 possible switching states defined below:

1. In the inserted (ON) state of the submodule, T_1 is on and T_2 is off. The output voltage (V_{SM}) of the submodule equals the capacitor voltage (V_C) and the capacitor charges if I_{SM} is positive and discharges otherwise.
2. In the bypassed (OFF) state, T_2 is on and T_1 is off, V_{SM} is zero and the capacitor voltage remains constant.
3. In the blocked state, both switches (T_1 and T_2) are off and the current can conduct only through the freewheeling diodes. The capacitor will be charged in the current I_{SM} is positive, but it cannot discharge (ideally).

The steady state equivalent circuit of the MMC shows in Fig H.2 is shown in Fig H.4. Here, the upper and lower arm valves are represented by controlled voltage sources for each phase. The expressions for the voltage across the valves in the upper arm are given by

$$V_{jU} = \frac{V_{dc}}{2} - e_j, j = a,b,c \tag{H.1}$$

Similarly, the voltages across the valves in the lower arm are given by

$$V_{jL} = \frac{V_{dc}}{2} + e_j, j = a,b,c \tag{H.2}$$

The above equations follow from the relations

$$V_{dc} = V_{aU} + V_{aL} = V_{bU} + V_{bL} = V_{cU} + V_{cL} \tag{H.3}$$

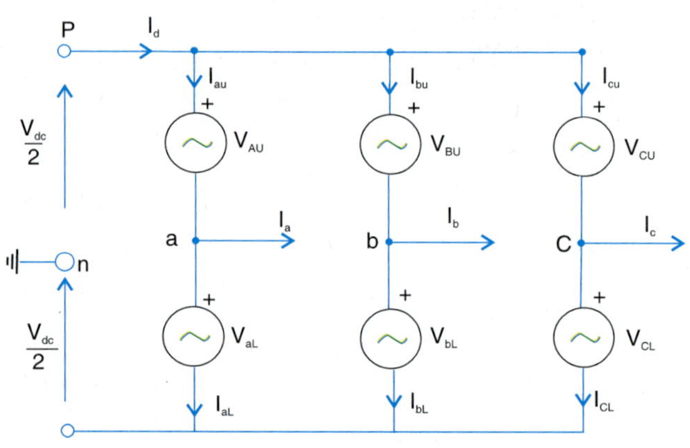

Fig. H.4: Steady state equivalent circuit of MMC

It is to be noted that in transient conditions, the above equation does not apply as the arm voltages are controlled. Balancing currents will flow between phases when equation (H.3) is not satisfied. The reactors shown in Fig H.2 are deployed to limit the balancing currents. These reactors also act to limit the increase in the currents in case of a DC fault, to enable the IGBTs to be turned off at current levels below the critical levels.

When the three phases impress the same voltage V_{dc} across a phase unit, and the converter is ideally symmetrical, the DC current will be divided equally between the three phases. Also, the AC current will split equally between the upper and lower arm in each phase. Thus, we get,

$$I_{jU} = \frac{1}{3} I_{dc} + \frac{1}{2} I_j, j = a,b,c \tag{H.4}$$

$$I_{jL} = \frac{1}{3} I_{dc} - \frac{1}{2} I_j, j = a,b,c \tag{H.5}$$

The DC components in the arm currents increase the conduction losses in the MMC compared to the conduction losses in a two-level converter. However, the total losses in the MMC are lower due to reduced switching losses compared to the two-level converter.

Advantages of MMC Topology

1. Modular construction in power section as well as in control and protection. As a result, the system has excellent scalability and flexible overall design.
2. In normal operation, no more than one level per converter arm switches at any given time. As a result, the AC voltages can be adjusted in very fine increments and a DC voltage with very little ripple can be achieved. This minimizes the generated harmonics and in most of the cases, completely eliminate the need for AC filters
3. The small voltage steps cause very little radiated or conducted high frequency noise.
4. The low switching frequency of individual devices results in low switching losses.
5. Standard AC transformers can be used
6. Due to elimination of AC filters and their switchgear, high reliability and availability can be achieved. Due to modular design, the project execution times can be reduced.
7. It is easy to replace the existing components by state-of-the art ones, since switching characteristics of each individual power module is independently determined. In contrast, in a two-level converter, the devices should have identical switching characteristics

It is to be noted that IGBT devices, similar to thyristors, fail by short circuit. This implies that in a high voltage IGBT valve, by providing redundant devices, it is possible to defer the replacement of failed devices to scheduled maintenance periods. With MMC topology, redundant power modules can be provided such that in case of failure of a power module, the remaining modules are not subjected to a higher voltage. The insertion of redundant power modules only increases the number if total power modules in a converter arm and are controlled to deliver zero voltage during normal operation. In the event of the failure of a power module, the fault is detected and the defective power module is shorted out by a reliable, high-speed bypass switch.

It is necessary to ensure, within certain limits, a uniform voltage distribution across individual capacitors of the multilevel converter. This is achieved by periodic feedback (at intervals less than 100 μs) of the capacitor voltages to a central control unit. Due to the fact that in the converter arms, current flows in both directions, charging and discharging of capacitors is feasible.

H.3 CONTROL OF MMC

The control action in a modular multilevel converter involves the determination of the converter arm voltages and synthesizing the required waveform across each converter arm by switching in or out the power (or sub) modules at desired instants.

The control of a voltage source converter (VSC) given in section 4.11 (chapter 4) is applicable in the determination of the desired sinusoidal AC output voltages for the three phases. From the knowledge of e_p and e_r in Fig 4.20, we can obtain the desired three phase voltages as,

$$e_a = \sqrt{\frac{2}{3}}\left[e_p \sin(\omega t + \theta) - e_r \cos(\omega t + \theta)\right] \tag{H.6}$$

$$e_b = \sqrt{\frac{2}{3}}\left[e_p \sin\left(\omega t + \theta - \frac{2\pi}{3}\right) - e_r \cos\left(\omega t + \theta - \frac{2\pi}{3}\right)\right] \tag{H.7}$$

$$e_c = \sqrt{\frac{2}{3}}\left[e_p \sin\left(\omega t + \theta + \frac{2\pi}{3}\right) - e_r \cos\left(\omega t + \theta + \frac{2\pi}{3}\right)\right] \tag{H.8}$$

From equations (H.1) and (H.2) we can determine the desired converter arm voltages.

It is to be noted that the output voltage in a VSC is controlled to regulate the desired active power or DC voltage and the desired reactive power. Typically active power is controlled at the rectifier and the inverter controls the DC voltage. The desired reactive power can be determined from the requirements of the load supplied or the magnitude of the AC voltage that may have to be regulated.

Power Module Control System

Assuming that there are N sub-modules in each phase unit, the converter can generate a phase voltage with $(N + 1)$ levels. At any instant, $n(t)$ modules are ON with n_U modules in the upper arm and n_L modules in the lower arm $(n(t) = n_U + n_L)$.

Assuming the voltages across the N capacitors are identical,

We have

$$e_j(t) = \frac{V_{dc}}{2} - n_U(t)V_c \tag{H.9}$$

$$= \frac{-V_{dc}}{2} + n_L(t)V_c \tag{H.10}$$

where $V_c = \dfrac{V_{dc}}{N}$

The total number of 'ON' state submodules in a phase unit,varies from zero to N during generation of one cycle of the converter output voltage waveform. Once the reference voltages for the converter arms are determined, the power module control (PMC) system is responsible for the generation of firing pulses sent to the power modules. The PMC must select the most appropriate combination of capacitors to be activated (inserted into the circuit by turning 'ON') during an interval of time. Several strategies are possible. A simple and yet efficient strategy is the selection of the PM (or SM) based on nearest level method.

The selection is primarily based on the arm current: for a positive current (see Fig H.3), the capacitors with the lowest voltages are brought in to service to build up the necessary arm voltage and to charge them up. In the case of a negative current, the capacitors with the highest voltages and selected since this will reduce their voltage. While this selection (of PMs) method is quite straight forward, there should also be mechanism to prioritize the already active (ON) modules over inactive (OFF) modules to reduce the switching frequency. It is to be noted that the optimal selection of active PMs is different at each time step. A strict priority based system of selection of PMs may increase the harmonic current while reducing the switching losses by reducing the switching frequency. Hence, the priority based mechanism is a tradeoff between harmonic distortion and switching losses. However, as the number of PMs per arm goes up, the overall harmonic distortion goes down, since the granularity of the waveform synthesizer is much higher.

As a thumb rule, adequate results are obtained with average switching frequencies around 2 to 4 times the fundamental frequency with 30 or more PMs per arm [5].

Remarks

1. The selection of PMs to be turned 'ON' during an interval helps to balance the capacitor voltages. Since the blocking voltages of the IGBTs in a PM is the capacitor voltage (across the PM), the capacitor voltage balancing is an important function of the PMC.
2. For a specified output voltage, (in a time interval), the sum of the inserted PMs in a phase unit is constant. Hence, inserting a PM in the upper arm (of the phase unit) is done simultaneously with bypassing one PM in the lower arm (and vice-versa)
3. In the Trans Bay Cable Project (2010) in California, there are 216 power modules in an arm, each containing two IGBT/diode devices. In contrast, there are only 38 SMs per arm in the HVDC Light project (by ABB) for a DC link voltage of ±320 kV. It is possible that a submodule may contain more than two IGBT and diodes. However, there will be only two switches in a SM with each switch containing one or more (series connected) IGBT/diode devices. Incidentally, the ABB term their multilevel converters as Cascaded Two Level (CTL) converter.
4. A selected list of VSC–HVDC projects since 2010 is given in Table H.1

Table H.1: Selected Multilevel VSC-HVDC Projects

Installation	Installed	Manufacturer	Rated Power (MW)	Converter Topology
TransBay Cable Project	2010	Siemens	400	Multilevel
BorWin2 (OWF)	2013	Siemens	800	Multilevel
HelWin (OWF)	2013	Siemens	576	Multilevel
DolWin1 (OWF)	2013	ABB	800	Multilevel
SylWin1 (OWF)	2014	Siemens	864	Multilevel
South-West Link	2014	Alstom	1440	Multilevel
HelWin2 (OWF)	2015	Siemens	800	Multilevel
DolWin2 (OWF)	2015	ABB	900	Multilevel

OWF = Offshore Wind Farm

Note: BorWin1 is also OWF project by ABB (400 MW, 2-level)

H.4 PROTECTION OF MMC

When a short circuit occurs between the DC terminals of the converter or along the transmission line, the current rises in the converter arms. The reactors in series with the arms limit the rate of rise of current. The IGBTs can be switched off within the few microseconds. As with any VSC topology, the currents flow from the three phase AC system through the free-wheeling diodes to the short circuit. To prevent this, one option is to open the AC breaker. This requires at least 3 cycles. The diodes should have the capability to withstand the fault currents until the AC breaker opens. If the surge current capability of the diodes is not adequate, thyristors can be connected in parallel with the diodes to protect the diodes. The thyristor is fired to relieve the overload on the diodes.

HVDC Breakers

HVDC breakers will be required to clear line faults when multi terminal DC (MTDC) systems will be introduced to form a HVDC grid. For a point to point power transmission with two terminals, a HVDC breaker can be used to protect against a line fault. While the technology using mechanical breakers is already available, it may take few tens of milliseconds to clear the fault (see section 6.6 in chapter 6 for further details).

In the case of DC grids with VSC converters, it is anticipated that the clearing time of the DC line faults should be in the range of 2 to 3 milliseconds. Recently, ABB has introduced a hybrid breaker using power semiconductor (IGBT) switching elements. The schematic diagram is shown in Fig H.5. During normal operation, the current flows through the bypass, consisting a fast acting (mechanical) disconnector and an auxiliary (IGBT based) DC breaker. There is no current flowing in the main DC breaker during normal operation. When a fault is detected the auxiliary DC breaker turns off, commutating the load current to the main DC breaker: The fast acting disconnector also opens simultaneously. The main circuit breaker is then triggered to interrupt the fault current. The residual DC current breaker acts to isolate the faulted line and to protect the arrester banks from thermal overload. At present, the major constraint is the upper limit on the fault current that can be handled by the breaker. A reactor is introduced to limit the fault current. Using Bimode Insulated Gate Transistor (BIGT), which is a new power semiconductor device that combines an IGBT and antiparallel diode into a single device, it is anticipated that the current ratings can be raised.

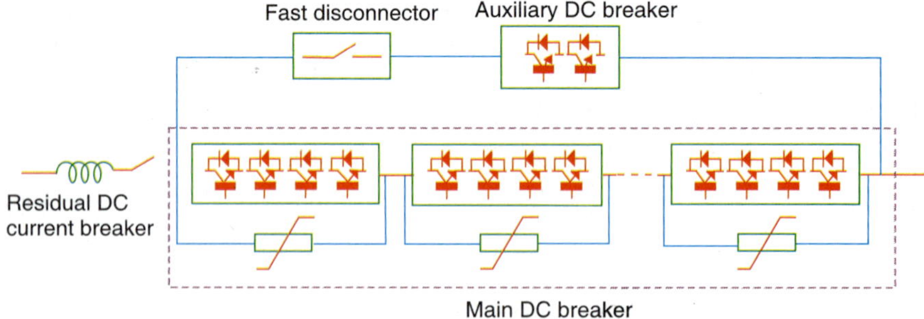

Fig. H.5: Schematic diagram of a hybrid HVDC breaker

The role of the auxiliary breaker is to limit the power losses during normal operation. This happens as the voltage rating of the auxiliary breaker is limited. Since no current flows in the main breaker during normal operation, it does not contribute to any power loss. The function of the fast disconnector is to protect the auxiliary breaker against overvoltages during the interruption of the fault current. The role of the auxiliary breaker is only to commutate the load current.

H.5 MULTILEVEL CONVERTER USING FULL BRIDGE MODULES

The concept of cascaded multilevel VSC converters was proposed by Lai and Feng [12]. The first application of this concept was made by GEC Alsthom [13] for STATCOM. They term it as 'chain-link' converter which is made up of series connection of Full Bridge (FB) modules in each phase. It is also possible to apply FB modules in VSC-HVDC and this can be compared with the Half-Bridge (HB) module used in the MMC by Siemens. (See Fig. H.6)

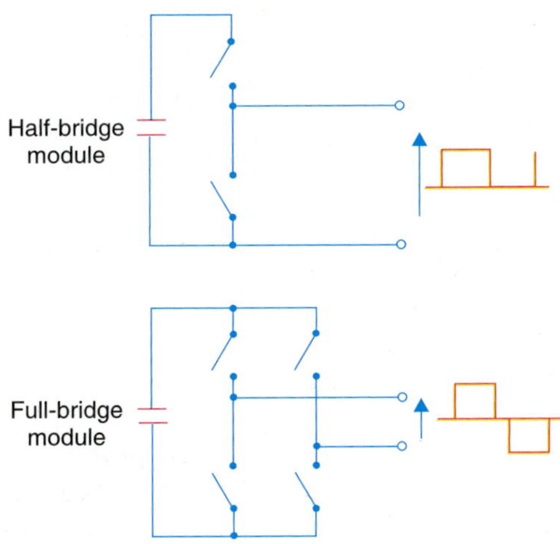

Fig. H.6: Half-bridge(HB) and Full-bridge(FB) switched capacitors

From HB MMC to FB MMC, the required number of semiconductors is increased by a factor of approximately two. (It is to be noted that from the two level converter to HB module, the number of semi conductors is increased by a factor of two for the same rated DC voltage). However, HB module can produce only unipolar voltage output, whereas the FB module can produce a bipolar voltage output, including a zero state. This implies that during a DC line fault, the fault current can be blocked if the capacitor voltages sufficiently high. The basic characteristics of the HB and FB MMC have been outlined in [8] with respect to the performance at deblocked and blocked conditions.

Alstom's MaxSine HVDC is similar to MMC HVDC where FB modules are combined with two level converter, thus introducing a new hybrid family of VSC-HVDC topologies. The hybrid converters fall into two main categories, series and parallel circuits [14].

VSC-HVDC has been applied for off shore platforms (either supply from or to onshore), city centre in-feed, interconnecting grids and energy trading. There is an ambitious and innovative project (Tres Amigas) on the anvil in USA to build 5000 MW interconnection between three interdependent networks (Western, Eastern and Texas grids). Located near the border of New Mexico and Texas, the super station will ultimately consist of six HVDC back to back (BTB) converter stations which will be configured to connect the three AC networks. The initial BTB stations, each rated at 750 MW will use VSC-HVDC. After strengthening the existing AC networks by 500 kV and 765 kV lines, three BTB LCC (Line Commutated Converter) stations of 920 MW each will be built. The objective is to export renewable energy (wind and solar) to load centres[15].

H.6 MODELLING MMC HVDC FOR TRANSIENT SIMULATION

For power flow and stability studies, simplified converter models can be used. However, for transient simulation involving electromagnetic transients (EMT), it is necessary to model every level independently, particularly when analyzing the failure of a module or its control system. Also, modeling of the diodes and losses requires detailed simulation. [4, 5, 6]

The basic approach is to represent each arm in a MMC converter by an equivalent circuit (Thevenin or Norton). The equivalent circuit for a submodule (or power module) is shown in Fig H.7. Here R_1 and R_2 are the resistances of the two switches (shown in Fig H.3). Each switch is modelled as a low resistance when it is ON and a high resistance, when it is OFF R_c is defined as

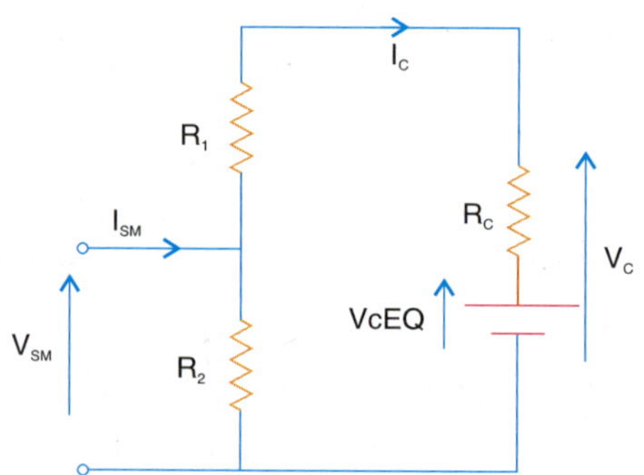

Fig. H.7: Equivalent circuit of a submodule

$$R_c = \frac{\Delta T}{2C} \tag{H.11}$$

where ΔT is the time step (typically 20 μs).

From the equivalent circuits of individual modules, it is possible to obtain Thevenin [4] or Norton [5] equivalent circuit of a multivalve (or arm).

It is to be noted that EMT simulation is feasible in real-time, using RTDS (Real Time Digital Simulator) which enable hardware – in - the loop (HIL) studies.

REFERENCES AND BIBLIOGRAPHY

1. Friedrich K., "Modern HVDC PLUS application of VSC in Modular Multilevel Converter Topology" Industrial Electronics (ISIE), IEEE International Symposium on, 2010, pp. 3807–3810.
2. Marquardt R. and Lesnicar A., "New concept for high voltage Modular Multi level Converter", IEEE PESC 2004 Conference, Aachen, Germany.
3. Jacobson B., Karlsson P., Asplund G., Harnefors L. and Jonsson T., "VSC-HVDC transmission with cascaded two – level converters", CIGRE B4-110, 2010.
4. Gnanarathna U.N, Gole A.M, Jayasinghe R. P, "Efficient modelling of Modular Multilevel HVDC Converters (MMC) on Electromagnetic Transient Simulation Programs", IEEE Trans. on Power Delivery, V.26, No. 1, Jan. 2011, pp. 316–324.
5. Le-Huy P., Giroux P. and Soumagne J.C, "Real-Time simulation of Modular Multilevel Converters for network integration studies", Int. Conf. Power Syst. Transients, Delft, 2011.
6. Gnanarathna U. et al., "Loss estimation of Modular Multilelvel Conveters using electromagnetic transients Simulation", Int. Conf. Power Syst. Transients(ICPT), Delft, 2011.
7. Callavik E.M, Lundberg P., Bahrman M P. and Rosenquist R.P, "HVDC technologies for the future onshore and offshore grid", CIGRE US National Committee 2012 Grid of the Future Symposium, Kansas City, USA, October 2012.
8. Jonsson T., Lundberg P., Maiti S. and Jiang-Hafner V., "Converter technologies and functional requirements for reliable and economical HVDC grid design", 2013 CIGRE Canada Conference, Calgary, Alberta, September, 2013.
9. Barker C.D, Davidson C.C, Trainer D.R. and Whitehouse R.S., "Requirements of DC-DC Converters to facilitate large DC grids", CIGRE, B4-204, 2012.
10. Dodds S. et al., "HVDC VSC (HVDC Light) transmission-operating experiences", CIGRE, B4-203, 2010.
11. Adapa R., "HVDC Technology: The State of the Art", IEEE Power & Energy Magazine, Nov-Dec. 2010, pp. 19–29.
12. Lai J and Peng F.Z, "Multilevel converters – A new breed of power converters, IEEE Trans. Ind. Appl. Vol. 32, no. 3, May/June 1996, pp. 509–517.
13. Ainsworth J.D. et al., "Static VAR compensator (STATCOM) based on single phase chain circuit converters", IEE Proc. Generation, Transmission and Distributions, Vol 145, No. 4, 1995, pp. 381–386.
14. Davidson C. and Trainer D., "Innovative concepts for hybrid converters for HVDC power transmission", 9th IET Conf. on AC and DC Power Transmission, 2010, pp.1–5.
15. Kirby C., Macleod N.M., Stidham D. and Reynolds M., "Application of high capacity Voltage Source Converter HVDC technology at the Tres Amigas Superstation", CIGRE 2012 San Francisco Colloquium, B4-12, 2012.
16. Flourentzou N., Agelidis V.G. and Demetriades G.D., "VSC-based HVDC power transmission systems: An overview", IEEE Trans. on Power Electronics,V.24, no.3, March 2009, pp.592–602
17. Glasdam J. et al., "Review on multi-level voltage source converter based HVDC technologies for grid connection of large offshore wind farms", IEEE Int. Conf. on Power System Technology (POWERCON), Auckland, 2012

INDEX

D

E

F

G

H

I

L

U

V

W

X